— PRAISE FOR —

Herbal Formularies for Health Professionals

"There are many herb books, but few address the fine points of clinical herbal prescribing. To remedy this deficiency, Dr. Jill Stansbury, ND, has written perhaps the best herbal formulary since the days of the renowned Eclectic physicians. Her first book in this set of five volumes covers the digestive and eliminatory systems including the liver and gallbladder, urinary system, and skin. Each section and page clearly shows the skill, experience, and expertise of the author and will help even the seasoned naturopathic physician or herbalist improve their clinical practice and success in treating their patients."

—DAVID WINSTON, RH (AHG), dean of David Winston's Center for Herbal Studies; founder of Herbal Therapeutics Research Library

"*Herbal Formularies for Health Professionals, Volume 1* is simply the best text available for the modern clinical practice of Western herbal medicine. It's likely to become an instant classic and essential book for any student of herbal medicine, as well as seasoned practitioners. Dr. Stansbury includes copious examples from her extensive clinical work with many kinds of patients to help students and practitioners alike find healing solutions to common ailments and symptoms."

—CHRISTOPHER HOBBS, PhD, LAc, fourth-generation herbalist

"Fabulous! This book is an outstanding resource. Dr. Jill Stansbury, one of the most respected herbalists and naturopathic doctors of our time, has created an up-to-date formulary that combines the best of traditional formulas with modern research and her own extensive clinical experience. This impeccably researched volume, the first in a set of five, is a master reference tool for herbalists, doctors, and other health care professionals seeking to create precise formulas based on the individual, not on the diagnosis."

—ROSEMARY GLADSTAR, herbalist; founder of Sage Mountain Herbal Retreat Center

"Esteemed clinical herbalist and naturopathic physician Jill Stansbury shares thirty years of experience with botanical medicines from all over the globe in this incredibly helpful text, which is part guidebook, part *materia medica*, and part recipe-based formulary, laced with an enthusiastic appreciation of all things herbal. In short, *Herbal Formularies for Health Professionals, Volume 1* is a well-researched and beautifully presented herbal resource. Because of its easy readability, logical organization, and wealth of information, it is an asset for students, newly graduated herbalists and doctors, and seasoned practitioners alike. It will no doubt be a frequently referenced text in the library of any herbal practitioner."

—TIMOTHY MILLER, ND, MAc, LAc, RA, founder of Naturopathic CE

"Accomplished naturopathic physician Dr. Jill Stansbury brings her vast clinical experience, deep knowledge, commitment, and passion for herbs to this exciting modern-day formulary. *Herbal Formularies for Health Professionals, Volume 1* is an extensive reference guide to the creation of classic herbal formulas using traditional herbal healing wisdom as well as modern science. This text should be a foundational part of herbal education. There is no other formulary as comprehensive or as accessible to the clinician; this book truly fills an empty space in the list of current herbal clinical education publications."

—DR. MARY BOVE, ND, author of *Encyclopedia of Natural Healing for Children and Infants*

"*Herbal Formularies for Health Professionals, Volume 1* delves deeply into herbal formulas for digestive, liver, urinary, and skin problems, along with descriptions and, where known, causes of these problems. In a chapter on the art of formulation, Jill Stansbury outlines example cases and provides appropriate formulas for each. Her considerations are fascinating. This refreshing work will be highly useful for both serious students and practitioners."

—HENRIETTE KRESS, herbalist; founder of Henriette's Herbal Homepage

"Dr. Jill Stansbury's *Herbal Formularies for Health Professionals, Volume 1* is a formidable accomplishment and historic contribution to the rapidly developing field of botanical medicine. In the tradition of the great pharmacopeias of the Eclectic and naturopathic physicians, this comprehensive manual contains a wealth of concisely detailed herbal formulations and easily readable contemporary research that address both the widespread illnesses of our time as well as specialized and less common conditions.

"*Herbal Formularies* is practical and immediately usable for anyone interested in the applications of herbal medicine. Its depth will satisfy senior clinicians, academic researchers, and medical educators, while its accessibility will greatly further the education of students and the serious lay person."

—DAVID CROW, LAc, founder of Floracopeia Aromatic Treasures

"Reading Dr. Stansbury's formulary has quickly supported us in updating and filling in gaps in our dispensary stock to better serve our customers. I believe this book will similarly inform the dispensing practices of practitioners of all experience levels looking to glean insight from one of the most experienced herbalist physicians in the United States.

"I'm thrilled to see this first of five volumes of herbal internal medical formularies being released to a wide audience. Previous efforts have been predominately geared toward licensed acupuncturists or naturopathic physicians, but these volumes will speak to a broader, dynamic audience. Clinicians will be surprised to discover how much groundwork has already been laid for future generations of health care professionals. This book offers mature templates for improvisation to treat a myriad of complex health conditions with botanical medicines."

—BENJAMIN ZAPPIN, LAc, herbalist; cofounder of Five Flavors Herbs

"Formulation is one of the key elements of a clinical herbalist's practice. It is also an aspect that intimidates many practitioners, especially those newer to herbal medicine. Jill Stansbury's book is a valuable guide to formulation that will lend a strong guiding hand for herbalists of all skill levels. Her language is clear, instructive, and easy to read. She gives multiple examples drawn from her extensive practice, and her explanations offer direction for working with individual patients. *Herbal Formularies for Health Professionals, Volume 1* is a comprehensive and practical book that I will recommend to my herbal students."

—7SONG, director of Northeast School of Botanical Medicine; Director of Holistic Medicine, Ithaca Free Clinic

HERBAL FORMULARIES FOR HEALTH PROFESSIONALS

VOLUME 1

DIGESTION AND ELIMINATION

INCLUDING THE GASTROINTESTINAL SYSTEM,
LIVER AND GALLBLADDER,
URINARY SYSTEM, AND THE SKIN

DR. JILL STANSBURY, ND

Chelsea Green Publishing
White River Junction, Vermont

Project Manager: Patricia Stone
Editor: Fern Marshall Bradley
Copy Editor: Nancy Bailey
Proofreader: Deborah Heimann
Indexer: Shana Milkie
Designer: Melissa Jacobson

Printed in the United States of America.
First printing February, 2018.
10 9 8 7 6 5 4 23 24 25 26 27

Library of Congress Cataloging-in-Publication Data
Names: Stansbury, Jill, author.
Title: Herbal formularies for health professionals. Volume I, Digestion and elimination, including the gastrointestinal system, liver and gallbladder, urinary system, and the skin / Dr. Jill Stansbury.
Other titles: Digestion and elimination, including the gastrointestinal system, liver and gallbladder, urinary system, and the skin
Description: White River Junction, Vermont : Chelsea Green Publishing, [2017] | Includes bibliographical references and index.
Identifiers: LCCN 2017044410| ISBN 9781603587075 (hardcover) | ISBN 9781603587082 (ebook)
Subjects: | MESH: Formularies as Topic | Phytotherapy—methods | Digestive System Diseases—drug therapy | Urologic Diseases —drug therapy | Skin Diseases—drug therapy
Classification: LCC RM666.H33 | NLM QV 740.1 | DDC 615.3/21—dc23
LC record available at https://lccn.loc.gov/2017044410

Chelsea Green Publishing
White River Junction, Vermont, USA
London, UK
www.chelseagreen.com

To the Standing Green Nation—
all plants everywhere as living, breathing, sentient beings.
May we learn to treat them with respect and care.

And to all the traditional peoples, cultures, and nations
who have lovingly entered into sacred relationship with
the earth and the plants, and contributed to the art
and craft of herbal wisdom shared in these pages.

— CONTENTS —

— INTRODUCTION —

Capturing the Wisdom

Formularies are long-standing traditions in medicine. One of the earliest known such documents is the *Ebers Papyrus*, dated to 1550 BCE, which was written on a scroll of paper made from papyrus leaves. The Alexandrian School of Medicine in ancient Greece, circa 330 BCE, used handwritten herbals as their "textbooks," which detailed the clinical use of plant-based medicines. At the advent of the Common Era, herbal formularies typically detailed actual formulas. The treasured leather-bound books were filled with pages edged in gold, and they offered exacting recipes for treating various diseases, based on a skilled herbalist's clinical experience. As pharmaceutical medicine emerged, however, such recipe books evolved to detail the most commonly used drugs, and offerings of "recipes" went out of fashion. While the earliest formularies were entirely based on natural medicines—primarily herbs, animal secretions or tissues, and mineral-based substances—they evolved to consist primarily of pharmaceutical agents, as modern medicine evolved away from natural products and toward synthetic substances. In the present era, such "formularies" are largely drug compendiums. Many countries and governmental agencies have compiled lists of their most valuable medicines, detailing drugs approved and sanctioned by the region's health care system. In the United States, the *United States Pharmacopeia* and the *National Formulary* (USP–NF) are published annually to this day, having evolved from earlier herbal pharmacopeias.

Herbalists have continued the tradition of producing recipe-based formularies, also known as "herbals," and such books are often among the most esteemed and well used in their beloved collection. Nearly all budding herbalists go through a stage of collecting effective recipes from their elders, possibly starting with a recipe for a surefire method of making a perfect salve or an effective cough syrup. As they advance their clinical skills, they may collect formulas for precise clinical situations, from eyewashes to a tea that someone with nausea and vomiting can keep down and from palatable medicines for toddlers to safe recipes that use potentially toxic herbs to ameliorate pain. In China, effective formulas may be highly guarded family secrets, passed down through generations, and may offer a lucrative livelihood for the owners. For many centuries, herbalists have prized formulas endorsed by a master clinician, and herbal formularies offering dozens of such recipes are considered treasure troves. Such information is so priceless that in ancient times, healers or scholars would travel for months to reach the location where such a book was housed and spend another month or two to copy the prized text by hand. For example, *De Materia Medica*, by Pedanius Dioscorides, is believed to have been in circulation for well over 1,400 years prior to the advent of the first commercial printing press, and individuals or professional scribes created handwritten copies. Because early European monasteries were frequently the repositories of books, monks often served as professional scribes and sometimes also acted as healers, brewers, and medicine makers due to being keepers of the wisdom housed on the premises. Traditional herbals were among the first books printed and were extremely popular in Asia, Europe, Egypt, and India. Among the essential wares of the first European immigrants to North America were their seeds—and their herbal formulary. Throughout the 1400s and up through the 1600s, such herbals contained a mix of magical and medicinal information about plants, but present-day herbals include a more scientific mix of biochemical, pharmacologic, and toxicologic information about plants. With the resurrection of herbal medicine as one arm of the alternative medicine field, herbal formularies may now include research-based evidence and mechanisms of action to serve a new generation of medical herbalists.

Having made rather arduous journeys to the Amazon basin nearly 50 times in my life to learn about plants and medicine directly from Indigenous communities, I can understand the passion that drove ancient herbalists and physicians to go to great lengths to seek the wisdom keepers. I know that I am not alone in cherishing my books and recipe file and in holding my elders in high respect. When they are willing to share their hands-on knowledge regarding the use of plants as healing allies, I listen up!

I have taught at hundreds of medical conferences over the last several decades, addressing audiences of skilled clinicians. Over the years I have learned that many of us attend such events not to reap a bounty of new material at each lecture, but rather in hopes of gleaning a few "clinical pearls." We may not take notes as a presenter reviews molecular pathways or pathology, but when an expert in the field offers therapeutic advice on how to better heal a patient, just watch as everyone reaches for their pencils. That's the pearl of clinical wisdom we know we can implement in our practice as soon as we return home. Similarly, many of us scour the published studies on herbs, even if the only studies available are animal studies or cell culture investigations, seeking any shred of guidance to better treat challenging cases. When a respected colleague shares real-life experience, the information is highly valuable and provides the basis for mealtime and evening discussions when herbalists gather.

About This Book

This text is the first in a set of five comprehensive volumes aimed at sharing my own clinical experience and formulas to assist herbalists, physicians, nurses, and allied health professionals to create effective herbal formulas. The information in this book is based on the folkloric indications of individual herbs, fused with modern research and my own clinical experience.

This volume and the others that will follow are organized to reflect logically related organ systems. I begin with digestion and elimination, drawing inspiration from the old medical term *emunctory*, which refers to an organ of elimination and can be applied to the bowels, the urinary system, and the skin. Digestion and elimination are considered the foundation of health. Optimizing the health and function of the emunctories will benefit many other organ systems and ameliorate many health problems outside of these emunctory organs themselves—from allergic reactivity to infectious illness to hormone balance to musculoskeletal inflammation and more.

Each volume offers specific herbal formulas for treating common health issues and diagnoses within the selected organ system, creating a text that serves as a user-friendly reference manual as well as a guide for budding herbalists in the high art of fine-tuning an herbal formula for the person, not just for the diagnosis. Each chapter includes a range of formulas to treat common conditions as well as formulas to address specific energetic or symptomatic presentations. I introduce each formula with brief notes that help to explain how the selected herbs address the specific condition. At the end of each chapter, I have provided a compendium of the herbs most commonly indicated for a specific niche, a concept from folklore simply referred to as *specific indications*. These sections include most herbs mentioned in the corresponding chapter and highlight unique, precise, or exacting symptoms for which they are most indicated. Please note that these listings do not encompass *all* the symptoms or indications covered by the various herbs, but rather only those symptoms that relate to that chapter—the indications for GI symptoms, indications for biliary issues, indications for skin complaints, and so on. You'll find certain herbs repeated in the specific indications section of all four system chapters in the book, but in each instance, the description will feature slightly different comments. Readers are encouraged to refer back and forth between the various chapters to best compare and contrast the information offered.

The Goals of This Book

My first goal in offering such extensive and thorough listings of possible herbal therapies is to demonstrate and model how to craft herbal formulas that are precise for the patient, not for the diagnosis. It is my hope that after studying formulas in this book and following my guidelines for crafting a formula, readers will assimilate this basic philosophic approach to devising a clinical formula. As readers gain experience and confidence, I believe they will find that they rely less and less on this book and more and more on their own knowledge and insight. That's what happened to me over the years as I read the research and folkloric herb books and familiarized myself with the specific niche-indication details of a wide range of healing plants. I now have this knowledge in my head, and devising an herbal formula for a patient's needs has become second nature, and somewhat intuitive. But from talking with my herbal students over several decades of teaching, I have come to understand that creating herbal formulas is one of the most challenging leaps between simply absorbing information and using it to treat real, live patients. Students often feel inept as they try to sift through all their books, notes, and knowledge and struggle to use "information" to devise a single formula that best addresses a human being's complexities. Thus, I felt that it was high time that I created a user-friendly book to help students refine formulation skills and to help all readers develop their abilities to create sophisticated, well-thought-out formulas.

Another goal I aim to achieve through this set of herbal formularies is to create an easy-to-use reference that practitioners can rely on in the midst of a busy patient day. In this "information age," it is not hard to track down volumes of information about an herb, a medical condition, or even a single molecule isolated from a plant. The difficulty lies in remembering and synergizing it all. While this text doesn't pretend to synergize the "art" of medicine in one source, I believe it will help health professionals quickly recall and make use of herbal therapies they already know or have read about by organizing them in a fashion that is easy to access quickly.

Naturopathic physicians are a varied lot. Add in other physicians and allied health professionals, and the skill sets are varied indeed. I rely on my naturopathic colleagues to inform me about the latest lab tests, my allopathic colleagues to inform me about new pharmaceutical options, and my acupuncture colleagues to inform me on what conditions they are seeing good results in treating. This text allows me to share my own area of expertise. I have included a large number of sidebars that feature some of the more in-depth research on the herbs and individual molecular constituents, helping to provide an evidence-based foundation for the present era of medical herbalism.

I realize that not all clinicians specialize in herbal medicine, even naturopathic physicians. I hope that this formulary will serve as a handy reference manual for those who can benefit from my personal experience, formulas, and supportive discussions.

Creating Energetically Fine-Tuned Formulas

Much like a homeopathic *materia medica*—another term from folklore referring to collected information on individual medical materials, in this case, the plants themselves—this set of formularies aims to demonstrate to clinicians how to choose herbs based on *specific indications* and clinical *symptoms* and *presentations*, rather than on diagnoses alone. For example, I do not offer a single one-size-fits-all formula for dermatitis. Instead, I've compiled more than a dozen specific dermatitis formulas, such as a Tincture for "Prickly Heat" and Itchy Skin in Hot Weather, a Tea for Chronic Eczema, Dermatitis, and Hives, and a Tincture for Dry Skin Concomitant with Hypothyroidism. Along with the formulas, I provide targeted lists to help readers begin to craft their own formulas. Budding herbalists can draw inspiration from lists such as "Herbs for Skin Eruptions of the Hands and Feet," "Herbs for 'Wet' Atopic Dermatitis," and "Herbs for Itching Skin." I include supportive research on herbs that helps to explain why a particular herb is chosen for a particular formula as well as endnote citations that provide details of specific studies for those interested. I also provide findings from research on individual herbs that are essential to the treatment of the various conditions featured in a chapter. To make the text as useful as possible for physicians and other clinicians, I also offer clinical pearls and special guidance from my own experience and that of my colleagues—the tips and techniques that grab attention at medical conferences year after year.

The Information Sourced in This Book

The source of the information in these volumes is based on classic herbal folklore, the writings of the Eclectic physicians, modern research, and my own clinical experience. Because this book is designed as a guide for students and a quick reference for the busy clinician, the sources and research are not rigorously cited, but enough so as to make the case for evidence-based approaches. When I offer a formula based on my own experience, I say so. I also make note of formulas I've created that are more experimental, due to lack of research on herbs for that condition or my lack of clinical experience with it.

My emphasis is on Western herbs, but I also discuss and use some of the traditional Asian herbs that are readily available in the United States. In some cases, formulas based on Traditional Chinese Medicine (TCM) are featured due to a significant amount of research on the formula's usage in certain conditions. I readily admit that TCM creates formulas *not* for specific diagnoses, but rather for specific energetic and clinical situations. However, I have included such formulas, perhaps out of context, but with the overall goal of including evidence-based formulas, with the expectation that readers and clinicians can seek out further guidance from TCM literature or experienced clinicians where possible. In reality, TCM is a sophisticated system that addresses specific presentation, and I have borrowed from this system where I thought such formulas might be of interest or an inspiration to readers. I admit that listing just one formula for a certain condition based on the fact there have been numerous studies on it is somewhat of a corruption of the integrity of the TCM system, which is aimed at precise patterns and energetic specificity. Nonetheless, I chose to do so with the goal of creating a textbook to help busy clinicians find information quickly, while still encouraging individualized formulas for specific presentations.

While I have endeavored to create herbal formulas to address as many different conditions and presentations as possible, this text purposefully avoids addressing specific types of digestive, liver, and renal cancers because to do the topic justice would require a textbook all its own. And frankly, there is not yet the evidence to cite, nor do I have the clinical experience in dozens of such cases to feel I could pose enthusiastic herbal formula suggestions. One exception in this volume is skin cancers, which are covered briefly in chapter 5. Some skin cancers are slow-growing and accessible, and the topical use of herbal therapies can be an important consideration in treating squamous cell lesions and actinic keratosis. Liver and bladder cancers are also mentioned due to the growing volume of research on complementary herbal therapies, although the information is not intended to replace an expert's care.

How to Use This Book

Each chapter in this book details herbal remedies to consider for specific symptoms and common presentations of various diagnoses. Don't feel that you must be a slave to following the recipes exactly. When good cooks create a food recipe, they are always at liberty to alter the recipe to create the flavor that best suits the intended meal—the big picture. The formulas listed should not be thought of as *the* formulas to make, but rather as a guide and example, inviting the clinician to tailor a formula for each individual patient.

To create an herbal formula unique to a specific person, the clinician should first generate a list of actions that the formula should perform (intestinal carminative, biliary antispasmodic, urinary antimicrobial, and so on), and then generate a list of possible herbal *materia medica* choices that perform the desired actions. If these ideas are new to you, you may want to begin by reading chapter 1, "The Art of Herbal Formulation," before you start generating lists.

Look to the formulas in chapters 2 through 5 that address specific symptoms for guidance and inspiration. (These formulas are grouped within the chapter by a general diagnosis, such as "Formulas for Constipation" or "Formulas for Dermatitis.") Regard the lists and formulas I have provided as starting points and build from there. In my commentary on the individual formulas and in sidebars that focus on specific herbs, I offer further guidance as to whether the formula or individual herbs are safe in all people, possibly toxic in large doses, intended for topical use only, or indicated only in certain cases of that particular symptom. Once herb and formula possibilities have been identified, the reader should then review the "Specific Indications" section at the end of the chapter to narrow in on choices of which herbs would be *most* appropriate to select and to learn more about how those herbs might be used. Herbalists can narrow down long lists of herbal possibilities to

Unity of Disease (Totality of Symptoms)

The concept that any given health issues a person may experience are actually one disease, as opposed to a number of disparate diagnoses to be treated individually, is a core tenet of naturopathic medicine and the philosophical underpinning of holistic medicine in general. Any one symptom does not provide the full story, and just because you can label the symptoms with a Western diagnosis and offer the established therapy for that diagnosis does not mean you are really helping a person to *heal.* A careful consideration of the sum totality of all symptoms is important to reveal underlying patterns of organ strength or weakness, excess or deficiency states, nervous origins versus nutritional origins, and, of course, a complex overlap of all such issues. The most effective therapies will address *all* issues in their entirety and involve an understanding of the entire energetic, mental, emotional, nutritional, hereditary, situational, and other processes creating a complex web of cause and effect—the unity of any given individual's "dis-ease."

just a few *materia medica* choices that will best serve the individual. In many cases, the reader/clinician will be drawing upon herbal possibilities from a number of chapters and organ systems as the clinical presentation of the patient dictates. Thus, you are not making a formula by throwing together all the herbs listed as covering that symptom or symptoms, but you are studying further and narrowing down the list of possibilities to consider based on the sum totality of all the symptoms. In some cases, you will rule out herbs on the list for a particular symptom after reading the specific description of that herb at the end of the chapter. In some cases, you might decide to put one herb in a tea and another in a tincture due to flavor considerations. In other cases, you might decide that you will prepare only a topical remedy. And in other urgent situations, you might come up with a topical, a pill, an herbal tea, *and* a tincture to address the situation as aggressively as possible. Aim to select the best choices, and avoid using too many herbs in one formula. Larger doses of just a few herbs tend to work better than smaller doses of many herbs, which can confuse the body with a myriad of compounds all at once. The use of three, four, or five herbs in a formula is a good place to start; this approach also makes it simpler to evaluate what works when the formula is effective as well as what is poorly tolerated, should a formula cause digestive upset or other side effect.

As you work with this book, you will discover that there is some overlap of information among the various chapters. For example, in chapter 5, "Creating Herbal Formulas for Dermatologic Conditions," you'll find a list of herbs appropriate for dermatitis due to underlying digestive and liver symptoms, and in chapter 2, "Creating Herbal Formulas for Gastrointestinal and Biliary Conditions," I've included a list of herbs appropriate for constipation associated with skin eruptions.

Learning from the Formulas in This Book

In reviewing the formulas in this book, notice how specific herbs are combined with foundational herbs to create different formulas that address a variety of energetic presentations. There are a handful of all-purpose immune modulators, all-purpose alterative herbs, and all-purpose anti-inflammatories that can be foundational herbs in many kinds of formulas. Such foundational herbs can be made more specific for various situations by combining with complementary herbs that are energetically precise. Notice how the herbs are formulated to be somewhat exacting to address specific symptoms and make a formula be warming, drying, cooling, or moistening and so on. Also, note how acute formulas may have aggressive dosages and include some strong herbs intended for short-term use, while formulas attempting to shift chronic tendencies are dosed two or three times a day and typically include nourishing and restorative herbs intended for long-term use. Also notice how some potentially toxic herbs are used as just a few milliliters or even a few drops in the entire 2-ounce formula. These dosages should not be exceeded, and if this is a clinician's first introduction to potentially toxic herbs, further study and due diligence are required to fully understand the medicines and how they are safely used. Don't go down the poison path without a good deal of education and preparation. I am able to prepare all of the formulas in these texts upon request, but I can only offer those containing the "toxic" herbs (*Atropa belladonna, Aconitum, Gelsemium, Hyoscyamus*, and so on) to licensed physicians.

It is my sincere hope that this book helps you in your clinical work and efforts to heal people.

DR. JILL STANSBURY

— CHAPTER ONE —

The Art of Herbal Formulation

Creating an effective and sophisticated herbal formula is somewhat of an art, and like all arts, it is difficult to put into step-by-step directions; however, this is my attempt to do just that. This book aims to explain how to create specific formulas for *presentations* rather than *diagnoses*—rather than offering a single formula or two for a general condition such as hypertension or irritable bowel syndrome, this text offers exacting formulations to best address the precise presentation of the person. I have personally seen many *different kinds* of hypertension, dyspepsia, cystitis, and other conditions that conventional medicine tends to treat with one across-the-board medication or therapy. My aim is to coach readers on how to create numerous finely tuned herbal options for treating the person, and not the diagnosis—a core tenet of natural medicine.

Creating effective herbal formulas and treatment plans requires many skills: knowledge of the herbs; herbal combinations best for a particular situation; the proportions to use in a formula; the starting dose, frequency of dose, and how long to dose; what form of medicine is best, such as a tincture, a tea, a pill, or a topical application; broader protocol options that may include diet, exercise, nutritional supplements, or referral to allied health professionals; and the follow-up plan.

Hippocrates said, "It is more important to know what kind of person has a disease than to know what kind of disease a person has." To know what sort of person has a disease requires careful listening and skilled and nuanced questioning in a safe and comfortable setting (see "Asking the Right Questions" on page 8). Listen for underlying causes, for overarching emotional tone, for what a person is able to do for themselves (sleep more, exercise more, eat better, brew a daily tea), and for what they are resistant to (sleep more, exercise more, eat better, brew a daily tea). Address underlying causes and start where people show some interest and capacity. Cheerlead to instill enthusiasm if required, so that people become better educated and thereby better motivated to make important changes or adopt valuable healing practices. Giving the right medicine is only one aspect of doing healing work with a person; creating a sacred space that invites truth and sharing is key to getting to the point where you know what sort of person has a disease.

Healing also stems from understanding pain, suffering, challenges, and unique situations, and from nonjudgmental listening to the stories so often linked to our physical ailments. It comes from giving encouraging words and sympathy, congratulating and complementing people's efforts and accomplishments, and giving people the tools, resources, and support to succeed. This kind of true caring and earnest effort to provide real support are among our best medicines.

The Importance of Symptoms

Naturopathic medicine has a different philosophical stance on physical symptoms than allopathic medicine, especially infectious and eruptive symptoms, and that view is worth briefly describing here for those who may be unfamiliar.

The Biochemical Terrain

The terrain of the human body invites microbes specific to the chemical composition of the ecosystem, and when infectious microorganisms have consumed those specific chemical substrates, the disease-producing microbes are no longer supported. The microbes themselves change the chemical composition of the tissue by consuming the "food" that invited them, and when those nutritional resources are exhausted, the biochemical makeup of the tissues is changed for the better. Antibiotics are not the best way to treat chronic infections. Instead, optimize the ecosystem to support the desired beneficial flora. As when bacteria start to stink up the compost bucket on the kitchen counter, we don't spray the compost with a germicide—we clean the bucket! This text stresses "opening the emunctories" as a means of treating infections and reducing inflammation. By changing the biochemical terrain, we are less susceptible to opportunistic infections. The emunctories are our organs of elimination: kidneys, liver, bowels, and skin, and stimulating the eliminative functions of these organs can often have far-reaching systemic effects in reducing infection and inflammation.

The Role of a "Healing Crisis"

Acute infections, diarrhea, vomiting, fever, skin eruptions, abcesses, and boils have been referred to as symptoms of a "healing crisis"—meaning the acute symptoms are actually a part of the body's attempt to heal itself, allowing the infectious organisms to consume the "morbid matter" and, thereby, restore a healthier ecosystem.

Naturopathic philosophy embraces the symptoms of a healing crisis as a triumph of the vital force. Our symptoms serve us and call attention to the imbalances requiring changes. Be thankful when the body has the vitality to manifest a healing crisis. Be concerned when infections, eruptions, and discharges stop, and allergies, autoimmunity, joint pain, ulcers, blood pathology, and so on emerge. These are not "healing crises," but rather signs that vitality is being damaged and pathology is becoming deeper and more serious.

Because the symptoms of a healing crisis are the way in which the body can heal itself, such symptoms should not be suppressed, but rather supported and made as tolerable as possible. Only when such symptoms are so severe as to threaten damage to the body should they be suppressed. Use herbs to help a fever do its job, use herbs to help vomiting or diarrhea eliminate infectious microbes—in other words, use herbs to palliate the discomfort involved in righting the ecosystem and to open the emunctories.

The Harm in Suppressing Symptoms

Habitual suppression of symptoms over time can force the body to give up its struggle for health. For example, laxatives may relieve constipation temporarily, but ultimately damage normal peristalsis. Stimulants may initially provide energy, but ultimately exhaust the adrenals and nervous system. For example, a pot of coffee may jolt an exhausted person awake, but it will not improve core energy status in the long run and will only further exhaust it. Antibiotics can kill infectious pathogens, but unless the underlying ecosystem that supports such microbes is significantly improved, the same pathogens will be supported a second, third, and fourth time and become resistant to antibiotics if they are given repeatedly, such as for chronic otitis, cystitis, or sinusitis.

Asking the Right Questions

Learning how to ask questions that will elicit relevant information is as much an art form as creating an herbal formal. Aim your questions to gather information in several categories. What follows here is a broad, but not exhaustive, list of questions that can yield helpful information.

Etiology

How long has this been happening?
How did it begin?
How has it evolved over time?
What else was going on in your life at the time this began?
Have any other symptoms, complaints, problems accompanied this complaint?
What is your health history?
Any previous episodes? Related pathologies? Family members with this complaint?
What is the predominant emotion associated with this complaint? Did the complaint begin during a period of grief? Anxiety? Ambition?
Did the complaint begin following hard labor or an injury, stress, eating a new food, traveling out of the country, starting a new medication (and so on)?

Quality and Occurrence of the Complaint: "PQRST"

Provocation. Does anything seem to bring on the complaint? Does anything alleviate it? What makes the complaint better or worse?
Quality. What is the character of the complaint: burning, throbbing, dull, sharp, shooting, aching . . . ?
Radiation. Does the pain travel? Does this symptom affect any other organ? Is this complaint associated with any other symptom?
Severity. On a scale of 1 to 10, how severe is this? Does it interfere with sleep, activities, work, sex, relationships, child rearing, creativity, and hobbies?
Time. Timing throughout the day, throughout the month (hormonal fluctuations?), throughout the year (seasonal allergies? seasonal affective disorder? Oriental or Native American concepts of seasons?). Timing may involve an association with eating

food or going without eating, an association with anxiety/relaxation, an association with menstrual cycle, or an association to a certain environment or allergen exposure. Is the complaint any better or worse with sleep?

Concomitant Symptoms

Do you feel hot or cold when this occurs?
Is there lethargy or anxiety, heart palpitations, or weakness?
Does it occur during times of stress and activity, during sleep, or after prolonged sitting?
Is there fear or fatigue, mania or depression, weakness or restlessness (and so on)?
Is there a desire to be consoled or to be left alone?

Constitution and Energetic Considerations

Note the constitution by asking the right questions, and observe the person to get additional clues:

Complexion. Pale? Yellow? Cyanotic? Flushed? Haggard? Dry? Damp? Inflamed? Quick or slow to perspire? Oily or dry skin?
Pulse. Strong or weak? Fast or slow? The quality? The variability?
Tongue. Large or small?
Muscles. Well developed, overdeveloped, underdeveloped? Spastic, atrophic? Soothed by pressure and massage or aggravated by pressure or massage?
Senses. Hyperacute includes sensitivity to odors, noises, bright light; pain; racing thoughts. Hypoacute includes loss of taste, hearing, sensation; poor ability to concentrate, remember, respond.
Diet and Appetite. Hungry, anorexic? Can eat large quantities or tolerate only small amounts? Hypoglycemic symptoms? Unusual cravings for a particular flavor or food? Unusual aversion or aggravation by particular foods or flavors? Are the symptoms better following meals, or between meals when stomach is empty? What is the general diet, nutrient intake, bowel habit, and quality of the stool?
Thirst. Large thirst versus small or no thirst? Thirst due to dry mouth? Thirst due to compulsion? Thirst for large gulps or thirst for small sips? Thirst for warm fluids or thirst for cold fluids?
Sex drive. High, low? Markedly cyclical?
Sleep. Requires more than 8 hours of sleep to feel rested, able to function with little sleep? Sleeps soundly all night, or wakes many times? Takes a long time to fall asleep, but then sleeps well, or falls asleep readily but wakes at a particular time unable to sleep further? Eating before bed disturbs or improves sleep? Restless during sleep or wakes with a jolt? Has nightmares or difficult dreams?

The "Triangle" Exercise

Once you have listened thoughtfully and well to a description of symptoms, I recommend starting with a traditional and simple method of thinking through the choice of the components of a formula: *the triangle.* Visualize a triangle with a horizontal base and two slanting sides that meet at a point at the top of the triangle. Like the triangle, your formula needs a *base* on which to *rest*—that's where you'll start—and it also needs two axes that point the formula in the right direction.

In general, the herb or herbs chosen as the base should be nourishing and nontoxic—something tonifying, restorative, alterative, adaptogenic, or nutritive to the main organ system, tissue, or issue of concern. For example, cardiovascular formulas might have *Crataegus* as a base herb on which to rest. A formula for insomnia with exhaustion might rest on *Withania*, and a skin condition formula might rest on *Calendula* or *Centella.* There is an old Wise Woman saying that all healing begins with nourishing. Everyone needs nourishment and tonification, so such herbs are always appropriate. Simple, right?

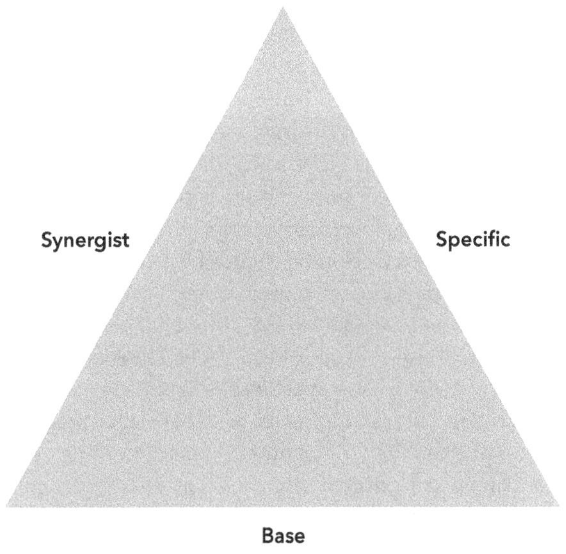

Figure 1.1. The Triangle—A Pragmatic Tool for Crafting Herbal Formulas

From there, herbs for the other two axes of the triangle are selected to "point" the formula in a more specific direction. The base herbs tend to be nonspecific and indicated in a very wide number of clinical situations, so to make your formula more specific for an individual patient, you will need to drive it in the right direction. Through the ages, herbalists have used varied vernacular to refer to such herbs as *synergists, specifics, energetics, kickers, directors*, and so on. The choice of names does not matter, but for the sake of discussion I use the words *synergist* and *specific* to apply to the other two axes of the herbal formula triangle. In Traditional Chinese Medicine (TCM), royal terms such Emperor, Minister, and Assistants may be used.

The choice of a specific requires a detailed understanding of *materia medica*, such that one or two energetically and symptomatically precise herbs can be selected. The lists of herbs offered in "Specific Indications" at the end of each chapter help provide exacting details of symptoms for which individual herbs are most indicated. If there is more than one specifically indicated herb, use them both! If one tastes good and one tastes bad, put the tasty one in a tea and put the less palatable option in a tincture or pill. The point is to identify one or two specifics that best address all symptoms possible as well as the constitution, energy, and the entire body's strengths and weaknesses.

As an example, if your case mainly involves hypertension, you may choose to rest the formula on *Crataegus*. The unique identifying symptoms are that the woman's blood pressure began going up during menopause and she is also experiencing episodes of tachycardia and heart palpitations. After reviewing the specific indications for various herbs, you might choose *Leonurus* as your specific to point or drive the formula in the right direction and to the right place. If, on the other hand, the patient is a long-term smoker with high cholesterol and vascular inflammation, your specific may be *Allium*. And if the hypertension formula is for a highly stressed Type A personality who also suffers from alarming hypertension accompanied by a throbbing headache, an appropriate choice of specific might be *Rauvolfia*.

A *synergist* is less specific but takes into account underlying contributors such as other organ systems involved and other energetic considerations that contribute to the overall case. To follow up on the hypertension examples above, the menopausal hypertension formula might include *Actaea* as a synergist to also offer hormonal and nervine support. The long-term smoker may benefit from *Ginkgo* as a synergist to combat circulatory stress, and the Type A severe hypertensive might benefit from *Piscidia* as a vasodilating synergist.

The "Specific Indications" in chapters 2 through 5 offer details about individual *materia medica* options to help choose specifics and synergists to best address the sum total of the symptoms. A synergist may be very "specific"—it's all semantics. The point is to take into consideration various individual factors and unique presentations that contribute to the case.

Returning to the example of the menopausal woman with hypertension and episodic heart palpitations, if she also was suffering from stress and poor sleep as underlying contributors, you might select *Withania* as a synergist. If, on the other hand, she was sleeping fine and did not feel particularly stressed but was having some difficulty with constipation and hemorrhoids, you might select *Aesculus* as the synergist. Or if she was having constipation, hemorrhoids, and episodes of heavy menses as part of the menopausal transition, you might select *Hamamelis* as the synergist. Review the "Specific Indications" sections at the end of each chapter to start developing a familiarity with each individual herb and its "personality" or specific symptoms that it best addresses.

Whatever the details you are presented with, you will use the model of the triangle and select at least three herbs: a nourishing base (tonic), a specific, and a synergist. There is no reason not to select more than one nourishing base or more than one specific or synergist if something very appropriate pops out. If, for example, the menopausal woman with hyptertension had a long-standing history of anxiety, stress, mood swings, and emotional lability well before the onset of the recent hypertension, then her nervous system may need nourishment and tonification as much as, or more so than, her cardiovascular system. In that case, you might choose two nourishing herbs upon which to rest the formula—one such as *Crataegus* for the vascular system and one such as *Avena* for the nervous system. Or if she had a history of these stress-related symptoms and they also tended to cause irritable bowel and diarrhea on many occasions, then you might choose *Crataegus* and *Matricaria*, which provide a nourishing base for the vascular, nervous, and digestive systems.

Therefore, when you reflect on all the details of a case, other symptoms often emerge that are not the chief complaint, but highly important accompanying considerations that should not be overlooked. A simple

More about Formula Components

Base. Also called the lead herb or the Director, the base should be a tonic that has a nourishing and restorative effect on the main organ system affected. This herb does not require a great amount of skill to choose as it is often among the herbs best known for having an affinity to a particular organ system.

Synergist. Also called the Adjuvant, the Balancer, or the Assistant. This component helps correct or complement the action of the lead herb and helps drive it to the desired tissues. The selection of these herbs requires an in-depth understanding of the case and person being treated in order to address underlying causes and give the formula the needed energetic specificity.

Specific. Also called the Kicker or the Energetic Specific. This component is selected not just for a specific condition or diagnosis, but for a specific quality, essence, or expression of any given disease or disorder. Such qualities include considering the pulse, the tongue, the affect, the pathology, the etiology, and the person themselves to guide the selection of the most specific medicine. The practice of learning and using specific medications is gleaned from careful study, observation, and clinical practice.

presentation of perimenopausal hypertension with episodic heart palpitations and no other symptom details may lead to the formula as described above: *Crataegus*, *Leonurus*, and *Actaea*. However, if you learn that this woman has had gallstones, a history of fat intolerance, and many digestive symptoms and that the heart palpitations are worse after meals, you might choose *Curcuma* or *Silybum* as synergists because biliary congestion may contribute to hypertension and, for that matter, may deter the processing of hormones and postprandial lipids. This is another example of an important underlying factor that should not be overlooked. Or if you find that she is a highly allergic person who sometimes has wheezing with chemical exposure, occasional hives, and chronic low-grade eczema on her hands that flares up after a day of heavy cleaning with exposure to a lot of water and cleaning products, you might choose *Angelica* as a second specific because it is specific for hypertension, asthma, eczema, and hives.

Mastering the Actions of Herbs

Herbal clinicians should have an excellent grasp of primary actions while gaining a solid knowledge of *materia medica* and specific indications. Clinicians should know herbs that are the best antispasmodics for a variety of situations, the best anti-inflammatories, the best vulneraries, the best nervines, the best antimicrobials, and so on. Such actions of herbs are also foundational considerations when creating an herbal formula or when considering which herbs to select as a tonic base, synergist, and specific of a formula triangle.

Another exercise that I often encourage my students to undertake is learning basic categories of actions of herbs. Actions include antispasmodics, antimicrobials, carminatives, alteratives, adaptogens, demulcents, vulneraries, and so on. I encourage my students to type up pages or create a "little black book" that helps to remind them of the hundreds of herbs they are learning, organized as to their categories of action. And from there, they can go deeper. For example, individual antispasmodic herbs might be categorized as having an affinity for a certain organ system or being best suited for a particular quality of spasm. Consider the following antispasmodics: *Lobelia* is especially indicated for respiratory smooth muscle and cardiac muscle spasms, *Dioscorea* is specific for twisting and boring muscle spasms about the umbilicus, *Piper methysticum* can allay acute musculoskeletal pain and urinary spasms, and *Viburnum opulus* or *prunifolium* is especially effective for spasms of the uterine muscle. Sidebars throughout this book and the volumes that will follow offer quick reference lists that summarize various actions.

It is best to avoid a "What herbs are good for intestinal cramps?" or "What herbs are good for bladder infections?" style of creating herbal formulas. When treating ulcerative colitis with gas, bloating, and intestinal cramps, for example, rather than asking what herbs are "good for ulcerative colitis," I encourage clinicians

Supporting Vitality Instead of Opposing Disease

Western medicine has its "differential diagnosis" where the presenting symptoms can generate a list of possible (differential) causes, ultimately leading to the diagnosis. Although this approach has some value, herbal medicine is less concerned with the formal diagnosis. Instead, herbal medicine aims to carefully consider organ system strengths and weaknesses, underlying causes (stress, toxicity, poor nutrition, poor sleep, circulatory weakness, inflammatory process, allergic hypersensitivity, and so on), and how to support and nourish basic organ function and systemic vitality. Medicines employed by herbalists are typically aimed at restoring function and supporting the innate recuperative powers of the body. The intelligent wisdom of the body to heal itself is sometimes referred as the "vital force," and herbalists aim to support the vital force more so than to oppose the symptoms of disease. Almost all herbs offer at least some nutrition, being more like foods than drugs, and by nourishing the body and stimulating the vital force, the body is supported in healing itself.

to ask themselves, "What *actions* do I need this formula to perform?" In this case, we need a formula that offers carminative and intestinal antispasmodic *actions*, anti-inflammatory *actions*, antiulcerative *actions*, and possibly hemostatic *actions*. Throughout this text, the comments that accompany each formula mention actions we are attempting to accomplish with the recipe. For example, "Mint is included here as a reliable intestinal antispasmodic. . . . " or "*Uva ursi* is one of the most well-known urinary antimicrobial herbs."

When thinking through an herbal formula, especially for difficult or complex cases, it is useful to write down what actions you wish the formula to perform and then list several herbs that perform this action. The next step is to consider any specific indications for the listed herbs to help narrow in on the best choices. And finally, the most nourishing herbs may be chosen as supporting the foundation of the triangle, a base on which the formula may rest. Other herbs from the action lists may then be chosen as specifics or synergists to best offer all the needed actions, while creating energetically specific, finely tuned formulas.

Energetic Fine Tuning

Using the triangle method and an awareness of the actions of herbs, as detailed above, will assist you in selecting three or more herbs that address a case in its entirety, and as such, the formulas are likely to be effective and successful. To fine-tune your formulas even further, an added tier of specificity is the energetic state of your patient. TCM philosophy often depicts the energetic state as a mixture of polar opposites in keeping with Taoist philosophy of yin and yang polarities. For example, is your patient hot or cold? Tight and constricted or loose and atrophic? Excessively damp in the tissues or excessively dry? Tired and lethargic or energized and manic—and so on. Ayurveda, the traditional medical system of ancient India, sets up a three-pronged system of doshas—vata, pitta, and kapha—rather than the two-pronged polar opposites of Taoism, but is similarly aimed at addressing differing constitutions and energetic presentations. The four-elements theory of ancient Western herbalism looks for symptoms or presentations categorized into earth, air, fire, and water related symptoms, with herbal therapies being chosen accordingly. Again, the precise system, vernacular, and approach do not matter, as long as you are aware of some sort of energetic presentation. Whether or not you take it upon yourself to learn the doshas, TCM, or four-elements thinking and prescribing, you can still begin to notice whether a patient is hot or cold, damp or dry, for example, and choose herbs based on the specific clues or symptoms that you discern through thorough questioning and from simple and obvious physical exam findings.

For example, an IBS patient may present with frequent loose stools and urgent diarrhea, and he tells you that he is bothered by fatty foods, acids, and spices, which make him feel hot and sweaty. Furthermore, you learn that these substances promote a burning discomfort in the bowels followed by an urgent need to race to the bathroom. You would most likely want

Pharmacologic versus Physiologic Therapy

Pharmacologic Therapy. Pharmacologic prescribing is the use of a potent medicine to force a rapid pharmacological response. The energy to catalyze changes, movement, and homeodynamic balance seems to come from the medicine itself. The chemical constituents have strong actions in the body and act as cardiosedatives, emetics, antibiotics, vasoconstrictors, antispasmodics, diuretics, and so on. Pharmacologic medicines can be used heroically and can save lives, but they don't build the vital force and restore organ function, and their use may be needed repeatedly. Most pharmaceutical drugs are pharmacologic in nature and are often suppressive to the body's vital force—for example, acetaminophen to suppress a fever, antibiotics to kill pathogens, or bisphosphonate drugs to halt bone cell turnover rates.

Some of the more toxic herbs can have pharmacologic activity, but in general, herbs are more like foods, offering nutrition and physiologic support. Pharmacologic medications do very little to nourish, tone, or deeply "cure" anything; the symptoms return as soon as the medication is removed. For example, steroids may suppress wheezing in the lungs or suppress eczematous skin lesions, but because they do not alter the underlying condition, wheezing and itching will recur as soon as the medications are stopped. At times, for serious and acute situations, an herb with a pharmacologic action may be chosen as a synergist or a specific in an herbal formula, but never as a lead herb upon which to base a formula.

Physiologic Therapy. Physiologic prescribing involves the use of gentle medicines over an extended period of time to nourish organs and restore normal tone and function. Physiologic medications do not have rapid or strong pharmacological actions and even if prescribed inappropriately would be unlikely to push the limits of homeostasis or cause undesirable side effects. Physiological medications balance, nourish, and tone, and they help restore optimal physiology through gentle support of assimilation, elimination, detoxification, metabolism, perfusion, and nerve function. Most herbs are physiologic in nature, capable of restoring normal functioning, organ tone, and homeostatic balance. In contrast to the above pharmacologic examples, herbs may be used to support a fever when needed to allow the body to fight infections, or to encourage the body to build new bone cells rather than halt all bone cell turnover. Because physiologic medications repair and restore tissue and organ function, they can usually eventually be stopped as the body becomes capable of maintaining balance without them. Herbs with nutritive physiologic actions should be those tonics upon which all formulas are based.

to choose a cooling soothing base for a formula such as *Althaea* or *Ulmus*, because demulcent herbs are cooling and calming to mucosal tissues and can allay the hot and overactive symptoms. Even though *Rumex* and *Taraxacum* are known to help digestion and especially liver function and the digestion of fats, it could be a problem to rest your formula on a significant quantity of these herbs because these alterative herbs are somewhat stimulating or warming to the digestive function. For an overstimulated, hot bowel, they are not the perfect choice energetically.

When I teach, I often present sample cases and lead discussion with my students to explore how to think through the choices of herbs for the base, synergists, and specifics, based on details of the particular patient. It can be helpful to work through two examples of people with the same "diagnosis," such as insomnia or acne, and discover how differently a formula evolves based on the patient's unique constitution, energetics, and symptoms. In the first sample case, I offer an extensive discussion that arrives at a specific formula. Following that, I present a more condensed sample case, with several options for base, synergist, and specific herbs. See how you do at finishing the selection process for the sample cases described below. There is no single right or wrong answer.

SAMPLE CASE: VARIATIONS OF INSOMNIA

Patient 1: This insomnia patient has been exhausted for years, with an accompanying history of chronic hay fever and occasional respiratory infections requiring medical attention. The insomnia is of recent onset and involves not being able to fall asleep for many hours, lying in bed very tired and exhausted, but awake. The person will finally fall asleep and wake in the morning groggy and still exhausted. She is cool with cold hands and feet in bed at night and even during the day. She experiences gas and bloating if she eats raw broccoli and onions. For such a patient, you might consider basing a formula on "energy or chi" tonics, such as *Panax*, *Eleutherococcus*, *Astragalus*, *Rhodiola*, *Ganoderma*, *Cordyceps*, or *Oplopanax*. You might settle on a base of *Panax* because it is more warming than some of the others. For the specific, you might select *Astragalus*, because it is also a chi and immune tonic, and it is also specific for allergies and respiratory infections that linger and exhaust. The synergist might be *Zingiber* because it enhances circulation in the cold extremities and improves digestion by warming and stimulating core organs. It could help drive the other ingredients in the formula where they need to go by being a heating, stimulating, and moving herb.

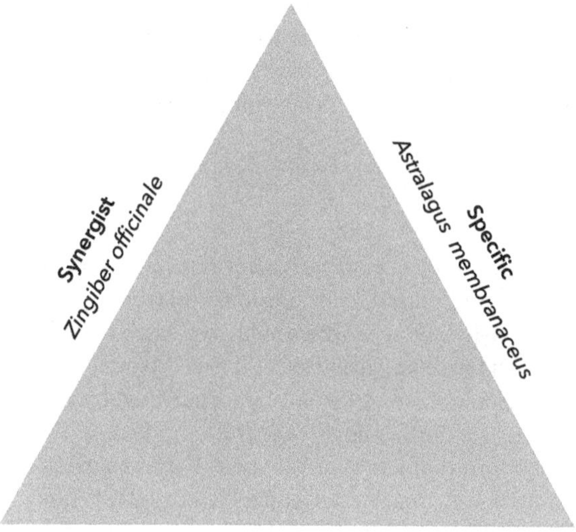

Figure 1.2. Formula for Insomnia Patient with Exhaustion

Patient 2: This second patient has suffered from restless sleep for many years, but it is presently worse. The pattern has been to fall asleep readily but to wake after a few hours and then fall back to restless sleep for the remainder of the night, waking briefly every half hour or so, tossing and turning with stiff muscles, and going back to sleep for short stretches only. She has occasional episodes of feeling too hot in bed and rolling away from her partner feeling uncomfortably warm, and taking a while to fall back asleep. The person has come for a consultation because recently, instead of frequently waking and rolling over and going back to sleep, she now lies awake for half an hour, sometimes several hours, before falling back to sleep. She also experiences back and joint stiffness at night in bed and for a short while in the morning upon waking. She reports having only a few minor colds per year, recovers quickly without needing treatment, and doesn't even feel terribly ill with them. She reports having oily skin, still prone to breakouts on the central face, especially premenstrually, and having some minor PMS symptoms, primarily emotional lability and crankiness.

There are many relevant details here. For this restless, stiff insomnia patient, who is excessive in movements, muscle tension, nervous tension, and heat, you want to rest a formula on something cooling and relaxing, particularly to the nervous and musculoskeletal systems, such as *Avena*, *Passiflora*, or *Scutellaria*. While *Valeriana* and *Piper methysticum* are some of the most powerful herbs specific for both sleep and muscle relaxation, you might choose not to "rest" the formula on them because they are rather hot and could be a problem long term or if put in the formula in too great a proportion. Not only are they heating, but they are not particularly nourishing and never listed in the folkloric literature as nervous system restoratives or daily long-term-use herbs. *Valeriana*, *Actaea*, and *Piper methysticum* could certainly be used in the formula, as long as care is taken to use them as synergists or specifics in smaller proportion than the other ingredients in the formula, while basing the formula on something more restorative to the nervous and musculoskeletal system and cooling to the body, such as *Avena* or *Scutellaria*. Another approach might be to consider adding *Valeriana* or *Piper methysticum* to a formula for short-term use or to be taken only before bed to reap the benefits of the more powerful muscle-relaxing effects of these herbs. In addition, you would create a separate formula to be used during the day and over the long term, attempting to restore the nervous and muscular tone so that the more powerful muscle relaxers and before-bed formula are no longer needed. *Scutellaria, Matricaria, Eschscholzia, Hypericum, Passiflora, Avena,* and *Melissa* are cooling, soothing, nourishing, and restorative to the nervous system, with *Scutellaria* and *Passiflora* having the greatest effects on muscle tension as well. Many people with insomnia, who are not particularly exhausted like the first insomnia patient, benefit from sedating relaxing herbs like *Valeriana,* while patients with insomnia and long-term exhaustion might only become further tired or even lethargic and depressed with strong sedatives like *Valeriana* or *Piper methysticum.*

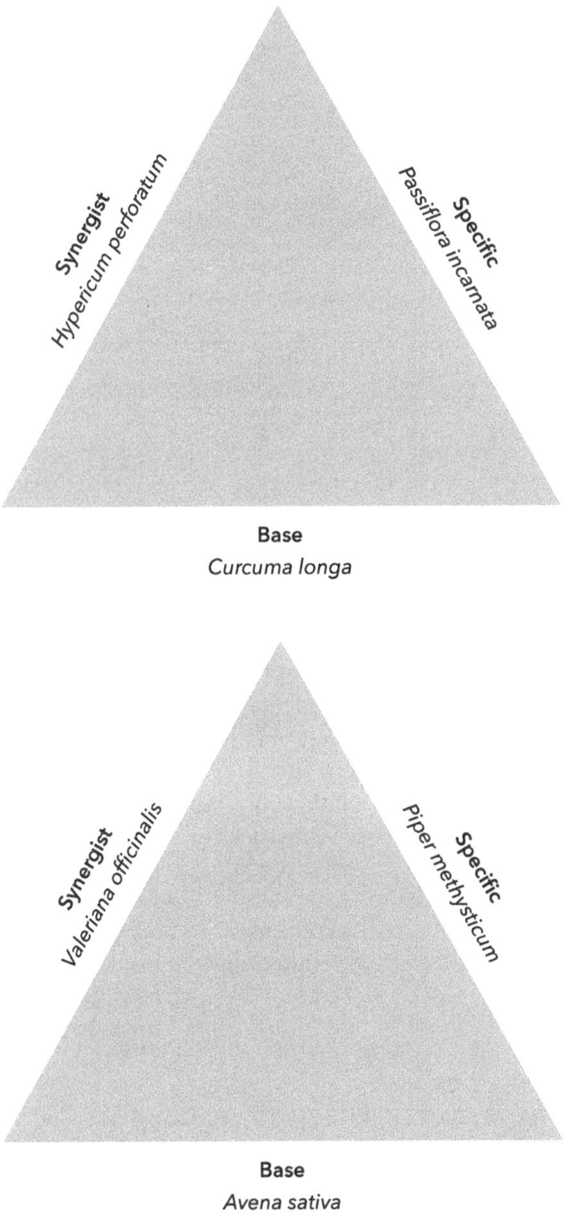

Figure 1.3. Two Formulas for Restless Insomnia

Other considerations or clues for the second insomnia patient are her oily skin, skin eruptions, aggravation by elevated premenstrual hormones, and hormone-related emotional tension. Although these are all very minor and very common in the general population, they are all heat symptoms. These symptoms also suggest that the liver and hormonal metabolism may be contributing underlying factors in this person's overall constitution and balance. The liver not only processes hormones and removes waste products from the blood that can otherwise contribute to acne and skin eruptions, but, in TCM, the liver is said to "rule" the joints and tendons. In many traditions, including TCM, liver herbs are often said to be specific for vague muscle stiffness that does not represent tendonitis, arthritis, fibromyalgia, or other condition. Therefore, a good synergist for this case might be a cooling liver herb noted to improve skin and hormonal balance. Some choices here might be a simple alterative such as *Taraxacum, Arctium, Silybum,* or *Curcuma.* Thus, for the second insomnia patient you might end up with a formula such as *Hypericum, Passiflora,* and *Curcuma,* to be taken multiple times per day for many months, with a before-bed formula of *Avena, Piper,* and *Valeriana.* Yes, there is some arbitrariness in the selection of herbs, but only to a degree. The arbitrariness stems from a somewhat capricious choice between herbs that have very similar energies and specific indications, or mixing and matching formulas in such a way that the overall base, and specific energy is the same. It is possible to create several variations of a formula having nearly identical action. For example, both *Taraxacum* and *Arctium* roots may be somewhat interchangeable as an alterative base in a formula. Either fennel or caraway seeds may offer interchangeable carminative effects in an IBS formula. *Valeriana* or *Piper methysticum* might both be effective in the above insomnia case. Or a synergist of *Curcuma* or *Silybum* might be logical to help the liver process hormones.

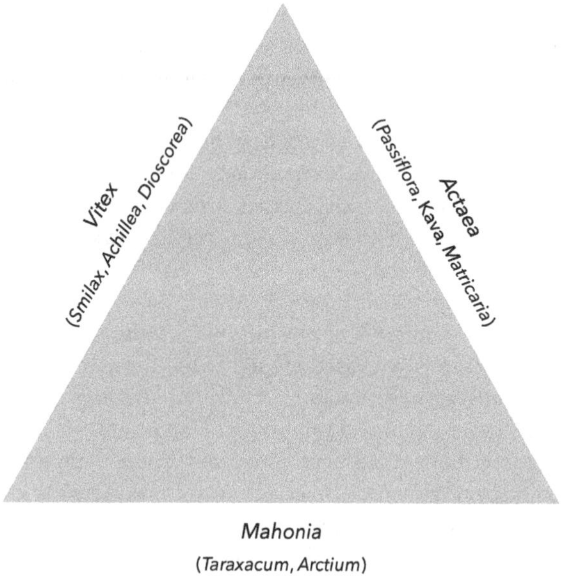

Figure 1.4. Formula Possibilities for Acne with Fiery Symptoms

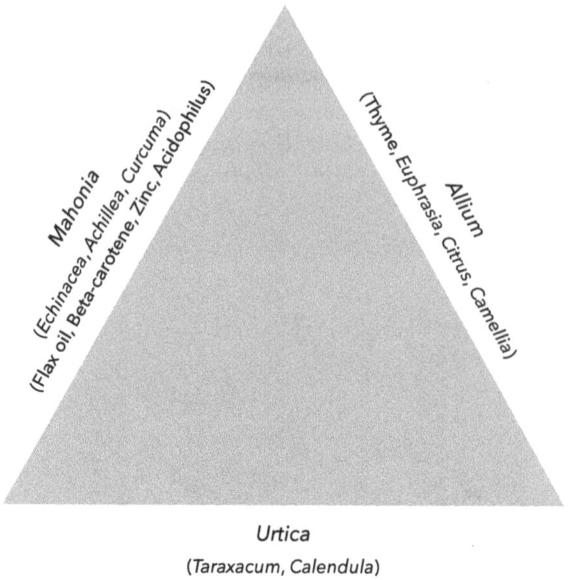

Figure 1.5. Formula Possibilities for Acne with Dampness and Infections

SAMPLE CASE: VARIATIONS OF ACNE

Patient 1: The first case is a 25-year-old woman with acne. History reveals frequent heavy menses, PMS including menstrual headaches, breast tenderness, and significant anxiety, along with chronic constipation as concomitant complaints. She rarely suffers from infections but when she does, they will come on quickly with painful sore throats, high fevers, and acute illnesses. She will recover quickly in a day or two. She also has frequent episodes of insomnia and muscle pain in the neck and upper back, most often premenstrually.

Figure 1.4 shows possible herbal combinations based on these considerations of the patient's constitution, action of herbs, and energetics.

4 Elements	Actions of Herbs	Energetics
Seek water to balance fiery symptoms	Seek alteratives, cholagogues	Seek cooling
Seek earthy therapies and herbs to ground the volatile tendencies	Seek hormonal-balancing agents	Seek sedating
	Seek nervines	Seek yin tonics
	Seek detox	

Patient 2: The second case is also a 25-year-old woman with acne, but her history reveals minor allergies, hay fever, frequent upper-respiratory infections, occasional yeast vaginitis, and tetracycline use for several years. She is chilly, has a damp constitution with much mucous production, tends toward loose stools, and has low-grade infections that linger a long time. Figure 1.5 shows possible formula combinations appropriate for this case.

4 Elements	Actions of Herbs	Energetics
Seek warming herbs to balance cold	Seek immune stimulants	Seek chi tonics
Seek drying herbs to balance damp	Seek antimicrobials	Seek yang tonic
Seek fiery herbs	Seek intestinal, vaginal, and respiratory astringents	Seek warm and drying tonics
	Seek antiallergy herbs	Seek to purge fluid and dampness
		Seek to move medicine to head, skin, upper respiratory tract

SAMPLE CASE: VARIATIONS OF RHEUMATOID ARTHRITIS

Patient 1: Our first example is a 50-year-old woman with a chief complaint of rheumatoid arthritis, primarily in the hands, wrists, neck, and shoulders. Onset was associated with conflicts and issues with children, loss of control in influencing children's lives, and disappointments. She reports minor anxiety and frequent bouts of insomnia. She also suffers from frequent constipation, a chronic cough due to a dry scratchy throat, and occasional brief episodes of cystitis. Her hands become red and swollen in acute episodes that are experienced as aching, burning, sore, and tender and then settle down over a month's time. Her neck becomes tight and spastic, with a throbbing headache and burning sensation on the skin. Symptoms wax and wane independently of diet or activity, but perhaps correlate to stress. She has a trim build, is often warm, is often thirsty, has a big appetite and a fast metabolism, and is very active. See figure 1.6 for formula possibilities for this case.

4 Elements	Actions of Herbs	Energetics
Seek to cool fire	Seek nervines	Seek to cool and moisten
Improve dryness, heat with watery, cooling herbs	Seek anti-inflammatories	Aim to soften, lubricate, and smoothe
Improve insomnia, restlessness, and worry with grounding, earthy herbs	Seek antispasmodics	Seek to quiet and calm energy
Aim to ground with moist earthy herbs	Seek tissue lubricants	Aim to move medicines to nerves
	Seek nerve and connective tissue tonics	

Patient 2: The second example is a 50-year-old woman with a chief complaint of rheumatoid arthritis—the arthritic pain is in multiple joints including the low back, hands, wrists, shoulders, and hips. There is a family history of rheumatoid arthritis and osteoarthritis. Constant mild to moderate stiffness is worst after sleeping or prolonged inactivity and also is aggravated with exertion. She has some minor arthritic and degenerative changes in the low back and some bony deformity beginning in the finger joints. The sensation is heavy, stiff, and aching and is better when resting the affected limb. She tends to be chilly, has chronic postnasal drip, is on hormone replacement therapy, retains fluid, has mild constipation, is slightly overweight, and reports low energy. Figure 1.7 shows formula possibilities for this case.

4 Elements	Actions of Herbs	Energetics
Seek to dry out excess water	Seek anti-inflammatories, anodynes	Seek warming and drying
Seek to warm, fire	Seek to move fluid, diurese	Seek stimulating, moving therapies
Seek to lift, lighten with airy or volatile compounds	Seek connective tissue tonic	
	Seek alterative detox therapies	

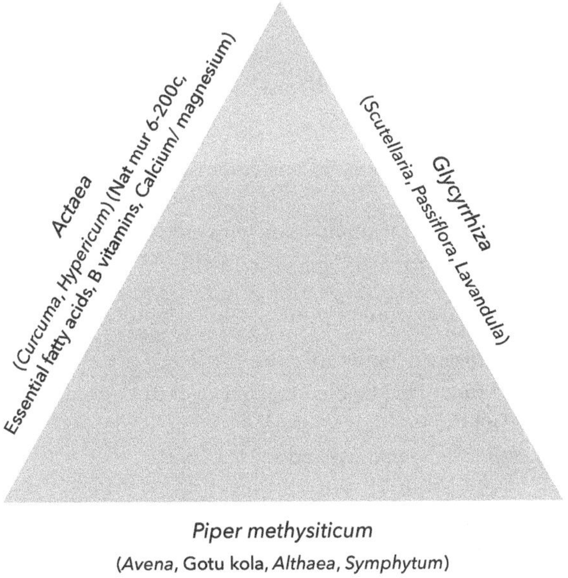

Figure 1.6. Formula Possibilities for Rheumatoid Arthritis with Stress

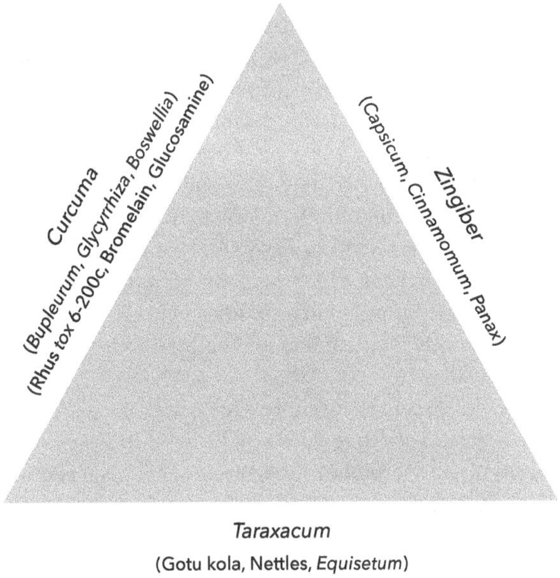

Figure 1.7. Formula Possibilities for Rheumatoid Arthritis with Cold Constitution

The Use of "Toxic" Herbs in Formulas

In the sample case of insomnia with exhaustion where *Zingiber* is used as synergist because of its heating and moving energetic qualities, *Zingiber* might be used in a lesser proportion than the other two herbs, in order to ensure the formula is overall nourishing, restorative, and tonifying. *Zingiber* is not particularly toxic, but due to its strong energy, there is a minor concern over its proportion in a formula. This is especially true for caustic, irritating, or outright toxic botanicals. Some such toxic or otherwise powerful herbs have strong moving or driving action: *Phytolacca* is a lymph mover, *Gentiana* is a bile mover, *Rhamnus* is an intestinal smooth-muscle mover, *Sanguinaria* is a tissue mover, and *Pilocarpus* is a secretory stimulant. All are used in formulas in a lesser quantity than the lead herbs, and yet their presence is equally contributory and as powerful as the other primary ingredients.

The case is the same when energy or motion is excessive, and a goal of the formula is to calm and soothe or to quiet down an excessively hot or hyperfunctioning action in a tissue or organ. *Veratrum, Aconitum, Atropa belladonna, Digitalis, Lobelia,* and *Conium* all have powerful relaxing, sedating, diminishing effects on nerve, cardiac, respiratory, and musculoskeletal tissues, appropriate only in small amounts in the overall formula.

In many cases, herbs with such extremely strong energies are added to formulas only when extremely strong imbalances are occurring. In general, the stronger and more potentially toxic or harsh the herb, the smaller the dose in the formula. And yet even that milliliter, or as little as five drops, drives the formula and contributes equally to the other herbs occurring at a dose of 10 or even 100 times the amount. The smaller the degree of atrophy/hypertropy or the more minor the degree of imbalance of hot/cold or excess/deficient status of the body, the less the need for any strong energy herbs, and the more the formula can be based on purely nourishing and restorative ingredients for the base, synergist, and specific alike.

In the previous example, the insomnia patient is so exhausted and so cold that a little bit of *Zingiber* to warm up our formula is appropriate. *Zingiber* is quite hot but overall nontoxic and safe, even long term. You would not include *Podophyllum* or *Iris* right off the bat to warm the bowels, and you would not choose *Thymus* or *Allium* (she is sensitive to onions and might react) to warm the lungs and reduce infections. None of those herbs are as indicated or specific as *Zingiber* to drive the formula into the proper locations. Thus, for this insomnia patient you might end up mixing a formula of 30 milliliters *Panax*, 20 milliliters *Astragalus*, and 10 milliliters *Zingiber* to fill a 60-milliliter or 2-ounce dropper bottle of tincture.

Further Guidance on Creating Warming and Cooling Formulas

The actions of herbs—both traditional folkloric and mechanistic—can be organized into simple categories of warming or cooling. Vasodilators, for example, and the hot spicy "blood movers" are generally warming, and demulcent herbs are generally cooling.

Warming stimulants. Here are some general guidelines for the use of warming stimulants in treating acute conditions.

- Use for colds, fever, and chills with acute onset
- Use for abundant mucous, phlegm in the throat and lungs
- Use as diaphoretic for those who fail to mount a useful fever in acute illness
- Use for many common infections of childhood in small doses
- Discontinue therapy when improvement is achieved
- Contraindicated in yang constitutions, for those who feel uncomfortably hot
- Contraindicated in hemorrhage, free sweating, or night sweats

In cases of chronic disease, observe these guidelines for warming stimulants.

- Use for circulatory enhancement, excessive clotting, blood stasis
- Use for fluid accumulation, poor circulation to organs
- Use for chronic inflammation with stiff and swollen character that feels better in hot weather or after a hot bath
- Use for cold hands and feet
- Use for atonic, feeble, sluggish constitutions
- Don't use for high fevers due to chronic debility, rather for weakness and sense of chill

Cooling remedies. These are guidelines for the use of cooling remedies in treating acute conditions.

- Use for acute mucosal inflammations, cool tissues with demulcent anti-inflammatories

Warming Stimulants

ACTION	HERBS
Peripheral vasodilators	*Achillea millefolium* *Allium sativum* *Cinnamomum* spp. *Ginkgo biloba* *Zingiber officinale*
Secretory stimulants	*Armoracia rusticana* *Capsicum* spp. *Iris versicolor* *Pilocarpus jaborandi*
Diaphoretics	*Achillea millefolium* *Capsicum annuum* *Zingiber officinale*
Warm antimicrobials	*Allium sativum* *Curcuma longa* *Origanum* spp. *Thymus vulgaris*
Chi tonics	*Eleutherococcus senticosis* *Panax* spp. *Withania somnifera*

- Use for burning sensations in the throat, intestines, or skin
- Use for tight, dry sensations in the mucous membranes or joints
- Use for acute infections with fever or sensation of heat, cool with antimicrobials and alteratives
- Use for acute toxicity states, joint pain, or headache; cool with bitters, alteratives, diuretics
- Contraindicated for acute disease with chills, abundant watery mucous, consolidation in the lungs

In cases of chronic disease, follow these guidelines when using cooling remedies.

- Use for dry, hot constitutions; use bitters, demulcents, alteratives, and diuretics
- Use for general yang states with warmth, redness, and heat in the body
- Use for tendency to acute fevers, infections, inflammation, and joint pain
- Contraindicated for chronic diseases associated with excessive dampness, coldness, deficiency, and fluid stasis

Cooling Remedies

ACTION	HERBS
Demulcents	*Aloe vera, A. barbadensis* *Althaea officinalis* *Ulmus fulva* *Verbascum thapsus*
Bitters/alteratives	*Arctium lappa* *Mahonia aquifolium* *Rumex crispus* *Taraxacum officinale*
Diuretics	*Apium graveolens* *Equisetum arvense* *Galium aparine* *Petroselinum* spp.
Cooling antimicrobials	*Echinacea purpurea* *Hydrastis canadensis* *Lomatium* spp. *Mentha piperita*
Astringents	*Geranium maculatum* *Hamamelis virginiana* *Rubus idaeus*
Energy dispersants	*Achillea millefolium* *Mentha piperita* *Iris tenax, I. versicolor* *Galium aparine*

Types of Herbal Medicines

Herbs are available as teas, tinctures, powders, encapsulations, syrups, solid extracts, and other forms. Following are pros, cons, and indications for the most common types of these herbal preparations.

Herbal Teas

Teas are especially indicated for individuals with digestive and bowel weakness to avoid irritation by pills or sometimes tinctures and to improve absorption of desired substances. Teas are also indicated for urinary ailments where substances in water quickly reach the urinary passages. Demulcents to soothe inflamed mucosal surfaces such as the esophagus, stomach, intestines, and bladder are best delivered via teas. Teas are also relatively inexpensive and when of good quality can be very effective. When the desired herbs are particularly unpalatable, tinctures or pills would be friendlier. Or when a high dosage of an herb, action, or chemical is desired, it may be easier to accomplish that with pills or tinctures.

Tinctures

Tinctures are especially handy when fresh plant juices are desired for preservation. For the home herbalist, tinctures are a practical way to stock the medicine chest with inexpensive, valuable medicines with a long shelf life. Tinctures will store for many years, whereas dried herbs age as the months go by and should be replaced about every two years to be of medicinal quality.

To any clinical herbalist, another important virtue of tinctures is that they can be combined into formulas as precisely indicated for a given individual, where commercially available pills are a one-size-fits-all formula. Tincture prescriptions can be formulated in an exacting manner to reduce the need to take many different medications to address many different organs, imbalances, and pathologies. Tinctures, when well thought out, can address many different pathologies, levels, and energies all at once.

Pills

Tablets and capsules can be a convenient and practical way to combine herbs with vitamins, minerals, amino acids, protomorphogens, bile, and numerous other diverse substances. Formulas designed for everything from prenatal nutrition to treating congestive heart failure or hypothyroidism are available, and when well indicated they are very helpful. However, unless you are producing your own encapsulations, you are usually limited to prescribing the pill combinations as the proprietor has formulated them. Because they might not be as specific for particular presentations as your own formulas, you might complement proprietary encapsulations with more specific herbs in a tea or tincture form as needed to round out the herbal prescription.

Pills are also useful when high dosages are being pushed for urgent health problems or when numerous different medications are being employed at once. Herbal pills may be simply dried and powdered plant material, or they may contain purified and concentrated plant constituents, such as curcumin, manipulated to boost its absorption and assimilation.

Knowing When to Use What Form of Medicine

When there are urgent circumstances, standardized or concentrated medicines might be desirable for their concentration and known potency, such as using standardized *Ginkgo* for serious ischemic disease or *Convallaria* for heart failure or silymarin concentrate for acute amanita toxicity. On the other hand, there is no reason to take expensive standardized *Matricaria* when a dollar's worth of good quality chamomile tea would likely do the trick for a flare-up of indigestion or irritable bowel.

When treating complex and chronic disorders, it is often appropriate to use many different herbs or supplements at once. For example, a protocol might call for ginkgo, garlic, milk thistle, passion flower, and echinacea, along with a multivitamin, flaxseed oil, and some additional beta carotene, zinc, magnesium, and calcium. A person on a budget (which is almost everyone) would not be able to keep up with such a program for very long if each item had to be purchased as a separate bottle of pills. Not to mention how difficult it is for many people to take handfuls of pills day in and day out. However, if the herbs could be combined in a tincture and a tasty tea and supported with the perfect vitamin and herbal/nutritional combination, the cost and convenience could both be improved.

Preparing Formulas for Use

Teas and tinctures are used throughout this formulary. With regard to herbal teas, as a general rule, delicate plant parts such as flowers are steeped rather than boiled, because the aromas and flavors can be lost or destroyed with vigorous boiling. Green leafy herbs such as nettles, alfalfa, and mint are also gently steeped, but harder, denser plant parts such as roots, barks, and seeds are best simmered to extract the medicinal components. Simply steeping herbs is referred to as preparing an infusion, while gently simmering herbs is referred to as a decoction. Unless otherwise specified, it is ideal to infuse teas for 10 minutes. For herbs that are decocted, they simmer gently for 10 minutes and then are removed from the heat to stand 10 to 15 minutes more. In a few cases in this text I recommend steeping or simmering for a longer period, when the intention is to liberate minerals or other compounds that are not readily released with simple infusions or decoctions. In some cases, it is recommended that vinegar or lemon juice be added to the water to best extract minerals. In other cases, it is recommended that mucilagenous herbs be macerated in cold water for many hours or overnight before bringing to a brief simmer to best extract mucilage.

A general dosage for a tea is a minimum of three cups a day. For an acute or urgent situation, such as trying to treat a bladder infection or a peptic ulcer, it is recommended that the tea be consumed as often as possible, such that there is a constant trickle of medicine passing through the irritated bladder or peptic mucous membranes.

Herbal tinctures are also used throughout this formulary with recipes given for the proportions and ratios. Because commercial tinctures are readily available, I do not go into detail discussing the making of tinctures. However, in some cases, herbal oils or vinegars that are not readily available are used in formulas, and in these cases, I offer brief details on how to prepare them. I also provide a few unique methods of preparing formulas, such as using solid extracts to thicken formulas to help them cling to the esophagus, or placing the formula in a spray bottle to use as a throat spray.

A general dosage strategy for tinctures is to use 1 dropperful (½ to 1 teaspoon) three times a day when treating chronic conditions. Most people have difficulty taking medicines any more than this, unless they are acutely ill or uncomfortable and motivation is high. In acute situations, tinctures may be taken as often as hourly for acute pain or infectious illnesses, or even as often as every 10 minutes. When tinctures or other herbal medicines are taken at this aggressive dosage, it is always for a very limited length of time, such as every 10 minutes for an hour, or every hour for a day, with instructions to reduce the frequency as symptoms improve.

Although the focus of the book is to create effective herbal formulas, in some cases, protocols are offered to create comprehensive therapies for chronic conditions such as biliary cirrhosis or renal failure. In such cases, I may suggest the use of a tea, a tincture, medicinal foods, and herbal encapsulations to treat such challenging conditions; additional examples are ulcerative colitis, interstitial cystitis, or autoimmune disorders. When herbal capsules are employed, they are often dosed as little as two per day or as much as two or three at a time, three times a day. While teas and tinctures can be taken with or without food, herbal capsules are most often taken with meals to enhance their digestion and to prevent them from being nauseating on an empty stomach.

Extending the Triangle Philosophy

As explained, herbal formulas exemplified throughout this book aim to use nourishing herbs as base ingredients in all formulas, complementing them with specifics and synergists. All the formulas include an introductory sentence or two explaining why the herbs are chosen, cite any research to support its use in specific conditions and situations, or comment on why the formula is appropriate for a specific presentation of the condition being discussed. I encourage readers to refer to the "Specific Indications" section at the end of each chapter to help master *materia medica* knowledge. Staying focused on the desired actions can help you avoid creating a formula that is too broad. With so many available herbs to choose from, knowledge of niche indications for various herbs helps prevent a "kitchen sink" approach. Instead of choosing every herb you know of that may be effective for bladder infections or indigestion, you can select the best choices with a solid knowledge of *materia medica.* It is a common mistake to place too many herbs with the exact same function in a formula and miss including herbs that address the underlying cause or other special considerations. For example, chronic bladder infections could be due to underlying diabetes where elevated sugar in the urine invites bacterial infection. The synergist should be something that addresses diabetes. In other cases, chronic bladder infections can be due to poor circulation through the kidneys, where stagnant circulation may allow bacterial infections, and the synergist should be something that increases renal perfusion. Once one has a solid knowledge of the *materia medica,* the specific indications of the herbs, prescribing and formulating become somewhat intuitive. This becomes easier, and even second nature, the longer one focuses on making formulas and the more experience one obtains.

Many herbalists believe that the plants are their teachers and that they learn from the plants themselves over time, as the plants teach us what they are capable of. Many herbalists are deeply aligned with nature and often have cultivated nature-based spirituality. Herbalists may embrace the notion that clinical intuition is evidence of a connection with the plants, and that this is the real root of herbal medicine as a healing discipline.

— CHAPTER TWO —

Creating Herbal Formulas for Gastrointestinal and Biliary Conditions

The role of intestinal bacterial flora to health, both locally in the gut and systemically, cannot be overemphasized, because bacterial flora has been shown to influence many diseases, such as inflammatory bowel disease, diabetes mellitus, obesity, nonalcoholic fatty liver disease, and autoimmune diseases. However, establishing a healthy intestinal ecosystem is not accomplished by simply giving probiotic species as a supplement, because many factors that allow these probiotic species to thrive must also be addressed, including diet, liver function, gastric motility, and fiber intake, to name just a few.

The most successful treatment protocols, regardless of the diagnosis, will take into consideration the entire person and all possible contributions to the condition. Frequently, dietary factors need to be addressed, such as excessive sugar that feeds yeast and undesirable intestinal bacteria, or insufficient fiber that creates an inhospitable environment for beneficial intestinal flora. Stress and anxiety may be deranging normal digestive physiology. Perhaps a history of heavy antibiotic use has altered the intestinal ecosystem. There may be biliary or pancreatic insufficiency. Hypothyroidism may slow down digestion; undiagnosed hyperthyroidism may cause diarrhea.

This chapter exemplifies formulas that take diverse presentations and situations into account. I present herbal treatment approaches and formula examples for common digestive disorders including digestive ulcers, irritable bowel disorder (IBD), diverticulitis, gastroesophageal reflux disease (GERD), pancreatic and digestive insufficiency, constipation and diarrhea, small intestinal bacterial overgrowth (SIBO), anorexia, halitosis, and leaky gut. Liver and gallbladder disorders are further discussed in chapter 3.

Rule Out Food Allergy and Intolerance

Allergic phenomena are one important contributor to diarrhea, nausea, vomiting, and irritable bowel syndrome (IBS) and should always be suspected in any of these complaints, especially when the person has evidence of other allergies such as hay fever, eczema, hives, migraines, or asthma. Wheat, dairy, nuts, and other allergens may not be suspected by an individual because they are eaten so frequently and the digestive inflammation is so severe and long standing. The correlation of the symptoms to the ingestion of a particular food can be obscured. Removing all common allergens from the diet as well as avoiding foods containing preservatives, artificial flavorings, and other chemicals, is helpful in these situations. The use of antiallergy herbs is recommended as well.

However, in a chicken-and-egg cycle, sometimes a GI disorder can lead to a food allergy. For example, pathogenic bacteria, dysbiosis, and small intestinal bacterial overgrowth (SIBO) can *cause* lactase deficiency[1] and dairy intolerance. Lactose intolerance depends not only on the expression of lactase but also on the dose of lactose consumed, the species composition of intestinal flora, gastrointestinal motility, and the presences of undesirable bacteria and SIBO, which generate gas and other fermentation products of lactose.[2] Gluten is another common allergen, and its ubiquitous presence in readily available foods makes it easy to overconsume. Gluten may cause not only intestinal damage, as in the case of celiac disease, but can also cause a plethora of systemic symptoms from dermatitis to arthritis.[3]

GI Formula Basics

A wide range of herbs have a direct effect on the gastrointestinal system, with the following general categories being among the most frequently used to treat common complaints.

Demulcents

Use demulcents to soothe burning pain, to soften hard stools, and to help heal ulcerative lesions. Demulcents are trophorestorative to the intestinal mucosa, supporting mucosal cell integrity and regeneration and helping to optimize bowel tone. Demulcents can be useful as base herbs in herbal formulas when ulcers or other mucosal lesions are present. Demulcent herbs that soothe the bowels, such as *Althaea officinalis*, *Glycyrrhiza glabra*, and *Ulmus fulva*, may be helpful—especially *Ulmus* when prepared into a porridge.

Antispasmodics

Use antispasmodics for intestinal cramps, acute gallbladder colic, excessive gurgling and rumbling in the GI tract, and excessive peristalsis such as urgent explosive diarrhea. *Mentha* spp. are some of the best intestinal antispasmodics and antinausea herbs, but can relax the gastroesophageal sphincter and worsen

Using Antimicrobials in GI Formulas

Antimicrobial herbs are indicated for intestinal infections such as gastritis, enteritis, stomach "flu," and traveler's diarrhea. Microbes such as *Helicobacter pylori* have been proven to contribute to GERD, ulcers, and many GI disturbances, and therefore antimicrobial agents are appropriate in formulas for these complaints, too. SIBO also requires herbal antimicrobials and avoiding foods that feed the problematic bacteria. Viruses are implicated in initiation of some cases of colitis and, of course, in infectious hepatitis. Viruses may trigger autoimmune diseases of the bowels such as Crohn's disease. Thus, antiviral and antibacterial herbs and general immune support may be useful in many formulas for GI health.

gastroesophageal reflux disease (GERD). There are numerous carminative herbs to help dispel gas and bloating. Many of them are aromatic herbs with pleasing flavors, and they improve the palatability of herbal teas. *Matricaria chamomilla, Dioscorea villosa,* and *Foeniculum vulgare* seeds are among the useful antispasmodics. When there is a great deal of uncomfortable flatulence and bloating, carminative herbs such as *Foeniculum* seeds, *Pimpinella anisum* seeds, *Matricaria* flowers, and *Mentha piperita* leaves are all comforting, often quite quickly. When there are intestinal cramps and spastic pain, *Dioscorea villosa* and *Mentha* can be very helpful.

Alteratives

Alterative herbs are appropriate in numerous digestive disorders because they help to eliminate wastes without harsh laxative effects. Alteratives also help to assimilate nutrients in cases of malnutrition, hypochlorhydria, and malabsorption. Alteratives make useful base or lead herbs in many gastrointestinal and biliary conditions. *Taraxacum officinale, Arctium lappa, Mahonia aquifolium* (also known as *Berberis aquifolium*), and *Rumex* spp. are among my favorite alterative herbs. Alteratives are appropriate in many formulas outside of GI complaints, too, because they may help reduce infections, improve the complexion and skin lesions, and improve overall health and energy. Alteratives might be featured strongly in a formula in cases where such symptoms accompany digestive complaints.

Digestive Stimulants

The use of a small amount of *Piper nigrum, Zingiber officinale,* or *Capsicum annuum* may be appropriate in those who have sluggish bowel function and poor circulation in the intestines. This problem occurs most often in the elderly but may also occur in younger people who have diabetes, hypothyroidism, or other complaints. Apple cider vinegar, bitters before meals, and lemon water before meals can also help those with hypochlorhydria and malabsorption.

Cholagogues

The use of cholagogues supports bile flow, fat digestion, and liver function and has warming and stimulating effects. Most of the alteratives listed in "An Overview of Alterative Herbs" on page 27 have gentle cholagogue action. *Hydrastis canadensis, Raphanus niger,* and *Chelidonium majus* are additional cholagogues for biliary insufficiency. See also the discussion on gallstones and acute cholecystitis in chapter 3.

Hemostatics

When bleeding is present, astringent hemostatic botanicals such as *Achillea millefolium, Capsella bursa-pastoris, Geranium maculatum,* and *Hamamelis virginiana* in capsule form or in tea blends are indicated. People with chronic bleeding should be monitored for anemia.

Secondary Support Herbs and Adjuvants

Antiallergy herbs, nervine herbs, adaptogens, and antimicrobials are all useful categories to add to GI formulas to provide secondary support.

"Bitters"

A bitter is, logically enough, an herb having a bitter flavor, including leaves and especially bitter roots. The reasoning for taking a "bitter" is nearly identical to the reasoning that supports using an alterative. Bitter flavors are noted to promote the flow of bile from the liver and gallbladder, referred to as a cholagogue action. Almost all alterative herbs have a bitter flavor.

The study of individual chemical compounds in plants will sometimes include bitter principles, which are not a particular molecular type such as flavonoids or glycosides, but rather any substance having a bitter flavor. Both arcane and modern herbal texts will speak of the use of "digestive bitters" as herbal tools for enhancing liver function, bile flow, digestion, and nutrition. The inclusion of bitters in the diet, from bitter melon to bitter greens such as mustard greens, is also a method of promoting bile flow and supporting digestion. Most herbalists and naturopathic physicians devote a significant amount of time to patient education and to evaluating and improving a patient's diet, including making an effort to incorporate many different flavors of plants in the diet, and guarding against eating only of a sweet or salty palette.

The Importance of Alterative Botanicals

Alteratives have a wide variety of indications but are particularly called for when conditions of general and chronic dysfunction and imbalance are present. Hypofunction of the digestive, eliminatory, and circulatory systems are obvious indications, and maldigestion, hypochlorhydria, constipation, flatulence, liver dysfunction, toxemia, acne, chronic infections, and many other ailments may all be improved through the use of alterative botanicals. For best results, it's appropriate to use alteratives for at least three months.

Alterative herbs should be avoided in cases of acute diarrhea, cholecystitis, cholelithiasis, peptic ulcers, ulcerative colitis, or other inflammatory conditions, as the stimulation of the digestive secretions may exacerbate these conditions. Similarly, alteratives should be avoided in cases of intestinal blockage, impaction, or other conditions where promoting peristalsis would be hazardous.

Dr. Stansbury's General Alterative Tea

This formula is my standard alterative decoction to gently promote digestive secretions, gastric and intestinal motility, the absorption of nutrients and elimination of wastes, and support beneficial intestinal flora. Such a formula will also support liver function, hormone

Teas versus Capsules versus Tinctures

Many of the formulas in this chapter emphasize the use of herbal teas over pills or tinctures, because teas have the benefit of offering direct surface contact with the GI mucosa. When the goal is to directly astringe or have demulcent effects on the tissues, it is necessary that the herbs actually touch the gastric or intestinal mucosa in order to be effective. *Ulmus* tincture, for example, would not have the same powerful effect on the mucosa as would *Ulmus* tea or gruel. Encapsulated *Ulmus* powder would also be more effective than *Ulmus* tincture.

In challenging and severe cases, treatment protocols may be more aggressive by using more than one prescription, such as using a tea along with a tincture or an encapsulated product. In such cases, herbs that work via their surface effects are best used in a tea. Herbs that have systemic effects, such as immune modulators or herbs with antiallergy effects, may be administered as tinctures because they work by affecting blood cells. The goal is to get them absorbed into the bloodstream, not to have them contact the GI mucosa.

Herbs for Relief of Digestive Pain

The following herbs are recommended for relief of digestive pain and inflammation associated with diabetes, old age, and circulatory inflammation.

Aesculus hippocastanum
Allium sativum and *Allium cepa*
Angelica sinensis
Collinsonia canadensis
Commiphora mukul
Curcuma longa
Ginkgo biloba
Hamamelis virginiana
Vaccinium myrtillus
Zingiber officinale

These herbs can be helpful to relieve symptoms related to infection and dysbiosis.

Achillea millefolium
Allium sativum
Astragalus membranaceus
Calendula officinalis
Echinacea purpurea
Glycyrrhiza glabra
Mahonia aquifolium
Origanum vulgare
Thuja occidentalis

These herbs can help relieve pain due to ulcerative lesions.

Calendula officinalis
Centella asiatica
Equisetum hymenale or *E. arvense*
Glycyrrhiza glabra
Hypericum perforatum

balance, and detoxification pathways, and will be useful as one prong of a formula in treating many conditions. A basic alterative tea such as this can be a useful way to begin therapy for not only digestive complaints, but also to treat PMS, acne, muscle stiffness, hypertension, high cholesterol, and diabetes.

Taraxacum officinale
Rumex crispus
Arctium lappa
Mahonia aquifolium
Cinnamomum verum or *C. zeylanicum*
Glycyrrhiza glabra
Citrus aurantium peels
Foeniculum vulgare seeds

Combine in equal parts. Simmer 1 teaspoon of the mixture per cup of hot water for 10 minutes. Remove from the heat, let stand covered for 10 minutes more, and strain. Drink 3 or more cups per day for many months.

An Overview of Alterative Herbs

Although there are a good number of herbs that could be considered to have an alterative effect, I feature my personal favorites—*Mahonia, Arctium, Rumex*, and *Taraxacum*—in many formulas due to their nutritive qualities, broad indication, and suitability in tea formulas. However, there are many other herbs suitable to exact alterative effects that are useful in other niche situations. As this list shows, some have affinity for different organ systems; refer to "Specific Indications," beginning on page 80, for further prescribing details for many of these herbs.

SPECIFIC AFFINITY	HERB
Gastrointestinal tract	*Arctium lappa* *Baptisia tinctoria* *Calendula officinalis* *Chionanthus virginicus* *Echinacea* spp. *Iris versicolor* *Mahonia aquifolium* *Petroselinum crispum* *Phytolacca decandra* *Plantago* spp. *Podophyllum peltatum* *Rumex* spp. *Smilax ornata* *Stellaria media* *Stillingia sylvatica* *Taraxacum officinale* *Trifolium pratense*
Liver	*Arctium lappa* *Chelidonium majus* *Cichorium intybus* *Curcuma longa* *Leptandra virginica* *Mahonia aquifolium* *Rumex* spp. *Silybum marianum* *Taraxacum officinale*
Spleen	*Ceanothus americanus* *Chelidonium majus* *Iris versicolor* *Silybum marianum*
Lymphatic system	*Aesculus hippocastanum* *Baptisia tinctoria* *Calendula officinalis* *Fouquieria splendens* *Hydrastis canadensis* *Iris versicolor* *Mahonia aquifolium* *Phytolacca decandra* *Scrophularia nodosa*
Endocrine system	*Arctium lappa* *Eleutherococcus senticosus* *Fucus vesiculosus* *Glycyrrhiza glabra* *Panax ginseng* *Rhodiola kirilowii* *Serenoa repens* *Smilax ornata* *Taraxacum officinale* *Turnera diffusa* *Withania somnifera*

Formulas for Dyspepsia

Dyspepsia is the diagnosis for digestive symptoms unassociated with underlying digestive lesions or pathology. As many as one-third, or even more, of all people referred to gastroenterologists are not found to have detectable pathologies. Patients may complain of epigastric pain, gas, bloating, and flatulence. There may be loss of appetite, rapid nausea following meals, or fat intolerance. Patients may experience sensitivity to spices, coffee, orange juice, or other foods and drinks previously tolerated, or they may have an unusual taste in the mouth. Food allergy can be the cause of many dyspeptic and vague GI complaints and is especially likely in people with a strong family history of the same as well as in people who are suffering from other allergies, eczema, asthma, or chemical sensitivity. Liver congestion, intestinal dysbiosis, and poor circulation to the digestive organs are other possible contributors to general dyspeptic states. Those who experience dyspepsia due to anxiety and stress symptoms, on the other hand, might be most improved by the use of nervine herbs and various relaxation therapies.

Tea for Gas and Bloating

This formula can be quickly comforting to nausea, cramps, and bloating when 2 or 3 cups are consumed over the course of an hour. Both *Matricaria* and *Mentha* can work immediately and can be a useful starting therapy for quick symptomatic relief.

Matricaria chamomilla 4 ounces (120 g)
Mentha piperita 4 ounces (120 g)

Steep 1 tablespoon of the herb mixture per cup in hot water and strain. Drink freely, 3 or more cups per day.

Tea for Dyspepsia Due to Anxiety

Since *Matricaria* is a nervine as well as a digestive antispasmodic and anti-inflammatory, it is a perfect choice for dyspepsia due to emotional causes. *Scutellaria* offers additional calming effects and is easy on the digestive system. A tea will be faster acting, but repeated doses of tincture will also be beneficial.

Matricaria chamomilla 4 ounces (120 g)
Scutellaria lateriflora 4 ounces (120 g)

Steep 1 tablespoon of the herb mixture per cup of hot water. Strain after 10 to 15 minutes. Drink freely, at least 2 cups in fairly rapid succession when symptoms are acute.

Tea for Stomach Pain with Extreme Distension

In folklore, *Dioscorea* is specific for twisting, boring, and spastic pain in smooth muscle. Modern studies show *Dioscorea* to reduce stomach acid while enhancing motility as well as supporting healthy intestinal flora.[4] *Lobelia* and *Mentha* are also smooth muscle relaxants, and *Mentha* can prevent any tendency for *Lobelia* to be nauseating. A tea will be faster acting, but repeated doses of tincture will also have a beneficial effect.

Mentha piperita 4 ounces (120 g)
Dioscorea villosa 2 ounces (60 g)
Lobelia inflata 2 ounces (60 g)

Steep 1 tablespoon of the herb mixture per cup of hot water. Strain after 10 to 15 minutes. Drink freely, at least 2 cups in fairly rapid succession.

Tincture for Dyspepsia Due to Low Stomach Acid or Digestive Enzymes

When food is not digested properly, fermentation ensues, which produces gas and, in many cases, digestive upset. This formula features bitter agents known to promote production of hydrochloric acid (HCL) and may be used as a stand-alone therapy or with digestive enzymes and HCL and/or bile supplements. Gentian promotes gastrin and somatostatin secretion.[5]

Taraxacum officinale 10 ml
Humulus lupulus 10 ml
Rumex crispus 8 ml
Gentiana lutea 2 ml

Place 3 drops of the combined tincture in a cup of warm water and sip slowly. Repeat 3 or more times daily, especially half an hour before meals.

Gentian and Wormwood Capsules for Hypochlorhydria

Gentiana (gentian) and *Artemisia* (wormwood) are a traditional combination used to stimulate digestive secretions and peristalsis. Research suggests that *Gentiana* promotes gastric emptying and intestinal propelling,[6] and that the duo increases vascular tone via sympathetic reflexes involving the vagus nerve within 5 minutes of ingestion. Digestion is supported due to hemodynamic effects[7] as well as possible secretory effects in the stomach. Gentiopicroside in *Gentiana* is one bitter compound credited with digestion-enhancing effects.

Gentiana lutea
Artemisia absinthinum

Blend equal parts of the powders and encapsulate the mixture. Take 2 capsules 15 minutes prior to eating.

Tincture for Dyspepsia Due to Sluggish Digestion

Similar in concept to the Gentian and Wormwood Capsules for Hypochlorhydria, this formula adds "hot" herbs *Capsicum* and *Zingiber* to enhance circulation, open up blood vessels to enhance absorption, and offer carminative and blood-moving effects.

Artemisia absinthium 15 ml
Hydrastis canadensis 4 ml
Picrasma excelsa 4 ml
Zingiber officinale 4 ml
Capsicum annuum 3 ml

Place 3 drops of the combined tincture in a cup of warm water and sip slowly. Repeat 3 or more times daily, especially a half hour before meals.

Formula for Dyspepsia Due to Low Stomach Acid

Bitter herbs generally stimulate stomach acid production. Taking bitters in the form of a tincture or tea, rather than capsules, allows patients to benefit fully from the bitter taste. See also the Fire Cider recipe on page 37.

Taraxacum officinale 14 ml
Rumex crispus 12 ml
Gentiana lutea 4 ml

Take bitters straight or prepare an aperitif using 1 teaspoon of the combined tincture in ½ cup of lemon water or ½ cup of diluted medical vinegar or shrub. Take half an hour before a meal with a second ½ cup right at mealtime.

Formula for Dyspepsia Due to Excessive Stomach Acid

High stomach acid is associated with ulcers, stress, and burning pain in the epigastric area. Any underlying stress should be treated directly, but this formula will help allay acute pain. Some people make excessive stomach acid when pancreatic enzymes or bile are insufficient. Consider liver support and/or digestive enzymes to try to normalize stomach acid production. (See also "Formulas for Ulcers" on page 62.)

Conventional antacids reduce stomach acid production, but this herbal formula supports the protective barrier of the stomach lining and leaves stomach acid intact, eliminating the rebound effect seen with pharmaceutical acid-blocking drugs and thus avoiding negative long-term consequences. This formula can be prepared as a tincture or a tea, with the latter likely being more effective.

Glycyrrhiza glabra
Ulmus fulva

For a tincture, combine equal parts of the two individual tinctures and take 1 or 2 dropperfuls at a time, as often as every 10 minutes for acute pain, reducing frequency as symptoms subside.

For a tea, first prepare *Glycyrrhiza* tea by steeping 1 tablespoon of the herb in 3 cups of hot water. Allow to cool to room temperature, then strain. Place 1 heaping tablespoon of *Ulmus* powder in a blender or large glass, and pour in the 3 cups of cooled tea. Blend or stir vigorously. For acute pain, it is helpful to sip such a tea constantly throughout the day so that the liquid is constantly soothing the lesion.

Daikenshuto— A Traditional Dyspepsia Remedy

This traditional Japanese formula is shown to improve constipation and relieve associated gas and bloating.[8] *Zanthoxylum* fruits are sometimes called Szechuan peppercorns due to their hot spicy flavor. Prickly ash, *Zanthoxylum clava-herculis*, is a possible substitute if the Asian species is not available.

Zingiber officinale 5 ounces (150 g)
Panax ginseng 3 ounces (90 g)
Zanthoxylum piperitum 2 ounces (60 g)

Decoct 1 teaspoon of the herb mixture per cup of water by simmering gently for 10 minutes and then letting stand in a covered pan, 20 minutes more. Strain the decoction. Drink 3 or more cups per day.

Angelica "Candies" for Dyspepsia and Constipation

Angelica archangelica is a traditional digestive tonic herb of old European herbals. Modern studies indicate that *Angelica* may promote ion flow in intestinal epithelial cells, enhancing chloride secretion in the colon,[9] improving peristalsis and intestinal dryness. This remedy can be prepared in any quantity you wish; I've given instructions here for a modest amount.

Angelica, about 3 or 4 fresh young stalks
Baking soda
Zingiber, a 4-inch piece of fresh ginger root
Honey

Harvest *Angelica archangelica* when the succulent hollow stalks are juicy and not yet pithy or stringy, usually no later than May Day. Cut into pieces 5 to 10 inches long and place them in a saucepan. Add measured quantities of water until the stalks are covered, and then add ½ teaspoon baking soda per pint of added water to help prevent color loss.

Simmer gently until just starting to soften. Strain and immerse in ice cold water. When the stalks are cool enough to handle, peel as much of the outer rind as possible and then cut the stalks into ½-inch- to 1-inch-long pieces and place them back in the saucepan.

Meanwhile, prepare a simple syrup of honey in a strong ginger tea. The amount of syrup you will need depends on the quantity of stalks, but you will likely need at least 2 cups. To make the tea, boil freshly chopped or grated ginger root in water for 10 minutes, and then strain. Combine 3 parts honey to 1 part of the freshly strained hot ginger tea.

Cover the drained *Angelica* roots with the *Zingiber* honey syrup, while still hot or after cooled, as convenient, and let stand for 24 hours, stirring 3 or 4 times over the course of the day. Each day for 4 days, drain off the syrup into a separate pan, boil it, and then pour it back over the *Angelica* pieces and stir well. On the fifth day, remove the *Angelica* pieces from the pan and transfer to a cookie sheet or dehydrating tray lined with wax paper. Let air-dry for 3 or 4 days, turning the pieces once or twice a day. When the syrup has crystallized on the surface of the stalks, sprinkle with powdered sugar or dried coconut to help absorb all possible remaining moisture. Store in small airtight containers. Eat several per day as part of a treatment for constipation and dyspepsia.

General Carminative Tea

A carminative is an agent capable of dispelling gas and bloating from the stomach and intestines. This category includes various mints, chamomile, and the Apiaceae family seeds, such as fennel, anise, and caraway. Whole seeds can be used in herbal teas but they will make the strongest medicine if crushed in a mortar and pestle prior to infusing. In patients with gas and bloating due to poor circulation or insufficient secretions, ginger and some of the hot culinary spices can have a carminative action, but can cause gas and bloating in those with sensitive stomachs or IBS. Simply ask people how they tolerate spicy food when considering if cinnamon or ginger may make them better or worse. Any of these can be effective as "a simple," but several are combined here.

Foeniculum vulgare powder or freshly crushed seeds 2 ounces (60 g)
Mentha piperita 2 ounces (60 g)
Matricaria chamomilla 2 ounces (60 g)

Steep 1 tablespoon of the herb mixture per cup of hot water. Drink 3 cups in a row to treat uncomfortable gas.

Formulas for Irritable Bowel Syndrome

Irritable bowel syndrome (IBS) is most common in women, with the onset commonly occurring during their twenties and thirties, but the condition may affect as much as 20 percent of the Western population overall. IBS is considered "functional," rather than being associated with any underlying pathology; however, research is emerging that associates the condition with deranged brain-gut signaling, hypersensitivity of visceral sensory afferent fibers, bacterial gastroenteritis, small intestinal bacterial overgrowth (SIBO), genetic alterations, and food sensitivity.[10]

IBS is frequently associated with personalities prone to nervousness, depression, and hysteria. Some people suffering from IBS may be fearful that they have cancer or a serious life-threatening disease. People often experience alternating bouts of constipation and diarrhea, with asymptomatic periods of various lengths in between flare-ups. There is often accompanying flatulence, belching, and stomach upset. On physical exam, the intestines may feel full, tense, and firm and may be tender to the touch. There is not typically blood or mucous, as there is with inflammatory bowel conditions, but the pain may be every bit as severe and bleeding may occasionally occur. When stress and nervous issues underlie digestive upset, nervine herbs including *Matricaria*, *Melissa officinalis*, *Verbena*, *Hypericum*, and *Passiflora incarnata* will complement herbs that act more directly on the digestive system.

Carminative Herbs for Intestinal Cramps and Bloating

These herbs are some of our best choices to include when there are significant intestinal cramps and twisting, spastic, boring pains in the abdomen. All are tasty enough to use in tea formulas for digestive complaints. Those available as essential oils can also be used topically. *Mentha piperita* (peppermint) is available in enteric-coated capsules to deliver large and concentrated doses to the colon.

Avena sativa (whole groats or rolled oats)
Cinnamomum zeylanicum or *C. verum*
Dioscorea villosa
Elettaria cardamomum
Foeniculum vulgare
Glycyrrhiza glabra
Humulus lupulus
Lavandula angustifolia
Matricaria chamomilla
Matricaria recutita
Mentha piperita
Mentha spicata
Nepeta cataria
Origanum vulgare
Pimpinella anisum
Trigonella foenum-graecum
Viburnum prunifolium
Zingiber officinale

Be Cautious with Bitter Alteratives in IBS Formulas

Since alterative herbs are stimulating to motility, bile flow, peristalsis, and gastrointestinal secretions, it best to start with a small dosage such as ¼ teaspoon of tincture twice daily for an IBS patient. Increase to ½ teaspoon and up to 1 teaspoon multiple times a day. Withdraw these herbs if significant inflammation, diarrhea, or bleeding occurs. Alterative herbs can be very beneficial in small doses, such as ¼ teaspoon of tincture or a ½ cup of tea between episodes or flare-ups to enhance digestive absorption, metabolism, and waste excretion, increasing as tolerated.

All-Purpose Teas for IBS

Teas are a good choice for IBS because they can have direct surface contact with the intestinal mucosa. This tea is tasty enough to truly enjoy. Encourage a patient to drink it in large quantities and to avoid coffee, soda, alcohol, and other irritating beverages. Chamomile offers anti-inflammatory and antispasmodic properties, and licorice helps restore and regenerate intestinal mucosa cells. Fennel is a fast-acting carminative for gas and bloating, and mint is one of the strongest intestinal antispasmodics that I have encountered. Consuming 2 or 3 cups of this tea in a row can alleviate acute symptoms; regular use can reduce the frequency and severity of bouts of IBS.

Matricaria chamomilla
Glycyrrhiza glabra
Foeniculum vulgare
Mentha piperita

Combine equal parts of the dry herbs. To make the tea, add 1 tablespoon of the herb mixture per cup of hot water, and steep for 10 minutes. Consume at least 3 cups per day; more, if possible, particularly during a flare-up.

Soothing demulcent herbs such as slippery elm (*Ulmus fulva*) can be a helpful complement to this tea. Stir several tablespoons of the raw powder into foods such as oatmeal or applesauce or into juices.

Formula for IBS with Painful Gas and Cramping

For very uncomfortable gas and intestinal cramps, this formula might be more powerful than the one above, because every herb in it has antispasmodic effects. This formula is most effective taken as a tea, but it could be prepared as a tincture to have on hand in the purse or desk drawer for emergency use.

Mentha piperita 2 ounces (60 g)
M. spicata 2 ounces (60 g)
Dioscorea villosa 2 ounces (60 g)
Foeniculum vulgare 2 ounces (60 g)
Angelica archangelica 2 ounces (60 g)

Steep 1 tablespoon of the herb mixture per cup hot water and then strain. Drink 3 or more cups a day. For acute or new onset symptoms, drink 3 cups in one hour.

If preparing as a tincture, choose only one of the mints, and combine 8 ml of each herb; this will fill a 1-ounce bottle. Take 1 dropperful of tincture every 5 to 10 minutes for an hour, reducing the frequency as symptoms improve in cases of acute intestinal discomfort.

Marijuana for GI Symptoms

Marijuana (*Cannabis sativa*) has been used since ancient times to treat nausea, vomiting, and other gastrointestinal disorders. Many of the medicinal actions of marijuana are credited to agonism of the endocannabinoid system, and there are extensive cannabinoid pathways in the digestive system. Cannabinoid pathways play roles in GI motility, secretion, intestinal secretions, and epithelial barrier functions. Cannabinoids also bind additional receptors in the gut including transient receptor potential channels (TRPV1), the peroxisome proliferator-activated receptor alpha (PPARα) and the G protein-coupled receptor 55 (GPR55), all known to play roles in bowel inflammation and colon cancer.[11]

Prickly Ash–Ginger Tincture for Painful Cramping

Zanthoxylum (prickly ash) and ginger are a traditional combination for intestinal cramps. Modern studies suggest the combination reduces inflammatory T-cell responses in animal models of enteropathy.[12]

Zanthoxylum clava-herculis 40 ml
Zingiber officinale 20 ml

Mix the tinctures and take 20 drops in a sip of warm water every 15 to 30 minutes as needed for bouts of IBS with severe cramping.

Enteric-Coated Peppermint Oil for Gas and Bloating

Mint is a fast-acting smooth muscle relaxant, useful for treating intestinal cramps, gas, and bloating,[13] and it can also be used prior to a colonoscopy to improve comfort. Mint oil delivered through stomach acid may also exert antimicrobial actions in the gut, helping to treat small intestinal bacterial overgrowth (SIBO).[14]

Enteric mint oil capsules (*Mentha piperita*)

Take 1 or 2 capsules 3 or 4 times daily, after meals, or as needed for digestive pain and discomfort.

Tea for IBS with Diarrhea Predominant

In many cases, diarrhea is the body's effort to eliminate irritating or allergenic substances. In these instances the use of agents that soothe mucosal irritation are indicated. Excessive peristalsis may also contribute to diarrhea in IBS. This formula uses the antispasmodics *Matricaria* and *Mentha; Ulmus* and *Glycyrrhiza* soothe the GI tract. This formula would be most effective taken as a tea to obtain direct surface contact with the intestinal mucosa. See also "Formulas for Diarrhea" on page 51.

Glycyrrhiza glabra
Ulmus fulva
Matricaria chamomilla
Mentha piperita

Combine equal parts of the dry herbs. Steep 1 tablespoon of the herb mixture per cup of hot water and then strain. Drink 3 or more cups a day. Drink 3 cups in an hour to relieve acute or new onset symptoms.

Tea for IBS with Blood and/or Mucous Diarrhea

Blood in diarrhea signals the need for both astringents and demulcents, and even though they are opposite in concept, they work well when used in tandem. *Achillea* is usually highly effective in reducing GI bleeding, but agents with a soothing and anti-inflammatory action are still needed. *Geranium* could be substituted in this formula as an alternate herb to check intestinal bleeding. This formula could be prepared as a tincture, but the tea will be much more soothing and thereby effective.

Ulmus fulva 3 ounces (90 g)
Glycyrrhiza glabra 2 ounces (60 g)
Matricaria chamomilla 2 ounces (60 g)
Achillea millefolium 1 ounce (30 g)

Steep 1 tablespoon of the herb mixture per cup of hot water for 10 minutes and then strain. Drink as much as possible; 6 to 8 cups a day is recommended for the short term. Decrease the dosage over several days as symptoms improve.

Tea for IBS with Constipation Predominant

Patients with IBS do not tolerate irritant or commercial over-the-counter laxatives, because IBS patients tend to quickly develop diarrhea. Even alteratives with their gentle laxative action need to be attempted cautiously; combine alteratives with carminatives and demulcents and keep the proportion in the formula low. Add greater amounts of alteratives incrementally once it is certain that the patient tolerates them well. Here *Taraxacum* is used as a very gentle alterative with minimal laxative action, at only 1 ounce in the ½ pound mixture that includes plenty of the demulcent herb *Ulmus*. If a patient tolerates this tea well, *Taraxacum* in tincture form may be added as a complementary therapy.

Ulmus fulva 4 ounces (120 g)
Elettaria cardamomum 2 ounces (60 g)
Curcuma longa 1 ounce (30 g)
Taraxacum officinale 1 ounce (30 g)

Steep 1 tablespoon of the herb mixture per cup of hot water and then strain. Drink 3 or more cups a day. Drink 3 cups in an hour to relieve acute or new onset symptoms.

Tea for IBS Stimulated by Stress

For patients whose IBS flares up with stress or who suffer concomitantly with anxiety disorders, it is important to include nervines in the formula. *Matricaria* is both a nervine and a GI trophorestorative, making it a good choice for this condition. *Glycyrrhiza* is both an adrenal tonic and a GI anti-inflammatory and mucosal tonic and complements *Matricaria* well to treat IBS. When combined with the additional nervines *Melissa* and *Avena*, this formula can have a substantial calming effect on the nervous system.

Matricaria chamomilla
Melissa officinalis
Glycyrrhiza glabra
Avena sativa

Combine equal parts of the dry herbs. Steep 1 tablespoon of the herb mixture per cup of hot water and then strain. Drink 3 or more cups a day. Drink 3 cups in an hour to relieve acute or new onset symptoms.

Tea for IBS in Allergic Individuals

Allergens in the bowels can trigger peristalsis, inflammation, and irritability and may occur more often in those with other allergies such as hay fever, eczema, asthma, or hives. This formula adds *Tanacetum* and *Curcuma*, which have anti-inflammatory and antiallergy effects, to the base of the standard soothing GI herbs *Glycyrrhiza* and *Ulmus*. A tea is the most effective vehicle for this formula, but it could be prepared as a tincture for people who dislike or have difficulty preparing teas.

Glycyrrhiza glabra
Tanacetum parthenium
Curcuma longa
Ulmus fulva

Combine equal parts of the dry herbs. Steep 1 tablespoon of the herb mixture per cup of hot water and then strain. Drink 3 or more cups a day. Drink 3 cups in an hour to relieve acute or new onset symptoms.

Herbs for Digestive Symptoms Related to Allergic Reactivity

These herbs have antiallergy effects and are used for asthma, hay fever, and migraines and are also valuable choices to include in formulas for IBS, allergic diarrhea, and many food sensitivities in atopic individuals.

Angelica sinensis	*Matricaria chamomilla*
Capsella bursa-pastoris	*Petasites hybridus*
Crataegus spp.	*Picrorhiza kurroa*
Curcuma longa	*Scutellaria baicalensis*
Glycyrrhiza glabra	*Tanacetum parthenium*

Tea for IBS in Association with PMS or Hormonal Fluctuations

Some women experience IBS flare-ups premenstrually, but this complication is seen less commonly than IBS triggered by stress and allergies. Adding alteratives may help the liver process estrogen, and adding *Dioscorea* may help the adrenals boost cortisol and progesterone, all serving to achieve hormonal balance. *Angelica* improves circulation in the pelvis and may reduce vascular congestion and hyperreactivity.

Matricaria chamomilla 2 ounces (60 g)
Angelica sinensis 2 ounces (60 g)
Glycyrrhiza glabra 2 ounces (60 g)
Taraxacum officinale root 1 ounce (30 g)
Dioscorea villosa 1 ounce (30 g)

Steep 1 tablespoon of the herb mixture per cup of hot water and then strain. Drink 3 or more cups a day. Drink 3 cups in an hour to relieve acute or new onset symptoms.

Formulas for Inflammatory Bowel Conditions

Inflammatory bowel diseases (IBD) include Crohn's disease (regional enteritis) and ulcerative colitis. Involving possible autoimmune components, Crohn's disease may be more difficult to treat than ulcerative colitis, but even Crohn's disease can be treated with herbal and natural therapies. Symptomatically they are similar. Both conditions can cause abdominal pain, blood and mucous in the stools, ulcerative and inflammatory lesions in the intestines, fever, anorexia, malaise, and weight loss. As with irritable bowel syndrome, dietary changes will not produce a cure, but they can lessen the frequency, severity, and duration of episodes. Anything that optimizes digestion and bowel function, such as probiotics, diet changes, stress management, and herbal and nutritional supplements, can have a positive effect on these conditions. Help a person identify any foods that aggravate their symptoms by conducting a food intolerance test or trying an elimination diet. Dairy, corn, wheat, citrus, sugar, caffeine, tomatoes, and soy are all common offenders. The FODMAP diet has become established as one effective therapy. On this diet, patients avoid fermentable oligosaccharides, disaccharides, monosaccharides, and polyols (FODMAP) for a month or more to help recover the intestinal ecosystem flora.[15] High-fiber diets sometimes help and sometimes aggravate this condition; it is best to individualize the diet. Sugars, fats, and refined carbohydrates should be greatly limited or cut out altogether. Fresh fruits and especially vegetables should be encouraged, taking into consideration individual sensitivities and preferences. Regular garlic consumption should also be encouraged; it seems helpful to many. Following an episode of bowel inflammation, it may be helpful to undergo a fast of 2 to 4 days, and add broths, steamed vegetables, and fresh fruits slowly over the next 5 to 7 days. Build up to raw foods and meats, in moderation only.

Herbs showing at least some efficacy in randomized controlled trials[16] include *Aloe vera* gel, *Triticum aestivum* (wheat grass juice), *Andrographis paniculata*, *Boswellia serrata*, *Plantago ovata* seeds, *Artemisia absinthium*, *Tripterygium wilfordii*, *Cannabis*, *Melissa*, *Taraxacum*, *Hypericum*, *Foeniculum*, *Calendula*, and *Silybum*. Additional animal models of colitis suggest that saponins in the roots of *Anemarrhena asphodeloides* are metabolized into sarsasapogenin in the gut, which is shown to inhibit many inflammatory mediators expressed in colitis and may help balance T-helper and regulatory T cells.[17] Clinical investigations have shown *Curcuma* to prevent chemical-induced colitis in animals.[18]

Because IBD increases the risk of colorectal cancer, herbs with immune-modulating action should be considered, even when patients are managed on pharmaceuticals. *Astragalus*, shiitake mushrooms, turmeric, ginger, and umbel family spices are all tasty enough to use in preparing soup stock for vegetable and other soups. Bone broths prepared with these herbs can also provide good nutrition when patients are unable to eat much solid food during a flare-up. *Scutellaria baicalensis* has numerous anti-inflammatory mechanisms on the bowel, and animal studies suggest the flavonoid baicalein may reduce colon cancer cell development. *Astragalus* is also shown to deter the proliferation of gastric cancer and is featured in the following formulas and recipes for inflammatory bowel disease.[19]

Basic Decoction for IBD

The *Dioscorea* compound methyl protodioscin exerts an anti-inflammatory effect on intestinal mucosa and supports healing and improves intestinal barrier function.[20] *Salvia miltiorrhiza* has an anti-inflammatory effect on the bowel.[21] Pomegranate rinds, *Punica granatum*, are traditionally reported to have astringent effects on the intestines and to be useful in treating diarrhea. Small clinical trials suggest utility in treating ulcerative colitis.[22] If *Punica* is not readily available dried, the rinds from fresh fruits may be used. This formula works well along with the Basic Tincture for IBD, because they contain differing herbs that complement each other.

Astragalus membranaceus 2 ounces (60 g)
Dioscorea villosa or *rotunda* 2 ounces (60 g)
Salvia miltiorrhiza 2 ounces (60 g)
Glycyrrhiza glabra 2 ounces (60 g)
Punica granatum 1 ounce (30 g)
Curcuma longa 1 ounce (30 g)

Decoct 1 teaspoon of the herb mixture per cup of water. Prepare 6 to 8 cups at a time: Simmer for 10 minutes, let stand 10 minutes more, and strain. Drink the entire 6 to 8 cups over the course of a day. Reduce intake over time as symptoms improve.

Basic Tincture for IBD

Artemisia absinthium and *Tripterygium wilfordii* were superior to a placebo in inducing remission in some clinical trials.[23] *Boswellia serrata* may match the efficacy of steroids in the treatment of ulcerative colitis.[24]

Artemisia absinthium
Tripterygium wilfordii
Boswellia serrata
Andrographis paniculata
Scutellaria baicalensis

Combine equal parts of the herbs and take 1 teaspoon of the combined tincture 5 or 6 times per day, reducing as symptoms improve.

Smoothie for Inflammatory Bowel Disease

Smoothies are an excellent choice for treating IBD because medicines and nutrients can be better absorbed in them than from pills, plus the direct surface contact can help allay pain and heal intestinal ulcers and lesions. Choose *Astragalus*, *Matricaria*, *Glycyrrhiza*, or other tea to act as a base.

2 cups (480 ml) herbal tea
1 tablespoon fresh wheatgrass juice
1 tablespoon *Aloe vera* gel
1 tablespoon glutamine powder
1 teaspoon apple pectin

Place the herbal tea in a blender. Add the wheatgrass juice, *Aloe* gel, glutamine powder, and apple pectin, and blend on medium high for 1 minute. Drink immediately.

Immune-Modulating Soup for Ulcerative Colitis

Lycium and *Astragalus* are widely used traditional medicines in China, particularly for "nourishing Yin" and "reinforcing Qi" in hot inflammatory states. Animal models of ulcerative colitis suggest the duo to offer nourishing polysaccharides, which help restore intestinal cell functions and physical integrity and may also improve intestinal mucosa barrier functions.[25] Coriander prevents inflammation in animal models of colitis.[26] Many Apiaceae family seeds are flavorful and credited with anti-inflammatory effects: Dill, fennel, anise, and caraway seeds might be used plentifully in the diet. Ergosterol is a vitamin precursor and phytosterol ubiquitous in mushrooms. Some of the anti-inflammatory and immune-modulating effects of shiitake (*Lentinula edodes*) and hoelen (*Poria cocos*) may be due to ergosterol.[27]

For the vegetable component, you can add a combination of onions, garlic, carrots, celery, and green beans, as you prefer. Dry beans and quinoa are also good additions.

1 cup (80 g) shredded or 6 slices *Astragalus membranaceus*
½ cup (40 g) *Centella asiatica*
¼ cup (50 g) *Curcuma longa*, fresh, if available, or ⅛ cup dry
¼ cup (50 g) freshly grated ginger root
2 tablespoons coriander
1 cup (50 g) *Poria cocos*
1 cup (100 g) *Lycium barbarum*
1 cup (50 g) sliced shiitake (*Lentinula edodes*)
4 cups (500–1,000 g) chopped vegetables
2 cups (200 g) dry beans or quinoa

Place the *Astragalus*, *Centella*, *Curcuma*, ginger root, and coriander in a stock pot filled three-quarters with water and simmer gently for 1 to 2 hours; strain. Return the liquid to the pot and add the *Poria*, *Lycium*, and shiitake, along with all vegetables plus beans or quinoa, as you wish. Add more water if needed to cover the ingredients fully. Simmer several hours more over low heat, until the beans are soft. Serve in soup bowls.

Leftover soup will store for 3 days in the refrigerator. Eat several bowls per day or one for each meal as a medicinal food for acute IBD and as a convalescent food to support recovery following an acute episode.

Herbal Capsules for Ulcerative Colitis and Crohn's Disease

Berberine acts as an antagonist at dopamine receptors, contributing to immune-modulating effects for colitis.[28] Because autoimmune bowel inflammation is typically recalcitrant, it is often best to use a tincture, a tea, and one or more types of capsules at once, because the therapy must often be aggressive. Berberine, *Curcuma*, or silymarin capsules are recommended complements to the teas and tinctures described in this chapter; take 2 or 3 capsules 3 times a day.

Tincture for Colitis with Fever

Severe colitis may generate a fever as part of the inflammatory and immune reaction. Treat the bowel symptoms directly with an herbal tea (one of the teas in this section or in "Formulas for IBS" on page 30), and consider this formula as a complementary tincture. The goal is not just to bring down the fever with an aspirin-like anti-inflammatory, but to help control the entire underlying immunologic reaction that is producing the fever.

Glycyrrhiza glabra 20 ml
Tanacetum parthenium 20 ml
Mentha piperita 20 ml

Take one dropperful every 15 to 30 minutes, reducing the frequency of dosage as symptoms improve. Hourly dosing may be required for a day, reducing to every 3 or 4 hours over the next several days, and down to 4 times a day thereafter.

Formula for Colitis with Bloody Diarrhea

When there is blood in diarrhea, astringents will often control the bleeding. Demulcents can soothe the irritated tissue. In this formula *Geranium* and *Quercus* act as intestinal astringents. *Centella* and *Glycyrrhiza* can help heal bleeding ulcers and have a restorative and healing effect on the intestinal mucosa. This formula can be prepared as a tea or a tincture.

Glycyrrhiza glabra 2 ounces (60 g)
Ulmus fulva 2 ounces (60 g)
Geranium maculatum 1 ounce (30 g)
Quercus rubra 1 ounce (30 g)
Centella asiatica 1 ounce (30 g)
Mahonia aquifolium 1 ounce (30 g)

For a tea, gently simmer 1 teaspoon of the herb mixture per cup of hot water. Prepare 6 to 8 cups at a time and sip the tea constantly throughout the day, decreasing the following day as the bleeding is controlled.

Formulas for Anorexia

Lack of appetite can occur as a result of anxiety and depression, of various diseases in and outside of the digestive system, of chemotherapy (as a side effect), and of other causes. A teaspoon or two of bitter herbs placed in water and sipped before meals can help overcome lack of appetite due to hypochlorhydria. Peppery species such as black pepper (*Piper nigrum*) and cayenne pepper (*Capsicum annuum* and other species) enhance the absorption of nutrients by enhancing circulation in the stomach and intestines and by affecting the microarchitecture of gap junctions and cell-to-cell adhesion. Simple home remedies such as apple cider vinegar or fresh squeezed lemon juice diluted in water and consumed before meals can help acidify the stomach and enhance digestion. When nervous system conditions have upset the appetite, the use of calming herbs in combination with digestive aids may be most effective.

Tea to Allay Nausea and Anorexia of Chemotherapy

In general, formulas for chemo-induced anorexia are aimed more at soothing irritated organs and tissues than at stimulating stomach acids and intestinal motility, and that is true of this tea. A tea is the best vehicle, because pills and tinctures can be nauseating themselves to a person on chemotherapy.

Mentha piperita
Matricaria chamomilla
Althaea officinalis
Foeniculum vulgare
Zingiber officinale

Combine equal parts of the dry herbs. Steep 1 tablespoon of the herb mixture per cup of hot water and then strain. Drink as much as possible in small frequent sips throughout the day. For severe nausea, some may respond well to mixing the tea with an equal amount of a carbonated mineral water. When nausea is severe and not even water is tolerated, try freezing this tea in an ice cube tray. Place the frozen cubes in a resealable plastic bag. Wrap a thick towel around the bag and crush the ice cubes into slivers using a hammer or kitchen mallet. The frozen tea slivers can be placed in a small teacup and taken by spoonfuls.

Tonic for Children with Poor Appetite or Failure-to-Thrive Pediatric Patients

Gentian is extremely bitter and will not be tolerated by most children; however, when balanced with something sweet and then highly diluted, it can be effective in stimulating children's appetites. Frequent dosing is best—at the very least 20 to 30 minutes before each meal. See also *Angelica* "Candies" for Dyspepsia and Constipation on page 29 and Licorice Syrup for Barrett's Esophagus on page 43 in this chapter for further instruction on preparing herbal syrups.

Gentiana lutea 15 ml
Mentha piperita syrup or other herbal syrup 15 ml

Essential oil of mint or other desired flavor such as citrus may be added to help mask the gentian.

Put 5 drops in a sip of water. Encourage the child to take a sip as often as possible, and especially before meals.

Tincture for Anorexia in Cold, Enfeebled, and Aged Persons

This formula combines the bitter alteratives *Rheum* and *Cinchona* with warming and circulatory enhancing agents for best effect in anorexia in elderly people with poor circulation, slow peristalsis, and diminished digestive acids and enzymes. The use of probiotics and actual digestive enzymes or HCL supplements will be complementary to this formula for many patients, selected on a case-by-case basis. *Gentian*, *Artemisia*, *Berberis*, *Hydrastis*, *Rumex*, *Acorus*, *Achillea*, *Cinchona*, and other bitter cholagogues can all be used in this manner.

Rheum palmatum 15 ml
Cinnamomun verum 15 ml
Cinchona officinalis 15 ml
Angelica archangelica 15 ml

Take 1 dropperful of the combined tincture 20 minutes before each meal.

Nourishing Herbs for Malnutrition

While preparing teas for chemotherapy patients or formulas for children with failure to thrive, consider adding some of the following herbs to provide nutrients. Teas prepared from these herbs can also be used to make soups and broths, as the cooking liquid for oatmeal, to thin down a smoothie, and in other creative ways to increase nutrient intake.

Avena sativa
Centella asiatica
Equisetum arvense, *E. hymenale*
Fucus vesiculosus
Medicago sativa
Urtica dioica

Fire Cider to Stimulate Digestive Acid and Secretions

Vinegars macerated with hot spicy herbs are sometimes referred to as fire cider due to the hot fiery flavor and the apple cider vinegar base. Fire ciders may include ginger, horseradish, turmeric, garlic, onions, and hot peppers macerated in apple cider or other quality vinegar (see page 74 for vinegar recipe). This formula combines fire cider with the digestive bitter *Artemisia* and a small amount of sweetener such as honey. Use as an aperitif before meals.

½ cup (120 ml) cold water or hot herbal tea
1 tablespoon apple cider vinegar (or a fire cider or a bitter tonic vinegar)
Maple syrup or honey, to taste
20 drops *Artemisia* tincture
Dash fresh ground black pepper (optional)

This beverage can be prepared in water or tea as desired. Add the vinegar, maple syrup, and *Artemisia* to the chosen liquid and stir well. Add pepper if it can be tolerated. Sip over a span of 10 to 15 minutes prior to all meals.

42 Cocktails

Not a drink, 42 Cocktails is a traditional naturopathic encapsulated formula aimed at stimulating hydrochloric acid and digestive secretions. The freshly ground powders are blended and encapsulated in size 00 capsules.

Gentiana lutea powder
Artemisia absinthium powder

Encapsulate equal amounts of the powders and take 1 capsule 10 to 15 minutes before each meal.

Tea for Chronic or Frequent Nausea

This tea formula can help chemotherapy patients as well as patients with multiple food sensitivities that are interfering with the appetite. The *Ulmus* and *Hydrastis* powders will cling to the leaves and flowers of the other herbs, allowing the blend to be easily infused.

Matricaria chamomilla 4 ounces (120 g)
Mentha piperita 2 ounces (60 g)
Ulmus fulva 1 ounce (30 g)
Hydrastis canadensis 1 ounce (30 g)

Steep 1 tablespoon of the herb mixture per cup of hot water and then strain. Drink by small cupfuls over the course of the day.

Tincture to Promote Appetite

A bitter tincture like this one can promote stomach acid and digestive secretions, increase circulation in the digestive mucosa, and stimulate peristalsis. The bitter herbs are placed in a carminative base of peppermint glycerite to avoid possible griping or overstimulation of secretions and intestinal peristalsis.

Mentha piperita glycerite 45 ml
Artemisia absinthium 30 ml
Rumex crispus 15 ml
Iris versicolor 15 ml
Zanthoxylum americanum or *Z. clava-herculis* 15 ml

Take 10 to 30 drops of the combined tincture straight or in a sip of water, repeating throughout the day, particularly a half hour before meal or snack times.

Supportive Tincture for Anorexia Nervosa

During the recovery phase, bulimics and individuals recovering from anorexia nervosa can benefit from this tincture, which supports appetite and digestive function and helps treat the inability to eat.

Rumex crispus
Zingiber officinale
Glycyrrhiza glabra
Althaea officinalis
Matricaria chamomilla

Combine equal parts of the herbs and take 20 to 60 drops of the combined tincture as frequently as every hour and at least 3 times daily.

Tea for Anorexia Nervosa

This tea complements the above tincture, aiming to support nutrition, the nervous system, and the digestive mucous membranes.

Avena sativa groats
Medicago sativa
Urtica urens
Matricaria chamomilla
Scutellaria lateriflora
Althaea officinalis
Glycyrrhiza glabra

Combine equal parts of the herbs, adding more licorice to taste. Steep 1 tablespoon of the herb mixture per cup of hot water and then strain. Drink as much as possible to help restore digestive mucosa and offer nervine and nutritional support.

Glycyrrhiza Simple for Oral Lesions Due to Bulimia

Licorice is excellent for promoting healing of erosive oral lesions, and as a solid extract, it will stick to the oral mucosa.

Glycyrrhiza solid extract

Place 1/4 teaspoon of the solid extract in the mouth and hold until dissolved and distributed in the saliva. Repeat 6 or more times a day.

Demulcent Lozenges for Lesions Due to Bulimia

Licorice solid extract can be combined with *Ulmus* (slippery elm) powder to create a simple paste appropriate to use for both acute and chronic oral and esophageal lesions.

1 tablespoon *Glycyrrhiza* solid extract
2 tablespoons *Ulmus* powder
Essential herbal oil (optional)

Place the licorice solid extract in a small bowl and whisk in the *Ulmus* powder briskly with a fork until a cookie dough–like consistency is achieved. If desired, add a few drops of an essential oil such as mint or citrus to enhance the flavor. Add additional solid extract or more *Ulmus* powder to achieve a dry dough consistency that

Glutamine Powder for Oral Lesions Due to Bulimia

Glutamine may improve mucous membrane barrier functions by helping to tighten the tight junctions of intestinal mucosal epithelial cells.[29] Glutamine powder is available by the pound or in lesser quantities, has a pleasant taste, and dissolves easily in water or herbal tea. It is also pleasant tasting enough to put directly in the mouth.

To treat oral lesions, place 1/4 to 1/2 teaspoon glutamine powder in the mouth and hold until dissolved and distributed in the saliva. Repeat 6 or more times a day.

can be rolled into jelly bean–size balls. Flatten each ball with the thumb to create a thin lozenge and allow the lozenges to dry for several hours on a piece of wax paper. Place in a small resealable plastic bag or plastic storage container or wrap individually in waxed paper to use throughout the day.

Formulas for GERD

Alternative therapies for gastroesophageal reflux disease (GERD), also called acid reflux, involve rebuilding and restoring the health of the digestive mucosa to tone and strengthen the esophageal sphincter. The removal of inflaming or offending foods and beverages is one, often invaluable, approach. Greatly limit or eliminate common food allergens such as wheat, citrus, dairy, or tomatoes, for example. Eliminate coffee, alcohol, and soda and avoid any agents known to be difficult to digest or that lead to mucosal inflammation such as fried foods, heavy meats and fats, or chemical irritants. Digestive health is also supported by using probiotic supplements, eating fiber-rich foods, and ensuring nervous system health. Stress activates the sympathetic nervous system, which overrides the parasympathetic control of healthy digestion. Deep breathing, meditation, and other relaxation techniques invoke the parasympathetic system and are to be encouraged.

Helicobacter pylori bacteria have been found to be associated with gastric ulcers and GERD. Antibiotics are often prescribed to address the "infections." However, the digestive ecosystem may allow, if not invite, this microbe to proliferate such that antimicrobial therapies are incomplete in and of themselves. Many herbs, including *Matricaria* and *Hydrastis*, have been found to deter *H. pylori* and may be superior to pharmaceutical antibiotics by simultaneously offering anti-inflammatory and digestion-enhancing effects as well as antimicrobial effects.

Herbs such as *Glycyrrhiza*, *Ulmus*, and *Matricaria* may be helpful in rebuilding the digestive mucosa and reducing the frequency and intensity of GERD. The demulcent effect of the mucilaginous herbs *Ulmus*, *Althaea*, or *Symphytum* is also helpful to alleviate the discomfort and heartburn sensation of GERD. Alterative herbs such as *Arctium*, *Taraxacum*, *Rumex*, and *Mahonia* (also known as *Berberis*) may help those where biliary insufficiency and insufficient digestive secretions have contributed to the derangement of the mucosa and health of the sphincter. *Mentha* has been noted to relax digestive smooth muscle, including relaxing the tone of the gastroesophogeal sphincter, and may worsen reflux. When a carminative herb is needed to allay gas, bloating, and distensive pain, *Foeniculum* and *Matricaria* may be better choices. Myrtle berries may minimize GERD inflammatory symptoms.[30] Lycopene—the well-studied anti-inflammatory flavonoid from tomatoes—protects against mucous membrane ulceration,[31] even though consuming whole tomatoes can aggravate symptoms for some people with GERD. Lycopene is added to some vitamins and nutritional supplements and is commercially available.

Aloe gel can improve reflux symptoms and should be included in teas and smoothies wherever possible. Glutamine, a gentle and decent-tasting amino acid, also supports intestinal mucosal cells and can be added to teas or stirred into the day's drinking water. These agents are mentioned as supportive therapies or complementary ingredients in the following recipes for GERD.

Sweet and Sour Alginate "Cordial" for GERD

Palatable herbs with mucoadhesive properties are placed in a base of magnesium alginate and used to prepare a warm drink to consume after meals. The alginate may be used alone or combined with pectin,[32] hyaluronic acid,[33] and locust bean/guar gum from *Parkia biglobosa*,[34] when available, as in this recipe. Polysaccharides from *Tamarindus indica* seeds yield a mucoadhesive polymer[35] and offer a sour flavor. Licorice solid extract has anti-inflammatory and ulcer-healing effects to mucous membranes and offers a sweet flavor.

1 tablespoon magnesium alginate
½ teaspoon pectin gel
¼ teaspoon tamarind paste
¼ teaspoon *Glycyrrhiza* solid extract
¼ teaspoon hyaluronic acid
1 cup (240 ml) hot water or tea

Combine all ingredients (as available) in the hot water or tea, stirring vigorously. Drink immediately. Can be taken with a digestive enzyme to offer further digestive support.

Inhibiting Gastric Acid Has Harmful Side Effects

Proton-pump inhibitors (PPIs) and histamine-blocking drugs, also known as H2 antagonists, such as cimetidine, are the mainstay of acid reflux treatment but may provoke SIBO (small intestinal bacterial overgrowth, a type of dysbiosis) as well as exacerbate nonsteroidal anti-inflammatory drug-induced small intestinal injury.[36] Acid-blocking drugs may increase the risk of peritonitis in some patients, due to dysbiosis,[37] and allow pathogenic bacteria, such as *Clostridium difficile*, to thrive.[38] In fact, children put on acid-blocking drugs are at a more than four-fold risk of developing *Clostridium difficile* infections.[39] PPIs may also exacerbate the damaging effects of aspirin and steroids on the intestinal mucosa.[40] Acid-blocking drugs may increase the risk of pneumonia in stroke patients,[41] impair semen quality,[42] exacerbate chronic obstructive pulmonary disease (COPD),[43] increase the risk of chronic kidney disease,[44] increase the risk of fracture in osteoporotic adults,[45] and increase the risk of lower respiratory tract infections in susceptible individuals.[46] It may be wise to treat GERD without such drugs in the first place, given how difficult *Clostridium* infections are to eradicate and how serious the complications. See also "Acid-Blocking Pharmaceuticals and Digestive Ulcers" on page 63 for further ideas on avoiding the H2 antagonist drugs.

Carminative Tea for GERD and *Helicobacter* Infection

Matricaria deters *Helicobacter pylori*, which may contribute to GERD, and is also an excellent carminative agent. Many people with GERD have a great deal of substernal pain and pressure that can be alleviated, at least in part, with the use of carminative anti-inflammatory agents. This formula can be prepared as a tea or a tincture, as desired. Although teas may be more effective, it is also useful to have a tincture on hand in the office or in the purse for occasions when symptoms flare up suddenly. This formula is based on agents noted to have activity against *H. pylori* and selected for palatability as a tea. The proportions may be amended to suit personal tastes.

Achillea millefolium finely ground flowers
2 ounces (60 g)
Matricaria chamomilla flowers
2 ounces (60 g)
Rosmarinus officinalis finely cut leaves
2 ounces (60 g)
Foeniculum vulgare whole seeds or powder
2 ounces (60 g)
Zingiber officinale finely cut dried root
2 ounces (60 g)

Steep 1 tablespoon of the herb mixture per cup of hot water, and drink freely, at least 3 cups per day. Consuming 2 or 3 cups over the course of a half hour may help allay acute discomfort. An equivalent tincture may be prepared from the combining equal parts of the herbs and taking 1 dropperful every 10 minutes for a half hour to treat acute symptoms.

Tincture for GERD, Especially with Sore Throat

The astringent herbs *Hamamelis* and *Quercus* may help tighten the gastroesophogeal sphincter while also providing anti-inflammatory effects. *Hydrastis* has effects against *Helicobacter pylori* infections and is astringent, and the folkloric literature emphasizes *Hydrastis* for excessive secretions from infected mucous membranes. *Matricaria* provides overall anti-inflammatory and trophorestorative effects on the gastric mucosa and all digestive mucous membranes and is a nervine for when stress contributes to excessive stomach acid production. This formula could also be prepared as a tea, but *Hydrastis* is challenging for many to drink. It could be omitted from the tea.

Hamamelis virginiana
Quercus rubra
Matricaria chamomilla
Hydrastis canadensis

Combine equal parts in a 2-ounce bottle. Take a dropperful of the combined tincture as often as hourly, reducing as symptoms improve. Use at a minimum of 3 times daily.

Matricaria and the GI

Matricaria is included in many of the formulas in this chapter due to its pleasant taste, lack of negative side effects, and numerous medicinal actions. *Matricaria* contains aromatic terpenes such as alpha-bisabolol and its oxidation products, the azulenes, including chamazulene. Matricin, a sesquiterpene lactone, is metabolized into chamazulene. The chamazulene that occurs in *Matricaria, Achillea,* and a few other members of the Aster family is noted to possess potent anti-inflammatory activity, particularly to the GI tract. These compounds have been shown to inhibit cyclooxygenase-2 (COX-2) enzymes[47] and to have platelet-stabilizing effects.[48] The flavonoids apigenin, quercetin, patuletin, and luteolin are also anti-inflammatory. Antispasmodic activity, especially in the GI, has been noted, along with an anti-inflammatory effect on GI mucous membranes. Antimicrobial activity, including activity against *Helicobacter,* has been demonstrated.[49] These compounds also have been shown to have gastroprotective effects via numerous mechanisms, including protecting against alcohol-induced ulceration via angiotensin-converting enzyme (ACE) and glutathione regulation.[50] and also via potassium channel effects.[51] Alpha-bisabolol itself is also credited with gastroprotective effects,[52] with one mechanism being making sulfur groups more bioavailable, helping to support glutathione detoxification and antioxidant pathways.[53]

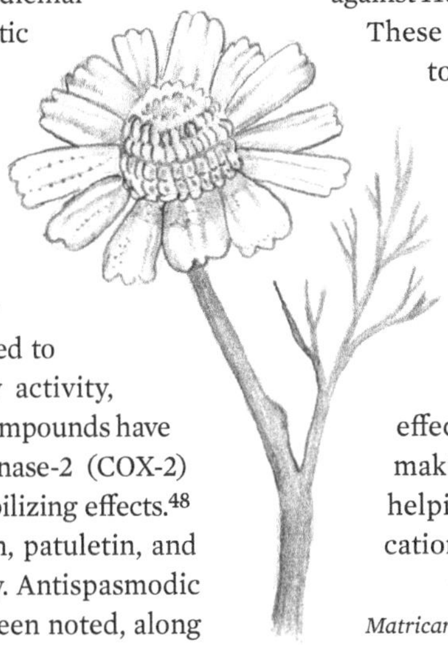

Matricaria chamomilla, chamomile

Antimicrobial Tincture for GERD and Helicobacter Infections

As with peptic ulcers, *Helicobacter pylori* infections have been implicated in many cases of reflux, making antimicrobial herbs a logical part of a formula for reflux. This formula is based on herbs shown to deter *H. pylori.* For best result, also offer patients mucosal restoratives, probiotics, and thoughtful dietary advice to permanently improve the GI mucosal health and bacterial flora.

Achillea millefolium 30 ml
Curcuma longa 30 ml
Mahonia aquifolium 30 ml
Rosmarinus officinalis 30 ml
Gentiana lutea 15 ml
Zingiber officinale 15 ml
Fennel essential oil 20 drops

This yields a 4-ounce bottle and a sufficient quantity to take 1 or 2 teaspoons of the combined tincture 5 or 6 times a day, reducing as symptoms improve.

Postprandial Gum Chewing to Reduce GERD

Chewing stimulates saliva flow and digestive motility and can reduce GERD and improve gastroparesis. Studies suggest that chewing alone may reduce reflux,[54] and companies now offer medicinal chewing gum for this purpose such as gum containing licorice and papain.[55] Gum chewing improves the clearance of refluxate from the esophagus, a benefit that endures for at least 3 hours after 1 hour of gum chewing. The chewing of xylitol gum has many benefits to dental health and may also help reduce childhood ear infections. Xylitol chewing gum is also shown to increase gastrointestinal motility recovery following major surgeries.

Carminative Tea for GERD

In addition to treating a possible *H. pylori* infection contributing to GERD symptoms, as with the above tincture, many people will appreciate a palliative they can use immediately for pain, bloating, pressure, heartburn, and acidic symptoms in the throat. Although mint is an excellent carminative, avoid it for GERD because it relaxes the gastroesophageal sphincter and may worsen reflux. The following carminatives and mucosal nutrients may work well to quickly palliate distensive pain and bloating.

Elettaria cardamomum 2 ounces (60 g)
Foeniculum vulgare 2 ounces (60 g)
Althaea officinalis 2 ounces (60 g)
Glycyrrhiza glabra 1 ounce (30 g)
Zingiber officinale 1 ounce (30 g)

Place 6 teaspoons of the ground herbs and seeds in a saucepan and soak in 6 ounces of cold water or overnight when possible. Bring to a very brief simmer for 2 or 3 minutes, remove from the heat, and cover the pan. Let stand for 15 minutes and then strain. Drink freely, as much as possible.

Seaweed Alginates for GERD

Alginates are polysaccharide compounds found in the cell walls of brown algae, including the kelps and bladderwracks. These seaweeds are currently the world's largest sea "crop" used in large quantity to produce stabilizers and emulsifiers for the food industry, to make bandages for wound care, and to fashion gums and gels used as vehicles to deliver various oral medicines.[56] Kelp (*Macrocystis pyrifera*) is used in alginate production in California, and *Laminaria japonica*, *Laminaria hyperborean*, and *Ascophyllum nodosum* are used in other regions of the world. The gel-forming physical nature of alginates can bind stomach acid at the gastroesophageal junction and create a barrier that helps resist reflux. Several clinical trials have shown alginates to help relieve symptoms in GERD patients.[57] Because stomach acid often floats on top of ingested food in the stomach, referred to as an "acid pocket," immediately consuming sodium or magnesium alginate at the close of all meals may bind the acid, cap off the gastroesophageal sphincter, and reduce reflux without antacids. Alginates impregnated with calcium carbonate are also being developed to remedy reflux,[58] and alginates impregnated with *Coptis chinensis* and *Euodia rutecarpa* are shown to protect the gastric mucosa from alcohol-induced injury.[59]

Demulcent Tea for GERD

When you want to brew a demulcent tea but soaking the herbs overnight is inconvenient, here is a simple alternative. Make sure that your *Symphytum* tincture is obviously mucilaginous.

Glycyrrhiza solid extract 20 ml
Symphytum tincture 40 ml

Pour the licorice solid extract into the bottom of a 2-ounce bottle. Top off with the *Symphytum* tincture and shake well. This formula will be so thick that it will barely move through a dropper. Cap the bottle with a regular cap, and uncap it to pour the dose into a spoon. Hold ¼ teaspoon in the mouth and throat as long as possible, allowing the formula to soothe the irritation caused by stomach acid. Repeat every 1 to 2 hours, reducing as symptoms improve.

Tea for GERD Due to Poor Mucosal Integrity

In some cases, esophageal reflux may be related to loss of mucosal integrity due to old age, intestinal dysbiosis, and inability of the mucosa to repair and regenerate itself. *Centella* is a tissue restorative and *Calendula* may help improve circulation to the mucosa. *Glycyrrhiza* and *Althaea* may provide both symptom relief and act as mucosal demulcents and anti-inflammatories. Offer this tea to patients along with probiotics, nutrients, and dietary advice.

Glycyrrhiza glabra
Centella asiatica
Althaea officinalis
Calendula officinalis

Combine equal parts of the dry herbs. Steep 1 heaping tablespoon of the herb mixture per cup of hot water and then strain. It's a good idea to add 1 to 2 teaspoons of glutamine powder and as much *Aloe* gel as tolerated to each cup. Drink 3 or more cups per day.

Restorative Agents for Mucous Membranes

Like the squamous epithelium of the skin, mucosal epithelial cells of the digestive system have a very rapid cell turnover rate. Infections, allergic reactions, inflammatory diseases, and other factors may impair the ability of the mucosa to keep up with the repair and regeneration of tissues. The following herbs and nutrients are noted to support the stomach and intestinal mucosal cells, helping to heal ulcers; improve cell-to-cell adhesion and thereby barrier functions; and improve the quality and quantity of mucous that the intestinal goblet cells produce. Such herbs can be mixed with antimicrobial agents to treat *Helicobacter*-induced GERD, combined with anti-inflammatory herbs to treat irritable bowel disease (IBD), or combined with antiallergy herbs to create formulas for those with multiple food sensitivities.

Calendula officinalis
Centella asiatica
Equisetum arvense, *E. hymenale*
Filipendula ulmaria
Glycyrrhiza glabra
Matricaria chamomilla
Ulmus fulva
Beta carotene
Glutamine
Probiotics
Zinc

Herbs for *Helicobacter* Infections

Helicobacter pylori has been shown to be highly associated with peptic ulcer diseases as well as proven to be a potent carcinogen associated with gastric carcinoma,[61] particularly gastric B-cell lymphoma. A long list of herbs has been shown to have activity against *H. pylori*, including the following.[62] Choose one or more such antimicrobial herbs when putting together formulas for *Helicobacter* and associated conditions based on specific indication and other properties of these herbs as listed in "Specific Indications," beginning on page 80.

Achillea millefolium
Azadirachta indica
Carum carvi
Coptis chinensis
Curcuma longa
Elettaria cardamomum
Foeniculum vulgare
Gentiana lutea
Hydrastis canadensis
Juniperus communis
Lavandula angustifolia
Mahonia aquifolium
Matricaria chamomilla
Melissa officinalis
Mentha piperita
Myristica fragrans
Origanum majorana
Passiflora incarnata
Pimpinella anisum
Rheum palmatum
Rosmarinus officinalis
Sanguinaria canadensis
Zingiber officinale

Licorice–Aloe Paste for GERD and Esophagitis

Aloe gel may improve heartburn and other digestive symptoms.[60] The gel can be prepared into syrups or pastes to help it cling to the esophagus as long as possible. Tamarind is a sour-flavored leguminous pod that is prepared into a syrupy paste for cooking and is shown to improve mucoadhesivity of medicines. *Glycyrrhiza* (licorice) is available as a solid extract that will also help the medicine cling to the throat and be slowly cleared from the esophagus. Both tamarind and *Glycyrrhiza* will improve the flavor of the *Aloe* gel.

Aloe vera gel 1 ounce (30 g)
Tamarindus indica paste 1 tablespoon
Glycyrrhiza solid extract 2 teaspoons
Glutamine powder 2 teaspoons

Place all ingredients in a small bowl and blend vigorously with a fork. Take 2 teaspoons 3 times a day, directly off the spoon or diluted in one of the herb teas presented in this chapter. Additional doses may be taken as needed for acute symptom relief. Use the entire batch of paste over the course of the day, and make fresh each day.

Licorice Syrup for Barrett's Esophagus

Barrett's esophagus is the gradual replacement of esophageal epithelium with cells more typical of the lower digestive tract and is associated with an increased risk for esophageal cancer. Insufficient saliva and mucosal secretions contribute to Barrett's esophagus and may be improved by prolonged mastication.[63] *Curcuma*, a well-established anticancer herb, is noted to offer protective effects from developing Barrett's esophagus.[64] *Panax ginseng* is not featured in the basic GERD formulas, but studies have shown the plant to ameliorate inflammatory changes that occur with chronic GERD,[65] and it might be especially chosen when GERD occurs with chronic stress and emotional symptoms. Berries and their flavonoids are known to protect vascular and other tissues in many research models of inflammatory stress, and they can be included in the delivery of this formula.

Glycyrrhiza solid extract 30 ml
Curcuma longa 10 ml
Panax ginseng 10 ml
Fucus vesiculosus 10 ml

Place the licorice solid extract in the bottom of a 2-ounce bottle and add the tinctures to fill the bottle. Shake vigorously to homogenize. The licorice solid extract can help the other herbs cling, at least momentarily, to the esophagus. Take 1 teaspoon 3 or 4 times a day, aiming to hold it in the mouth for as long as possible, which may prolong direct contact with the esophagus. This syrup can also be stirred into berry powder such as *Crataegus* or *Vaccinium* powders to make a paste that can be simply swallowed off the spoon.

Formulas for Esophageal Disorders

Esophagitis is most commonly due to reflux disease (GERD), but it may also occur due to chronic irritation from alcohol, coffee, or other mucosal irritants, including pharmaceuticals. Allergic phenomena can contribute to eosinophilic esophagitis, which can be chronic and sometimes poorly understood.[66] Elimination of all common food allergens is warranted when treating esophageal disorders not obviously associated with GERD. Esophagitis may cause heartburn, dysphagia, and uncomfortable symptoms. It should always be taken seriously, because the condition may progress to involve substantial hemorrhage, the formation of strictures, and full malignant transformation in some cases, such as with Barrett's esophagus described on page 43.

Agents and therapies that improve the health of the entire digestive ecosystem are required to improve the tone of the gastroesophageal sphincter. Simple demulcent mucosal anti-inflammatories are appropriate for symptom relief and to help restore the esophageal mucous membrane. Teas, glycerites, or syrups are the best vehicles, as they provide the most prolonged surface contact with the esophagus. *Althaea*, *Ulmus*, *Symphytum*, and *Glycyrrhiza* are appropriate.

Formula for Esophageal Motor Weakness

Muscular and neurologic diseases can result in dysphagia that has no reliable pharmaceutical or other remedies. Herbs that strengthen the muscles stand the best chance of improving this type of dysphagia, though they will not remedy Lou Gehrig's or other muscular dystrophic pathology. This remedy may be prepared as a tea or a tincture.

Panax ginseng
Lepidium meyenii
Salvia miltiorrhiza

Combine in equal parts for either a tea or tincture. Gently simmer 1 teaspoon of the herb mixture per cup of water for 15 minutes. Remove from heat and let stand, covered, for 10 minutes more, and strain. Take large doses such as 4 or 5 cups of tea or 4 dropperfuls of tincture every day for at least 6 months. If there is any improvement at all from such a formula, it would be worth continuing for a year or more.

Tincture for Globus Hystericus

Globus hystericus is a type of dysphagia most associated with mental-emotional disturbances and may be a type of conversion reaction, also referred to as globus sensation. Globus hystericus may also relate to the Plum Pit sensation in Chinese medicine. This formula is often helpful, as it contains muscle-relaxing nervines that help to address the underlying stress and *Mentha*, which is very specific for relaxing the smooth muscles of the digestive tract.

Mentha piperita 30 ml
Passiflora incarnata 15 ml
Matricaria chamomilla 15 ml

Take 1 dropperful of the combined tincture every 3 to 6 hours, reducing as symptoms improve.

Decoction for Globus Hystericus Based on Classic Chinese Formula

Human clinical trials of this decoction have suggested efficacy for Globus hystericus[67] and for dysphagia in Parkinson's patients.[68] This formula can also be prepared into a throat spray to use during meals for those with stroke-related dysphagia.[69]

Magnolia officinalis
Poria cocos
Zingiber officinale
Perilla frutescens
Pinellia ternata

Decoct equal parts of the herbs, gently simmering 1 teaspoon of the herb mixture per cup of hot water for 10 minutes. Strain and consume 3 or more cups per day.

Formula for Eosinophilic Esophagitis

Also called allergic or reactive esophagitis, eosinophilic esophagitis may be treated in much the same manner as GERD and gastritis, but emphasizing immune modulators. This formula is intended to be a very thick paste, so that it will cling to the esophagus. This formula would be complemented by probiotics, dietary changes, and the removal of all irritants and allergens possible.

Matricaria tea ½ to 1 cup (120 to 240 ml)
Ulmus powder 1 heaping tablespoon
Glycyrrhiza solid extract 1 teaspoon
Aloe vera gel 1 tablespoon
Blueberry or elderberry powder 1 teaspoon
Curcuma powder 1 teaspoon
Astragalus powder 1 teaspoon

Prepare the *Matricaria* tea and set aside. Pour 1 tablespoon cold water over the 2 tablespoons of *Ulmus* (slippery elm) powder and stir vigorously, and let stand. Place the licorice, *Aloe* gel, and three powders in a small bowl and blend. Once the *Ulmus* powder has absorbed the liquid, stir it into the other herbs in the bowl. The resulting paste can be thinned down with the *Matricaria* tea and then taken by the spoonful, or it can be taken straight. Place a small ½ teaspoon in the mouth and allow to slowly dissolve into the saliva and be swallowed as slowly as possible, attempting to prolong the surface contact. Repeat every 15 to 30 minutes, and make a new batch whenever needed. This formula should begin to provide relief within 1 to 2 days' time, and the frequency of dosing can be reduced.

Tincture for Esophageal Varices

Esophageal varices are usually associated with serious liver diseases including cirrhosis, hepatic carcinoma, and portal hypertension. Therefore, liver herbs and circulatory-enhancing herbs are the basis of this formula, which should be used as part of a comprehensive and aggressive protocol to treat any underlying portal hypertension and liver disease.

Silybum marianum
Curcuma longa
Angelica sinensis
Ginkgo biloba
Aesculus hippocastanum

Combine equal parts and take 2 dropperfuls every 2 to 3 hours.

Throat Wash for Esophagitis and Fungal Infections

Fungal infections of the throat and esophagus usually occur only in diabetics with poor glucose control and in those with immunodeficiency states. Hypochlorhydria may also be a factor if the digestive environment is not sufficiently acidic to kill off naturally occurring yeast and fungus. This formula combines powerful antimicrobials *Allium* and *Origanum* with immune-modulating *Astragalus* and digestive-enhancing *Mahonia*.

Allium sativum
Origanum vulgare
Astragalus membranaceus
Mahonia aquifolium

Prepare as a tea or a tincture using equal parts of the herbs. For throat infections, gargling with a tea or the tincture (placed in water) would be ideal, but for throat infections, small frequent sips of tea may be the best technique. Prepare a large batch of tea by steeping 1 tablespoon per cup of hot water, sip as often as possible throughout the day. The frequency may be reduced over time as the infection improves.

Antiviral Throat Wash

Viral esophageal infections usually occur only in immune-compromised individuals, where *Herpes simplex* and cytomegaloviruses may take hold. Those who have had local irradiation and those with Sjögren's syndrome may also be afflicted because damaged mucosal tissue and insufficient saliva allow for opportunistic infections. This formula combines antiviral herbs with immune-enhancing herbs.

Glycyrrhiza glabra
Origanum vulgare
Hypericum perforatum
Astragalus membranaceus

Prepare as a tea or a tincture using equal parts of the herbs. Teas may be prepared by steeping 1 tablespoon of the herb mixture per cup of hot water. If the throat is also affected, gargling with the tea may be helpful. For esophageal infections, small frequent sips of tea may be the best technique, consuming 5 or 6 cups of tea over the course of the day.

Formula for Esophageal Ulcers

Esophageal ulcers may result from bulimia and other situations of repetitive vomiting or reflux. Failing to fully wash down iron supplement pills, alendronate (Fosamax), and other medications can also result in ulcerated esophageal tissue. Take care to eat a bit of solid food after ingesting any problematic medicines to ensure that the pills are pushed all the way into the stomach. Crohn's disease may affect the esophagus and should be considered in such patients who are experiencing upper-gastrointestinal symptoms in addition to the more typical bowel symptoms.[70] This formula is similar to ones for peptic or intestinal ulcers.

Glycyrrhiza glabra
Centella asiatica
Matricaria chamomilla
Symphytum officinale

Combine equal amounts of the dry herbs and prepare as a tea, using 1 tablespoon of the herb mixture per cup of hot water. Even better, combine 1 cup tea prepared from dried *Matricaria* and *Centella* with ½ teaspoon of *Glycyrrhiza* solid extract and 1 teaspoon *Symphytum* tincture, or ½ cup of a long-soaked cold *Symphytum* infusion. This creates the thickest liquid possible to help the formula adhere to the esophagus and prolong surface contact. Prepare a large batch each morning or evening and sip frequently throughout the day.

Neem Formula for Ulcers Due to *Helicobacter pylori* or Dysbiosis

Azadirachta indica (neem) bark has potent antisecretory and antiulcer effects,[71] and clinical trials have shown 30 to 60 milligrams of dried bark powder to be highly effective in treating ulcers when taken daily for 10 weeks.[72] Neem has also been found to prevent gastric and hepatic carcinogenesis and is often combined with garlic for this purpose.

Azadirachta indica
Allium sativum

Tinctures, where available, can be taken 1 or 2 dropperfuls at a time 3 to 6 times daily. Or you can add the combined tincture to other formulas in this chapter as appropriate. Neem and garlic powders are also available, as are oils. However, the oral ingestion of neem oil must be limited due to safety concerns and is best for topical application. Powders of neem may be combined in equal amounts with garlic powder, encapsulated by hand, and consumed orally to treat intestinal worms, dysbiosis, *H. pylori* infection, and gastric and esophageal ulcers. Take 2 such capsules twice a day.

Formulas for Constipation

Slow movement in the intestines can be due to insufficient fiber intake, low thyroid function, intestinal spasm, insufficient secretions, intestinal dryness, or intestinal lesions that obstruct the bowels, including colon and rectal cancers. Constipation may be an early symptom of pathologies outside of the digestive tract as well. For example, ovarian cancer, renal disease, or liver disease may involve constipation or other changes in bowel habits. A thorough workup to rule out pathology is in order. In cases where constipation has been long-standing, muscle tone in the intestines and peristalsis can be lost, necessitating the use of irritant laxatives such as *Rhamnus purshiana* or *Cassia* (also known as *Senna*) to promote smooth muscle contraction.

In most cases, however, the use of irritants is to be avoided, reserving them for only the most severe and unresponsive cases when peristalsis has been lost. For most cases of simple constipation, the alterative herbs are some of the best choices. Alteratives such as *Rheum*, *Taraxacum*, *Arctium*, *Mahonia*, and *Rumex* may be taken multiple times per day for many months to help restore normal intestinal function. These may be combined with carminative agents such as *Cinnamomum*, *Foeniculum*, and *Zingiber* as well as demulcents that act as intestinal lubricants and stool softeners. For young children, stewing prunes in a tea of *Rheum* and *Cinnamomum* may be effective.

For chronic constipation, it is also helpful to remove food allergens and difficult-to-digest foods such as fried foods, meat, dairy products, breads, and other processed grain products and replace them with large quantities of fresh fruits and vegetables. Encourage patients to eat a raw salad every day, along with at least five other servings of vegetables and several liters of water. Exercise is also helpful for constipation—even gentle walks of a mile or two will be helpful to many people. Too much sitting and fiber-deficient diets are both leading causes of chronic constipation.

Alterative Tincture for Constipation

Many alteratives have mild and gentle laxative effects and, unlike irritant laxatives, will not cause diarrhea or cramping. Be cautious about using any laxative agents, even gentle ones, in patients with IBS because laxatives can quickly promote diarrhea that may take weeks to subside. This formula could also be prepared as a tea.

Taraxacum officinale 15 ml
Rumex crispus 15 ml
Foeniculum vulgare 15 ml
Glycyrrhiza glabra 15 ml

Take ½ teaspoon of the combined tincture 4 times daily.

Triphala for Constipation

My colleague and friend, Ben Zappin, L.Ac., urged me to include triphala in the section for chronic constipation. Triphala can be translated as "three fruits" and is a traditional Ayruvedic herbal combination prepared from Amalaki (*Emblica officinalis*), Bibhitaki (*Terminalia belerica*), and Haritaki (*Terminalia chebula*) and used as a bowel tonic. Triphala may improve absorption and elimination without creating dependency. The blend is well-known and readily available in the US marketplace. The fruits used in the blend are also credited with powerful antioxidant and tissue-protective effects.

Triphala powder

Stir a teaspoon of the powder blend into water and take at night before bed. Triphala can also be combined with a teaspoon of bulking fiber, such as apple pectin, psyllium, or ground flaxseeds, stirred vigorously, and consumed promptly.

Smoothie for Constipation in Children

Children with constipation are often not eating enough fiber or drinking enough water. In other cases, children may be intolerant of a particular food that is poorly digested. In all cases, a dietary consult will often resolve the problem. This drink recipe often helps ease constipation during the period of adjustment to a new diet. Use the child's favorite fruit juice to make this smoothie.

1 cup (240 ml) of fruit juice or soy, rice, or almond milk
½ cup (120 ml) prune juice
½ cup (120 ml) alterative herb tea (such as *Taraxacum or Arctium*)
1 teaspoon ground psyllium seeds
1 teaspoon apple pectin

Put all ingredients in a blender and blend on high speed until smooth. Drink promptly.

Irritant versus Bulk Laxatives

Irritant laxatives, as the name implies, irritate the intestines to the point where peristalsis is promoted as the body attempts to move the offending substances out. Irritant laxatives are only appropriate in cases where peristalsis is weak to nonexistent and must be combined with antispasmodic carminative herbs to prevent them from being too harsh and causing diarrhea, cramping, and sudden, urgent loose stools. *Rhamnus purshiana* and *Cassia* (also known as *Senna*) pods are some of the most common irritant laxatives available.

Bulk laxatives are indigestible fibrous substances found in foods such as oatmeal, whole flaxseeds or ground meal, and psyllium seed husks or powders that absorb water and swell inside the intestines, stimulating stretch receptors and promoting peristalsis. These are appropriate for everyone, are safe and gentle, and are most effective taken by the tablespoonful and stirred into water or juice and promptly consumed, except for oatmeal and oat bran, which are simply eaten as breakfast porridge. It is important that ample water be consumed any time bulk laxatives and fiber supplements are being used.

IRRITANT LAXATIVES

Aloe barbadensis
Cassia senna
Juglans nigra
Podophyllum peltatum (use in small doses only)
Rhamnus frangula
Rheum officinale
Ricinis communis (oil, never beans!)

BULK LAXATIVES

All raw fruits and vegetables
Cereal grains, such as bran, oats, brown rice
Fruit pectin
Linum usitatissimum (flax)
Plantago spp. seed heads ("bran")

Irritant Laxative Tea for Loss of Peristalsis

This formula uses the irritant laxative *Rhamnus* to promote peristalsis in those with weak or atonic bowels. Cinnamon is carminative, which will reduce any potentially cramping effects of the *Rhamnus*, and is also warming; it stimulates circulation to further promote

Folkloric Herbs for Constipation

The herbs most indicated for constipation in traditional herbal folklore include all of the alterative herbs; all of the high-fiber herbs such as psyllium and flaxseeds, especially ground into powders and taken in food and drink; and irritant laxative herbs.

CATEGORY	HERBS
Alteratives	*Arctium lappa* *Mahonia aquifolium* *Rumex* spp. *Taraxacum officinale*
Bulk laxatives	*Avena sativa* *Linum usitatissimum* *Malus* spp. *Plantago psyllium*
Irritant laxatives	*Cassia senna* *Juglans nigra* *Podophyllum peltatum* *Rhamnus purshiana* *Rheum palmatum* *Ricinus communis* (castor oil)

peristalsis. *Glycyrrhiza* offers some protective demulcent effects and is anti-inflammatory. This formula could also be prepared as a tincture.

Taraxacum officinale 4 ounces (120 g)
Rhamnus purshiana 4 ounces (120 g)
Cinnamomum verum 4 ounces (120 g)
Glycyrrhiza glabra 4 ounces (120 g)

Combine the dry herbs to yield 1 pound of the blend. Gently simmer 1 tablespoon of the herb mixture in 3 cups of water for 15 minutes in a covered pan and then strain. Drink the entire amount over the course of a day. Continue daily for several weeks to several months.

Irritant Laxative for Colonic Inertia or Loss of Peristalsis

Reserve this powerful tincture for those with atonic bowel. *Rhamnus* and *Podophyllum* are both irritant laxatives and are too harsh to use on their own. In order to avoid cramping and diarrhea, combine these strong herbs with carminatives and demulcents to balance the formula. The ginger and fennel help dispel gas and bloating to prevent any griping effects from the irritants, and the *Ulmus* provides a soothing base.

Foeniculum vulgare 8 ml
Ulmus fulva 8 ml
Taraxacum officinale 4 ml
Rhamnus purshiana 4 ml
Podophyllum peltatum 4 ml
Zingiber officinale 2 ml

Take 1 dropperful of the combined tincture 2 times a day. Continue for 2 or 3 days while evaluating the results. If you decide to increase dosage, do so cautiously because *Podophyllum* may be slow to act. If there are no positive results over 5 days' time, try slowly increasing the amount of *Rhamnus* in the formula.

Irritant Laxative–Free Laxative Tincture

Patients with irritable bowel syndrome (IBS) do not tolerate irritant laxatives, even when severely constipated. It is better to use alterative herbs with a specific stimulant. *Capsicum* is used here to bring more heat to the bowels, but other herbs to consider are *Zingiber*, *Piper nigrum*, or *Allium*.

Rumex crispus 28 ml
Taraxacum officinale 28 ml
Capsicum spp. 4 ml

Take 1 dropperful 4 or 5 times a day, increasing to 2 dropperfuls at a time if symptoms do not improve in 48 hours.

Demulcent Tea for Intestinal "Dryness"

When the stools are hard and dry and passed with difficulty, demulcent agents such as *Ulmus* may be combined with laxatives, such as the *Rumex* in this formula.

Glycyrrhiza glabra
Ulmus fulva, shredded bark
Rumex crispus

Combine equal parts of the dry herbs. Use 1 teaspoon of the herb mixture per cup hot water, simmer for 5 minutes, and then strain. Drink 3 to 6 cups per day.

Cascara Stewed Prunes

This formula is tasty and usually fast acting. It can be used for toddlers or anyone who would prefer it.

Rhamnus purshiana
Dried prunes

Simmer 1 tablespoon of dried *Rhamnus* bark in 2 cups of water for 10 minutes. Let stand 10 minutes, and strain, returning the tea to a small saucepan. Add 2 or 3 prunes and simmer gently until the prunes swell and soften. Eat just half a prune, and wait 8 hours before eating the second half. If no bowel movement occurs within 24 hours, eat another entire prune, and repeat if needed for further instances of constipation.

Apple Pectin Truffles for Constipation

Increasing dietary fiber stimulates intestinal stretch receptors and stimulates peristalsis. Fiber also supports the beneficial microbes in the digestive tract.

Cashew butter or other nut butter
Carob syrup, yacon syrup, agave nectar, or honey
Apple pectin
Rheum palmatum powder

Combine roughly equal parts of nut butter and sweetener in one bowl and blend equal parts of apple pectin and *Rheum* powder in another. Blend the nut butter and sweetener into a sticky mixture, and add in just enough of the apple pectin and *Rheum* powder to create a cookie dough consistency. Roll the dough into balls and store in a small storage container in the refrigerator. Aim to eat 2 or 3 per day to increase dietary fiber.

This recipe can be amended in a myriad of ways, such as by adding cocoa powder, dried coconut, chia seeds, or other favorite cookie ingredients. Be certain to drink ample water to boost the efficacy of the pectin in the truffles.

Formulas for Hemorrhoids

Hemorrhoids are usually related to increased pressure in the intestines such as occurs with chronic constipation or portal congestion in the liver. Treating these underlying disorders is necessary in treating chronic hemorrhoids. Blood vessel weakness may also predispose to hemorrhoids and may be especially suspected in those with easy bruising and poor wound healing. Many red, purple, and blue pigments in plants improve blood vessel integrity via effects on collagen.[73] Liberal consumption of colorful fruits and vegetables such as berries, beets, carrots, and grapes over a lifetime can have many health benefits including stabilizing vascular tissue, protecting it from loss of elasticity. Herbs most noted to speed the resolution of painful hemorrhoids are *Hamamelis*, *Aesculus*, and *Ruscus aculeatus*. All may be used both internally and topically. The use of sitz baths can be very comforting to those with acute hemorrhoidal pain. Regular exercise, a high-fiber diet, and ample water intake are other important measures to alleviate intestinal pressure.

Tincture for Hemorrhoids

In addition to these herbs specific for hemorrhoids, *Hypericum* is included in this formula as a source of procyanidins, flavonoids that can strengthen venous connective tissue integrity.

Hamamelis virginiana 10 ml
Taraxacum officinale 8 ml
Hypericum perforatum 8 ml
Ruscus aculeatus 4 ml

Take ½ to 1 teaspoon of the combined tincture 4 times a day for 3 months or longer to help improve the chronic tendency to hemorrhoids and address any underlying constipation or portal congestion.

Sitz Bath for Hemorrhoidal Pain

This simple formula features *Hamamelis*, which has anti-inflammatory and astringent effects on swollen blood vessels. It can be pain-relieving after a single treatment, but for best effects should be used regularly over the course of a week or more, while also using other supportive remedies internally.

Hamamelis virginiana powder 2 ounces (60 g)
Epsom salts 2 ounces (60 g)
2 gallons of water
Rubber tub large enough to sit in

Bring the water to a gentle simmer in a large (or several smaller) stock pans. Remove from the heat and add the *Hamamelis* powder and Epsom salts and stir with a wooden spoon to help dissolve. Place the rubber tub in a bathtub and add the hot brew. When the water is cool enough, sit in the sitz bath for 15 to 20 minutes, and then rinse off in the shower. Pour the sitz bath brew onto a lawn or garden outdoors to avoid putting too much particulate down the household drain. Witch hazel compresses can be applied to the anus between treatments, as below. Repeat one or two times daily, or as often as possible.

Simple Topical for Hemorrhoidal Pain

Witch hazel is traditional for vascular congestion, and commercial products such as Tucks pads offer astringent effects with topical use. The pads are available in a small tin, convenient to carry in a purse or keep in the bathroom for use instead of toilet paper after a bowel movement.

Hamamelis virginiana

Place a pad soaked with home-prepared tea or a commercial witch hazel product against the anus 3 or more times daily to speed resolution of acute hemorrhoidal swelling and to offer pain relief.

Castor Mint Oil for Hemorrhoidal Pain

The cooling sensation of mint essential oil can help alleviate hemorrhoid pain when applied topically.

Castor oil 20 ml
Mint essential oil 10 ml

Combine the oils in a 1-ounce bottle and use the blend to moisten a cotton ball to apply topically to the anus for pain relief. Repeat 3 or 4 times through the day.

Tincture for Portal Congestion with Bleeding Varices

Portal congestion may result from liver congestion, organomegaly, or portal hypertension. *Aesculus* is specifically indicated for portal congestion and combines well with the vascular astringents *Quercus* and *Geranium* to control bleeding in the esophagus or bowels. Beta-aescin from *Aesculus* has a gentle vasoconstrictive effect,[74] contributing to its ability to treat venous insufficiency, varicose veins, and hemorrhoids. Rutosides, a group of flavonoids found in citrus rinds and other sources including *Aesculus hippocastanum*, may also reduce edema associated with vascular congestion,[75] contributing to the ability of the plant to support venous and lymphatic return.[76] *Curcuma* and *Silybum* offer further support in decongesting the liver to help reduce varicosities. Using *Silybum* encapsulations and lipotropic formulas for the liver along with this tincture may produce good results.

Aesculus hippocastanum 15 ml
Quercus spp. 15 ml
Geranium maculatum 10 ml
Curcuma longa 10 ml
Silybum marianum 10 ml

Take 1 to 2 teaspoons of the combined tincture 4 times daily, continuing for many months. Follow with an alterative tea and further liver support, perhaps indefinitely.

Herbs for Vascular Congestion and Hemorrhoids

Vascular congestion may result from chronic constipation, obesity, and sedentary habits. Such conditions can impede circulation, venous return, and lymphatic flow and result in backaches, toxicity symptoms, vascular congestion, and hemorrhoids, sometimes referred to as pelvic stagnation by herbalists.

Angelica sinensis
Asclepias tuberosa
Ceanothus americanus
Collinsonia canadensis
Ginkgo biloba
Hamamelis virginiana
Hypericum perforatum
Juglans nigra
Quercus spp.
Schisandra chinensis
Scrophularia nodosa
Silybum marianum
Taraxacum officinale
Trifolium pratense

Formulas for Halitosis

Halitosis is most often due to digestive issues—from poor dental hygiene, caries, and loose crowns or cracks in the teeth harboring bacteria capable of producing a foul odor to intestinal dysbiosis and chronic constipation. In other cases, lung tumors or respiratory diseases may also cause halitosis. The use of bitter and alterative agents including *Taraxacum, Rumex, Arctium, Silybum, Curcuma, Artemisia,* and *Gentiana* for many months may improve liver function, bile flow, and digestion and thereby improve halitosis. Dietary changes may also be corrective for some people with a slow digestive transit time.

Tincture for Halitosis

This formula aims to offer alterative, liver, and antimicrobial support when toxemia and intestinal dysbiosis underlie halitosis. "Toxemia" is a state of intestinal dybiosis that can result from poor diet or altered intestinal motility. It is a concept commonly embraced by traditional herbalists and naturopathic physicians, who will often address shifts in microbial ecosystems away from beneficial species and toward more pathogenic species with alterative and liver supportive therapies.

Taraxacum officinale 20 ml
Curcuma longa 20 ml
Mentha piperita 20 ml

Take 1 teaspoon of the combined tincture 3 or 4 times a day for at least 3 months.

Mouthwash for Halitosis Due to "Deranged Stomach"

Based on a traditional formula, this homemade tincture is part medicine and part mouthwash. This formula treats the breath and supports oral health and should be combined with probiotics and an alterative tea or encapsulation to best support intestinal flora and optimize digestion.

Sherry 4 ounces (120 ml)
Cinnamon powder 4 tablespoons
Caraway powder 4 tablespoons
Clove powder 1 tablespoon
Nutmeg powder 1 tablespoon
Essential oils of mint, lavender, and rose

Put the sherry and the powders in a small-lidded canning jar, and let the powders macerate in the sherry for 2 to 4 weeks, shaking vigorously each day. Strain and add 10 drops of each of the three essential oils. Use by the ½ teaspoon as a gargle each day by diluting with a small sip of water. Be sure to take other steps to support digestive function also.

Formulas for Diarrhea

Diarrhea has many possible etiologies, each needing to be treated accordingly. Diarrhea can have infectious causes, in which case the diarrhea is part of the healing mechanisms of the body to dispel the pathogens as quickly as possible. In these cases, pharmaceutical agents aimed at stopping the diarrhea are illogical and oppose the healing efforts of the body. Instead, herbal agents that are antimicrobial and astringent to the intestines are indicated in such cases; these include *Mahonia, Hydrastis, Agrimonia,* or *Artemisia.*

Diarrhea may also be inflammatory and triggered by allergens or exposure to irritants and chemicals, particular foods, alcohol, or a heavily spiced meal. In such cases, simple, soothing, and anti-inflammatory demulcent agents are indicated, such as *Ulmus* and *Althaea.* Anti-inflammatory herbs such as *Glycyrrhiza* or *Matricaria* are also helpful here and make palatable teas.

Diarrhea may also be triggered by emotional causes. In such cases *Matricaria, Scutellaria,* and other nervines may be most helpful. *Avena* may be nerve-calming, nourishing, and demulcent to the gastrointestinal tract.

Diarrhea can occur due to hypermotility that may occur as part of hyperthyroid excessive stimulation of all metabolic functions. The treatment of this type of diarrhea involves normalizing thyroid and endocrine balance.

When it is necessary to halt excessive fluid losses, astringents such as *Rubus* leaves, *Vaccinium* leaves, *Quercus* bark, *Agrimonia* leaves, and *Hamamelis* bark are all helpful in reducing diarrhea. When diarrhea is due to atony of the intestines, agents that are slightly stimulating without being laxative or irritating are appropriate and include *Zingiber, Cinnamomum,* and *Mahonia* (also known as *Berberis*).

Both constipation and diarrhea can occur due to intestinal lesions including growths, tumors, and cancer, so any unusual presentation or unresponsive cases require a diagnostic workup to rule out serious underlying pathologies.

Traveler's Diarrhea

Bacteria are responsible for at least 50 percent of all acute cases of traveler's diarrhea, with viruses being the second most likely causative organisms, and protozoa being to blame less than 10 percent of the time.[77] Of the causative bacteria, *Escherichia coli* strains are the most common culprit, but *Salmonella, Campylobacter,* and *Shigella* are also possible causes of traveler's diarrhea. Enterotoxigenic *E. coli* may be isolated from around 50 percent of people afflicted.[78] Enterotoxigenic *Bacteroides fragilis* is another less well-known potential bacterial pathogen recently identified as being associated with traveler's diarrhea. *Arcobacter* is another bacterial pathogen similar to *Campylobacter.*[79] In some locales, intestinal viruses including *Arbovirus* and the Norwalk viruses,[80] also referred to as *Norovirus,* are associated

Types of Diarrhea

This summary of the "textbook" types of diarrhea is to provide insight into the many underlying causes of diarrhea to consider when choosing herbs for formulas. In this book, I do not offer detailed formulas for every type of diarrhea. Instead, I encourage readers to note any underlying cause of diarrhea that can be identified: infectious agents; irritants; allergic reactivity; excess bile; excess ingestion of xylitol, mannitol, or other unusual carbohydrates; the presence of intestinal lesions; and hypermotility due to hyperthyroidism.

Secretory diarrhea. This is a type of diarrhea where the intestines secrete excessive fluid and electrolytes rather than absorbing them. This condition occurs due to the presence of bacterial endotoxins or viruses in the intestines as well as the presence of excessive fats or bile acids in the intestines, occurring secondary to hepatic or biliary conditions. Hence, the treatment for secretory diarrhea may be symptomatic, such as astringent herbs (such as *Rubus, Geranium, Quercus, Agrimonia*) to reduce excessive secretions, as well as antimicrobial herbs (such as *Hydrastis, Coptis, Matricaria*) or biliary supportive herbs (*such as Curcuma, Silybum, Taraxacum*) as specifically indicated. *Croton lechleri* (Sangre de Drago, Dragon's Blood) has also been shown to reduce secretions in even severe diarrhea, such as that occurring in cholera. *Rhodiola*[81] and resveratrol[82] may have stabilizing effects on stimulated ion channels. Resveratrol is a naturally occurring blue-purple phenolic compound credited with potent antioxidant effects. Resveratrol occurs in grapes and red wine, berries, dark chocolate, and the red papery husk of Spanish peanuts.

Osmotic diarrhea. This is a type of diarrhea where the ingestion of sugar substitutes such as xylitol and mannitol remain in the large bowel, unable to be absorbed, and osmotically retain water in the intestinal lumen. The main treatment for osmotic diarrhea is, of course, ceasing to ingest the offending agents.

Exudative diarrhea. This type of diarrhea results from inflammation or lesions in the mucous membrane where excessive mucous, blood, and serum proteins are released. Also referred to as exudative enteropathy, this type can be caused by ulcerative colitis, IBS, carcinomas, and other mucosal lesions. Herbs and nutrients that help repair the mucosal health and integrity are most valuable for this diarrhea. Demulcents such as *Ulmus, Symphytum,* and *Glycyrrhiza* are appropriate, as is the amino acid glutamine, noted to help repair gastrointestinal ulcerative lesions.

Accelerated transit time. This is a type of diarrhea where accentuated peristalsis moves intestinal contents too rapidly for adequate absorption, also referred to as hypermotility. This condition may occur due to endocrine or metabolic imbalances such as hyperthyroidism or emotional stress in some susceptible individuals. It may also result from to the presence of intestinal irritants, such as laxatives or drugs, or in other cases foods that act as allergens. They trigger rapid peristalsis as the body finds them objectionable and attempts to move them out as fast as possible. Alcohol, coffee, and certain foods are notorious for such irritation, as might be any food in an allergic person. The most common offenders are dairy products and wheat. Botanical agents that reduce mucosal irritation, such as *Matricaria, Ulmus,* and *Glycyrrhiza,* may allay such types of diarrhea.

with 7 percent of diarrhea-related deaths in the United States and identified in over 15 percent of Europeans returning from travel abroad.[83] Microsporidiosis may also be the culprit. Microspores such as *Enterocytozoon bieneusi* and *Encephalitozoon* are obligate intracellular microorganisms and were previously classified as protozoans, but recently reclassified as fungi.

Traveler's diarrhea may be multifactorial and a reaction to the introduction of many microbial strains foreign to the intestinal ecosystem all at once. The bacteria mentioned above are very harmful to the intestinal cells, causing inflammation and damage, but other factors may weaken or strengthen intestinal mucosal cells' ability to resist shifts in the overall ecosystem.

Essential oils are among our most powerful tools for acute infections, and essential oil of *Origanum vulgare* (oregano) has been shown to be one of the strongest and broadest acting of the common and inexpensive essential oils. Oregano has shown activity against *Bacillus subtilis, Enterobacter cloacae, Escherichia coli, Micrococcus flavus, Proteus mirabilis, Pseudomonas aeruginosa, Salmonella enteritidis, Staphylococcus epidermidis, Staphylococcus typhimurium*, and *Staphylococcus aureus*.[84] While many chemical constituents in the common Lamiaceae (mint) family essential oils are active against these microbes, carvacrol has shown the highest antibacterial activity.[85] *Apium graveolens* (celery), *Foeniculum vulgare* (fennel), and umbel family plants also have antimicrobial effects[86] and may alleviate gas, cramping, and nausea as well.

Some intestinal pathogens, including *Pseudomonas aeruginosa, Staphylococcus, E. coli*, and *Candida albicans*, may form biofilms that deter the efficacy of antibiotics.[87] For this and other reasons, pharmaceutical antibiotics are not usually indicated. They have not been shown to be highly effective and have even been noted to foster antibiotic resistance and promote carrier states of some enteric pathogens. Herbal formulas may prove more effective in the long run.

Prophylaxis for Traveler's Diarrhea

Begin taking one or more of the following 1 month prior to departure for overseas travel.

- Probiotics
- Turmeric capsules, teas
- Glutamine powder in water
- *Aloe vera* gel in water, pineapple juice, or herbal tea

Be sure to take some of these remedies daily while you are traveling.

- Oregano oil capsules
- Culinary herbs, especially garlic, ginger, and turmeric (liberal amounts)
- Teas made with *Astragalus*, medicinal mushrooms
- Bromelain supplements (500 milligrams or more 3 times daily)
- Glutamine powder (½ to 1 teaspoon in water several times per day)
- *Aloe vera* gel (mix into pineapple juice with glutamine)
- Papaya seeds (readily available in tropical countries)
- Berberine pills, tinctures, powders
- Potassium supplements and electrolyte powders
- Charcoal capsules
- Gentle fiber supplements such as oat bran (prepare as food)
- Huang Lian Su tablets
- Mint essential oil
- Fennel essential oil

Tincture for Exudative Diarrhea

In cases of exudative diarrhea, an herbal formula should astringe the excessive secretions, such as with *Geranium* and *Hydrastis* in this formula, while simultaneously treating the underlying inflammation contributing to the lesions, as with the *Ulmus* and *Glycyrrhiza* used here.

Ulmus fulva 8 ml
Glycyrrhiza glabra 8 ml
Hydrastis canadensis 7 ml
Geranium maculatum 7 ml

Take ½ to 1 teaspoon of the combined tincture as often as hourly, reducing each day as symptoms improve.

Tea for Exudative Diarrhea

Herbs and nutrients that help repair the mucosal health and integrity are most valuable for exudative diarrhea. Demulcents such as *Ulmus, Symphytum*, and *Glycyrrhiza* are appropriate, as is the amino acid glutamine, as all are noted to help repair gastrointestinal ulcerative lesions. This formula could also be prepared as a tincture, although the tea will be more powerful and faster acting. When there is blood in the stool, include one of the digestive styptic herbs listed in "GI Astringent Herbs" on page 54, such as the *Geranium* included in this formula. Patients with blood in the stool require further investigations to rule out colon cancer or other lesions.

Ulmus fulva 2 ounces (60 g)
Artemisia absinthium 1 ounce (30 g)
Mahonia aquifolium 1 ounce (30 g)
Quercus alba or *rubra* 1 ounce (30 g)
Glycyrrhiza glabra 1 ounce (30 g)
Matricaria chamomilla 1 ounce (30 g)
Geranium maculatum 1 ounce (30 g)

Add 1 teaspoon of the herb mixture per cup of water and simmer gently for several minutes in a covered pan. Remove from the heat, let stand 15 minutes, and strain. Drink freely. For acute diarrhea and blood in the stool, aim to drink as much as possible, and sip constantly to keep the liquid flowing through the intestines all day.

GI Astringent Herbs

Astringent herbs can tighten swollen and congested mucous membranes. When taken orally, they can help reduce excessive secretions and bleeding from the stomach and intestines and help control diarrhea. Many such herbs are high in tannins, molecular compounds capable of astringing blood vessels and boggy tissues. Note how these herbs are included in formulas for diarrhea, blood in the stool, or mucous and undigested food in the stool. When high tannin herbs have a particular ability to control bleeding, they are also referred to as *styptics* or *hemostatics*. Patients with blood in the stool, visible or occult, require further workup to rule out intestinal lesions and colon cancer.

Agrimonia eupatoria	*Geranium maculatum*
Cinnamomum verum	*Hamamelis virginiana*
Collinsonia canadensis	*Quercus* spp.
Filipendula ulmaria	*Rubus idaeus, R.* spp.

Herbs for Parasites and Dysbiosis

These herbs all have mild to moderate to powerful antiparasitic effects and can be used in teas, tinctures, and encapsulations to treat intestinal worms, parasites, and pathogenic intestinal bacteria.

Allium sativum	*Matricaria chamomilla*
Artemisia annua, A. vulgaris, A. absinthium	*Melaleuca alternafolia*
	Origanum vulgare
Azadirachta indica	*Rumex acetosella*
Citrus paradisi (seed extract)	*Spilanthes acmella*
Coptis chinensis	*Tabebuia impetiginosa*
Hydrastis canadensis	*Thuja occidentalis*
Juglans nigra	*Thuja plicata*

Jungle Tea for Secretory Diarrhea or Cholera

It may take some searching to find the Amazonian jungle herbs used in this formula, but Sangre de Drago bark can be highly effective for controlling severe, infectious diarrhea, particularly when combined with guava bark. I gained experience with these plants due to extensive time I spent in the Amazon, where all the indigenous communities with whom I worked use these herbs for diarrhea, bleeding from the bowels, and intestinal lesions and infections.

Croton lechleri (Sangre de Drago)
Psidium guajava (guava)

Combine equal parts of the dry barks, roughly 1 cup of each, simmer in 8 cups water, and strain. Drink as much as possible throughout the day, decreasing as symptoms subside.

Tincture for Infectious Diarrhea

The "stomach flu," traveler's diarrhea, food poisoning, and ameobic dysentery may all respond to this formula. It covers the bases: *Glycyrrhiza* and *Ulmus* provide soothing demulcent effects; *Mentha* allays gas, cramping, and bloating; and *Glycyrrhiza, Mahonia,* and *Allium* offer antimicrobial effects. This formula is a tincture but could also be prepared as a tea, omitting the *Allium*. A tincture may be less effective than a tea for diarrhea, but it is still helpful and may be more convenient while traveling or in other circumstances.

Glycyrrhiza glabra 15 ml
Mahonia aquifolium 15 ml
Zingiber officinale 7.5 ml
Achillea millefolium 7.5 ml
Mentha piperita 7.5 ml
Ulmus fulva 7.5 ml

Take 1 to 2 teaspoons of the combined tincture every 30 to 60 minutes, reducing frequency as symptoms abate.

Pills for Food Poisoning and Traveler's Diarrhea

Berberine and similar compounds can be purified and concentrated out of *Coptis chinensis* and are available from Chinese apothecaries as small yellow tablets and sold under the name of Huang Lian Su. They are easy to travel with, as are large bottles of berberine capsules. Take 2 pills at the first hint of dysbiosis, and repeat 3 or 4 times daily until symptoms are controlled. Encapsulated *Hydrastis candensis* and *Mahonia aquifolium* (also known as *Berberis aquifolium*) may also do the trick.

Allium sativum pills are also versatile, inexpensive, and easy to travel with. Garlic pills may be taken prophylactically while traveling, as much as two or three at a time, 4 or 5 times a day, when needed to treat food poisoning and infectious diarrhea.

Tea for Food Poisoning

Pathogenic microbes underlie diarrhea resulting from tainted food, which necessitates the inclusion of antimicrobials in the formula. *Mahonia* (also known as *Berberis*) is active against many microbes and is astringing to the GI mucosa, serving several functions in this formula.

Therapies for Intestinal Parasites

Consuming certain supplements and foods can help with recovery from intestinal parasites. Use papaya seeds in smoothies, for example, or dry and grind them to use like black pepper. Pumpkin and sunflower seeds are both available in their hulls. Chewing 2 or 3 tablespoons of such seeds results in woody slivers of the hulls that pass through the intestines and discourage pinworm and possibly other parasites. Choose from the following supplements and foods:

Bromelain
Probiotics
Raw pineapple
Fresh pineapple and fresh papaya juice
Raw grated papaya (where available)
Papaya seeds
Fiber supplements
Whole sunflower seeds or pumpkin seeds, with the hulls intact

Avoid sugar during the recovery period, because sugar favors intestinal dysbiosis and parasites.

Glycyrrhiza glabra
Mahonia aquifolium
Mentha piperita
Ulmus fulva
Perilla frutescens

Combine equal parts of the dry herbs. Steep 1 tablespoon of the herb mixture per cup of hot water for 10 minutes, and then strain. Sip this tea constantly throughout the day.

Tincture for Intestinal Parasites

Juglans nigra and *Artemisia absinthium* are traditional herbs used to treat amoebic and intestinal parasites. The use of carminative herbs high in volatile oils featured in this formula adds both antiparasitic effects and anti-inflammatory actions.

Juglans nigra 15 ml
Artemisia absinthinum 15 ml
Allium sativum 10 ml
Foeniculum vulgare 10 ml
Zingiber officinale 10 ml

Take 1 or 2 dropperfuls of combined tincture at a time, 3 to 6 times a day. Garlic capsules may be used in conjunction with this formula, along with oregano oil and/or bromelain capsules.

Escherichia coli–Induced Diarrhea

Escherichia coli (*E. coli*) and other gut pathogens attach to enterocytes, collapse the microvilli of infected cells, and alter numerous absorptive and other cellular functions. Secretory diarrhea results as these injured cells release numerous small molecules through the gap junctions between the intestinal mucosa cells.[88] The pre-existing strength of the gap junctions between mucosal cells may affect susceptibility to *E. coli* and other enteric pathogens. Glutamine and curcumin from *Curcuma longa* (turmeric) may help tighten gap junctions between intestinal epithelial cells and improve intestinal health prophylactically.[89] Glutamine powder supplementation may decrease the ability of *E. coli* to adhere to the intestinal mucosa[90] and may reduce endotoxin-induced mucosal injury.[91] Glutamine occurs naturally in significant amounts in *Aloe vera* and in small amounts (1⁄10th of what's in *Aloe vera*) in *Ananas comosus* (pineapple).[92] *Aloe vera* leaf products and pineapple juices, especially fresh, may be consumed as preventives when traveling. Glutamine powder is inexpensive and easy to travel with, making it useful as a preventive measure. Stir ½ teaspoon to 1 teaspoon into water or herbal tea; this may be taken 3 times daily starting as early as a month prior to international departure. The proteolytic enzyme bromelain from pineapples has been shown to affect enterocyte receptors for infectious strains of *E. coli* in a manner that inhibits the ability of the bacteria to bind intestinal cells.[93]

Antimicrobial Seeds and Spices

A folkloric remedy for intestinal infections and intestinal parasites was to dry papaya seeds and grind them into a powder for use as a condiment in foods. *Carica papaya* (papaya) seed extracts have been shown to have activity against *Escherichia coli*, *Staphylococcus aureus*, *Salmonella typhi*, and *Pseudomonas aeruginosa*.[94] Including liberal amounts of the antimicrobial culinary spices *Allium* (garlic), *Zingiber* (ginger), and *Curcuma* (turmeric) in the diet is also highly encouraged when traveling, and they offer numerous other health benefits in the diet in general.

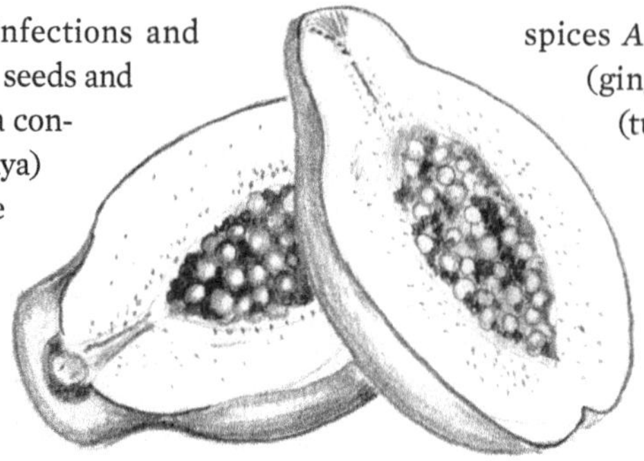

Carica papaya, papaya

Berberine Alkaloids for Infectious Diarrhea

The berberine alkaloids are a group of isoquinoline alkaloids that are among our best herbal therapies for infectious diarrhea. The berberine family of alkaloids includes berberine in *Mahonia*, hydrastine in *Hydrastis*, phellodendrine in *Phellodendron*, and coptine in *Coptis*, and all have an affinity for secretory infections of mucous membranes. Several studies have identified wide antimicrobial properties for these herbs.[95] *Coptis* (also known as *Coptidis*) has been used in China for secretory diarrhea and gastroenteritis for more than a thousand years. *Coptis* roots contain at least 12 protoberberine alkaloids, which recent research has reported are transformed by phase I and phase II metabolism, and are ultimately excreted via the urine.[96] The plant also goes by the name of Huang Lian Su available in pill form from vendors of traditional Chinese medicines, and it is easy to travel with the small vials even for backpackers. *Berberis* species have also shown activity against the amoeba *Entamoeba histolytica*, as have *Tinospora*, *Terminalia*, and *Zingiber*.[97]

Tincture for Explosive Diarrhea

Urgent and explosive diarrhea is evidence of extremely rapid peristalsis and necessitates the inclusion of antispasmodic herbs to control the hypermotility. Very watery stools with the presence of undigested food is another sign of hypermotility. *Atropa belladonna* is a powerful antimotility agent, but should be prescribed only by skilled clinicians and is best used in small doses. (See also the Tincture for Stress-Related Excessive Peristalsis, below.) Belladonna has both drying and spasmolytic actions on the intestines. An alternative would be *Mentha* essential oil in tincture and separately as enteric-coated capsules. This formula could act quickly for severe intestinal cramping.

Mentha piperita 15 ml
Foeniculum vulgare 15 ml
Dioscorea villosa 15 ml
Matricaria chamomilla 11 ml
Atropa belladonna 4 ml

Take 1 dropperful of the combined tincture every 15 to 30 minutes for several hours, reducing as symptoms improve.

Tincture for Stress-Related Excessive Peristalsis

When a patient's diarrhea seems entirely related to stress rather than exudative lesions or pathogens, the use of bowel-soothing nerve tonics is appropriate. This formula adds the prescription-only herb belladonna, which is highly effective in quickly halting excessive motility.

Omit it from the formula if you are not trained in the use of this druglike herb, and if you do include it, remove it from the formula once the acute problem is resolved. Belladonna can cause dry mouth, facial flushing, and visual disturbances, even hallucinations, so do not use more than 4 to 6 ml per ounce of the tincture formula.

Matricaria chamomilla 14 ml
Scutellaria lateriflora 14 ml
Withania somnifera 14 ml
Mentha piperita 14 ml
Atropa belladonna 4 ml

Take 1 dropperful of combined tincture every 30 minutes to 3 hours, reducing frequency as bowel function normalizes.

Tincture for Excessive Peristalsis Associated with Hyperthyroidism

Hyperthyroid patients may suffer from diarrhea unrelated to bowel health, but induced by hypermotility. This formula uses *Atropa belladonna* to reduce excessive motility, along with thyroid-balancing herbs, *Leonurus* and *Lycopus*. *Atropa belladonna* can cause a dry mouth, flushed face, feverlike heat but without perspiration, visual disturbance, and possibly hallucinations at dosages greater than the level in this tincture. Remove belladonna from the formula at the least hint of any such symptoms. *Atropa belladonna* is available only to licensed health care professionals and should only be used by them.

Ulmus fulva 20 ml
Leonurus cardiaca 20 ml
Lycopus virginicus 16 ml
Atropa belladonna 4 ml

Take 1 dropperful of the combined tincture every 30 minutes to 3 hours, reducing frequency as bowel function normalizes.

Tea for Severe, Electrolyte-Depleting Diarrhea

Food poisoning and other pathogens can cause severe diarrhea that leads to dehydration and electrolyte imbalance in some cases. This formula helps treat the infection (*Mahonia* and *Glycyrrhiza*), astringe the tissue (*Mahonia* and *Quercus*), and reduce the inflammatory processes (*Matricaria* and *Glycyrrhiza*). The addition of honey and salt boosts the restorative effects on blood sugar and electrolytes.

Mahonia aquifolium
Quercus spp.
Matricaria chamomilla
Glycyrrhiza glabra
Salt
Honey

Combine equal parts of the dry herbs. Steep 1 tablespoon of the herb mixture per cup of hot water and then strain. Drink as much as possible, adding a bit of salt and honey to each cup.

Universal Electrolyte Replacement Beverage

When severe and of significant duration, vomiting and diarrhea can lead to electrolyte imbalance, vascular collapse, and renal failure. Sodium depletion may lead to metabolic acidosis, and potassium depletion may lead to cardiac excitability. Excessive mucous secretions may contain large amounts of potassium. Magnesium depletion is also possible and is associated with nerve excitability and risk of tetany and cardiac abnormalities. Infants and the elderly or otherwise debilitated individuals are especially susceptible to acute dehydration and resulting electrolyte imbalance complications.

1 quart (960 ml) spring water or coconut water
1 teaspoon salt (sea salt if possible)
1 teaspoon salt substitute (KCl/potassium chloride)
2 teaspoons honey or agave nectar
2 teaspoons fruit juice or beet juice concentrate

Bring the water to a boil and add the salt and salt substitute, honey or nectar, and fruit or beet juice concentrate. Stir to dissolve. Drink hot or cold, taking only small sips. This natural electrolyte replacement beverage may also be frozen in ice cube trays and then crushed into slivers and placed in a bowl to consume by the spoonful for those who are nauseated by drinking or who vomit all fluids taken. Consume as much as possible until diarrhea is controlled.

Whole Food Electrolyte Replacement Broth

Less quantifiable than the previous recipe in terms of knowing how much potassium the medicine contains, this broth, however, is better-tasting. Miso and honey in a veggie broth make a delicious clear soup to help provide salt, potassium, chloride, and glucose. Homemade vegetable broth is ideal and is prepared by simmering leftover kitchen scraps such as carrot tops and peels, onion skins, and other veggies in water, but a quality organic commercial product would also serve.

1 quart (960 ml) vegetable broth
2 tablespoons miso
2 teaspoons honey

Bring the vegetable broth to a simmer. Add the miso and honey and stir to dissolve. Serve hot in a bowl to eat with a spoon or in a mug to sip slowly. Continue consuming throughout the day until the quart is gone, and the diarrhea is controlled. Prepare a second or even third batch as needed.

Tea for Allergic Diarrhea

When food allergens trigger diarrhea, or when multiple factors have made the bowels hyperreactive, the addition of antiallergy herbs such as *Perilla*, *Tanacetum*, and *Astragalus* to soothing herbs such as *Glycyrrhiza* and *Foeniculum* may help control the diarrhea as quickly as possible. This formula can be prepared as a tincture, although the tea may be more effective and faster-acting.

Foeniculum vulgare
Glycyrrhiza glabra
Tanacetum parthenium
Astragalus membranaceus
Ephedra sinica
Perilla frutescens

Combine equal parts of the dry herbs. Steep 1 tablespoon of the herb mixture per cup of hot water and then strain. Drink as much as possible, sipping constantly throughout the day.

Topical for Diarrhea with Intestinal Cramps

This topical formula can complement a tea or tincture for intestinal pain, cramping, and bloating. It can be a first-line formula for infants with colic and toddlers with upset tummies.

Mentha piperita essential oil 10 drops
Foeniculum vulgare essential oil 10 drops
Castor oil or other oil 1 teaspoon

Add the essential oils to the castor or other oil in a tiny dish or shot glass and stir to mix. Rub the herbal oil onto the abdomen. Cover with a heat pack or hot wet towel.

Tincture for Diarrhea Due to Digestive and Intestinal Atony

Most cases of diarrhea are "excess" conditions, due to causes such as excess allergic reactivity or pathogens, or excessive motility. In some rarer cases, however, diarrhea may result from poor digestion higher up in the GI and an inability of the colon to absorb the liquid back out of the stool. The use of alterative herbs to stimulate digestive and biliary secretions, mucosal restoratives such as *Matricaria*, and a small amount of a stimulant, such as *Zingiber* used here, may promote circulation and altogether remedy the situation. This formula may also be prepared as a tea.

Mahonia aquifolium 28 ml
Matricaria chamomilla 28 ml
Zingiber officinale 4 ml

Take 1 dropperful of the combined tincture every 1 to 2 hours.

Tea for Diarrhea Due to Intestinal and Digestive Atony

Similar in concept to the preceding tincture, this tea has gentle alterative and stimulating effects for atony, poor circulation, and diarrhea related to digestive insufficiency.

Zingiber officinale 2 ounces (60 g)
Matricaria chamomila 2 ounces (60 g)
Mahonia aquifolium 2 ounces (60 g)
Centella asiatica 2 ounces (60 g)

Combine the dry herbs and steep 1 tablespoon of the herb mixture per cup of hot water and then strain. Drink a minimum of 3 or 4 cups per day, preferable sipping constantly throughout the day, and reducing the amount as bowel function normalizes.

Rubus Simple for Diarrhea in Infants

Raspberry leaves are a gentle and nutritive astringent agent suitable for infants. Prepare them as a simple or combine with *Matricaria*.

Rubus idaeus leaves

Steep 1 heaping tablespoon in 2 cups of water, strain, and aim to give 3 to 6 teaspoons per dose every 1 to 2 hours for a day, continuing a second or third day if necessary.

Formulas for Gastritis, Enteritis, and Gastroenteritis

Gastritis is an inflammation of the stomach lining and enteritis is an inflammation of the intestinal lining, and both may be acute in onset. In many cases, both the stomach and intestines are affected, which is referred to as gastroenteritis. There is overlap between this condition and traveler's diarrhea because infectious agents may be the cause. Gastroenteritis can also overlap with GERD due to low-grade, chronic *Helicobacter pylori* infections. Gastroenteritis is sometimes referred to as the "stomach flu," even though the influenza virus does not cause the condition. Other viral infections are the most common cause of gastroenteritis, including rotavirus, adenovirus, astrovirus, and calicivirus, all associated with poor sanitation. Children are more susceptible to these pathogens and often experience significant vomiting. Contaminated food is another common cause of gastroenteritis due to the presence of *Salmonella, Shigella*, or *Campylobacter* bacteria. In developing nations *Escherichia coli 0157* and *Listeria monocytogenes* are common, and potentially fatal cholera infections may occur in epidemics following natural disasters and widespread sanitation issues. Eating thoroughly cooked foods when traveling and avoiding raw salads and exposure to tap water can help reduce gastroenteritis. *H. pylori* can also cause gastritis, usually of a more chronic type, having a unique ability to colonize the extremely acidic environment of the stomach. Persistent presence of *H. pylori* changes the gastric mucosa over time, inducing low-level gastritis that progresses to atrophy of the mucosa, and may further progress to cause metaplastic cell changes and finally adenocarcinoma.

Some of the formulas included in this section are indicated for atrophic gastritis in contrast to the acute symptoms seen with other types of infectious gastroenteritis. Noninfectious causes of gastroenteritis include alcohol, nonsteroidal anti-inflammatory drugs (NSAIDs), and corticosteroids, which may induce gastritis in particular, as these agents harm and even ulcerate mucosal cells. Gastritis may also be chronic and related to stress or may result from regular ingestion of irritants such as alcohol. Toxins and allergens capable of inducing the inflammation include mushrooms, potatoes, seafood, or dairy as well as drug reactions and accidental ingestion of heavy metals or other poisons. Other situations that increase the risk of developing gastroenteritis include liver and renal disease, cocaine use, being on a respirator, and autoimmune diseases that affect the stomach, such as Crohn's disease.

Infectious gastroenteritis can cause significant symptoms for a few days, but is often self-limiting, even without treatment in many cases. However, for those with underlying diseases and very young or very old

Hot Gastritis versus Cold Gastritis

Gastritis is classified medically as being erosive (ulcerative) or nonerosive (inflammatory), distinctions that typically require a biopsy. Clinically, herbalists may classify gastritis as being "excessive/hot" or "atrophic/cold," and formulas can be crafted accordingly. Hot gastritis presents with a sense of burning, excessive stomach acid, and/or deficient mucosal barrier and often in a younger person and concomitant with stress, alcohol, or pathogens. Cold gastritis may present with sensations of pressure or cramping rather than burning and is associated with deficient stomach acid and/or enzymes and atrophy of cells and mucosa, and most often occurs in older people and those with long-standing *Helicobacter pylori* or other irritant.

Hot gastritis might necessitate demulcents and anti-inflammatories in the formula, while cold gastritis would benefit from cholagogues, bitters, and warming herbs such as *Zingiber*, *Cinnamomum*, or *Capsicum*. Energetically neutral herbs such as *Mentha*, *Matricaria*, and *Glycyrrhiza* would be appropriate for both hot or cold gastritis. Alterative herbs and bitters are always appropriate for cold gastritis and can often, but not usually, be an appropriate complement to the primary herbs in a formula for hot gastritis. Antimicrobial herbs are important to include in formulas for infectious diarrhea and *H. pylori* and are listed in "Antimicrobial Herbs for Infectious Gastroenteritis" on page 61.

patients, the condition can be fatal due to severe dehydration and electrolyte depletion.

Gastritis can induce mild to severe pain that is often intermittent and, when chronic, may be associated with a constant dull aching sensation. A low fever may occur with infectious gastritis, along with fatigue and general malaise. There may be indigestion and loss of appetite or frank nausea and vomiting, and when the intestines are affected, diarrhea occurs. Passing black stools or vomit tinged with blood or with an appearance resembling coffee grounds indicates significant bleeding of the digestive mucosa and is reason to seek immediate medical attention. Chronic gastritis should be addressed as thoroughly as possible, even if the symptoms are mild, because the chronic inflammation is associated with increasing the risk of gastric cancer. Chronic gastritis due to long-standing irritation and/or insufficient circulation and secretions in the GI can lead to atrophy and low-grade inflammation and is less commonly discussed than the infectious and acute types.

Medical workups may include testing for *H. pylori*, tests for fecal blood, an endoscopy to visually assess the condition of the mucous membranes, and a gastric biopsy to rule out gastric cancer. Treatments may be specific, from treating infections when present, to simply avoiding irritants. Electrolyte and fluid replacement are necessary when vomiting and diarrhea are persistent. A bland diet such as simple soups and steamed vegetables are recommended during recovery, avoiding heavy fats, strong spices, and other irritants. Allopathic medicine may offer antacids and proton pump inhibitors to treat the symptoms and antibiotics to address *H. pylori* when present, while herbal medicines can effectively combine agents that allay symptoms, while addressing infection, irritation, or other underlying causes. Identifying the specific pathogen is not a priority in acute infectious gastritis, but treating the symptoms and preventing dehydration are important. Stress reduction may be an important part of a treatment protocol in many cases.

Using botanical demulcent agents such as *Ulmus*, *Althaea*, and *Glycyrrhiza* and eating oatmeal may help soothe irritated mucosal surfaces and allay discomfort. Berberine, *Baptisia*, *Commiphora myrrha*, *Hydrastis*, oregano oil, neem, garlic, and other antimicrobial herbs may deter *Helicobacter* and other pathogens. *Matricaria*, *Mentha*, *Foeniculum*, *Filipendula ulmaria*, or *Carum* can relieve nausea and help stop vomiting when present. Electrolyte replacement formulas, discussed elsewhere in this chapter, may be necessary. Avoid antidiarrheal agents in cases of gastroenteritis, because diarrhea is the body's method of eliminating the infectious or irritating agent, and halting this eliminative effort only hampers the healing effort. Activated charcoal and clay slurries may be offered to help absorb excessive fluids without artificially halting diarrhea. Hot packs and mint oil rubbed into the abdomen may help when stomach and intestinal cramps are present; emphasizing *Mentha*, *Matricaria*, and *Dioscorea* in teas; or including *Viburnum* or small doses of *Atropa belladonna* can also help relieve painful cramping. When there is bloody diarrhea, *Achillea*, *Quercus*, *Hamamelis*, or *Geranium* should be included in the formula to stop the bleeding. Probiotic supplements may help reestablish healthy intestinal flora and be part of a convalescent therapy.

Tincture for Hot Infectious Gastritis

This tincture combines the antimicrobial and astringing *Hydrastis* with soothing and cooling base herbs. For best results, this formula should be used along with the Tea for Hot Gastritis (below), slippery elm porridge or pills, and a thoughtful diet.

Matricaria chamomilla 15 ml
Ulmus fulva 15 ml
Glycyrrhiza glabra 15 ml
Hydrastis canadensis 15 ml

Take 1 dropperful of the combined tincture every several hours, decreasing frequency as symptoms improve.

Tea for Hot Gastritis

This tea is good-tasting and uses *Symphytum* as a soothing and cooling demulcent agent. *Althaea* or *Ulmus* would be alternatives to *Symphytum* to lend a very cooling energy to the neutral herbs.

Symphytum officinale root or root powder 2 ounces (60 g)
Glycyrrhiza glabra 2 ounces (60 g)
Matricaria chamomilla 2 ounces (60 g)

Steep 1 tablespoon of the herb mixture per cup of hot water for 10 minutes and then strain. Drink as much as possible. A minimum of 3 cups a day are needed to see results, with double that being even better.

Tincture for Cold (Atrophic) Gastritis

This formula is warming and stimulating due to *Artemisia* and *Gentiana*. The presence of the nourishing base herbs *Matricaria* and *Glycyrrhiza* keep it in balance and prevent it from being irritating. This formula is not quite tasty enough to prepare as a tea, unless a patient tolerates bitter flavors well.

Matricaria chamomila 30 ml
Glycyrrhiza glabra 30 ml
Taraxacum officinale root 30 ml
Artemisia absinthium 15 ml
Gentiana lutea 15 ml

Take a dropperful of the combined tincture 20 minutes before meals to improve digestion and food tolerance. Take as often as 5 or 6 times a day, reducing frequency as symptoms improve.

Tea for Atrophic Gastritis

Although *Artemisia* is an extremely bitter herb, when it is combined with the other herbs in this tea, the result is palatable. *Artemisia*, *Zanthoxylum*, and *Zingiber* are all warming and stimulating and help provide the energetic heat needed. *Artemisia* stimulates secretions and has antimicrobial effects. *Zanthoxylum* may improve circulation in the stomach and intestines.

Glycyrrhiza glabra 4 ounces (120 g)
Matricaria chamomilla 2 ounces (60 g)
Zingiber officinale 2 ounces (60 g)
Artemesia absinthium 1 ounces (30 g)
Zanthoxylum americanum 1 ounce (30 g)

Place 6 tablespoons of the herb mixture in a saucepan containing 6 cups of water. Bring to a gentle simmer and turn off the heat immediately. Cover the pan, let steep for 15 minutes, and then strain. Drink 3 or more cups a day. Sipping all day long to keep a constant trickle running through the stomach is advised.

Tincture for Infectious Gastroenteritis

Antimicrobial herbs are needed in formulas for acute infectious gastroenteritis. This tincture should be complemented by the use of a soothing tea for symptom relief as well as garlic or *Artemisia* pills where available. The tincture allows for easier consumption of unpalatable herbs and also allows for an aggressive protocol to be put together using multiple forms of medicine.

Allium sativum 15 ml
Artemisia absinthium 15 ml
Mahonia aquifolium 15 ml
Matricaria chamomilla 15 ml

Take 1 dropperful of the combined tincture every hour, reducing frequency as symptoms improve.

Antimicrobial Herbs for Infectious Gastroenteritis

Allium sativum
Artemisia annua, *A. vulgaris*
Azadirachta indica
Coptis chinensis
Hydrastis canadensis
Juglans nigra tincture or alcohol free
Mahonia spp.
Matricaria chamomilla
Melaleuca alternifolia
Origanum vulgare
Rheum palmatum
Rumex acetosella
Spilanthes acmella
Tabebuia impetiginosa
Thuja occidentalis, T. plicata
Thymus vulgaris

Tincture for *Clostridium difficile* Enteritis

Clostridium difficile (also known as *C. diff*) is a gram-positive anaerobic bacteria that is a leading cause of hospital-acquired nosocomial bacterial infection. It is associated with antibiotic use and dysbiosis and is one of the most deadly enteric pathogens in the United States.[98] Therapies to support healthy intestinal microbes range from simple probiotic supplements to fecal transplants[99] as well as naturally occurring polysaccharides shown to support optimal diversity in the gut microbiome.[100] *Clostridium* metabolizes polysaccharides in intestinal mucous, and *C. diff* patients display altered mucous composition,[101] suggesting that mucous-enhancing herbs may benefit barrier functions. Epigallocatechins from green tea inhibit *C. diff* virulence and protect the colon from damage.[102] Antibiotics should be avoided where possible, but when severe and refractory, *C. diff* infections may respond to doxycycline or the related tigecycline.[103] Research on botanicals against *C. diff* is lacking, but the following tincture formula combines some of our strongest digestive antimicrobials.

Hydrastis canadensis
Allium sativum
Azadirachta indica
Andrographis paniculata
Curcuma longa

Combine equal parts of the herbs and take 2 or 3 dropperfuls of combined tincture 8 times a day for acute and symptomatic infections, or 1 dropperful 5 times a day for less severe presentations. Due to the tenacious nature of *C. diff*, consider putting together the most aggressive protocol possible to bolster this tincture, including garlic pills, oregano oil capsules, and berberine supplements as well as probiotics, glutamine powder, and bromelain.

Tea for Infectious Gastroenteritis with Severe Diarrhea

This formula combines the astringents *Rubus* and *Agrimonia*, to help stop diarrhea, with the antimicrobial agents *Mahonia* and *Matricaria*. All are palatable enough to blend into a tea.

Matricaria chamomilla 2 ounces (60 g)
Rubus idaeus 2 ounces (60 g)
Mahonia aquifolium 2 ounces (60 g)
Glycyrrhiza glabra 1 ounce (30 g)
Agrimonia eupatoria 1 ounce (30 g)
Filipendula ulmaria 1 ounce (30 g)

Steep 1 tablespoon of the herb mixture per cup of hot water for 10 to 15 minutes and then strain. Drink 3 or more cups per day. Sipping constantly throughout the day may yield the best result.

Tea for Infectious Gastroenteritis with Vomiting

Mentha species and *Zingiber* are some of the most effective herbs for alleviating nausea and vomiting. Carbonated liquid is also very helpful, and this tea may be combined with mineral water or ginger ale.

Zingiber officinale
Matricaria chamomilla
Mentha piperita
Glycyrrhiza glabra

Combine equal parts of the dry herbs. Steep 1 tablespoon of the herb mixture per cup of boiling hot water and then strain. Drink as is after straining, or combine the strained tea with mineral water or ginger ale. Take small sips every few minutes; drinking a cupful may be nauseating until the herbs take effect. When no liquids at all are tolerated, freeze the prepared tea into ice cubes, crush, and take small spoonfuls. Smelling mint oil and rubbing mint essential oil over the abdomen can also be helpful when not even soothing liquids can be tolerated.

Formula for Allergic Gastroenteritis

Less common than infectious or NSAID-induced gastritis, allergic gastritis may also result from hypersensitivity reaction in the stomach or intestines. When a lapse in the diet, a new medication, or other exposure triggers gastroenteritis, antimicrobial agents are not necessary. Demulcent and anti-inflammatory agents, such as the herbs in this formula, can be effective for diarrhea and pain and can soothe irritated tissues.

Ulmus fulva powder or finely shredded bark 4 ounces (120 g)
Matricaria chamomilla 2 ounces (60 g)
Glycyrrhiza glabra 2 ounces (60 g)

Place 3 tablespoons of *Ulmus* powder or bark in the bottom of a saucepan and cover with 4 cups of water. Let stand overnight, or at least 3 hours. Place 2 tablespoons each of *Matricaria* and *Glycyrrhiza* in a separate saucepan, cover with 4 cups of water and bring to a simmer. Turn off the heat immediately and let stand, covered, for 5 minutes. Pour the hot brew into the pan with the *Ulmus* and water. Let stand 10 minutes more and then strain. Drink as much as possible over the course of the day, and repeat the following day, or for several days as needed.

Formulas for Ulcers

Optimally, the amount of stomach acid occurs in balance with the amount of gastric mucous secreted to protect the stomach tissues from the acid. When acid is excessive or the protective mucous barrier is deficient, ulcers may result. Many factors may contribute to the loss of this balance, including stress and the use of aspirin, nonsteroidal anti-inflammatory drugs (NSAIDS), steroids, or alcohol. If ulcers are present, it is important to reduce, if not eliminate altogether, all such substances. Agents that enhance the production of mucous, such as *Glycyrrhiza*, or that have the capacity to coat the stomach lining, such as demulcents, are foundational ingredients in ulcer formulas. As with GERD, *Helicobacter pylori* bacteria are noted to be associated with many cases of ulcers, and therefore antimicrobial herbs are usually appropriate in ulcer formulas.

General anti-inflammatories such as *Curcuma*, *Bupleurum*, and *Zingiber* are appropriate in the form of teas or tinctures. *Foeniculum vulgare* and *Pimpinella anisum* teas are noted to have ulcer-healing effects.[104] *Curcuma*, long used for all manner of inflammatory processes, is noted to affect inflammatory enzymes and mediators. It also has antiproliferative, antiangiogenic, and chemopreventive effects, all of which may be therapeutic to ulcers. *Curcuma* may be used in a foodlike way as well as in medicinal preparations.[105] *Centella asiatica* has been shown to promote the healing of digestive ulcers.[106] *Azadirachta* (neem) preparations have also demonstrated broad antimicrobial and antiulcer effects.[107]

Acid-Blocking Pharmaceuticals and Digestive Ulcers

Pharmaceutical acid-blocking treatments for ulcers are not preferred by herbalists or naturopathic physicians. The use of antacid products containing aluminum salts (such as Maalox) unnecessarily introduces aluminum into the body and should be avoided. The popular H2 receptor blocking drugs—such as cimetidine, ranitidine, and their over-the-counter relatives—also seem illogical to some practitioners. These drugs block the production of stomach acid. However, as soon as the stomach senses that acid is low, it responds by producing more acid. Furthermore, as previously stated, ulceration involves lack of an adequate mucosal barrier every bit as much as excessive acid production. Therefore, it may be more effective to restore the stomach's mucosal barrier than to just block the acid production over and over. The H2 blocking drugs appear to be of little real help in changing the ulcer phenomenon in any deep way and contribute to intestinal dysbiosis and other negative side effects when used over time. *Glycyrrhiza*, *Ulmus*, glutamine, and bismuth may support the regeneration of a healthy gastric mucous membrane so effectively that pharmaceuticals for band-aid relief of symptoms gradually becomes unnecessary.

The amino acid glutamine supports gastric mucosal cell synthesis and has been reported to promote the healing of ulcers.[108] Glutamine powder may be taken by teaspoonful stirred into water or in any of the tea formulas for ulcers and colitis; dosage is several times a day. The mineral bismuth, of Pepto-Bismol fame, has ulcer-healing effects and has long been used to soothe dyspeptic symptoms. Bismuth inhibits the enzyme urease,[109] which is involved in converting urea into ammonia and carbonic acid and creating a digestive ecosystem hospitable to *H. pylori*. Therefore, the inhibition of urease by bismuth has therapeutic effects for ulcers and all conditions associated with *Helicobacter* infections. Bismuth-based medicines may support eradication of *H. pylori* when it is persistent following antibiotic therapy.[110] *Fucus* spp. (kelp) and *Lepidium meyenii* (maca root) contain tiny traces of bismuth.

Lloyd Brothers' Formula for Gastric Ulcer

The Lloyd Brothers manufactured concentrated herbal products in the 1920s, particularly for the Eclectic physicians, a group of natural-medicine–inclined MDs of the era. This formula was their commercial recipe for treating gastric ulcers.

Echinacea angustifolia
Mahonia aquifolium
Collinsonia canadensis
Hamamelis virginiana

Combine equal parts of the herbs. Take the combined formula by the teaspoonful, 3 to 6 times daily.

Robert's Formula for Ulcers

This is a classic formula, rumored to have been named after an early American sailor named Robert, who promoted this combination to heal stomach ulcers. One common name for a species of wild *Geranium* is Herb Robert, due to its use in this formula. The traditional formula contained the demulcent okra, but because okra as an herbal powder may be hard to locate, this formula uses *Ulmus* powder instead. *Hydrastis* was in the traditional formula and is now known to have activity against *H. pylori*. As the flavor of *Hydrastis* may be challenging for some, *Mahonia* (also known as *Berberis*) is a possible more palatable substitute. The classic formula may be encapsulated, and there are various commercial versions on the market, but it also can be made "in-house" by herbalists. Some traditional formulas included cabbage powder—a known source of glutamine. Drinking a quart of cabbage juice per day is another traditional folkloric cure for ulcers.

Echinacea angustifolia finely chopped root 2 ounces (60 g)
Hydrastis canadensis powder or chopped root 2 ounces (60 g)
Phytolacca decandra chopped root 2 ounces (60 g)
Geranium robertianum chopped root 2 ounces (60 g)
Ulmus fulva powder, or finely shredded 4 ounces (120 g)
Glycyrrhiza glabra powder 4 ounces (120 g)

This formula yields 1 pound of dry herb blend. To prepare it, gently simmer 1 teaspoon of the herb mixture per cup of water for 10 minutes and then strain. For best results, prepare a quart or more each day, and drink the entire volume. Glutamine powder could be a valuable addition, stirring 1 teaspoon into each cup upon consumption.

Tea for Gastric Ulcers Associated with Gas and Bloating

Tea is especially effective for treating ulcers because it can achieve direct surface contact with the stomach lining. The carminative herbs in this formula can quickly allay acute pain and bloating. *Matricaria* is noted to have activity against *Helicobacter* infections, and *Foeniculum, Glycyrrhiza,* and *Ulmus* are all noted to promote healing of ulcers.

Matricaria chamomilla
Mentha piperita
Foeniculum vulgare
Glycyrrhiza glabra
Ulmus fulva

Combine equal parts of the dry herbs. Steep 1 tablespoon of the herb mixture per cup of hot water, allowing it to stand covered for 10 to 15 minutes, and then strain. Drink as much as possible, a minimum of 3 cups per day. Keeping a constant trickle of the brew moving through the stomach will be the most effective.

Mucosal Restoratives

The following herbs are all listed in folklore as having the ability to treat ulcers, soothe mucosal irritation, and help rebuild and repair digestive mucous membranes.

Aloe vera	*Glycyrrhiza glabra*
Althaea officinalis	*Plantago* spp.
Calendula officinalis	*Symphytum officinale*
Centella asiatica	*Ulmus fulva, U. rubra*
Equisetum spp.	Seaweeds

Tea for Bleeding Ulcer

Gastrointestinal bleeding can be occult, noticed only with stool testing, can be frank blood in the stool, and can even be a hemorrhagic emergency. When bleeding is not so severe as to necessitate a trip to the emergency room, this formula may be offered to ease bleeding. It combines the intestinal hemostatics *Geranium* and *Quercus* with the demulcent anti-inflammatories also used in other formulas in this chapter to treat blood in the stool. This formula is most effective prepared as a tea.

Geranium maculatum, G. robertianum
Quercus spp.
Foeniculum vulgare
Glycyrrhiza glabra
Ulmus fulva

Combine equal parts of the dry herbs. Gently simmer 1 heaping teaspoon of the herb mixture per cup of hot water for 5 minutes. Let stand 10 minutes more in a covered pan, and then strain. Drink as much as possible, aiming to keep a constant trickle moving through the stomach.

Simple Tincture for Bleeding Ulcer

The above tea formula would be more effective in controlling digestive bleeds as quickly as possible, but this tincture could be started promptly while preparing the tea or tracking down ingredients. A simple *Geranium* tincture could also complement the above tea and could be used in tandem.

Geranium maculatum, G. robertianum

As a simple, take ½ teaspoon hourly. Reduce frequency as the symptoms improve.

Tincture for *H. pylori*

Based on modern research, the herbs in this formula have been shown to be effective against *Helicobacter pylori.* Many of these herbs are not tasty, and thus the formula is prepared into a tincture. However, this formula contains no demulcents or restoratives like the classic Robert's Formula for Ulcers (page 63). It would be important to pair this tincture with a nourishing trophorestorative tea such as the Tea for Gastric Ulcers on page 66.

Glycyrrhiza and *Ulmus* as Key Herbs for Digestive Ulcers—Molecular Research

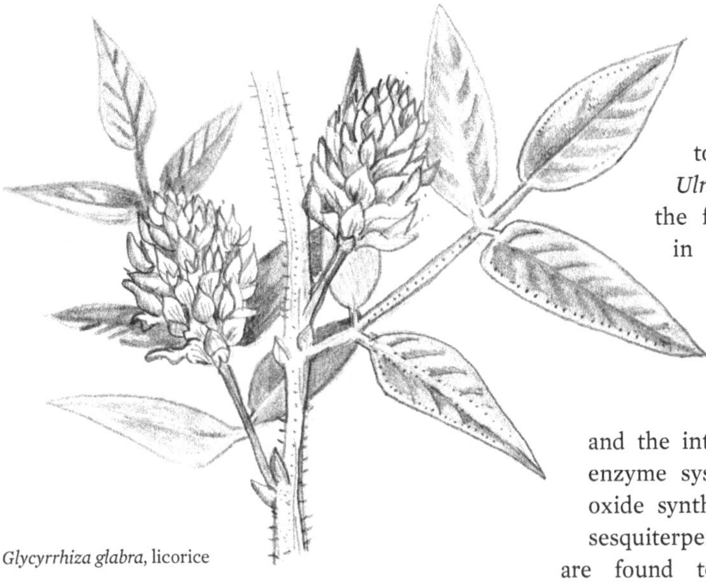

Glycyrrhiza glabra, licorice

Glycyrrhiza has noted ulcer-healing and mucosal-building effects. Flavonoids in *Glycyrrhiza* are noted to bind to histamine receptors and have antagonistic effects, thereby reducing stomach acid.[111] In addition, *Glycyrrhiza* is noted to increase the number of mucous-secreting goblet cells in the stomach, enabling the digestive mucosa to better protect itself from normal digestive acid.[112] A major component of *Glycyrrhiza glabra*, 18β-Glycyrrhetinic acid attenuates *H. pylori*-driven gastric inflammation, reducing mucosal erosions and inflammatory hyperplasia.[113]

Ulmus is noted to reduce reactive oxygen species[114] and is a long-standing folk medicine for digestive ulcers and burning epigastric pain. *Ulmus* and other mucilaginous herbs such as *Symphytum* and *Asparagus* contain glycoproteins that appear to support animal glycosaminoglycan molecules, commonly referred to as GAGS. The glycoproteins in *Ulmus* have been shown to inhibit the formation of inflammatory lesions in intestinal mucosa.[115] Researchers report that *Ulmus* ingestion leads to improvements in the levels of reactive oxygen species, including nitric oxide, the cytokines tumor necrosis factor (TNF), and the interleukins as well as inflammatory enzyme systems including COX-2 and nitric oxide synthase (NOS).[116] Three guaiane-type sesquiterpenes are also present in *Ulmus*; they are found to inhibit the lipopolysaccharide enzyme system and reduce excessive NOS formation.[117] Large, frequent doses are most effective, such as preparing a slippery elm beverage or porridge to fully coat and fill the stomach, once or twice each day.

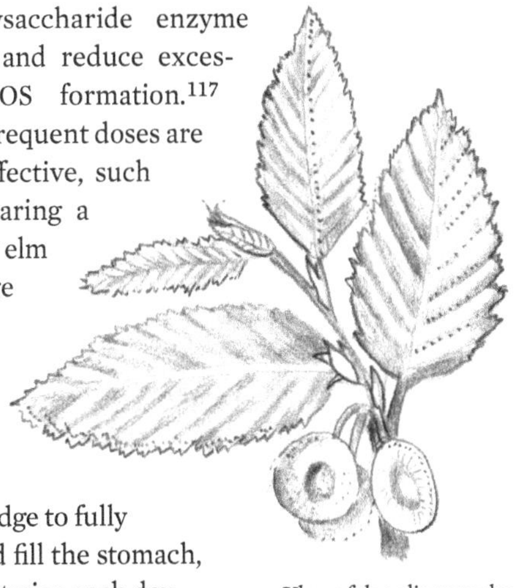

Ulmus fulva, slippery elm

Mahonia aquifolium
Echinacea angustifolia, E. purpurea
Thymus vulgaris
Artemisia absinthium

Combine equal parts. Take 1 or 2 dropperfuls of the combined tincture, 4 or 5 times a day, reducing week by week as symptoms improve. The ulcer should be healed in 3 months or less and the tincture discontinued at that time.

Tincture for Painful Gastric Ulcers

While teas and gruels may be more effective for ulcers than tinctures, a tincture is handier to use during the workday for many people. A pain-relieving tincture such as this can also complement a tea and help relieve acute symptoms until the tea can be brewed or obtained.

Foeniculum vulgare 15 ml
Glycyrrhiza glabra 15 ml
Mentha piperita 15 ml
Symphytum officinale 15 ml

Take the combined tincture by the dropperful every 5 to 15 minutes until symptoms are relieved.

Tea for Gastric Ulcers

This tasty tea contains trophorestoratives and pain-relieving ingredients. *Centella* and *Glycyrrhiza* are both noted to heal ulcers.

Matricaria chamomilla
Mentha piperita
Foeniculum vulgare
Glycyrrhiza glabra
Ulmus fulva
Centella asiatica

Combine equal parts of the dry herbs and blend well. Steep 1 tablespoon of the herb mixture per cup of hot water and then strain. Drink freely. A minimum of 3 cups per day, optimally more, is best for healing a digestive ulcer.

Slippery Elm Porridge for Acute Ulcer or Gastritis Pain

Historically this formula has been referred to as a "gruel" rather than a porridge, but most modern patients relate better to the word porridge. It can be extremely useful in alleviating acute burning pain of ulcers. The traditional formula uses plain water, but I prefer and have substituted chamomile tea.

1 cup (60 g) slippery elm powder
1 cup (240 ml) chamomile tea, cooled to a tepid temperature
¼ teaspoon fennel, cardamom, ginger, or cinnamon powder (optional)
½–1 teaspoon honey (optional)
Banana slices, chopped dried fruit (optional)

Place the slippery elm powder in a cereal bowl and slowly drip in the warm chamomile tea while whisking vigorously with a fork (like making gravy) to create a sticky oatmeal-like mass. The porridge is bland, but so pain-relieving that, once tried, its value becomes obvious to patients. To improve the flavor, aromatic spices, honey, and fresh or dried fruit may be added, if tolerated. It is best to prepare a simple porridge at first and then amend bit by bit.

Althaea and *Glycyrrhiza* Cold Infusion for Acute GI Pain

This is another effective way to relieve acute GI pain. It relies on a long cold maceration to best extract the mucilage out of *Althaea*.

Althaea officinalis 4 teaspoons
Glycyrrhiza glabra 2 teaspoons
Water 6 cups (1.4 l)

Cover the *Althaea* and *Glycyrrhiza* with the water and let sit 4 hours or overnight. Bring to a very brief gentle simmer of only 1 minute. Turn off the heat, cover the pan, and let stand for 10 to 15 minutes. Pour the entire brew into a muslin bag (jelly bag) or a muslin-lined stainless steel wire mesh strainer and allow it to drain into a bowl, and then transfer to a drinking mug. The tea should be thick and mucilaginous. Drink as much as possible when cool enough to drink.

Formulas for Gastric and Intestinal Bleeding

Intestinal bleeding is not a diagnosis unto itself, but rather can result from infection, irritation, or ulceration; from the lesions of Crohn's and other autoimmune diseases; or from colon cancer, polyps, and other lesions. Therefore, blood in the stool requires a medical workup to determine the underlying cause. *Shigella* and *Campylobacter* infections are particularly likely to induce intestinal bleeding, and erosive lesions can cause chronic blood loss that leads to anemia. The following formulas are examples of using herbs to treat various situations of intestinal bleeding other than those due to colon cancer.

Tea for GI Bleeding with Watery Diarrhea

These high tannins herbs are most effective when prepared as an herbal tea to accomplish direct surface contact with ulcerated or irritated intestinal mucosa. Because tannins taste bitter and astringent, *Glycyrrhiza* is used both as an antiulcerative agent and to improve the flavor of the tea.

Achillea millefolium
Geranium maculatum
Hamamelis virginiana
Glycyrrhiza glabra

Combine equal parts of the dry herbs and blend well. Gently simmer 6 heaping teaspoons of the herb mixture in 6 cups of hot water for 10 minutes in a covered pan, and then strain. Sip constantly throughout the day.

Tea for GI Bleeding with Pain and Ulceration

Similar to the above formula, this tea adds agents such as the demulcent *Ulmus* to quickly soothe pain. This formula has a greater anti-inflammatory effect due to the *Filipendula* and a greater antispasmodic effect due to the *Matricaria*. This formula can be prepared as a tincture, but a tea would be more effective.

Filipendula ulmaria 1 ounce (30 g)
Geranium maculatum 1 ounce (30 g)
Hamamelis virginiana 1 ounce (30 g)
Ulmus fulva, shredded 2 ounces (60 g)
Matricaria chamomilla 2 ounces (60 g)

Gently simmer 6 heaping teaspoons of the herb mixture in 6 cups of hot water for 10 minutes in a covered pan, and then strain. Sip constantly throughout the day.

Tea for GI Bleeding with Intestinal Infection

When intestinal bleeding is due to an intestinal infection such as *Shigella* or *Campylobacter* or due to traveler's diarrhea, antimicrobials are indicated. Many herbs have at least some antimicrobial activity, but the herbs in this tea are particularly specific for infectious diarrhea. Unfortunately, they are not particularly tasty. Patients should be coached to drink the formula as a tea anyway, for the surface contact benefits. For serious infections, this tea might be complemented by berberine capsules, *Artemisia* capsules, or a commercial product formulated for intestinal parasites.

Artemisia absinthium
Hydrastis canadensis
Mahonia aquifolium
Glycyrrhiza glabra

Combine equal parts of the dry herbs, or use more *Glycyrrhiza* as desired for taste. Gently simmer 2 heaping tablespoons of the herb mixture in 6 cups of hot water for 10 minutes in a covered pan, and then strain. Sip constantly throughout the day.

Formulas for Appendicitis

Appendicitis is most common in teenagers and young adults, although it may occur at any age from infants to the elderly. Investigations suggest that many cases are preceded by constipation, and that a shift in intestinal flora from healthy species to the more pathogenic may predispose. Fluid stasis or pelvic congestion associated with constipation may also predispose. While surgical treatment is almost always necessary, herbal therapy is appropriate as a follow-up. Historical literature of the early Americas purports that appendicitis may be a sign of a lymphatic or "phlegmatic" constitution, and that lymphatic herbs are helpful in convalescence formulas. *Calendula* and *Galium* are two lymphatic herbs to use in teas and tinctures for those recovering from an appendectomy. *Iris* supports liver function and reduces fluid stagnation in the body and is especially indicated when appendicitis is preceded by poor digestion, constipation, or liver complaints. While these herbs may not be useful to treat acute appendicitis, they may be used in convalescent and recovery formulas to help improve lymphatic circulation.

Tincture for Acute Appendicitis

Echinacea's lymphatic-enhancing properties are often overlooked, but *Echinacea* is emphasized in the folkloric herbals as being specific for organ inflammation and decay, including appendicitis. For best results from this tincture, it's advisable to use it in tandem with bromelain, an anti-inflammatory enzyme from pineapple. Bromelain is recommended before and after all surgical procedures because it is noted to reduce swelling, speed healing, and reduce the risk of hematomas and blood clots. Bromelain is widely available in capsules and 500 to 1,000 milligrams may be taken every several hours (at least 3 times a day).

Echinacea angustifolia, E. purpurea
Phytolacca decandra
Curcuma longa
Boswellia serrata

Combine equal parts. Take 1 dropperful of the combined tincture every 20 to 30 minutes, along with 1,000 milligrams of bromelain every 4 hours.

Post-Appendectomy Tincture

Acute appendicitis is generally a surgical emergency, but herbal medicines are appropriate to promote GI recovery following surgery. *Echinacea* is recommended in historical literature for organ breakdown and decay and, because the appendix may have become inflamed due to congestion, *Echinacea* is especially indicated for appendicitis recovery. *Calendula* is well-known to improve wound healing and support surgical recovery, and *Curcuma* is antioxidant and antimicrobial and also supportive to wound healing and connective tissue regeneration. *Galium* may treat lymph congestion that predisposes to infections and inflammation in lymphatic organs.

Curcuma longa 30 ml
Calendula officinalis 30 ml
Galium aparine 30 ml
Echinacea angustifolia 30 ml

Take 1 teaspoon of the combined tincture every 2 hours, beginning immediately following surgery. Reduce frequency to every 3 or 4 hours the following day, and then to 4 or 5 times a day thereafter until all of the tincture has been used up.

Formulas for Acute Pancreatitis

Acute pancreatitis is an excruciatingly painful condition. From 30 to 50 percent of severe cases can be fatal if not treated promptly, because large amounts of pancreatic enzymes are released, which may lead to circulatory and respiratory failure. Therefore, acute pancreatitis usually requires hospitalization due to the many possible complications such as systemic infection and organ failure, respiratory distress, and blood coagulation. Several species of peony contain paeoniflorin in the roots, which is shown to protect the kidney from injury in situations of acute pancreatitis, and *Paeonia lactiflora* and *suffruticosa* have been used in TCM for a variety of pancreatic and digestive diseases.[118] Biliary tract disease and alcoholism predispose to developing acute pancreatitis; drugs, vascular disease, hyperlipidemia, and a hereditary tendency are other possible etiologic contributors. Chronic pancreatitis may be associated with autoimmune reactivity.[119] Statistics suggest that older patients with severe acute pancreatitis may present with atypical clinical symptoms.[120] Both acute and chronic inflammation of the pancreas may lead to fibrosis of the organ and may possibly transition to ductal adenocarcinoma as pancreatic stellate cells become activated. Fibrotic infiltration of the pancreas can cause pancreatic exocrine insufficiency and is one long-term consequence, leading to malnutrition.[121] Herbs that limit fibrosis are indicated in all cases of pancreatitis and the use of polyphenols from plants may inhibit fibrogenesis and inhibit degeneration of the pancreas, with rhein, emodin, curcumin, and resveratrol the most studied.[122] Resveratrol, a component of red wine, skins of Spanish peanuts, cocoa, and other plants may protect the pancreas and intestinal barriers by increasing superoxide dismutase (SOD) and malondialdehyde (MDA) levels.[123]

Consider supplementing pancreatic enzymes in patients with steatorrhea (oily or fatty stools), diarrhea, weight loss, and lassitude and help educate patients to avoid acid-blocking agents to treat dyspeptic symptoms. Enteric-coated supplements and enzymes may be superior to conventional enzymes; another option is to open up capsules and mix the enzymes directly into food. Patients will be at risk for small intestinal bacterial overgrowth (SIBO) and an effort must be made to offer the teas and tinctures discussed in this section of the chapter.

Despite the grave nature of pancreatitis, there are few effective pharmacologic interventions. Steroids are not effective and surgical removal of the pancreas is incompatible with life. Thus, herbal anti-infectives, anti-inflammatories, and cholagogue herbs are of great value and some hospitals in China are exploring the efficacy of herbs in a hospital setting. Because the condition is extremely painful and does not often respond to pharmaceuticals, anodyne herbs are appropriate, and topical agents may be helpful. The following formulas, while inspired by what little research on herbal medicines for pancreatitis exists, represent my best effort to create supportive therapies that may help prevent fibrotic and inflammatory reactions.

Tincture for Acute Pancreatitis

Dioscorea is a traditional herb for digestive spasms and inflammation, and it contains the saponin diosgenin, which is shown to prevent pancreatic cell death in animal models of acute pancreatitis.[124] *Ginkgo biloba* may be added to protect the lungs in patients with pulmonary complications, as animal models of acute pancreatitis show *Ginkgo* to prevent lung inflammation and tissue changes by reducing activation of inflammatory responses.[125] *Phyllanthus emblica* is widely used in India and in Ayruvedic medicine as a broad-acting antioxidant and anti-inflammatory, actions that may extend to protecting the pancreas.[126] *Echinacea* may reduce the risk of necrosis, and although there is not research to support the hypothesis, it certainly will not hurt.

Dioscorea villosa
Ginkgo biloba
Rheum palmatum
Phyllanthus emblica
Echinacea angustifolia, E. purpurea

Combine in equal amounts. Take 1 to 2 teaspoons of the combined tincture per hour for several days, reducing as symptoms improve. Take the tincture along with a resveratrol supplement, *Curcuma* capsules, and the most aggressive protocol possible. Patients should fast, followed by at least 2 weeks of a light diet consisting of nourishing teas and broths.

Tea for Acute Pancreatitis Featuring Chinese Herbs

Rheum (Chinese rhubarb) is a source of emodin, and studies suggest that emodin and related compounds may naturally concentrate in the acutely inflamed pancreas,[127] that they may exert significant antifibrotic effects in animal models of pancreatitis,[128] and that they may significantly attenuate severe, acute pancreatitis,[129] encouraging phagocytosis of inflamed cells.[130] *Rheum is* commonly used in China for various gastrointestinal pathologies including acute and chronic pancreatitis. One clinical trial showed that the combination of *Rheum* and red peony root supported recovery following severe acute pancreatitis in human subjects.[131] One human clinical trial showed *Rheum* to help resolve acute abdominal pain and to speed resolution of fever, elevated white blood cells, and other lab indicators in patients with severe acute pancreatitis, compared to a control group.[132] *Panax ginseng* is shown to protect the pancreas from necrotic and fibrotic changes in animal models of pancreatitis.[133] *Salvia miltiorrhiza* (red sage, dan shen) roots have been shown to have numerous vascular benefits, including improving microcirculation in inflamed organs. Due to its antioxidant effects and many beneficial effects on inflammatory mediators, *S. miltiorrhiza* may also protect the pancreas, speed recovery, and prevent fibrotic changes.[134] *Rehmannia glutinosa* contains the iridoid glycoside catalpol shown to reduce inflammatory markers of pancreatitis such as amylase, interleukins, and nuclear factor kappa β.[135]

Rheum palmatum
Paeonia suffruticosa, P. lactiflora
Panax ginseng
Salvia miltiorrhiza
Rehmannia glutinosa
Glycyrrhiza glabra

Combine equal parts of the dry herbs. Prepare a pot of tea by gently simmering 2 tablespoons of the herb mixture in 8 cups of water. Continue simmering in an open pan until the tea is reduced to 6 cups, and then strain. Drink the entire 6 cups over the course of the day for at least 1 to 2 weeks, possibly as long as 2 to 3 months.

Pain-Relieving Tincture for Acute Pancreatitis

Since pancreatitis is acutely painful and unresponsive to steroids or pharmaceutical pain medication, anything capable of allaying pain would be most appreciated by the patient. This formula features general herbal anti-inflammatories and anodynes that it wouldn't hurt to try. This formula may be taken while fasting. The formula contains *Aconitum*, a potentially toxic herb that suppresses nerve centers, and is available to licensed physicians only. In small doses, however, *Aconitum* may help deaden pain. Bradycardia is a sign of aconite toxicity and a warning to stop dosing and remove the herb from the formula. Do not use aconite at all unless you are an experienced herbalist.

Piscidia piscipula (also known as *P. erythrinia*) 10 ml
Glycyrrhiza glabra 10 ml

Bupleurum falcatum 10 ml
Dioscorea villosa 10 ml
Echinacea angustifolia 10 ml
Hypericum perforatum 10 ml
Aconitum napellus 10 drops

This formula will fill a 2-ounce bottle. Omit the aconite if you are not a skilled and licensed practitioner. Take ½ teaspoon of the combined tincture every 15 to 30 minutes for a 24-hour period, reducing frequency as symptoms improve.

Convalescent Tincture for Pancreatitis

This formula is appropriate for low-grade chronic pancreatitis and as a recovery tonic following acute pancreatitis. These herbs are chosen for being general anti-inflammatories acting by a wide variety of mechanisms to provide a multipronged approach. *Bupleurum* is emphasized in the literature for pain and inflammation in the liver and other internal organs. *Harpagophytum*, Devil's Claw, is a traditional pain remedy in South Africa, most commonly used for musculoskeletal pain, but included here for its broad anti-inflammatory effects. Harpagoside is credited with some of the antioxidant and anti-inflammatory actions of *Harpagophytum*.[136]

Salvia miltiorrhiza 10 ml
Matricaria chamomilla 10 ml
Bupleurum falcatum 10 ml
Curcuma longa 10 ml
Echinacea purpurea, E. angustifolia 10 ml
Glycyrrhiza glabra 5 ml
Harpagophytum procumbens 5 ml

Take 1 teaspoon of the combined tincture in water or a digestive tea 3 times daily.

Convalescent Tea for Pancreatitis

It would work well to offer this tea with the previous Convalescent Tincture for Pancreatitis and liver-supportive capsules such as milk thistle or curcumin capsules. Lipotropic formulas typically contain milk thistle and curcuma and can be a useful combination product. Such measures may support full recovery from acute pancreatitis and limit fibrotic and degenerative changes in the pancreas and associated biliary and intestinal tissues.

Glycyrrhiza glabra 4 ounces (120 g)
Rheum palmatum 4 ounces (120 g)
Paeonia suffruticosa 4 ounces (120 g)
Taraxacum officinale 4 ounces (120 g)
Cinnamomum verum 2 ounces (60 g)
Crataegus oxyacantha berries 2 ounces (60 g)
Angelica sinensis 2 ounces (60 g)
Quercus alba, Q. rubra 2 ounces (60 g)

This formula yields a pound and a half batch of herbs, which should last many months stored in a large resealable plastic bag or airtight container. Simmer 1 teaspoon of the herb mixture per cup of hot water for 10 minutes. Cover and let stand 10 minutes more and then strain. Drink freely, a minimum of 3 cups per day.

Antifibrotic Formula for Chronic Pancreatitis

As discussed above, chronic pancreatitis leads to fibrosis of the pancreas and loss of pancreatic enzyme production and may predispose to pancreatic cancer. These herbs may prevent fibrosis by a number of mechanisms and would complement the use of lipotropic capsules and the tea formulas for pancreatitis in this section. This formula for a tincture is my best guess to prevent fibrosis; these herbs are noted to reduce fibrosis in other inflammatory conditions in the liver and kidneys and will certainly do no harm.

Curcuma longa 40 ml
Silybum marianum 40 ml
Rheum palmatum 40 ml

This recipe will fill a 4-ounce bottle. Take 1 teaspoon of the combined tincture 3 or 4 times per day. Continue long term, and consider complementing with resveratrol and curcumin capsules.

Antifibrotic Tea for Chronic Pancreatitis

This formula for a tea is my best guess to prevent fibrosis as these herbs are noted to reduce fibrosis in other inflammatory conditions in the liver and kidneys and will certainly do no harm.

Glycyrrhiza glabra 8 ounces (240 g)
Ganoderma lucidum 4 ounces (120 g)
Bupleurum falcatum 4 ounces (120 g)
Centella asiatica 4 ounces (120 g)
Ginkgo biloba 4 ounces (120 g)
Salvia miltiorrhiza 4 ounces (120 g)
Rheum palmatum 4 ounces (120 g)

This formula will yield a 2-pound blend, which should be stored in a glass jar in a cool dark place. Prepare by gently simmering 1 teas poon of the herb mixture per cup of water over low heat for 15 minutes. Let stand in a covered pan for 15 minutes more, and then strain. Drink 3 or more cups per day, at least 5 days per week, long term.

Liu-He-Dan Ointment for Pancreatitis

Emodin, a polyphenol found in *Rheum*, may even be useful when applied topically. Emodin is shown to increase in the plasma following topical application and is associated with reduced inflammatory markers in animal models of pancreatitis.[137] An ointment containing *Rheum* called Liu-He-Dan is used topically for acute pancreatitis and numerous papers have been published on its use to reduce inflammation in both acute and chronic pancreatitis. This simplified version may be prepared in the office and used to complement teas, pills, and tinctures for pancreatitis.

Rheum palmatum powder
Castor oil

Place several cups of *Rheum* powder in a glass jar and pour in enough castor oil to fully cover the powder; there should be at least an inch of oil on top of the powder. Stir thoroughly with a wooden spoon or chopstick once a day for 4 to 6 weeks, then strain the mixture several times through a muslin cloth or bag. Rub the strained oil topically into the abdomen 3 or more times daily.

Formulas for Diverticulitis

Diverticuli may occur in more than one-third of all people over the age of 50 and may become even more common with advancing age. Diverticuli may be "silent" and asymptomatic, noticed only with a routine sigmoidoscopy. In other cases, diverticuli are irritable and easily inflamed by certain rough-textured foods. They can become infected, inducing fever, worsening pain, and systemic symptoms. The symptoms may be nonspecific, such as simple gas and bloating or minor alterations in bowel habits, but diverticuli symptoms may also be acute. Acutely inflamed diverticuli may rupture causing hemorrhage or peritonitis.

The etiology of diverticuli is thought to involve low fiber intake, which causes an increase in intestinal pressure such that a ballooning of the mucosa through the muscular layer of the bowel occurs. Some patients and even physicians are under the impression that a bland, low-fiber diet is indicated for diverticuli, but this may actually contribute to the worsening of diverticulosis over time. Only foods that have been observed to be a problem for a particular patient need to be avoided, and in general oat bran, raw vegetables, and other easily digested fibers are encouraged.

Acute diverticulitis with abdominal pain and tenderness, bowel symptoms, fever, and leukocytosis can usually be managed at home with bed rest, fasting, and ample fluid intake. Simple demulcent teas, pills, or tinctures are appropriate to soothe the inflamed bowel, and anti-inflammatory antimicrobial herbs such as *Zingiber*, *Glycyrrhiza*, and/or *Curcuma* can be prepared into teas to sip constantly throughout the day. When fever and other symptoms of an infection are present, antimicrobial herbs such as *Echinacea*, *Allium*, *Origanum*, *Mentha*, *Hydrastis*, and *Mahonia* may be added to simple soothing formulas. When dosed aggressively, most patients with infected diverticuli will not require pharmaceutical antibiotics. The use of probiotics, hydrotherapy, and topical castor oil and essential oils are complementary to herbal therapies.

Tea for Simple Diverticulosis

This formula is aimed at improving intestinal health and pressure by using *Taraxacum* as an alterative to alleviate constipation and encourage desirable probiotic species. *Matricaria* and *Mentha* discourage cramping and increased intestinal pressure, and *Ulmus* has soothing effects on the intestinal mucosa, all decreasing the risk that diverticuli will become acutely inflamed or infected. This formula is most effective as a tea, but could also be prepared as a tincture if preferred.

Taraxacum officinale root 2 ounces (60 g)
Matricaria chamomilla 2 ounces (60 g)
Ulmus fulva 2 ounces (60 g)
Mentha piperita 2 ounces (60 g)

Simmer 1 heaping teaspoon of the herb mixture per cup of hot water for several minutes in a covered pan. Let stand 10 minutes more, strain and drink freely, at least 3 cups per day for several months.

Tea for Diverticulitis Based on TCM

Traditional Chinese Medicine (TCM) employs the following herbs for acute intestinal inflammation to improve digestive pain, constipation, and inflammation. Additional flavoring herbs such as licorice or fennel seeds may be added if desired.

Rheum palmatum root
Paeonia suffruticosa root
Prunus persica seed

Combine equal parts of the herbs. Gently simmer 1 teaspoon of the herb mixture per cup of water for 10 minutes. Strain and drink 3 or more cups per day.

Tincture for Diverticulitis with Fever or Infection

When diverticuli are inflamed and infected, antimicrobials are needed as well as palliative tools to allay the pain and irritation. *Curcuma* is both antimicrobial and anti-inflammatory, with *Echinacea* providing further antimicrobial support. For acute bouts of diverticulitis, use this formula as well as the above tea. *Mentha* helps with the gas and bloating typical of diverticulitis and *Ulmus* has direct soothing effects.

Echinacea purpurea, E. angustifolia 15 ml
Ulmus fulva 15 ml
Mentha piperita 15 ml
Curcuma longa 15 ml

Take 1 or 2 dropperfuls of the combined tincture 3 or 4 times daily.

Formulas for Malabsorption

The failure to adequately break down, absorb, and assimilate ingested food may have several underlying etiologies. Malabsorption may occur slowly and insidiously over many years' time as part of the aging process or as a congenital or acquired insufficiency of bile from the liver and gallbladder, HCL from the stomach, or pancreatic enzymes. Less commonly, genetic defects in digesting particular substances may occur, such as gluten intolerance in celiac disease or milk intolerance in lactase deficiency. There may be a congenital abnormality in the digestive organs or intestines such as biliary atresia or short bowel syndrome. The elderly are particularly likely to have hypochlorhydria and deficient pancreatic secretions. The nutrients commonly deficient are calcium, zinc, magnesium, vitamin B_{12}, and folic acid[138] as well as trace minerals. Hypochlorhydria also makes the elderly susceptible to small intestinal bacterial overgrowth (SIBO).

The symptoms of malabsorption are weight loss, poor wound healing, and digestive gas and bloating. Specific nutrient deficiencies may manifest in these characteristic ways.

- Glossitis
- Nervous symptoms such as paresthesias with B vitamin deficiency
- Muscle cramps and spasms with mineral and electrolyte imbalances
- Edema
- Lack of luster and integrity of hair and fingernails with protein deficiency
- Bruising with vitamin C, bioflavonoid, and vitamin K deficiency

Calcium deficiency is associated with tetany, muscle cramps, and bone pain. Anemia may ensue for patients absorbing iron and B vitamins poorly.

The underlying pathologies are of course treated as specifically as possible. Biliary insufficiency may be suspected with a history of liver disease, intolerance of fatty foods, and right upper quadrant pain. It may be improved by botanical liver support such as *Taraxacum*, *Silybum*, *Curcuma*, and *Chelidonium*. Pancreatic insufficiency may be suspected by oily or fatty stools (steatorrhea) that float in the toilet or are particularly sticky and malodorous, and difficult to flush away. Large malodorous stools also occur with celiac disease. Belching fairly immediately after eating, heartburn, and stomach pain are most typical of hypochlorhydria high up in the digestive system. Intestinal cramping and flatulence are most typical of milk intolerance or other food allergen aggravating the intestines. If the symptoms and diagnostic tests do not hint at a specific underlying cause, a simple trial and error approach may be appropriate. Patients who are hypochlorhydric usually respond readily to HCL supplements. Likewise, patients respond to pancreatic enzymes if there is enzyme insufficiency, and to biliary support if there is insufficient bile. Such therapies may simply be attempted for a week or two, and results evaluated.

Bitter herbs and alteratives such as *Artemisia*, *Juglans*, *Curcuma*, *Taraxacum*, *Rumex*, *Arctium*, and many others stimulate HCL, bile, and pancreatic enzymes, and so they are are appropriate in most cases of malabsorption. Liver herbs and the B vitamin relatives, choline and inositol, are noted to specifically improve bile quantity and quality and may help numerous patients, including those with fat intolerance and the elderly with general digestive insufficiency. When patients are suspected to have intestinal mucosa inflammation due to ingestion of food allergens, the addition of demulcents and anti-inflammatories are indicated, along with specific dietary changes. Teas, pills, and tinctures of *Ulmus*, *Glycyrrhiza*, *Althaea*, and the amino acid glutamine may help. When patients have become malnourished, nutritional supplements may help rebuild and restore the body. Take care, however, to not offer too many pills, because nutrients in pill form are unlikely to be well utilized or assimilated. Liquid nutrients such as teas and tinctures or the addition of liquid nutrients to smoothies or juices will be the easiest to absorb. High mineral herbs such as *Equisetum*, *Medicago*, *Centella*, and *Symphytum* are appropriate to help rebuild and restore connective tissue, skin, hair, and nails. Additions of small amounts of hot spicy herbs such as *Zingiber*, *Capsicum*, and *Piper nigrum*, may improve the absorption of nutrients due to local vasodilation in the intestines, especially in the elderly and those with a cold, deficient constitution. *Piper nigrum* (black pepper) has been shown to make intestinal mucosal tight junctions *less* permeable,[139] and yet *P. nigrum* has been found to increase the absorption of many beneficial nutrients by 100 fold. *P. nigrum* may enhance the absorption of nutrients through vasodilatory effects on the submucosal vasculature by a yet-to-be-explained mechanism that increases the length and girth of intestinal microvilli and thereby increases the absorptive surface area.[140] Preparation of these herbs in vinegar is a useful vehicle for promoting digestive function.

Bitter Tea for Hypochlorhydria

Bitter tea may be challenging for some patients to consume, while others may prefer it over tinctures because tea is less expensive and alcohol-free. The use of cinnamon in this tea both improves the flavor and acts as a carminative stimulant.

Achillea millefolium flowers
Rumex crispus root, finely chopped
Matricaria chamomilla flowers
Cinnamomum verum, small chips

Combine equal amounts of the herbs. Steep 1 to 2 teaspoons of the herb mixture per cup of hot water, and then strain. Drink 1 to 2 cups before each meal. If desired, add a teaspoon or two of fresh-squeezed lemon juice to each cupful.

Tincture for Malabsorption Due to Hypochlorhydria

Bitter herbs stimulate secretion of bile, HCL, and digestive enzymes, but all on their own, bitter herbs might be nauseating and too stimulating. Bitter herbs are combined with the digestive tonic and carminative *Matricaria* and *Zingiber* in this formula.

Artemisia absinthium 14 ml
Juglans regia, *J. nigra* 14 ml
Rumex crispus 14 ml
Matricaria chamomilla 14 ml
Zingiber officinale 4 ml

Take ½ to 1 teaspoon of the combined tincture on an empty stomach 15 to 20 minutes before meals. Another option is to prepare an aperitif using 1 to 2 teaspoons of the tincture, the juice from a lemon slice, and ¼ cup of water or chamomile tea to sip before meals.

Tincture for Malabsorption in the Elderly

Some patients, especially the elderly, may have both hypochlorhydria and poor circulation in the digestive organs contributing to malabsorption. Note how this formula uses half bitter agents and half circulatory-enhancing herbs.

Ginkgo biloba 14 ml
Gentiana lutea 14 ml
Artemisia absinthium 14 ml
Panax ginseng 14 ml
Zingiber officinale 4 ml

Take ½ to 1 teaspoon of the combined tincture on an empty stomach 15 to 20 minutes before meals. Another option is to prepare an aperitif using 1 to 2 teaspoons of the combined tincture, the juice from a lemon slice, and ¼ cup of water or chamomile tea to sip before meals.

Therapies for Malabsorption and Maldigestion

These agents help digest food and are often found in digestive formulas and enzyme blends. Such substances make excellent complements to the teas and tinctures featured in this section.

Ox bile
HCL
Pancreatic lipase, amylase, and proteases
Bromelain
Choline, inositol

Herbal Vinegar for Malabsorption

Vinegar alone is great for digestion, and especially one with appropriately chosen bitter and stimulant herbs, as in this example. Patients with maldigestion can be taught how to make their own herbal vinegar at home inexpensively, or you can prepare something like this recipe for them.

2 tablespoons *Artemisia annua* leaves
2 tablespoons *Urtica* spp. (nettle) leaves
1 small cayenne pepper, coarsely chopped
2 tablespoons coarsely chopped fresh *Zingiber* (ginger) root
Apple cider vinegar

Place the herbs, pepper, and ginger root in a blender. Pour in enough apple cider vinegar to cover, and puree until it becomes a small particulate consistency. Transfer to a glass jar. Shake the jar daily for several weeks and then strain the vinegar through a fine strainer into a glass jar. Use finished vinegar on steamed vegetables and to prepare salad dressings. The vinegar may also be prepared as an aperitif by adding it to water or tea. The vinegar will store at room temperature for one to two years.

Digestive Vinegar

This version of classic fire cider adds bitter herbs to stimulate digestion. The recipe can be amended for taste and purpose, making hotter or milder by changing the quantities of the ginger, pepper, garlic, and onion. Use larger or smaller quantities of bitter herbs as desired. Fruits such as mangoes, pineapples, and papayas, including papaya seeds, can also be blended into the vinegar, both to make the blend sweeter as well as to promote hydrochloric acid and help treat underlying digestive insufficiency and dysbiosis.

1 quart (960 ml) apple cider vinegar
1 cup (200 g) chopped fresh ginger root
1 cup (200 g) fresh habañero peppers, chopped
1 bulb peeled garlic cloves
1 onion, chopped
1 cup (200 g) chopped turmeric root
1 lemon, zest and juice
½ cup (15 g) dry *Medicago*
½ cup (60 g) dry chopped *Mahonia* root
½ cup (60 g) dry chopped *Arctium* root
½ cup (15 g) *Artemisia*

Place all ingredients in a blender and liquefy as finely as possible. Transfer the liquid to a large canning jar and shake daily for 6 weeks. Strain the herbs and place in individual bottles. The finished vinegar can be taken by the spoonful, used to prepare salad dressing, or used to prepare "sipping vinegar" by placing 1 tablespoon in the bottom of a tall glass along with a squeeze of lemon and filling with sparkling water.

Herbs for Malabsorption, Biliary Insufficiency, and Hypochlorhydria

Alteratives, bitters, and warming stimulant herbs are all capable of increasing digestive secretions and enhancing digestion and intestinal absorption. The more nutritive options, such as *Arctium*, *Mahonia*, *Rumex*, and *Taraxacum*, can act as tonifying bases in tinctures and teas. The potently bitter *Gentiana*, *Artemisia*, and *Iris* can act as synergists in formulas where more heat, circulation, and glandular secretions are needed for those with dry stool, hypochlorhydria, or cold dry constitutions. See "Specific Indications," beginning on page 80, for further guidance in the use of these herbs in various clinical situations.

Arctium lappa
Artemisia absinthium
Chelidonium majus
Cinnamomum verum
Gentiana lutea
Iris versicolor
Juglans nigra
Mahonia aquifolium
Podophyllum peltatum
Rumex crispus
Stillingia spp.
Taraxacum officinale
Zingiber officinale

Tincture for Poor Digestion Following Surgery or Illness

When patients have undergone surgery, been hospitalized, or been bedridden due to an illness, the digestive system can be weakened and may benefit from a "jump start." The same is true for patients who are taking many prescription medications. The use of chi tonics, bitters, alteratives, and stimulants in this formula may quickly improve appetite, bowel function, and digestion.

Curcuma longa 20 ml
Panax ginseng 20 ml
Taraxacum officinale 15 ml
Zingiber officinale 5 ml

Take ½ to 1 teaspoon of the combined tincture on an empty stomach 15 to 20 minutes before meals. Another option is to prepare an aperitif using 1 to 2 teaspoons of the tincture, the juice from a lemon slice, and ¼ cup of water to sip before meals.

Treating SIBO

Small intestinal bacterial overgrowth (SIBO) involves excessive and unbalanced bacteria in the small intestine, causing bloating, pain, gas, and diarrhea. Weight loss may also result as diarrhea and impaired appetite persist. The optimal small intestinal bacteria become replaced with colonic species[141] and possibly shift to support species that are more pathogenic than usual. SIBO can result in impaired absorption of minerals and nutrients, which can cause systemic complications such as osteoporosis and macrocytic anemia. Gut inflammation may interfere with gene expression involved with mucous secretion, linking SIBO to cystic fibrosis, irritable bowel syndrome, and chronic abdominal pain. SIBO is not a condition covered by name in older herb books. However, the symptoms of SIBO, including stomach pain, vomiting of undigested food, feeling hungry but losing one's appetite after only a few bites, and other dyspeptic symptoms, are addressed in folkloric writings.

The diagnostic method is not entirely established but many clinicians utilize glucose and lactulose breath tests, small intestinal aspiration and culture, and an assessment of the average digestive transit time. Some clinicians simply attempt a 2-week course of a broad-spectrum antibiotic such as rifaximin when SIBO is suspected, and if the patient is unresponsive, repetitive cycles of antibiotics are given.[142] Promotility drugs such as motilin agonists, dietary modifications, and other approaches are also warranted. There is growing evidence that acid-suppressing drugs used for reflux disease are associated with SIBO dysbiosis.[143] Proton-pump inhibitors (PPIs), the mainstay of acid reflux treatment, may promote SIBO as well as exacerbate nonsteroidal anti-inflammatory drug-induced small intestinal injury.[144] Gastroparesis and hypothyroidism also predispose to SIBO due to impaired GI motility; gastroparesis is covered later in this chapter.[145] As the dyspeptic symptoms can be very similar to those of dysbiosis of the colon, some patients may be offered probiotic supplements or prebiotics such as inulin and other fructooligosaccharides known to support colonic flora. However, patients with SIBO will often not react favorably because the colonic flora are overgrown in the small bowel, and therefore increasing or nourishing such bacterial overgrowth can worsen symptoms. In fact, many authorities recommend avoiding fermentable oligosaccharides in the diet altogether, and currently suggest the FODMAP diet, in which fermentable oligosaccharides, disaccharides, monosaccharides and polyols (FODMAP) are avoided. This diet is extremely restrictive, eliminating most carbohydrates, including sugars and breads, of course, but even otherwise healthy carbohydrates from legumes to fruits and many vegetables. Because the diet is so restrictive, most clinicians will recommend the FODMAP diet for just several months while taking herbs and possibly antibiotics to help shift the intestinal microbiota. For more specific guidance on a FODMAP diet, see various websites such as www.katescarlata.com/lowfodmapdietchecklists.

SIBO is implicated in the pathogenesis of irritable bowel syndrome,[146] and some of the therapeutic ideas posed in the "Formulas for IBS" section (page 30) may also be useful in treating SIBO. SIBO has also been associated with nonalcoholic steatohepatitis and Parkinson's disease.[147] Because the altered bacterial flora can produce endotoxins that alter gut motility and activate immune responses, motility-enhancing and immune-modulating herbs may be helpful. Because liver disease can impair small bowel motility and contribute to SIBO,[148] liver herbs may be appropriate.

Fiber for Intestinal Health

The necessity of high-fiber intake to promote bowel health and to be one prong in the management of gastrointestinal diseases can not be overemphasized. Guar gum, apple pectin, and psyllium can significantly modify intestinal microbiota and exert prebiotic effects, encouraging population of the gut by beneficial intestinal probiotic species. Fiber supplementation may help treat SIBO, constipation, irritable bowel syndrome, and other complaints.[149] Fiber supplementation may boost the efficacy of antibiotics in treating SIBO.[150] Consumption of fresh fruits and vegetables provides good amounts of fiber and is essential to bowel health and supporting the beneficial aerobic flora.

Therapies may include peppermint and oregano essential oils, both readily available in encapsulations, and berberine, also available in concentrated pill form. These capsules may be taken in tandem with the FODMAP diet mentioned above and the tea and tincture formula detailed below.

Tea for SIBO

All of the following herbs have substantial antimicrobial effects[151] and can be part of the broader protocol for SIBO.

Mentha piperita 4 ounces (120 g)
Ulmus fulva 4 ounces (120 g)
Origanum vulgare 2 ounces (60 g)
Thymus vulgaris 2 ounces (60 g)
Artemisia absinthium 2 ounces (60 g)
Glycyrrhiza glabra 2 ounces (60 g)

This recipe will yield a pound of tea for regular use. Steep 1 tablespoon of the herb mixture per cup of hot water and then strain. Drink 3 or more cups per day for 3 months.

Tincture for SIBO

The motility-enhancing herbs such as *Rheum* may be useful, and motility-enhancing formulas are also offered in "Formulas for Gastroparesis" (below). Gentle antimicrobial agents such as *Curcuma* and *Hydrastis* may be helpful, and when diarrhea is present, astringents such as *Agrimonia* may be included. This tincture would be best combined with oregano and peppermint essential oil capsules, a berberine supplement, and other therapies to create an aggressive protocol.

Curcuma longa
Rheum palmatum
Agrimonia eupatoria
Hydrastis canadensis

Combine equal parts of the herbs and take ½ teaspoon of the combined tincture as often as every 3 or 4 hours, reducing as symptoms improve.

Formulas for Gastroparesis

Gastroparesis is a condition of delayed gastric emptying; it causes pain, gas, and bloating but no actual obstruction and predisposes to dysbiosis and SIBO (see page 75). Those afflicted may feel full after eating just several bites or may vomit undigested food several hours after eating. Gastroparesis may occur in metabolic disorders such as diabetes,[152] particularly in cases of advanced autonomic neuropathy. Histologic study of gastric tissue in severe gastroparesis shows various changes including neuronal, smooth muscle, and interstitial cell abnormalities.[153]

Motility-enhancing agents include ghrelin (the hunger hormone) agonists and gastric electrical stimulation.[154] Other prokinetic drugs being investigated to improve gastric emptying include serotonin 5-HT4 receptor agonists, motilin agonists, dopamine D2 antagonists, muscarinic antagonists, and acetylcholinesterase inhibitors.[155] Caprylic acid is a medium-chain saturated fatty acid (MCFA) with metabolic supportive properties. It is shown to bind ghrelin, the only peptide hormone with an orexigenic effect.[156] Carnitine may promote gastric secretion,[157] and because the amino acid has also been established to support liver function and fat metabolism, the use of a carnitine supplement may support herbal formulas in treating poor digestion.

The motility issues of Parkinson's disease can affect the nerves of the entire gastrointestinal tract and cause gastroparesis, constipation, and SIBO. A traditional Japanese formula called the rikkunshito formula reduces L-dopa–induced impairment of gastric motility.[158]

Tincture for Functional Dyspepsia and Gastroparesis

This formula is aimed at enhancing intestinal motility and is based on current research and traditional herbal folklore. Anethole, a volatile oil in fennel and anise seeds, has been shown to improve dyspeptic symptoms as well as improve gastric emptying.[159] The anti-emetic effects of *Zingiber* may be helpful, but it may be best to avoid mint because it acts as a gastric relaxant. *Iris* is a folkloric secretory stimulant and may enhance digestion and motility.

Commiphora mukul 20 ml
Foeniculum vulgare 20 ml
Iris versicolor 10 ml
Zingiber officinale 10 ml

Take 1 or 2 dropperfuls of the tincture 3 or 4 times a day and complement with one of the tea or decoction formulas for gastroparesis.

Tincture for Slow Motility with Constipation

Based on a commercial formula shown effective for functional dyspepsia and gastric symptoms[160] and shown to stimulate ghrelin activity,[161] this combination may be prepared as a tincture or a tea to treat impaired gastric motility and is formulated here as a complex tincture. *Iberis*, or candytuft, is a Brassica family herb and may not be readily available. *Raphanus sativus* var. *niger* (black Spanish radish) is a possible substitute.

Iberis amara or *Raphanus sativa*, variety *niger* 20 ml
Angelica archangelica 20 ml
Chelidonium majus 20 ml
Matricaria chamomilla 20 ml
Foeniculum vulgare 20 ml
Silybum marianum 10 ml
Melissa officinalis 10 ml
Mentha piperita 10 ml
Glycyrrhiza glabra 5 ml
Zingiber officinale 5 ml

This formula will fill a 4-ounce bottle. Take the combined tincture by the teaspoonful 3 or more times per day.

Rikkunshito for Gastroparesis

Rikkunshito is a traditional Japanese formula used to treat upper gastrointestinal disorders, such as functional dyspepsia, gastroesophageal reflux,[162] and gastric motor function, via enhancing ghrelin production and response.[163]

Glycyrrhiza glabra
Zingiber officinale
Atractylodes lancea
Ziziphus jujuba
Citrus aurantium
Panax ginseng
Pinellia ternata
Poria cocos

Combine equal parts of the herbs. Decoct 1 teaspoon of the herb mixture per cup water, simmering for 10 minutes. Strain, and consume 3 or more cups per day.

ZhiZhu Decoction for Gastroparesis

ZhiZhu decoction is a traditional Chinese formula used for dyspeptic symptoms. This decoction adapts the traditional formula,[164] adding several other herbs traditional for gastroparesis.

Pinellia ternata
Glycyrrhiza glabra
Poria cocos
Panax ginseng
Codonopsis pilosula
Citrus aurantium
Atractylodes lancea
Zingiber officinale

Combine equal parts of dry herbs. Simmer 1 teaspoon of the herb mixture per cup of hot water for 10 minutes. Let stand in a covered pan, and then strain. Drink 3 or more cups per day.

Simple Oolong Tea for Gastroparesis

Chin-shin oolong tea is sometimes called tea ghrelin because it has been shown to bind growth hormone receptors as ghrelin does.[165] Oolong is a semifermented green tea, especially from mountainous regions of Taiwan.

Oolong tea

Steep 1 tablespoon oolong tea per cup of hot water, and then strain. Drink multiple cups throughout the day, especially after each meal.

Decoction from Ping Wei San

This formula is based on a classic Chinese formula for a powder to "Calm the Stomach." It is specifically indicated for digestive symptoms with a full heavy sensation, mucous congestion, and dampness. It may improve appetite and sense of taste as well as relieve GERD, vomiting, and nausea.

Atractylodes lancea 4 ounces (120 g)
Magnolia officinalis 3 ounces (90 g)
Citrus aurantium 2 ounces (60 g)
Glycyrrhiza glabra 2 ounces (60 g)
Ziziphus jujuba 2 ounces (60 g)
Zingiber officinale 1 ounce (30 g)

Decoct ¼ cup of the herb mixture in 8 cups of water, simmering gently uncovered down to 6 cups, and then strain. Drink over the course of a day.

Formulas for Leaky Gut Syndrome

Leaky gut involves a breakdown in the integrity of intestinal barriers resulting from various factors including the following:

- Altered bacterial flora
- Use of steroids, nonsteroidal anti-inflammatory drugs, and other drugs
- Use of alcohol
- Malnutrition
- Immune and inflammatory assaults

Inflammatory processes including an excess in cyclooxygenase, nuclear factor kappa β, and nitric oxide, among other inflammatory mediators, may occur following emotional distress, strenuous exercise, and viral or bacterial infections.[166] Small coils of polysaccharide-based tissue just below the brush border of the intestinal epithelium are believed to be involved in the manufacture of mucous-related secretions and hence are also sometimes called *mucopolysaccharides*. Using herbs that support mucopolysaccharides is one approach to improving intestinal barrier functions and making the cells less "leaky."

Polysaccharides from plants may help repair the intestinal cells as well as stimulate immune responses in the mucopolysaccharide network[167] Peyer's patches in the gut are affected by polysaccharides in plants and may be one immune-modulating action of the so-called immune polysaccharides, a group of large molecular weight starches occurring naturally in *Echinacea*, *Astragalus*, *Ganoderma* (reishi mushroom), and other immune-modulating herbs.[168] Allantoin, a well-known mucopolysaccharide found in *Symphytum* and *Aloe vera* is a sulphur-containing mucopolysaccharide; sulfated carbohydrates are particularly noted to improve connective tissue integrity.[169] Seaweeds are some of the richest sources of sulfated polysaccharides and credited with immune effects due to direct activity of blood cell glycoproteins,[170] such as ulvan, a sulfated polysaccharide from green algae.[171] *Hizikia fusiforme* is a Korean brown seaweed shown to improve intestinal tight junction integrity,[172] a mechanism that may deter metastasis of cancer as well as carcinogenesis in the colon. Bromelain and garlic both contain sulphur, and among many mechanisms of action, these agents may support regeneration of sulfated polysaccharides in the gut. Toxic *E. coli* strains and other pathogenic bacteria damage intestinal tight junctions and allow pathogens to spread through paracellular spaces. Daidzein, bromelain, and allicin have all been shown to inhibit the spread of pathogens.[173] Daidzein is a phytosterol in legumes that is credited with numerous medicinal benefits, and beans and leguminous herbs including *Astragalus*, *Pueraria*, and *Glycyrrhiza* can be included in protocols for leaky gut.

Daily Tea for Leaky Gut

This tasty tea combines herbs known to support the repair of mucous membranes. It is intended for long-term daily use.

Astragalus membranaceus 4 ounces (120 g)
Pueraria montana 2 ounces (60 g)
Hibiscus sabdariffa 2 ounces (60 g)
Calendula officinalis 2 ounces (60 g)
Centella asiatica 2 ounces (60 g)
Symphytum officinale 2 ounces (60 g)
Glycyrrhiza glabra 2 ounces (60 g)

This formula yields 1 pound of tea; it can be amended to suit individual tastes. Gently simmer 1 tablespoon of the herb mixture per cup of hot water for 10 minutes. Remove from the heat, let stand 10 to 15 minutes more, and then strain. Drink 3 or more cups each day.

Promotion of Hyaluronic Acid to Treat Leaky Gut

Hyaluronic acid is a prominent component of the extracellular mucopolysaccharide matrices in bones, skin, intestinal lining, and connective tissues, improving water retention and supporting numerous cell integrity and barrier functions. Human breast milk contains hyaluronic acid and helps to protect infants from gastrointestinal infections through enhancing barrier functions and supporting defense against intestinal microbes.[174] Hyaluronic acid increases the height and surface area of intestinal microvilli and helps proliferate epithelial growth in general. The blockade of hyaluronic acid leads to hypoplasia of the intestinal mucosa.[175] Endogenous polysaccharides create a cytoskeleton network and matrix that hold water and electrolytes and create electromagnetic fields. Agents that promote hyaluronic acid may thereby improve the all-electromagnetic currents moving through the cytoskeleton web. Hyaluronic acid, alginic acid, and pectic acid electrostatically attract bivalent ions such as calcium and other naturally occurring minerals. The presence of such salts in the polysaccharide matrix improves the ability to bind proteins and give tissues strength.[176] Hyaluronic acid can improve digestive and respiratory inflammatory disease and probably is most effective when direct surface contact can be accomplished. Reductions in GERD and chronic respiratory infections may result from the use of hyaluronic acid.[177]

Leguminous plants, including *Astragalus membranaceus*[178] and *Pueraria montana*,[179] may stimulate cellular synthesis of hyaluronic acid.[180] Many flavonoids have been shown to support hyaluronic acid. The *Epimedium pubescens* flavonoid icariin may protect the extracellular mucopolysaccharide matrix.[181] *Vaccinium myrtillus*, another high flavonol plant, may enhance hyaluronic acid and glycosaminoglycan expression,[182] and yet another flavonoid in *Hibiscus sabdariffa* may stimulate the synthesis of sulfated glycosaminoglycans.[183] Curcumin, a primary flavonoid in *Curcuma longa*, may help repair tight junctions.[184]

Daily Smoothie to Treat Leaky Gut

Naturally occurring phospholipids and phosphatidylcholine may enhance the intestinal barrier, helping cells resist permeation by potentially harmful compounds.[185] Excessive breakdown of phospholipids may increase permeability.[186] One of the many beneficial mechanisms of probiotic species is protection of intestinal phosphatidylcholine.[187] Although this formula doesn't contain probiotics, it may help support beneficial flora via the phosphatidylcholine paste.

1 cup (60 g) fresh pineapple, coarsely chopped
1 tablespoon (15 ml) *Aloe* gel
1 tablespoon (15 g) glutamine powder
1 teaspoon (5 g) phosphatidylcholine paste or 1 tablespoon soy lecithin granules
2 cups (480 ml) water, sparkling water, or herbal tea, as desired, chilled

Place the pineapple, *Aloe* gel, glutamine, and phosphatidylcholine or lecithin in the blender and puree to homogenize. Add the water or tea and blend again, or if sparkling water is chosen, blend by hand rather than in the blender. Drink one such smoothie each day for a month or two.

Dietary Supplements for Leaky Gut

These readily available supplements may be used to repair leaky gut and are complementary to the formulas and recipes in this section of this chapter. Although all are helpful, it is not necessary to take everything on this list at once. For severe or stubborn cases, one might select several to take at one time, continuing until the supply is used up. Then rotate through other options over several months' time.

Probiotics
Liver-supportive herbs
Digestive enzymes
Glutamine powder
Whey powder
Hyaluronic acid
Guar gum
Apple pectin
Berberine
Bromelain
Turmeric

Diet for Leaky Gut

Intestinal health is foundational to all of the conditions listed in this chapter, and these general diet guidelines are applicable to many digestive pathologies besides "leaky gut" proper. FODMAP refers to fermentable oligosaccharides, disaccharides, monosaccharides, and polyols, which are a category of carbohydrates (also discussed in "Treating SIBO" on page 75).

Avoid

Bread, gluten, pasta, crackers, pretzels, and all flour products
Alcohol
Sugar and sugary snacks
Fruit juices, jam, and dried fruits
Carbohydrate-rich fruits such as apples, pears, cherries, plums, and watermelon
Carbohydrate-rich vegetables such as potatoes, corn, and peas

Enjoy

Fermented food such as sauerkraut, kimchi, miso, and apple cider vinegar
Seaweeds (in broths, salads, condiments)
Low FODMAP veggies including cabbage, green beans, arugula, spinach, zucchini, squash, turnips, carrots, and bell peppers
Low FODMAP fruits including berries, melon, pineapples, citrus, grapes
Low FODMAP staples including lentils, quinoa, nuts, cheese, quality meats

Seaweed Broth for Leaky Gut

Seaweed offers sulfated polysaccharides with immune-modulating and mucosa-restoring properties. Bones, when simmered gently for a prolonged period (especially with vinegar added to the cooking water), release collagen peptides and other glycosaminoglycan molecules that support the repair and regeneration of mucosal membrane cells and barriers.

Several pounds of soup bones (900–1,000 g) (from as clean and organic a source as possible)
1 cup (20 g) dried *Ganoderma lucidum* slices
½ cup (10 g) seaweed (*Hizikia*, *Fucus*, and so on)
8–10 large dried *Astragalus membranaceus* slices
1–2 cups (240–480 ml) apple cider vinegar

Place all in a slow cooker and fill the pot three-quarters full with water. Simmer gently for 24 hours, adding a few more cups of water if evaporation occurs. Strain the entire contents of the slow cooker to collect the liquid; discard the bones and herbs. Use the broth to prepare soups, or simply return to a saucepan, add garlic, onions, and ginger to season, and consume as desired.

Specific Indications: Herbs for Gastrointestinal Conditions

Many herbs have a particular affinity for the gastrointestinal system, and this section of the chapter offers unique or specific indications for which individual herbs would be best used. Some plants may be niche herbs to treat specific diagnoses, such as constipation or ulcerative lesions. Others are noted to be particularly indicated for specific presentations, be it a unique sensation or a noteworthy underlying situation, such as stress or allergies that contribute to the digestive issue. There is some overlap between the herbs for gastrointestinal conditions and herbs for liver and gallbladder conditions. Thus I encourage you to also refer to the corresponding section of chapter 3 (the liver and gallbladder chapter) as you consider the best herbs to choose for specific presentations and situations.

Achillea millefolium • Yarrow

The leaves and flowers are valuable as a bitter alterative agent, a diaphoretic and circulation-enhancing agent, and a broad-acting antimicrobial agent. *Achillea* may be included in formulas for intestinal inflammation, IBS, liver congestion, or skin lesions associated with digestive symptoms. Due to fairly reliable hemostatic effects, *Achillea* is specific for bleeding hemorrhoids, blood in the stool, and passive hemorrhage associated with atony of the mucosal tissues.

Aesculus hippocastanum • Horse Chestnut

The processed nuts of *Aesculus* may improve poor digestion when associated with venous stasis and portal congestion. Other specific symptoms that indicate the need for *Aesculus* are a full sensation in the liver with tenderness in the right upper quadrant and a sense of weight in the stomach with gnawing and aching pain. *Aesculus* is also a traditional remedy for hemorrhoids with sticking, sharp, or shooting pain and for swelling of the rectal mucous membranes with pain and soreness in the anus. Due to its specificity for portal and venous congestion, *Aesculus* is also a leading remedy for varicose veins.

Allium sativum • Garlic

Allium bulb preparations may assist in correcting intestinal dysbiosis. It can act as an antimicrobial agent in the treatment of infectious gastroenteritis and amoebic dysentery and is a safe preventive agent when traveling. *Allium* is a warming, stimulating herb best for cold damp constitutions, catarrhal states, constipation, or slow peristalsis.

Aloe vera • Aloe

The rind of the *Aloe* fruit has laxative effects and promotes bile flow and peristalsis. Therefore, dried *Aloe* rind powder is indicated for chronic constipation. The gel and juice may be used to soothe intestinal pain, heal ulcers, and reduce inflammation of digestive mucous membranes. Because *Aloe* juice contains immune polysaccharides, consider *Aloe* also for bowel cancers and dysplastic changes.

Angelica sinensis • Dong Quai

Angelica's area of action is mainly on blood cells and cytokines giving it "blood-moving" properties, antiallergy effects, and an ability to enhance perfusion to various organs. *Angelica* root medications may be included in gastrointestinal formulas when vascular congestion, pelvic stagnation, menstrual cramps, or allergies are present and contributory.

Arctium lappa • Burdock

Arctium roots are an important alterative herb and cholagogue that can be included in formulas for intestinal dysbiosis, malabsorption, and dyspepsia. Systemic symptoms that specifically indicate its use include hyperlipidemia, acne and skin disorders, hyperestrogenism, and general malaise related to toxicity.

Artemisia absinthium • Wormwood

Wormwood leaf preparations are a folkloric standard herbal theray for treating intestinal worms, hence the name. The leaves are extremely bitter and will also stimulate peristalsis and intestinal secretions in cases of hypochlorhydria and digestive insufficiency. *Artemisia absinthium* is to be used with caution, in small doses of no more than 5 milliliters per ounce of formula, due to the high concentration of thujone, which is known to have central nervous system toxicity and to act as neuro-excitant capable of inducing seizures. Alcohol extracts such as absinthe prepared from the plant have been outlawed in many countries due to toxicity concerns. The plant has gone by numerous other species names over time including *A. majus*, *vulgare*, and *officinale*. Those with seizural disorders should avoid.

Artemisia annua • Sweet Annie

Artemisia annua leaves are specific for parasites, malabsorption, and amoebic dysentery. *Artemisia* may be used as a preventive agent when traveling abroad. The constituent artemisinin may deter malarial and other infections, and this species of *Artemisia* is especially useful for protozoan and amoebic infectious issues.

Artemisia vulgaris • Mugwort

These bitter leaves are specific for insufficient digestive secretions and for parasites, and have long been used as a culinary spice in fatty meat dishes to improve digestion. Because mugwort, also called common wormwood, contains potentially toxic volatile oils including thujone, only small, short-term doses should be used, and the essential oil should never be consumed orally. Avoid in those with seizural disorders.

Asclepius tuberosa • Pleurisy Root

Ascelpius is most noted for its effects on the lungs and is used to treat painful coughing with respiratory congestion. The roots are also useful to include in formulas for treating flatulence and dyspepsia, malodorous stools, and mucousy diarrhea.

Atropa belladonna • Belladonna

Belladonna is a potentially toxic herb used in small doses and is available only to licensed clinicians. Leaf preparations can be valuable to include in formulas for spastic colon and mucous colitis because it will quickly reduce intestinal secretions when excessive, and it will slow peristalsis by relaxing smooth muscles. Belladonna leaves may be processed in oil and may be used topically over the abdomen to help remedy intestinal spasms or prepared into ointments to relieve pain when applied topically to hemorrhoids and rectal fissures. Omit it from the formula if you are not trained in the use of this druglike herb, and if you do include it, remove it from the formula once the acute problem is resolved. Belladonna can cause dry mouth, facial flushing, and visual disturbances, even hallucinations, so do not use more than 4 to 6 milliliters per ounce of tincture formula.

Avena sativa • Oats

Although it is not specific for any particular gastrointestinal complaint, *Avena* may be included in GI formulas as a supportive and nutritive base. *Avena* groats in teas can be restorative for patients suffering from nervous exhaustion, debility, chronic disease, or digestive derangements associated with alcoholism or drug abuse. Oat straw tea is also an excellent source of minerals for those with malnutrition or digestive disorders associated with stress and anxiety.

Azadirachta indica • Neem

Neem is entering Western herbalism from Ayruvedic traditions, and a growing amount of scientific research shows the medicine to be useful for lice and parasites when used topically and for intestinal infections when consumed orally. Consider neem leaf products and neem oil preparations for difficult intestinal ulcerations and infections such as *Helicobacter pylori* or *Clostridium difficile.*

Berberis aquifolium • see *Mahonia aquifolium*

Capsella bursa-pastoris • Shepherd's Purse

The aerial parts of *Capsella* contain a classic hemostatic agent with astringent effects on the vasculature. *Capsella* can be included in formulas for intestinal bleeding and hemorrhoids.

Capsicum annuum • Cayenne Pepper

Capsicum can be used as a synergist in cold constitutions, in the elderly with exhaustion, and in the case of atonic dyspepsia with much flatulence. It brings more blood into the stomach and intestines, increases digestive secretions, and promotes peristalsis. It is best used in tinctures in small amounts combined with alteratives, cholagogues, and carminatives or demulcents to prevent any possible irritating effects. This species is also known as *C. frutescens.*

Carum carvi • Caraway

Caraway seeds, like most umbels, are pleasant-tasting carminative agents used as teas for cramps and distension as well as a corrigent with digestive stimulants and laxatives. Caraway seeds can be included in teas for IBS or other types of intestinal cramping, while caraway's essential oil can be applied topically over the abdomen to help treat intestinal cramps.

Ceanothus americanus • Red Root

Ceanothus is specific for liver congestion, pelvic and portal congestion, splenomegaly, vascular congestion, and hypertension. It is traditionally asserted to have an affinity for the lymphatic system, alleviating vascular congestion via enhancing entry of interstitial fluid into the vasculature and enhancing venous return. There has been little to no modern scientific research on this plant.

Centella asiatica • Gotu Kola

A highly nutritious and vulnerary plant, *Centella* can be included in teas to support convalescence and postsurgical recovery as well as to support patients recovering from malnutrition. *Centella* promotes hyaluronic acid. The entire plant can be dried to consume in teas, dried powders can be used in capsules, smoothies, and other oral preparations, or may be prepared into a tincture. It may assist in healing after surgical procedures, reduce scarring, and limit fibrosis in chronic inflammatory issues in the GI ranging from chronic pancreatitis to hepatitis to autoimmune diseases of the bowels.

Chelidonium majus • Celandine

Both the aerial parts and roots of this bitter plant are valuable cholagogues and liver herbs specific for pain or fullness in the right upper quadrant, pain that radiates to the right shoulder, jaundice, biliary disease, and gallstones. *Chelidonium* may be therapeutic to nausea and stomach pain that is related to poor digestion and biliary insufficiency. The plant is specifically indicated when there is a large flabby tongue with indentations of teeth on lateral margins, a thick or yellowish coat on the tongue, and constipation with dry hard stools. *Chelidonium* is also specific when there are abnormal stools, such as bright yellow stools, clay-colored stools, or light-colored stools that float, all indicative of biliary insufficiency. *Chelidonium* is also specific for skin disorders associated with digestive and liver disorders. Early American physicians regarded *Chelidonium* as one of the best remedies for biliary and hepatic congestion, though it is best avoided in acute inflammations of the liver.

Chelone glabra • Turtlehead

The bitter leaves have alterative and cholagogue properties used for liver congestion, dyspepsia, and recovery from infectious illness where the appetite and digestion have been affected. *Chelone* is specific for GI debility accompanied by jaundice and for dyspepsia following febrile diseases and exhaustive illnesses.

Chenopodium ambrosioides • Wormseed

A classic remedy for intestinal worms, this extremely bitter herb was given on sugar cubes, in syrup, or in castor oil several times a day for 5 days to a week. There are no such products on the market, but dried leaves may be used in teas or prepared in such products for oneself. The powdered seeds have also been used as a traditional vermifuge. The volatile oils in the plant have antiparasitic effects and can also stimulate digestion and motility. Goosefoot is an alternate common name for this herb.

Chionanthus virginicus • Fringe Tree

Chionanthus root bark is a classic liver herb specifically indicated for liver pain and fullness and for frothy or clay-colored stools. *Chionanthus* was highly regarded by the Eclectic physicians for portal congestion and hepatic enlargement. The plant is also recommended for infantile jaundice, for glucosuria due to faulty glycogen production in the liver and prediabetic states, and for long-standing digestive complaints including chronic gastritis and IBS, when poor digestion and liver function is contributory.

Cinchona officinalis, C. pubescens • Peruvian Bark

The source of quinine, the bark of this plant is a folkloric staple for poor digestion with poor appetite and is traditional for the intermittent "periodic fevers" typical of malaria. It is also called quinine bark and cinchona. As a digestive aid, *Cinchona* may act like other bitters and can be taken in water prior to meals to promote appetite and support digestive secretions. Tonic waters containing quinine can be the base of preprandial digestive aids or the base liquid for convalescent preparations. *Cinchona* tincture or tea is used in small doses only, because larger amounts can irritate the stomach rather than improve digestion. *Cinchona* and quinine may also be helpful for night sweats and debilitating hot flashes. The plant is astringent, drying, and cooling, so it may be used as a supportive ingredient in formulas for febrile illnesses.

Cinnamomum zeylanicum, C. verum • Cinnamon

Cinnamon bark is a palatable ingredient in teas and is useful when astringent and hemostatic effects are desired. Cinnamon is also carminative and is warming and stimulating to circulation in the intestines. Teas and tinctures are most specific for those with cold constitutions, for those with excessive intestinal secretions, or for those with blood in the stool. Use cinnamon in small amounts only or it can have an opposite effect, stimulating secretions and worsening diarrhea or loose stool. When cinnamon is cold infused, it helps to create a mucilaginous brew to boost the demulcent effect.

Collinsonia canadensis • Stoneroot

This little-used herb is specific for rectal tightness, hemorrhoids, and a congested feeling in the perineum. The root and whole plant are purported to improve vascular congestion in the pelvis. *Collinsonia* is a traditional remedy for all manner of rectal complaints including fissures, proctitis, straining with bowel movements, and fistulas.

Commiphora mukul • Guggul

Resin from guggul trees is traditional for diabetes, obesity, and high cholesterol, and the plant is a relative of the myrrh gum tree, *Commiphora myrrha*. Guggul tincture or other commercial products may be included in digestive formulas when there is a sense of weight or dragging in the pelvis. Guggul is a mucous membrane stimulant and antiseptic and is also specific for excessive discharges.

Coptis trifolia • Goldthread

The roots of *Coptis* contain berberine-like alkaloids, so like *Hydrastis* and *Mahonia*, *Coptis* is valuable for intestinal infections and for excessive intestinal secretions. *Coptis* is a broad-acting antimicrobial agent used internally for intestinal dysbiosis and sluggish digestion and to reduce chronic infections in diabetics.

Curcuma longa • Turmeric

Curcuma is antioxidant and anti-inflammatory and provides hepatic support. It is a general alterative and antimicrobial that can assist in hormone clearance via the liver. Use it in formulas to improve lipid and carbohydrate metabolism in cases of diabetes and hyperlipidemia. It also helps to improve intestinal dysbiosis and decrease propensity to fungal and other infections.

Cynara scolymus • Artichoke

Cynara can be included in digestive formulas when liver congestion and/or elevated cholesterol and hormonal imbalance accompany. See also the *Cynara* entry on page 104.

Dioscorea villosa • Wild Yam

The tuberous roots of wild yam can relieve colicky pains in abdominal organs including those caused by menstrual cramps, poor digestion, and flatulence. It also helps with other types of pain: liver and gallbladder pains that radiate to the shoulder or right nipple, twisting and boring pains about the umbilicus, and digestive pain that radiates through the abdomen to the spine. *Dioscorea* can also be useful in formulas for sudden urgent stools and for hemorrhoids with pains that radiate upward.

Echinacea spp. • Coneflower

Echinacea roots are specific for septic conditions with emaciation and debility. *Echinacea* may be included in digestive formulas with liver herbs in cases of hepatitis, cirrhosis, and appendicitis. *Echinacea* may be useful combined with astringents in cases of infectious diarrhea, food poisoning, and traveler's diarrhea.

Elettaria cardamomum • Cardamon

The seedpods are a pleasant-flavored carminative appropriate for digestive gas and bloating, distensive pain, belching with peptic ulcers, and flatulence with IBS.

Eleutherococcus senticosus • Siberian Ginseng

Eleutherococcus root preparations can be included as a synergist in formulas for chronic digestive irritability associated with nervous affectations and emotional instability. *Eleutherococcus* is a valuable adrenal tonic useful in treating long-term stress and GI issues accompanied by fatigue and general malaise. *Eleutherococcus* is also specific for endocrine imbalances and immune insufficiency, especially when related to overwork and nervous debility.

Equisetum arvense, E. hymenale • Horsetail

Equisetum is specific for intestinal ulcerations because the plant is traditionally emphasized as a nutritional tonic to help rebuild connective tissues and mucous membranes. In teas *Equisetum* can provide minerals and is indicated for GI complaints associated with malnutrition and malabsorption. A long maceration and/or a long gentle simmer of the dried herbs would best extract the minerals. To enhance this process, add weak acids in the form of lemon juice or apple cider vinegar while macerating teas.

Eschscholzia californica • California Poppy

The latex rich leaves and young seedpods are traditional remedies for pain, anxiety, muscular tension, and poor sleep. *Eschscholzia* may be a complementary component in formulas for digestive disorders that are associated with stress and anxiety.

Filipendula ulmaria • Meadowsweet

Filipendula leaves and young buds are indicated for inflammation of the intestines and mucous membranes and for loose or watery stools. *Filipendula* contains salicylates and tannins credited with anti-inflammatory and astringent effects.

Foeniculum vulgare • Fennel

Foeniculum seeds are helpful for relieving gas and bloating, peptic distension causing fullness and discomfort, burping, and cramping and gurgling in the intestines. Women who are nursing can drink fennel tea liberally to help ease their babies' colic. Whole seeds, seed powder, tincture, and essential oil are all readily available. Fennel essential oil may be useful topically, rubbed over the abdomen or right upper quadrant to help reduce biliary and intestinal spasms.

Fucus vesiculosus • Bladderwrack

Fucus may be used as a mineral tonic in cases of malnutrition. The crude powder of the dried seaweed may be added to food by the teaspoonful where the salty, fishy flavor is appropriate, or it may be taken in capsules.

Gentiana lutea • Gentian

The roots of this extremely bitter herb are best for atonic situations in the digestive tract and may be given before meals to stimulate the appetite in cases of anorexia. *Gentiana* can help recover enfeebled digestion following prolonged illnesses. Folkloric sources purport that *Gentiana* is specifically indicated when fatigue and mental lethargy accompany the physical symptoms.

Glycyrrhiza glabra • Licorice

Glycyrrhiza roots are excellent as demulcents and are anti-inflammatory and antiviral with wide utility in gastritis, digestive ulcers, liver disease, and hepatitis. Due to immune-modulating effects, *Glycyrrhiza* is also very effective for inflammation in digestive mucosa membranes due to allergic reactions. Licorice is available in a syrupy molasses-like preparation known as a solid extract. Licorice solid extract can help coat the oral and esophageal mucosa for pain, irritation, or ulceration of these tissues. Licorice teas are excellent in cases of gastric and intestinal ulceration or inflammation, and the powder and solid extract can be used as palatable "glue" in the making of hand-rolled herbal pills or balls. *Glycyrrhiza* may also be included in formulas when digestive and inflammatory symptoms are related to stress, due to the plant's restorative effects on the adrenal gland and hypothalamus-pituitary-adrenal (HPA) axis imbalances.

Hamamelis virginiana • Witch Hazel

Hamamelis is specific for pelvic congestion associated with poor circulation and profuse discharges and may be used as a styptic for intestinal or gastric bleeding. *Hamamelis* is also specific for hemorrhoids and venous engorgement, easy profuse bleeding of hemorrhoids, and soreness of the anus.

Humulus lupulus • Hops

Use *Humulus* as a digestive bitter and nervine in cases of stress, anxiety, and insomnia. It is useful as a complementary ingredient in adrenal and stress formulas in those with GERD symptoms. *Humulus* is also specific for digestive ailments in alcoholics when there is wakefulness, excitability, giddiness, and muscular twitching.

Hydrastis canadensis • Goldenseal

Hydrastis is an antimicrobial and drying agent useful in cases of gastritis, digestive ulcers, and bowel cancer. The plant is believed to tone and tighten damp, boggy, and atonic digestive tissues and is recommended for intestinal infections, dysbiosis, traveler's diarrhea, and atonic dyspepsia. It is also indicated for catarrhal states of the stomach and intestines where there are mucous discharges and mucous threads in diarrhea. *Hydrastis* is also appropriate for rectal prolapse and anal fissures. Specific indications for *Hydrastis* include sticking pain in the rectum with stool that persists for a long time, a bitter taste in the mouth, pulsations in the stomach, and an "all-gone" feeling.

Hypericum perforatum • St. Johnswort

Use *Hypericum* in gastrointestinal disorders when a connective tissue tonic is needed to improve vascular fragility and chronic hemorrhoids. *Hypericum* may also be used as a synergist in formulas for GI pain, vascular engorgement, and inflammation associated with anxiety and nervous disorders.

Iris versicolor • Iris

Iris roots are potentially caustic, but small amounts are used as an overall glandular stimulant, specific for promoting liver and digestive secretions as well as for stimulating the spleen and thyroid and moving congested fluids. *Iris* increases the secretions of salivary and digestive glands and is useful as a complementary herb in cases of digestive insufficiency related to hypothyroidism and biliary insufficiency where fat intolerance and fatty or oily stools (steatorrhea) are present. As *Iris* is a warming stimulating herb, small doses of only a few drops combined with other herbs are sufficient to gently stimulate the glands.

Juglans nigra • Black Walnut

Juglans is one of the most widely used folkloric herbs for intestinal parasites. *Juglans* is also indicated for chronic acne and skin eruptions associated with digestive upset. Medicines are prepared from the resinous inner rind of various walnut species.

Leptandra virginica • Culver's Root

The roots are used as a traditional medicine, and the plant is named after Dr. Culver, an early American pioneer physician who used the fresh root as a purgative. Such use is not recommended, as a high dose is required for a strong laxative effect and can be dangerous, as it may induce violent vomiting and bloody diarrhea. *Leptandra* is safer when the dried roots are used to make medicines and may help treat constipation associated with liver and gallbladder disease. *Leptandra* is specific for right upper quadrant pain especially when associated with drowsiness, dizziness, mental depression, lassitude, headache, nausea, diarrhea, and thickly coated tongue. *Note:* Some taxonomical systems have altered the Latin name of Culver's root to *Veronicastrum virginicum*.

Mahonia aquifolium • Oregon Grape

The inner bark of *Mahonia* roots are valuable as an alterative and liver- and digestion-supportive herb with wide application in teas, tinctures, and encapsulations. *Mahonia* is also a broad-acting antimicrobial appropriate for everything from infectious hepatitis to dysbiosis to traveler's diarrhea and food poisoning. *Mahonia* is specific for liver congestion with tenderness and slow digestion, coated tongue, and skin eruptions due to poor liver and digestive health. This plant and many others in the genus are also known as genus *Berberis*.

Matricaria recutita, M. chamomilla • Chamomile

Chamomile flowers are very helpful for digestive symptoms due to emotional upset and dyspepsia with gas, bloating, stomach pain and pressure, nausea, and burping. *Matricaria* is an excellent base herb in formulas for irritable bowel syndrome, diarrhea, intestinal ulcerations, colitis, intestinal cramping, flatulent colic, GERD, and burping with bitter or foul taste made worse by drinking coffee. *Matricaria* is specifically indicated for cranky babies and toddlers, for children with diarrhea, especially those with green watery stools, and for bouts of inappropriate sweating associated with digestive symptoms. A small spoonful may be given to infants and ¼ cup doses given to toddlers.

Medicago sativa • Alfalfa

Medicago leaves may be added to teas and taken as capsules as a mineral and nutritional support in cases of malnutrition and malabsorption and to support recovery after surgeries and debilitating illnesses.

Mentha piperita, M. spicata • Mint

The leaves of peppermint (*M. piperita*) and to a lesser degree spearmint (*M. spicata*) are among the best herbs for queasy stomachs and can be very valuable in formulas for nausea and bloating, colic in infants, and digestive upset with a large amount of painful gas, burping, rumbling, and flatulence. It can be included in tinctures and teas and used topically as an essential oil for colic and distensive or spastic pain in the stomach and intestines.

Myrica cerifera • Wax Myrtle

Myrica leaves are specific for liver disease and hepatic congestion, for biliary insufficiency with nausea, and for a bitter taste in the mouth and halitosis. *Myrica* is also indicated for loss of appetite and long-lasting stomach discomfort after eating; for digestive symptoms that are improved with consuming acids such as lemon juice or vinegar; or for a craving for acids such as pickles, sauerkraut, or vinegar-based dressings. Liver inflammation with jaundice, right upper quadrant pain, a constant sense of fullness, and clay-colored stool may respond to *Myrica*. Historic literature reports *Myrica* to be specific for a frequent urge for stool but only passing much gas.

Nepeta cataria • Catnip

Nepeta leaves are a carminative tonic for colic and intestinal cramps, children's digestive upset, and flatulence. A leading historical use of *Nepeta* has been for cranky babies and toddlers who appear to be suffering intestinal distress.

Origanum vulgare • Oregano

Oregano leaves are useful for flatulence and digestive upset and effective against intestinal parasites and dysbiosis. *Origanum* is also effective for childhood infections with digestive symptoms. Purified oregano essential oils have broad antimicrobial activity and encapsulated products can be part of therapies for *Helicobacter pylori*.

Paeonia lactiflora, P. suffruticosa • Peony

Red peony is the whole root of *P. lactiflora*, white peony is the peeled and processed root, and both are used medicinally. *P. suffruticosa* is called tree peony. Peony is considered to be a blood tonic in TCM, meaning that it supports body fluids, builds the blood, and may promote circulation in the intestine. Peony is traditional for pelvic pain, especially when due to blood stasis. It is also specifically indicated for blood deficiency that includes the symptoms of anemia such as pallor, weakness, and dizziness, and for some liver disorders associated with fever, restlessness, excessive sweating, and pain.

Panax ginseng • Ginseng

Panax may be used as a supportive herb in formulas where stress or long-standing illness is present or when exhaustion or debility accompany a digestive complaint. *Panax* is often a lead herb and chi tonic in TCM formulas.

Picrasma excelsa • Quassia

Quassia bark is a bitter stomach tonic said to combine well with vinegar or lemon juice. *Picrasma* is often seen in old formulas for digestive complaints of chronic alcoholics.

Pimpinella anisum • Anise

The seeds are an excellent and palatable carminative agent to help alleviate flatulence, cramps, and nausea. The tea is appropriate for infants with colic and may also be given to a nursing mother to be passed in the breast milk. *Pimpinella* can also be used as a corrigent to balance any potentially harsh effects of laxatives, cholagogues, and stimulants. In addition to seeds, *Pimpinella* is widely available as a tincture and essential oil; the latter can be used topically over the abdomen for cramps, distension, and nausea.

Punica granatum • Pomegranate

Pomegranate bark and fruit rinds are a traditional remedy for intestinal parasites and most often used as a tea. The fruit and rind are high in polyphenols and have been shown to have activity against drug-resistant bacteria.

Quercus alba, Q. rubra • Oak

Q. alba is white oak; *Q. rubra* is red oak. *Quercus* bark is an excellent digestive astringent recommended for swollen atonic digestive passages with excessive mucous discharges and when portal congestion contributes to digestive discomfort. *Quercus* is high in tannins, giving it significant astringency, and it combines well with mint or cinnamon for diarrhea. It has been traditionally recommended to improve gastric tone where there is fluid stasis due to alcoholism or liver disease.

Raphanus sativus var. *niger* • Black Spanish Radish

Raphanus tuber preparations can be included in GI formulas where biliary insufficiency contributes to chronic dyspepsia and constipation. *Raphanus* can help relax acute biliary spasm.

Rhamnus purshiana • Cascara

Bark from cascara trees contains the irritant laxative agent emodin, and teas or tinctures are most used for loss of peristalsis, constipation, and atrophy of intestinal muscles. A carminative agent is required in formulas that include *Rhamnus;* otherwise patients may experience intestinal cramps and explosive bowel movements.

Rheum palmatum • Turkey Rhubarb

The roots of this garden rhubarb relative are indicated for insufficient digestive secretions and slow or impaired gastric and digestive motility. Also called Chinese rhubarb, *Rheum* is specific for colicky digestive pains centered about the umbilicus, for bowel movements that are passed with much straining, for a feeling of overfullness even after eating only a small amount, and for a sour smell to body. *Rheum* was an important ingredient in formulas for dyspepsia, often referred to as a "neutralizing cordial" and combined with cinnamon, mint, and potassium bicarbonate. **Caution:** The leaves of *Rheum rhabarbarum,* or common garden rhubarb, are not to be consumed due to the high oxalic acid content, which can be toxic and even fatal when ingested in high amounts.

Ricinus communis • Castor

Castor oil is produced from ripe castor beans and is most often used topically over inflamed and congested organs, but it may be taken internally as a laxative in small doses. It is said to improve the efficacy of vermifuges when used in tandem and to be a valuable remedy for gastroenteritis, especially in children. A single dose of a teaspoon in children or a tablespoon for adults may improve chronic colicky bowel movements with gray, sticky, or other poor-quality stool. Combining castor oil with peppermint essential oil and licorice tea or solid extract can improve the flavor and reduce the thick, sticky quality. Although castor oil is produced from castor beans, the beans are *not* to be consumed in any form due to the presence of the highly toxic ricin. Because ricin is not fat soluable, castor oil contains very little and can be consumed orally in very small amounts. Castor oil is most commonly used topically in herbal medicine.

Rubus idaeus • Red Raspberry

Raspberry leaves may be used as a simple nutritive agent in teas for malnutrition. The flavor is pleasant and provides mild astringent effects on the intestinal mucosa.

Rumex acetosella • Sheep Sorrel

The diminutive sheep sorrel is used as an alterative for weak appetite and digestion and to allay nausea and vomiting. Several traditional detoxification and anticancer formulas have included sheep sorrel as an ingredient.

Rumex crispus • Yellow Dock

Rumex roots are a bitter alterative and can be added to formulas for hypochlorhydria, malabsorption, constipation, and digestive insufficiency. *Rumex* is specific for skin eruptions secondary to digestive insufficiency, biliary insufficiency, and poor elimination with toxicity.

Salix alba • White Willow

The bark from young willow branches is used as an intestinal astringent and anti-inflammatory; it's also useful in formulas for infectious diarrhea. Willow is a traditional febrifuge owing to its high salicylate content. Older texts mention willow bark teas for persistent hiccups.

Sanguinaria canadensis • Bloodroot

Sanguinaria roots are used as a synergist in formulas when digestive symptoms are associated with burning sensations and for hot flashes and vasomotor symptoms.

Only small amounts are used internally, because the high resin content can make herbal preparations potentially irritating and even caustic.

Scrophularia nodosa • Figwort

Scrophularia is used as a synergist in formulas for digestive complaints that have a full congested character and are associated with enlarged glands and lymphatic congestion as well as in formulas for painful hemorrhoids. The succulent aerial parts are used to prepare tinctures.

Silybum marianum • Milk Thistle

The seeds are very helpful for all matters of hepatic and biliary disorders and associated jaundice and digestive difficulty. *Silybum* is specifically indicated for infectious or auto-inflammatory ailments of the liver, gallbladder, pancreas, spleen, and kidneys. *Silybum* can also improve altered blood composition due to faulty digestion such as hyperlipidemia and hyperglycemia.

Smilax ornata • Sarsaparilla

The roots are used as an aromatic alterative agent with additional hormonal and adrenal supportive and slightly warming properties. *Smilax* is specifically indicated for poor digestion and also for fatigue and muscle weakness.

Stillingia sylvatica • Queen's Root

The roots of many *Stillingia* species are high in warming resins that may be used in formulas to treat dental and oral pain; these resins also have stimulating effects on digestion. Medicinal preparations of queen's root are specific for chronic constipation associated with atrophic and congested issue or secondary to liver disease. Small doses can have well-tolerated laxative effects; however, high doses can induce nausea, vomiting, and diarrhea. Older texts report that *Stillingia* is effective for pale or anemic children with no appetite and may reduce chronic ear infections by improving tissue congestion.

Syzygium aromaticum • Cloves

The dried aromatic flower buds act as digestive stimulants, promoting digestive secretions, stimulating the appetite, and strengthening peristalsis. Unlike irritant laxatives, *Syzygium* is carminative and antimicrobial, can relieve nausea and vomiting in cases of infections, and relieve flatulence, cramping, and distension. *Syzygium aromaticum* was formerly known as *Eugenia aromatica*.

Tabebuia impetiginosa • Pau d'Arco

Tabebuia has broad antimicrobial properties and can be included in teas and tinctures for intestinal infections.

Taraxacum officinale • Dandelion

Use the root for gastric headaches, biliary insufficiency, jaundice, coated tongue, pain in the right upper quadrant of the abdomen, hepatic torpor and constipation, and night sweats due to liver disease. The leaves are nutritious and can be eaten as spring greens or dried to include in teas for malnutrition.

Ulmus rubra, U. fulva • Slippery Elm

The inner bark of several *Ulmus* tree species yield a very valuable intestinal demulcent, useful in teas, capsules, and tinctures. Cold infusions are best used to extract the most mucilage and to yield the greatest pain-relieving effects on the digestive mucous membranes. The light fluffy powder can be prepared into a porridge historically referred to as "slippery elm gruel," which can provide welcome relief for burning pain, reflux, and ulcerative pain.

Viburnum prunifolium • Blackhaw

Bark from the shrub can be used to make teas or tinctures having an antispasmodic effect on the intestines, bile ducts, and uterus. *Viburnum* may be included in digestive formulas when there are cramps and spasms or in formulas for biliary colic.

Zanthoxylum clava-herculis • Prickly Ash

The bark is a warming, stimulating remedy that brings heat and blood to the stomach, increasing function. *Zanthoxylum* increases circulation and secretions in cases of digestive debility and insufficiency and is best for those with cold constitutions, weakness, lethargy, and poor circulation. *Zanthoxylum* has a carminative and antispasmodic action and is a mild appetite stimulant in cases of dyspepsia.

Zingiber officinale • Ginger

Zingiber is a warming, stimulating carminative in cases of dyspepsia and flatulent colic. *Zingiber* is also useful as an anti-inflammatory in cases of alcohol- or irritant-induced gastritis and for diarrhea due to atony of the bowels. *Zingiber* also has broad activity against numerous types of microbes and is well tolerated in teas, tinctures, and encapsulations.

— CHAPTER THREE —

Creating Herbal Formulas for Liver and Gallbladder Conditions

There is a saying in natural medicine, "If in doubt, treat the liver." The liver and digestion can be at the root of so many health issues, from skin disease to digestive difficulties, high cholesterol, diabetes, and hormonal imbalances, and other disorders. Because the liver processes hormones, drugs, nutrients, toxins, carbohydrates, and fats, supporting liver function can benefit a broad array of systemic conditions.

Liver disorders are a major health concern worldwide. Environmental chemicals, prescription drugs, over-the-counter drugs, hormones (such as birth control pills), chemicals that taint the food supply, and alcohol abuse put great burden on the liver. Reducing exposure to such hepatotoxins may offer some protection against liver disease. Nitrosamine, acetaminophen, doxorubicin, carbon tetrachloride, hexane, benzene, and aflatoxin are among the well-known chemical offenders that affect the liver. In fact, carbon tetrachloride is one of the most widely used hepatotoxic research tools. It is used to induce oxidative stress, inflammation, and fibrosis in the liver in animal or tissue culture research in order to test for herbs or compounds that may protect against such liver damage.

Short of being exposed to these severe hepatotoxins, simply being exposed to artificial ingredients and synthetic chemicals in food, too much fatty food, and environmental toxins contributes to liver burden and congestion. In some cases, bile flow is impaired, contributing to liver inflammation and gallbladder disease. When the liver has difficulty processing fats and cholesterol, blood lipids may elevate, and abnormally thick and poor quality bile may be passed to the gallbladder. Formation of gallstones, acute gallbladder pain, and gallbladder inflammation can result.

Chronic inflammation in the liver can lead to fibrotic/cirrhotic changes and loss of liver function as healthy hepatic stellate cells undergo transformation into fibroblast-like cells. If and when the irritating influences are entirely removed, the transformed stellate cells will frequently revert back to a normal state as the entire organ recovers, as long as the fibrotic changes have not progressed too far. Inflamed liver cells are resistant to many chemotherapy drugs because the high levels of reactive oxygen species interfere with cellular defenses and induce proteins that allow for cancer progression. Infectious agents such as hepatitis viruses will also induce inflammatory damage to the liver, especially when chronic, as can noninfectious causes of hepatitis. Thus, herbal formulas for liver disease might include antiviral herbs, immune modulators, and antifibrotic herbs as basic components.

An Overview of Hepatoprotective Herbs

Curcuma longa (turmeric), *Tinospora cordifolia*, *Eclipta alba*, *Silybum marianum*, *Rubia tinctorum*, *Andrographis paniculata*, and other herbs are being explored for their effects on liver diseases and have been shown to protect the liver from hepatotoxic substances.[1] Many of the hepatoprotective and regenerative effects of these herbs appear to take place via antioxidant and anti-inflammatory actions.[2] Alterative herbs can be complementary trophorestoratives for liver and biliary complaints; they can be included as nutritive tonics and base herbs in formulas and include *Taraxacum officinale*, *Arctium lappa*, and *Mahonia aquifolium* (also known as *Berberis aquifolium*). *Mentha piperita*, *Matricaria chamomilla*, and *Dioscorea villosa* have antispasmodic effects on the biliary system and may offer pain relief in acute gallbladder colic formulas. *Curcuma* has broad hepatoprotective effects and can make a gentle base herb in a variety of liver formulas.[3]

There are numerous published studies on *Panax* saponins protecting the liver against chronic alcohol consumption.[4] *Codonopsis lanceolata* has been traditionally used in Asia to treat hepatitis.[5] *Stellaria media* has shown hepatoprotective activity.[6] *Cyperus rotundus* roots are another traditional Chinese medicine for hepatitis.[7] *Scutellaria baicalensis* (huang qin) roots contain baicalein, which is shown to protect against cell death in cases of cirrhotic endotoxemia and to offer a protective effect on the intestines in cases of hepatitis.[8] *Phyllanthus niruri*, a folkloric plant for hepatitis, contains lignans shown to have activity against hepatitis B and to have hepatoprotective effects.[9] *Artemisia capillaris* (capillary wormwood) inhibits hepatitis B DNA replication.[10] *Terminalia macroptera* (and *T. catappa*) is a traditional hepatitis remedy, and although the root and tree bark are more potent, the leaves are more environmentally sustainable and simply require a higher dose.[11] *Terminalia arjuna*, *T. bellirica*, and *T. chebula* are also used medicinally in different parts of the world.

Herbs Used in TCM for Hepatitis Treatment

Numerous herbs are used in Traditional Chinese Medicine (TCM) for treating hepatitis. Many published studies examine their hepatoprotective, antifibrotic, and antiviral mechanisms of action. These are some of the most common.

Angelica sinensis	*Paeonia suffruticosa*
Artemisia annua or *A. capillaris*	*Phyllanthus niruri*
Astragalus membranaceus	*Pinellia tematae*
Atractylodes spp.	*Polygonum cuspidatum*
Bupleurum chinense	*Poria cocos*
Citrus aurantium peels	*Rheum palmatum*
Gardenia jasminoides	*Salvia miltiorrhiza*
Gentiana lutea	*Scutellaria baicalensis*
Glycyrrhiza glabra	*Zingiber officinale*
	Ziziphus jujuba

Liver Formula Basics: Actions of Herbs for Liver Complaints

In many cases, herbs are chosen for inclusion in a formula not because of a particular diagnosis, but for a specific desired action. Herbal actions that are often indicated for liver diseases are alteratives, cholagogues, biliary antispasmodics, and antimicrobial agents. The specific symptoms that lead to the inclusion of herbs having these actions are as follows:

Alteratives

Alterative herbs can be useful in a wide variety of liver complaints—from gallstones to biliary insufficiency to hepatitis—that typically involve digestive disturbance. Alterative herbs enhance digestion, particularly of fat, and are especially appropriate when the stools are gray in color and float in water (steatorrhea). Alterative herbs are also indicated for blood sugar imbalances because the liver plays important roles in glucose and glycogen storage and release. Alterative herbs will also improve cholesterol metabolism and can improve blood lipids. In addition, alteratives can improve conjugation and clearance of endogenous estrogens and can improve premenstrual syndrome (PMS) and numerous complaints related to hyperestrogenism, such as acne, breast tenderness, and dysmenorrhea. Alterative herbs such as *Taraxacum*, *Rumex* species, *Mahonia*, and *Arctium*, therefore, have wide applicability for skin, hormonal, and toxicity issues in addition to the ability to improve biliary and liver function in general.

Cholagogues

Cholagogues are agents that promote bile flow and are indicated for biliary insufficiency and poor fat digestion. While cholagogues will improve the tendency to develop gallstones by improving bile quality and quantity, they are contraindicated for acute biliary colic because promoting bile flow when the passages are blocked could aggravate pain or even rupture the gallbladder. *Chelidonium majus*, *Leptandra virginica*, and *Iris versicolor* are among the stronger cholagogue herbs.

Biliary Antispasmodics

Muscles of the biliary tree may spasm due to the presence of gallstones or simply because of thick poor-quality bile, and biliary antispasmodic herbs are essential to include in formulas for acute biliary colic or less severe episodes of pain in the right upper quadrant. Some of the best biliary antispasmodics include *Mentha piperita*, *Raphanus sativa* var. *niger*, *Matricaria*, and *Dioscorea*.

Liver Antimicrobials

Viral hepatitis will require agents to address underlying viral infection such as *Curcuma*, *Taraxacum*, and many others (see "Herbs Used in TCM for Hepatitis" on page 90). For example, turmeric enhances the synthesis and levels of hepatocellular proteins and inhibits viral replication,[12] and *Taraxacum* has shown activity against hepatitis viruses.[13]

Liver Antifibrotics

Chronic inflammation of the liver can lead to fibrotic and fatty transformation and degeneration. Include at least one or two antifibrotic herbs in all formulas aimed at treating chronic inflammation in the liver caused by conditions such as hepatitis C, alcoholism, or chronic exposure to toxins. *Curcuma*, *Silybum*, *Bupleurum chinense* or *B. falcatum*, *Angelica sinensis*, *Glycyrrhiza glabra*, and *Salvia miltiorrhiza* are among many herbs noted to have hepatoprotective effects against a variety of antigens and irritants that can trigger liver inflammation and lead to fibrosis.

Formulas for Hepatitis

Hepatitis is an inflammatory condition of the liver that can be due to viral infections or exposure to hepatotoxins such as drugs, poisons, and excessive alcohol. Over 4 million people are living with chronic hepatitis in the United States alone. Viruses that can cause hepatitis include the well-known hepatitis A, B, and C viruses as well as cytomegalovirus and Epstein-Barr virus. Some bacteria, fungi, and parasites may also cause hepatitis, especially in tropical regions of the world with poor sanitation. Hepatotoxins must be considered when viral screens are negative. Some patients may not be aware of exposures or may not consider substances such as lawn pesticides or workplace chemicals as particularly offensive to health. Furthermore, the onset of chronic hepatitis can be insidious. The symptoms seem flu-like, with no significant liver pain or jaundice, and the disease can progress unnoticed.

Typical symptoms of the various types of hepatitis are digestive upset, including nausea and vomiting, changes in bowel habits, and loss of appetite. Smokers may inexplicably experience a distaste for smoking. Acute and severe hepatitis may also include clay-colored stools, fever, and jaundice as well as pain in the right upper quadrant. Some cases of low-grade chronic hepatitis are caught when routine blood screening shows elevated liver enzymes indicating the presence of liver inflammation. Although many cases of hepatitis can be treated once the virus is eradicated from the body or the offending hepatotoxins are removed, other cases can be so long-lasting or so severe as to lead to cirrhotic damage of the liver and to liver failure and even death.

Infectious hepatitis is common and can afflict otherwise healthy people with competent immune function. Those with immunodeficiency such as HIV-positive patients may be at increased risk for the more severe and long-lasting types of viral hepatitis. As many as 10 percent of HIV-positive individuals are also found to carry hepatitis B. This book does not provide a deep discussion of the types of hepatitis, but many readers will be aware that such viruses may be transmitted via oral-fecal routes or via blood and bodily fluid exposure. Some community outbreaks of infectious hepatitis cases are associated with eating in a restaurant, and others can be associated with IV-drug use or sexual activity. Other cases of infectious hepatitis, however, such as hepatitis C, are complex and multifactorial, and the virus typically becomes chronic. Hepatitis C can range from mild and asymptomatic to fulminant cases with severe liver damage.

In many medical circles, therapy for acute infectious hepatitis is most often simple and supportive; most herbalists and naturopathic physicians will offer antiviral, hepatoprotective, and other types of herbs specific to the case and presentation. In my opinion, offering *Silybum*, *Glycyrrhiza*, *Curcuma*, and other appropriate herbs is always indicated for all types of hepatitis. Even in cases of severe or HIV-associated hepatitis, such herbs should be given along with antiviral drugs if chosen. Patients with chronic hepatitis must avoid further insults to the liver and avoid alcohol, acetaminophen, and exposure to household and other chemicals as much as possible. Such patients should also avoid food chemicals in the form of artificial flavors and preservatives and choose organic foods wherever possible.

Hepatitis Decoction Formula I, for Acute Viral Hepatitis

There are many formulas used traditionally to treat acute and chronic hepatitis,[14] including the traditional Chinese formula yin chen. This formula draws on some of the herbs most commonly used in such formulas. Chlorogenic and phenolic acids in *Artemisia scoparia* and *A. capillaris* have activity against hepatitis viruses.[15] *Paeonia lactiflora* has been used for liver disease for thousands of years, and modern research suggests that paeoniflorin promotes liver glutathione detoxification pathways, preventing cholestasis and tissue damage.[16] This decoction might be used for acute hepatitis, due to the bitter *Artemisia*, while Hepatitis Decoction Formula II may be better for chronic hepatitis due as much to inflammation as to the presence of hepatitis viruses.

Artemisia scoparia, A. capillaris, or possibly other species
Bupleurum falcatum
Atractylodes ovata
Paeonia lactiflora

Combine equal parts of the dry herbs. Decoct the herb mixture using 1 teaspoon per cup of water, simmering gently for 15 minutes. Let stand for 20 to 30 minutes and then strain. Drink 3 or more cups per day for several weeks to several months.

Hepatitis Decoction Formula II, for Chronic Hepatitis

This tea consists of four herbs commonly used in TCM for hepatitis and is appropriate for chronic situations and long-term use outside of acute hepatitis A or B infection. These herbs are selected for their good flavor and can all be decocted well together.

Poria cocos
Angelica sinensis
Paeonia lactiflora
Glycyrrhiza glabra

Combine equal parts of the dry herbs and gently decoct 1 teaspoon of the herb mixture per cup of water. Strain and drink as much as possible, such as 3 to 6 cups per day for several months' time. As liver function improves and symptoms abate or blood levels of liver enzymes improve, consumption of the tea may be reduced to just 2 cups per day, allowing some days or weekends off.

Simple Capsules for Acute or Chronic Hepatitis

Some Western herbs are readily available as capsules to complement a chosen tea, and their use can help make an herbal protocol more aggressive.

Artichokes and Beets for the Liver

Artichoke (*Cynara scolymus*) leaves are a traditional liver remedy, and the leaves may be juiced and consumed to treat liver and gallbladder disease as well as to help lower cholesterol and blood sugar.[17] Human clinical investigations suggest artichoke preparations to alleviate dyspeptic pain and improve lipids in patients with nonalcoholic steatohepatitis.[18] Artichoke dips, salads, and soup purees are all delicious ways to consume *Cynara*.

Betaine from beet (*Beta vulgaris*) roots may reduce elevated homocysteine, assisting liver methylation and clearance[19] as well as prevent fat accumulation in the liver and treat steatosis in animal models of nonalcoholic steatohepatitis.[20] Beets are also shown to protect against alcohol-induced hepatotoxicity[21] and hepatic carcinogens.[22] Offer patients a variety of beet recipes, from raw grated beets and cooked beets as side dishes to borscht, fermented Kvass beverages, and sauerkraut combinations.

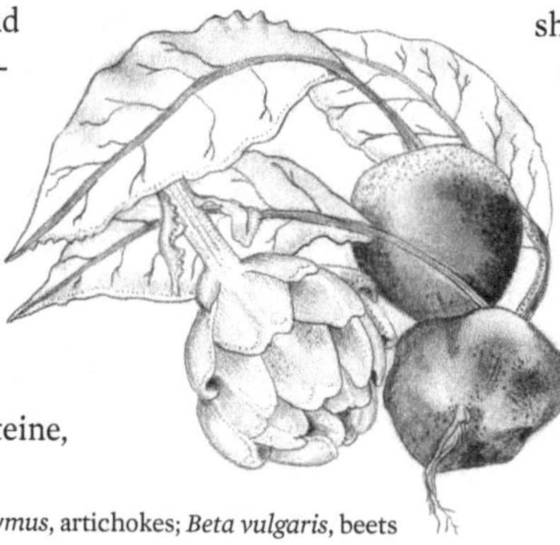
Cynara scolymus, artichokes; *Beta vulgaris*, beets

Silybum marianum capsules
Curcuma longa capsules
Lipotropic formula capsules

Choose one or all three of the above, and take two pills of each, 3 or even 4 times daily, for at least 2 weeks, and possibly for as long as 1 month, to address acute infectious hepatitis. *Curcuma* is better absorbed when taken with meals and especially a bit of fat. See also Golden Milk for Liver Disease and Inflammation on page 97.

Decoction for Acute or Chronic Hepatitis

While the previous formulas feature Chinese herbs that have had more scientific and clinical research, this formula is based on the Western herbal tradition that, with the exception of *Silybum*, has not yet been researched. Nonetheless these herbs are reported in the folklore and by modern clinicians to improve liver function.

Silybum marianum powder 4 ounces (120 g)
Taraxacum officinale 4 ounces (120 g)
Glycyrrhiza glabra 2 ounces (60 g)
Arctium lappa 2 ounces (60 g)
Foeniculum vulgare 2 ounces (60 g)
Citrus aurantium peel 2 ounces (60 g)

Combine and blend the dried herbs to help distribute the *Silybum* powder evenly. The overall mixture will be about 1 pound. Simmer 2 tablespoons of the herb mixture in 6 cups of water for 25 minutes. Remove from the heat, let stand 15 minutes more, and then strain. Drink in its entirety each day for at least several weeks.

Liver Lovin' Spice

In addition to crude powder, seeds of *Silybum* (milk thistle) are available whole and can be toasted or placed raw in a coffee grinder and ground to use as a condiment. The ground seeds can be combined with salt and spices of choice—in this case, herbs with liver benefits. The flavor will be more nutty and delicious when made for one's self in this manner.

Silybum marianum freshly ground powder
Salt
Curcuma longa powder
Zingiber officinale powder

Combine the ground milk thistle seeds, salt, and spices and use in salad dressings or marinades or simply sprinkled on foods at the dinner table.

Syrup for Nausea and Diarrhea in Hepatitis

Acute infectious hepatitis can involve digestive pain, nausea, vomiting, and diarrhea. This palliative formula could complement any of the other teas or capsules for hepatitis.

1 cup (60–90 g) coarsely chopped *Zingiber officinale* fresh root
2 tablespoons *Glycyrrhiza glabra* dried root powder

Nutraceutical Support for Liver Disease and Hepatitis

These herbal and nutritional compounds are all readily available and can be part of a treatment protocol for treating liver disease. All support the liver's innate detoxification pathways and can offer anti-inflammatory effects, improve liver function, and help protect against fibrosis.

s-adenosyl methionine (SAMe)[23]
Phosphatidylcholine
Lecithin
Berberine capsules, from *Mahonia aquifolium* or other sources
Lipotropic formulas containing cholagogues and other agents that promote bile flow and support liver function
Curcuma longa capsules
Silybum marianum capsules
Taurine
Glutamine
Betaine

Molecular Constituents against Hepatitis

These isolated compounds occurring naturally in medicinal plants have all shown activity against one or more strains of hepatitis viruses. Published research about each of these compounds suggests promise for hepatoprotection and for further investigation in the management of severe or chronic hepatitis.

Andrographolide
Artesunate
Astragalosides
Baicalein
Chrysophanol
Curcumin
Glycyrrhetinic acid
Kushenin
Phyllanthin
Picroside
Saikosaponins
Silymarin
Wogonin

Herbal Specifics for Hepatitis and Liver Infections

These are herbs other than the alterative herbs discussed elsewhere that are indicated in traditional herbalism as being specific for the various symptoms of hepatitis.

HERB	SPECIFIC CONDITIONS
Andrographis paniculata	Frequent infections, poor recovery from infections, fatigue
Astragalus membranaceus	Frequent infections, fatigue, lassitude
Ganoderma lucidum	Frequent and/or lingering infections, postviral fatigue and malaise
Glycyrrhiza glabra	Digestive upset, diarrhea, acute viruses, postviral fatigue and malaise
Grifola frondosa	Liver pain or congestion, frequent infections, fatigue, lassitude
Matricaria chamomilla	Diarrhea, loose stools, nausea, vomiting
Podophyllum peltatum	Chronic digestive colic, jaundice, liver disease, "bilious vomiting," enlargement of the liver, portal congestion and tendency to hemorrhoids, upper abdominal pain, heartburn, gagging and retching
Silybum marianum	Liver pain, jaundice, portal congestion, liver problems due to chronic alcohol abuse, gastric ulcers, varicosities due to portal congestion, rectal prolapse, fluid retention and dependent edema due to liver disease, stools yellowish or clay-colored and passed with difficulty, nausea and bitter taste in the mouth, coated tongue, nausea, retching, vomiting of green acid fluid, swollen enlarged gallbladder

Juice and fresh zest of an organic orange
1 cup (240 ml) honey
20 drops pure mint essential oil

Place the chopped *Zingiber* root in a pan with 4 cups water and the *Glycyrrhiza* root, and bring to a very gentle simmer for about 10 minutes. Remove from the heat, add the orange zest, and let stand covered for 20 minutes. Strain, and return the liquid to a small saucepan, and add the juice of the orange and the honey, and heat on the lowest setting to blend. Transfer to a glass bottle and add the mint essential oil. Take by the tablespoon for nausea or use to sweeten and flavor any of the teas featured in this chapter.

Glycerite Formula for Elevated Liver Enzymes

Elevated liver enzymes can result from birth control pills, exposure to hepatotoxins, a high fat burden in the liver, and many other assaults or irritants. When lab testing has ruled out hepatitis viruses, general alterative and liver-supportive herbs can be helpful. This formula combines the herbs most studied and used for liver inflammation and prepared in a glycerine form in order to avoid alcohol. Only a handful of herbs are commercially available in the form of glycerites, but the ingredients in this formula should be obtainable. When glycerites are not available, tinctures may be employed, but take care to evaporate the alcohol by stirring the tincture into a cup of hot tea before consuming.

Silybum marianum glycerite
Curcuma longa glycerite
Glycyrrhiza glabra glycerite
Taraxacum officinale glycerite

Combine equal parts. Take 1 teaspoon of combined glycerite 5 or 6 times a day for 3 months and then repeat blood work to evaluate success.

Tea to Prevent or Treat Cirrhosis

Fibrosis of the liver, which can occur when the liver is acutely inflamed, involves the activation of hepatic stellate cells. *Salvia miltiorrhiza*, *Cordyceps*, and other herbs have been shown to have hepatoprotective effects on the stellate cells and prevent their activation.[24] *Curcuma* and the curcumin it contains have been shown to similarly help the inflamed liver recover normal hepatic stellate cell status.[25] Human clinical studies suggest that *Panax ginseng* may decrease fibrotic transformation in hepatitis B patients.[26]

Silybum marianum for the Liver

Silybum marianum (milk thistle) seeds have hepatoprotective effects, reducing inflammation and activation following exposure to chemicals and toxins.[27] A 2014 review of Phase III Milk Thistle clinical trials report silymarin to be the best available medication for nonalcoholic fatty liver disease.[28] Much of the clinical and animal research on the use of milk thistle in various liver diseases shows antifibrotic[29] and anticirrhotic[30] actions. Milk thistle has been shown to offer hepatoprotection and, in a few cases, protection of other tissues against the toxic effect of acute cocaine-induced toxicity,[31] aflatoxin,[32] hydrogen sulfide gas,[33] iron overload,[34] high lipid load,[35] ethanol,[36] acetaminophen,[37] manganese,[38] chemotherapy drugs,[39] methotrexate,[40] doxorubicin,[41] vincristine,[42] statin drugs,[43] carbon tetrachloride,[44] *Schistosoma* infections, schistosomiasis drugs,[45] and other substances. Parenteral milk thistle preparations have become established as a viable treatment to save the liver following acute ingestion of hepatotoxic mushrooms.[46]

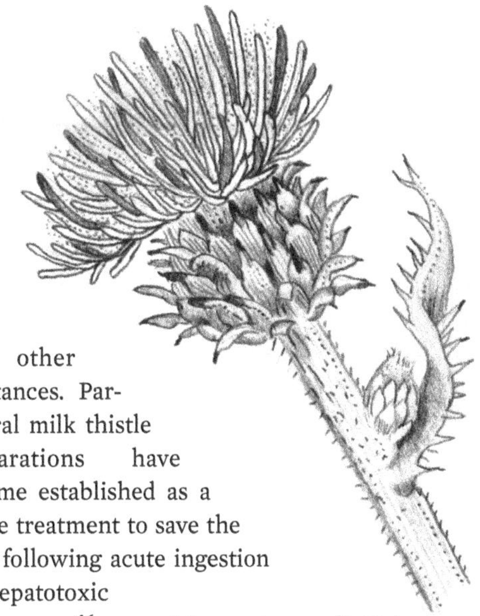
Silybum marianum, milk thistle

Curcuma longa
Salvia miltiorrhiza
Cordyceps sinensis
Panax ginseng

Combine equal parts of the dry herbs. Gently decoct using 1 teaspoon of the herb mixture per cup of hot water for 10 minutes. Strain and drink 3 or more cups each day for many months or even many years, taking a week, or at least a weekend, off every one to two months. As an alternative, the powders of these herbs may be obtained, blended, and put in smoothies or in Golden Milk for Liver Disease and Inflammation (see page 97).

Tincture for Acute Hepatitis

Hepatitis may be due to infectious, inflammatory, or toxic causes, and each underlying cause can be addressed specifically. The herbs in this formula are all-purpose liver anti-inflammatories, with antiviral and trophorestorative properties, appropriate for all underlying causes of hepatitis.

Silybum marianum
Glycyrrhiza glabra
Phyllanthus niruri or *P. amarus*

Combine in equal parts. Take 1 or 2 dropperfuls of the combined tincture 3 times per day, or purchase a similar encapsulated product and take 2 capsules, 3 times per day. The quantity and frequency of dosing may be reduced over several weeks' to several months' time.

Formula for Autoimmune Hepatitis

Autoimmune hepatitis is a chronic liver disease of unknown etiology, but associated with hypergammaglobulinemia. Immune-mediated hypersensitivity may also contribute to primary biliary cirrhosis and primary sclerosing cholangitis. This formula may be prepared as a tea or a tincture.

Angelica sinensis
Glycyrrhiza glabra
Bupleurum falcatum
Scutellaria baicalensis
Astragalus membranacous
Salvia miltiorrhiza

Combine equal parts of the dry herbs and decoct 1 teaspoon of the herb mixture per cup of water. Simmer gently for 10 minutes, let stand 10 minutes more, and then strain. Drink 4 to 6 cups per day. Or prepare as a combined tincture and take 2 dropperfuls at a time, 3 to 6 times daily.

Formulas for NASH, NAFLD, and Cirrhosis of the Liver

Chronic hepatitis and other long-term inflammatory conditions of the liver can lead to fatty and fibrotic degeneration and infiltration of functioning liver cells. Nonalcoholic fatty liver disease (NAFLD) is an umbrella term for fat deposition in the liver that is not associated with long-term alcohol use nor with an acute viral infection of the liver. Nonalcoholic steatohepatitis (NASH) involves progressive accumulation of fat in the liver and is often asymptomatic and noted only when routine blood tests show elevated liver enzymes. NASH may occur due to a combination of contributors including genetic predisposition, high cholesterol, and exposure to hepatotoxic drugs or environmental chemicals. Metabolic syndrome, including insulin resistance, diabetes, and obesity may also contribute to the development of NASH. Controlling all such factors can help prevent NAFLD and NASH from progressing to cirrhosis, liver failure, and a shortened life span. Herbal medicines, diet, and exercise can all be very valuable in preventing the fatty and fibrotic changes from progressing and interfering with liver function.

Reducing body weight and percent of body fat will reduce the progression and can even reverse a fatty liver. Intermittent fasting may be an effective weight-loss method. For example, 2 or 3 days each week, eat dinner at 6 pm, abstain from any evening snacks, and hold off eating breakfast the following morning until 10 or 11 am—a 16- to 17-hour period of fasting. This along with increased exercise are important complements to oral therapies for NAFLD. Probiotics, fiber, and glutamine may also reduce fat deposition in the liver and can be included, along with herbal therapies offered in this chapter, in protocols for treating fatty liver disease. *Curcuma*, *Glycyrrhiza*, and *Silybum* are featured in the formulas that follow due to their significant hepatoprotetive effects. *Tinospora*, known as guduchi, is another herb from Ayurvedic traditional medicine in India and is shown to have hepatoprotective effects.[47]

Formula for Fatty Liver, NAFLD, or NASH

The ingredients in this formula have all been subjects of some amount of research showing them to prevent fibrotic and fatty changes in the liver when due to viral infection and chronic liver inflammation. The ingredients are safe and gentle to use long term. Various *Silybum* (milk thistle) combination capsules for liver support may have similar formulas and are convenient and readily available. Isolated silibinin at a dose of 200 milligrams/kg shows good results in animal models of NASH.[48]

Silybum marianum 15 ml
Mahonia aquifolium 15 ml
Cynara scolymus 15 ml
Taraxacum officinale 15 ml

Take 1 or 2 dropperfuls of the combined tincture, 3 to 6 times daily. Or find a similar encapsulation and take 1 or 2 pills, 3 times a day.

Lloyd Brothers' Formula for Cirrhosis of Liver

The Lloyd Brothers manufactured concentrated herbal products in the 1920s, particularly for the Eclectic physicians, a group of natural medicine–inclined MDs of the era. This formula was their commercial formula for treating cirrhosis. *Echinacea* is often overlooked as being a valuable herb for reversing decay and degeneration.

Echinacea species
Chionanthus virginicus
Phytolacca decandra
Mahonia aquifolium

Combine in equal parts. Place 2 or 3 dropperfuls of the combined tincture in a cup of hot water to evaporate the alcohol. Drink the water over 2 hours' time and repeat 3 or more times per day.

Supplements and Nutrients Helpful against Fatty Liver

All of these supplements may be useful in protocols to treat liver disease. Some may be found in supplements crafted for the liver, such as lipotropic formulas, and others may be found as single nutrients.

Probiotics
Resveratrol
Vitamin E
Green tea
N-acetyl cysteine
Alpha-lipoic acid
Pantethine
Omega-3 fatty acids
Berberine
Choline
Curcumin
Glutamine
Guar gum

Silybum Simple for Liver Pathology Due to Iron Overload

Iron overload is damaging to the liver and can result from hemochromatosis or from thalassemia (when patients are treated with chronic blood transfusions). Silymarin is shown to improve the efficacy of the iron chelator desferrioxamine.[49]

Silybum marianum pills

Take 2 pills twice daily.

Golden Milk for Liver Disease and Inflammation

Curcumin, the active compound in *Curcuma longa* (turmeric), has poor bioavailability,[50] but it can be improved by placing the dry powder in fat or milk and adding a dash of black pepper. Piperine, an alkaloid found in *Piper nigrum* (black pepper), inhibits hepatic and intestinal glucuronidation, and thus black pepper enhances the assimilation of many important nutrients. It may increase the absorption of curcumin as much as 2,000 fold.[51]

2 cups (480 ml) nondairy milk, such as almond, soy, or hemp milk
1 teaspoon phosphatidylcholine or 1 tablespoon soy lecithin
1 teaspoon to 1 tablespoon turmeric powder (as desired, to taste)
¼ teaspoon freshly ground black pepper

Warm the milk gently in a saucepan, then transfer to a blender. Add the phosphatidylcholine or lecithin, turmeric, and pepper. Blend to homogenize and pour into a glass to drink promptly. High-quality commercial ground black pepper can be substituted for freshly ground pepper. This recipe can be amended in many ways: Add dates or other dried fruit for sweetening, add nut butter, add cinnamon, or add other medicinal herbs such as *Silybum* (milk thistle) powder as desired.

Herbal Formulas for Biliary Dysfunction

Biliary disorders include motility disorders, also known as biliary dyskinesia; issues with the quality and quantity of bile, known as biliary insufficiency; and acute spasms of the bile ducts, known as biliary colic. When the bile becomes so thick or flows so poorly that solids precipitate out, gallstones may result. And when grit in the bile and frank gallstones obstruct the bile ducts, acute inflammation of the gallbladder known as cholecystitis may result. Chronic inflammation of the biliary ducts can lead to biliary cirrhosis.

The treatment for all of these related diagnoses is somewhat similar in terms of herbal therapies. The goal is to optimize the flow and quality of bile via support of liver health. In my experience and opinion, many patients are quickly encouraged to undergo cholecystectomy, when improving biliary function would be preferable. In fact, a large percentage of patients who have their gallbladders out suffer from the same symptoms postsurgically as they did presurgically, a condition referred to as postcholecystectomy syndrome. Only when large stones are obstructing the common bile duct, threatening to rupture the gallbladder or inflame the pancreas, should immediate cholecystectomy be advised. In most other cases, alternative approaches to improve liver and biliary function can be highly effective and help to avoid unnecessary surgeries.

In addition to herbal agents that improve liver cellular function and bile flow, reducing the workload on the liver is appropriate for biliary disease. Through making dietary changes, avoiding hepatotoxins, and taking liver-supportive herbs, many cases of biliary dyskinesia, insufficiency, stones, and colic can be remedied without surgery and, in many cases, more effectively than with surgery. One human clinical study showed *Tinospora* to improve outcomes for patients undergoing biliary drainage surgeries to treat sepsis due to malignant obstruction of liver, gallbladder, and general biliary flow.[52]

Tincture for Biliary Insufficiency

Curcuma has an antispasmodic action on the bile ducts via effects on calcium channels,[53] promoting bile flow by as much 50 percent[54] and helping to reduce liver congestion. Curcuminoids in *Curcuma* help protect hepatic endothelial cells from hepatotoxins, improving sinusoid flow and suppressing hepatic microvascular inflammation.[55] In this formula, *Curcuma* is combined with other powerful bitter herbs. Due to strong cholagogue effects, this formula is contraindicated in acute biliary spasms.

Curcuma longa
Artemisia absinthium
Gentiana lutea
Rheum palmatum
Glycyrrhiza glabra

Combine in equal parts and place a dropperful of the combined tincture in a shot glass of water. Take a dose 3 or 4 times a day, reducing over time as symptoms improve.

Tincture for Cholestasis

Cholestasis is a cardinal manifestation of liver disease, and because effective pharmaceutical therapies are limited to nonexistent, herbal approaches are extremely valuable. Cholestasis may present with pruritus and elevated serum transaminase. This formula is more stimulating and has a greater cholagogue action than the Tincture for Biliary Insufficiency (page 97). Consider a formula such as this for biliary stasis, poor digestion, constipation, and a history of gallstones. Omit the *Chelidonium* for acute gallstones and right upper quadrant pain.

Silybum marianum
Chelidonium majus
Mahonia aquifolium
Chionanthus virginicus

Combine in equal parts. Take 1 teaspoon of the combined tincture 3 to 4 times a day for many months.

Simple for Cholestasis of Pregnancy or High Hormonal Load

The high hormonal loads of pregnancy can promote cholestasis of pregnancy. Cholestasis during pregnancy is a serious concern and associated with premature delivery and fetal distress and may cause stillbirth. Because the safety of herbs during pregnancy is often uncertain, the use of bile acids, such as ursodeoxycholic acid here, may be one alternative. Milk thistle may actually be a safer choice, although definitive research is lacking. Because the long-term use of bile acids may increase the risk of colon cancer and possibly other negative outcomes,[56] milk thistle deserves further investigation. The use of birth control pills may provoke cholestasis as well and elevate liver enzymes; bile acids and milk thistle may be appropriate in this situation, too.

Ursodeoxycholic acid

Take 500 milligrams twice daily.

Tincture for Cholelithiasis

This tincture can be used as part of the overall Protocol for Cholelithiasis, below.

Taraxacum officinale
Coptis chinensis
Dioscorea villosa
Silybum marianum

Combine in equal parts and take 2 dropperfuls of the combined tincture 3 or 4 times daily. In many cases, therapy will be needed for many years.

Protocol for Cholelithiasis

Environmental, dietary, infectious, and genetic factors contribute to gallstone susceptibility. Gallstones are comprised of cholesterol and bile pigments, and other cholesterol-related disorders including atherosclerosis, hyperlipidemia, and obesity are associated with an increased risk of gallstones, as are diabetes and metabolic syndrome. Individualized herbal formulas should address any contributors directly.

Supporting Bile Flow

Bile acids are synthesized from cholesterol in the liver and act as physiological detergent molecules, breaking down lipids and lipid-soluble nutrients into tiny spheres, improving their assimilation and metabolism. In cholestasis, impaired bile flow leads to accumulation of bile acids in the liver and induces inflammation and damage of hepatocytes. Ongoing cholestasis can lead to fibrotic and cirrhotic transformation and can eventually lead to liver failure as functional hepatocytes are gradually replaced by fatty and fibrous cells or to hepatocellular carcinoma as chronic inflammation allows oncogenic changes. Therefore, traditional cholagogue herbs that promote bile flow can help reduce liver damage; they include *Chelone, Chionanthus, Raphanus, Ceanothus, Curcuma, Silybum,* or *Cynara*. Herbs that protect against liver inflammation and fibrosis are complementary to cholagogues.

Epidemiologic studies suggest that coffee drinking offers some protection against gallstones.[57] Ursodeoxycholic acid is a naturally occurring bile acid that may be able to reduce bile cholesterol supersaturation and help dissolve cholesterol gallstones without significant side effects. Berberine supports bile production and flow and is found in *Coptis* but may also be supplemented in capsule form.[58]

For this protocol, take the Tincture for Cholelithiasis along with lipotropic formulas that contain choline and betaine, *Curcuma* capsules, mint oil capsules, berberine capsules, and possibly ursodeoxycholic acid. Continue the protocol for at least 3 months, at which time the protocol may be simplified and/or the dosages reduced.

Tincture for Acute Biliary Colic

The agents in this formula all have antispasmodic effects on the bile ducts and can be strong enough to relieve acute pain in cases of a biliary spasm and right upper quadrant pain. *Raphanus* (black radish) can be difficult to find but is so effective for acute biliary spasms that it is worth the effort. *Dioscorea* has relaxing effects on biliary spasm and can help allay pain in cases of acute biliary colic. Strong cholagogues should be avoided during gallbladder attacks, and this formula contains only antispasmodic and anti-inflammatory agents, making it appropriate for acute situations. Mint essential oil may also be used topically, by itself or with castor oil, applied over the right upper quadrant and covered with heat. It is also included in tincture form in this formula for oral use.

Dioscorea villosa
Raphanus sativus var. *niger*
Viburnum prunifolium, V. opulus
Mentha piperita

Combine equal parts. Take 1 teaspoon of the combined tincture every 5 to 15 minutes for acute gallbladder attacks, reducing as symptoms improve. This tincture may be complemented by also taking Spanish black radish pills and mint essential oil capsules, 1 or 2 of each at a time, every ½ hour until symptoms abate.

Black Spanish Radishes as a Cholagogue for Biliary Disease

Black Spanish radish (*Raphanus sativus* var. *niger*) has the same, or even hotter, spicy kick as its smaller red and pink relative, the common salad radish (*Raphanus sativus* var. *rubra*). Larger than pink radishes, black radishes are about the size of a small turnip. *Raphanus* is mentioned in old herbals as a cholagogue and as an effective plant for relaxing spastic bile ducts. I frequently include *Raphanus* in formulas for patients with gallstones and especially for people suffering from acute biliary colic. Modern animal studies have confirmed that *Raphanus* significantly increases bile flow. *Raphanus* is also credited with a liver-detoxifying effect due to induction of various detoxification enzyme systems.[59] Glucosinolates are found in radishes and other crucifers (plants in the Brassicaceae family, formerly Cruciferae), including broccoli and brussels sprouts, two plants that have been previously reported in the literature for promoting liver processing of hormones and other compounds. Several glucosinolate compounds unique to radishes, including glucoraphanin and glucoraphasatin, have been shown to promote phase II enzymes.[60] Radish sprouts are reported to have nearly four times more glucosinolate and eight times more isothiocyanate than mature black radishes.[61] Strong heavy metal chelation and antioxidant properties have also been credited to black radishes.[62] Significant amounts of catechin, the antioxidant of green tea fame, have also been identified in radishes.

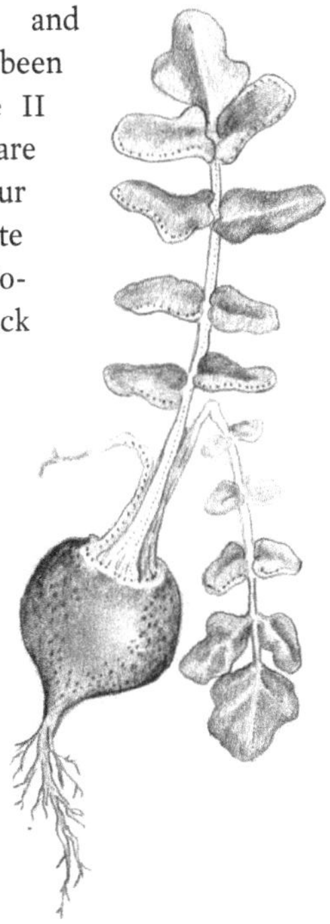

Raphanus sativus var. *niger*, black Spanish radish

Herbal Specifics for Biliary Pain

HERB	SPECIFIC CONDITIONS
Bupleurum chinense	Organomegaly, abdominal pain
Ceanothus americanus	Portal congestion, splenomegaly, lymphatic tenderness and swelling
Chelidonium majus	Swollen bile ducts, right upper quadrant pain that radiates to the shoulders, tongue that appears full and pale
Dioscorea villosa	Poor digestion with much pain and bloating; acute gallbladder colic; abdominal pain, radiating to back, chest, and arms, which can be lessened by maintaining a very erect posture
Mentha piperita	Digestive colic and bloating, biliary colic and acute gallbladder spasms with much painful gas
Raphanus sativus var. *niger*	Liver and splenic pain; painful incarcerated flatulence; distended, tympanic hard abdomen; periumbilical cramping and pain; loose, frothy, profuse stool passed with much pain and flatulence; putrid eructations; retching and vomiting; loss of appetite
Scutellaria baicalensis	Systemic inflammation, organomegaly, hepatic and splenic pain
Viburnum opulus, V. prunifolium	Cramps and colicky abdominal pains; abdominal and umbilical tenderness; large, hard stools, passed with cutting pains, that cause soreness of the anus

Tincture for Primary Sclerosing Cholangitis

Primary sclerosing cholangitis is a chronic cholestatic liver disease characterized by inflammation and destruction of the extrahepatic and/or intrahepatic bile ducts, resulting in biliary cirrhosis, the possible need for liver transplantation, and reduced life expectancy. *Centella* is used here for its ability to prevent fibrotic degeneration of various tissues. *Salvia* supports liver circulation, and *Silybum* and *Raphanus* are specific for improving bile flow.

Centella asiatica
Salvia miltiorrhiza
Silybum marianum
Raphanus sativus var. *niger*

Combine equal parts. Take 1 or 2 dropperfuls of the combined tincture, 3 to 6 times daily. This is a rather aggressive dosage to keep up long term, but because the condition is life-threatening, such an approach is warranted for at least a few months. The effects may be maintained by less frequent doses of 2 or 3 times a day thereafter.

Raphanus Simple for Biliary Atresia

Biliary atresia is a congenital obstructive fibrosis of the biliary tree resulting in cholestasis and icterus in newborns. It is a fairly rare condition occurring in only 1 out of every 15,000 to 20,000 births in the Western world, with reports of greater frequency in Asia. Without surgical intervention, the fibrotic condition will progress to full cirrhotic liver failure and death within the first two years of life. I know of one very serious case where the use of *Raphanus sativus* var. *niger* (black Spanish radish) appeared to be the key factor that saved the life of an infant with biliary atresia. This formula employs black radishes in whatever form they are available and is made as palatable as possible to give to a newborn, a difficult task in all cases.

Raphanus sativus var. *niger*, fresh juice of roots or sprouts, or powder
Infant formula, milk, or breast milk
Sweetener, such as maple syrup, if needed

Blend the radish juice, radish sprout juice, or radish powder (open up capsules to obtain the powder) into the infant formula or milk and give to the baby, drop by drop, as often as possible. This won't be easy due to the flavor, but it is worth the effort.

Treating Liver Cancer

Hepatocellular carcinoma (HCC), the most common liver cancer, is also one of the most lethal types of liver disease and has worldwide prevalence. Although most HCC cases are reported in the developing countries of Asia and Africa, there has been an alarming increase in HCC cases in Western Europe as well as in the United States.[63] Unlike other forms of hepatocellular carcinoma, HCC induced by hepatitis B virus (HBV) infection shows a poor prognosis with conventional therapies.[64] *Curcuma longa* (turmeric) has shown chemopreventive effects in an HBV-related in vitro model of HCC.[65] Viral hepatitis, chronic liver diseases, alcoholism, and dietary carcinogens such as aflatoxins and nitrosamines may all contribute to HCC; the influence of all of these factors may be attenuated by turmeric. One animal study showed turmeric to help regenerate damaged liver tissue, deter hepatitis virus gene expression, and prevent pathologic progression to hepatocellular carcinoma.[66] Curcumin has shown activity against important human viruses including the influenza virus, adenovirus, coxsackievirus, human immunodeficiency virus (HIV), and hepatitis viruses.[67]

Specific Indications: Herbs for the Liver and Gallbladder

There is some overlap between herbs that are indicated for biliary conditions and those for other digestive conditions. As you review the herbs listed here, it may also be helpful to browse "Specific Indications" in chapter 2 for ideas for gastrointestinal herbs. Many of the herbs discussed there can be selected as appropriate synergists and assistant complementary herbs in formulas to treat liver and biliary conditions.

Achillea millefolium • Yarrow

The flowering tops of yarrow can be prepared into a tea and a tincture to be used in formulas to treat irritable bowel syndrome, especially when liver congestion and atony of the tissues occur in those with cold constitutions, because *Achillea* has a warming and even diaphoretic effect. *Achillea* is also traditional for treating skin lesions associated with digestive symptoms and to reduce bleeding from the bowels.

Aesculus hippocastanum • Horse Chestnut

Aesculus may improve poor digestion when associated with venous stasis and portal congestion and presenting as a full sensation in the liver with tenderness in the right upper quadrant, and a sense of weight in the stomach with gnawing and aching. The ripe nuts are fresh tinctured, and sometimes prepared into oils and salve for topical use.

Aloe vera • Aloe

Aloe has mild laxative effects when products contain the *Aloe* rind and not just the mucilaginous pulp from the succulent leaves. *Aloe* will also promote bile flow and peristalsis; therefore, dried rind powder is indicated for chronic constipation, more so than the gel or juice, which is more specific for ulcerative lesions and mucous membrane inflammation. *Aloe* rind may actually worsen mucosal inflammation with diarrhea due to the stimulating and laxative action. In chronic constipation and biliary congestion, however, this effect is desirable.

Andrographis paniculata • King of Bitters

Andrographis is a traditional Ayruvedic herb, and both the root and leaves are used for jaundice, as a health tonic for the liver and cardiovascular health, and as an antioxidant. *Andrographis* is a liver protectant, immunostimulant, anti-infective, and antitoxin useful in formulas for hepatitis and elevated liver enzymes.

Arctium lappa • Burdock

Arctium is an important alterative herb and cholagogue that can be included in formulas for gallstones and biliary disease in general. Burdock root medications may improve skin disorders via improving biliary function and optimizing the elimination of toxins and the absorption of fat-soluble vitamins, including those important to skin health such as vitamins A, E, D, and K. *Arctium* is also indicated for liver disease and fatty liver due to its safe and gentle cholagogue and anti-inflammatory effects. Systemic symptoms that specifically indicate the use of *Arctium* include hyperlipidemia, acne and skin disorders, hyperestrogenism, and general malaise related to toxicity.

Research on *Curcuma* for the Liver

Curcumin, a bright yellow flavonoid found in turmeric, is a powerful antioxidant credited with anticancer effects and has been the subject of more than 5,000 scientific and clinical studies over the last decade.[68] The group of compounds related to curcumin in turmeric are referred to as curcuminoids and have many mechanisms of action for the liver, including promoting liver glutathione detoxification pathways and hepatoprotection.[69]

Research has also investigated bisabolene-type sesquiterpenes and saikosaponins in turmeric that are reported to have antifibrotic effects on the liver. Additional active molecules in turmeric include the essential oils, a group of aromatic sesquiterpene compounds including elemene, turmerin, turmerone, furanodiene, curdione, bisacurone, cyclocurcumin, calebin A, and germacrone.[70] Elemene is an approved anticancer agent in China, and it is shown to retard cell cycle arrest, induce apoptosis, and inhibit metastasis and tissue invasion.[71] Turmeric has been shown to protect the liver from hepatotoxins including carbon tetrachloride,[72] benzene,[73] aflatoxins,[74] alcohol, and concanavalin A.[75] It is also protective against other conditions and other agents[76] that damage the liver, including acetaminophen toxicity,[77] iron overload, biliary stasis, doxorubicin treatment,[78] cisplatin toxicity,[79] hepatitis virus,[80] and high-fat diets.[81] Many of these are liver toxins that one may be realistically exposed to in daily life, and many of them are also used in animal and tissue culture research, helping to demonstrate hepatoprotective actions of curcumin and other natural products when used prior to or immediately after these well-established liver toxins. Curcumin may reduce the incidence of liver cancer and fibrosis by protecting against hepatotoxic substances as well as decreasing cancer initiation through antimicrobial, anti-inflammatory, antioxidant, antiviral, and antimetastatic actions.[82] Turmeric extract may also improve the liver detoxification functions, even in acute inflammatory states that otherwise impair or delay excretion of wastes and toxins.[83] A randomized controlled trial investigated the effects of 3 grams/day of fermented turmeric powder on elevated alanine transaminase (ALT), a liver enzyme, compared to a placebo for 12 weeks and reported the levels to be significantly reduced compared to the control.[84]

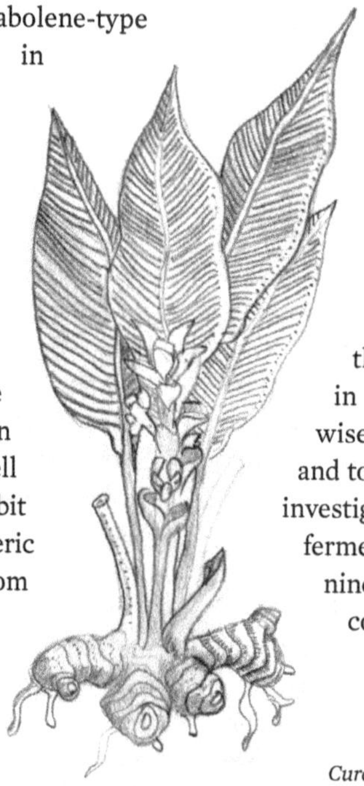

Curcuma longa, turmeric

Artemisia annua • Sweet Annie

The bitter leaves of this and other species of *Artemisia* are specific for parasites, malabsorption, and amoebic dysentery but may also have stimulating effects on all GI secretions for insufficiency and atony of the liver and biliary systems.

Artemisia capillaris • Capillary Wormwood

This species of wormwood has delicate finely divided leaves and is used in Traditional Chinese Medicine as an herb that helps to clear damp heat, and is an ingredient in the yin chen formula used to treat hepatitis with fever. It is recommended that the dose not exceed 30 grams per day because numbness, tingling, and other overdose symptoms may result. Contact dermatitis may also occur in some sensitive individuals following exposure to the fresh plant.

Artemisia vulgaris • Mugwort

The leaves of *A. vulgaris* are a bitter herb specific for insufficient digestive secretions and for parasites. Due to potentially toxic volatile oils, only small, short-term doses should be used, and the essential oil should never be consumed orally. Avoid in those with seizural disorders.

Astragalus membranaceus • Milk Vetch

Astragalus roots are especially indicated for frequent infections, fatigue, and lassitude. *Astragalus* is commonly combined with *Paeonia* and *Salvia miltiorrhiza* in China to treat liver disease. Immune-modulating and anti-inflammatory actions of *Astragalus* make it appropriate to include in formulas for hepatitis and to help prevent cirrhotic degeneration of the liver.

Berberis aquifolium, B. nervosa • See *Mahonia*

Beta vulgaris • Beet

Beets and the betaine found in the taproot are supportive to liver detoxification pathways and can be included liberally in the diet or in various beverages. Kvass, a traditional fermented beverage prepared from beets, is also useful and may be made at home or may be commercially available.

Bupleurum chinense, B. falcatum • Chinese Thoroughwax

Bupleurum, also known as chai hu, is a traditional Chinese medicine widely used to treat fever, hepatitis, jaundice, nephritis, and dizziness. The bright yellow bitter roots are used, and when they are baked with vinegar, the effect is especially directed at treating liver-related disease. *Bupleurum* is specific for organomegaly, especially hepatomegaly, and abdominal pain associated with liver and digestive disease. *Bupleurum* is often combined with *Paeonia* to treat liver congestion and disease in Traditional Chinese Medicine.

Ceanothus americanus • Red Root

The roots and root bark of red root, also known as New Jersey tea, are specific for liver congestion, pelvic and portal congestion, splenomegaly, vascular congestion, and hypertension. *Ceanothus* preparations are traditionally asserted to have an affinity for the lymphatic system, alleviating vascular congestion via enhancing entry of interstitial fluid into the vasculature and increasing venous return.

Chelidonium majus • Celandine

The roots and leaves of this bitter plant are a valuable cholagogue and liver herb specific for pain or fullness in the right upper quadrant, pain that radiates to the right shoulder, jaundice, biliary disease, and gallstones. *Chelidonium* may be therapeutic to nausea and stomach pain when related to poor digestion and biliary insufficiency. The plant is specifically indicated when there is a large flabby tongue with indentations of teeth on lateral margins, a thick or yellowish coating on the tongue, and constipation with dry hard stools. *Chelidonium* is also specific when the stool is abnormal; bright yellow, clay-colored, or light-colored stools that float are all indicative of biliary insufficiency. *Chelidonium* is also specific for skin disorders associated with digestive and liver disorders. The Eclectic physicians regarded *Chelidonium* as one of the best remedies for biliary and hepatic congestion, though it is best avoided in acute inflammations of the liver. This poppy family plant is high in isoquinoline alkaloids, and the dose should be kept on the small side.

Chelone glabra • Turtlehead

The bitter leaves have alterative and cholagogue properties used for liver congestion with jaundice and dyspepsia and to help recover from infectious or exhaustive illnesses where the appetite and digestion have been affected. *Chelone* is also specifically indicated for GI debility accompanied by jaundice.

Chionanthus virginicus • Fringe Tree

This woody plant is a traditional medicine for jaundice and hepatitis. The root bark is a classic liver herb specifically indicated for liver pain and fullness and for frothy or clay-colored stools. *Chionanthus* was highly regarded by the Eclectic physicians for portal congestion and hepatic enlargement. The plant is also recommended for infantile jaundice, for glucosuria due to faulty glycogen production in the liver, and for pre-diabetic states. *Chionanthus* is also specific for chronic digestive complaints, including chronic gastritis, and for irritable bowel syndrome when poor digestion and liver function are contributory.

Cnicus benedictus • Blessed Thistle

Many thistle family herbs, including *Silybum* and *Cynara*, are folkloric remedies for liver inflammation. *Cnicus* is said to be specific for sluggish liver function and is useful as an alterative herb to improve the liver's processing of hormones, carbohydrates, and lipids.

Curcuma longa • Turmeric

Curcuma is a popular and well-studied plant whose bright yellow roots offer antioxidant, anti-inflammatory, and hepatic supportive effects, assisting the liver to clear toxins and improve hormone clearance. Use of *Curcuma* may improve lipid and carbohydrate metabolism in cases of diabetes and hyperlipidemia. *Curcuma* may be used to improve intestinal dysbiosis and the propensity to fungal and other infections. *Curcuma* and related species can be used in formulas for liver and bladder disorders including gallstones, jaundice, hepatitis, and liver cancer as well as for hyperglycemia, obesity, diabetes, diabetes-related liver disorders, and numerous other issues. *Curcuma* is specific for cholestasis, biliary pain, joint pain, inflammation, and general maldigestion.

Cynara scolymus • Artichoke

Artichoke leaves are nutritive and can make a good base tonic ingredient in liver formulas. The leaves and leaf ribs can be prepared as a vegetable, and the plant has multiple effects, including liver anti-inflammatory, hepatobiliary tonic, antioxidant as well as the lowering of cholesterol. *Cynara* improves cholesterol and hormonal imbalances by supporting the hepatic processing of these substances, and it improves glucose and lipid processing, making it useful for hyperglycemia and hyperlipidemia.

Dioscorea villosa • Wild Yam

The underground tubers of wild yam are indicated for poor digestion accompanied by pain and bloating; acute gallbladder colic; abdominal pain radiating to the back, chest, and arms; and digestive pain that is ameliorated by holding a very erect posture or leaning the torso backward.

Eclipta alba, E. prostrata • Bhringraj

Also known as false daisy and yerba de tago, *Eclipta* is a weedy tropical herb whose leaves are both edible and medicinal; it is traditionally used as an anti-inflammatory, liver tonic, kidney tonic, and blood sugar–balancing agent as well as a hair tonic, preventing loss and graying. At high doses, *Eclipta* is harmful to the liver; at small doses, *Eclipta* is indicated for infective hepatitis, liver cirrhosis, liver enlargement, and gallbladder disease.

Ganoderma lucidum • Reishi Mushroom

Reishi mushrooms are woody and inedible but revered in Asia for thousands of years as a vitality and longevity tonic. Reishi is especially indicated in cases of frequent and/or lingering infections, postviral fatigue, and malaise. It may be considered in teas and tinctures for hepatitis C and chronic liver inflammation.

Gentiana lutea • Gentian

The extremely bitter roots of gentian are indicated for atonic situations in the digestive tract. *Gentiana* can help recover enfeebled digestion following prolonged illnesses and increase bile flow in cases of biliary insufficiency. Folkloric sources purport that *Gentiana* is specifically indicated when fatigue and mental lethargy accompany the physical symptoms.

Grifola frondosa • Maitake

Grifola is an edible mushroom prized in Japan and other Asian countries. Because the mushroom can grow as large as 100 pounds, it is also sometimes referred to as "King of Mushrooms." *Grifola* is available as a dry powder or small slices, can be prepared into a tincture, and is commonly found in encapsulated products for use as an immunostimulant, antiviral, and supportive measure for hepatitis and for inclusion in convalescence and recovery formulas following liver disease or gallbladder surgery.

Hydrastis canadensis • Goldenseal

Hydrastis may improve dyspepsia due to liver disease and can be included in formulas for jaundice and liver tenderness. *Hydrastis* is also specific for morning nausea and vomiting in chronic alcoholics and for anorexia due to hepatic ailments.

Iris versicolor • Blue Flag

Blue flag is a type of wild *Iris* whose roots are a lymphagogue and secretory stimulant that may be used in formulas for liver congestion with deficient bile flow, poor fat digestions, and steatorrhea. *Iris* is specific for liver enlargement with a sick headache and nausea, burning sensation in the stomach and intestines, liver pain, nausea and vomiting, flatulent colic and diarrhea, greenish loose stools, and skin eruptions with itching. As *Iris* is a warming stimulating herb, only small doses of a few drops at a time are appropriate when used as a simple. *Iris* is typically a supportive rather than a lead herb in formulas for liver complaints.

Leptandra virginica • Culver's Root

The bitter roots of *Leptandra* (may also be classified as *Veronicastrum virginicum*) are specific for pain in the right upper quadrant, typical of liver and biliary disorders, especially when associated with drowsiness, dizziness, mental depression, lassitude, headache, nausea, diarrhea, and a thickly coated tongue.

Mahonia aquifolium • Oregon Grape

The yellow bark of the roots of *Mahonia* is high in berberine and useful in treating liver congestion and biliary insufficiency. *Mahonia* is also a broad-acting antimicrobial appropriate for everything from infectious hepatitis and dysbiosis to traveler's diarrhea and food poisoning. *Mahonia* is specific for liver congestion with tenderness and slow digestion, a coated tongue, and skin eruptions due to poor liver and digestive health, chronic catarrh, weakness and emaciation from chronic disease, digestive derangements, and malnutrition. This plant and many others in this genus are also known as genus *Berberis*.

Mentha piperita • Peppermint

Mint leaves and purified menthol in mint oil can have fairly powerful antispasmodic effects on digestive smooth muscle. Peppermint is specific for digestive colic and bloating, biliary colic, and acute gallbladder spasms with much painful gas. Mint tea, concentrated mint oil capsules, and mint essential oil applied topically over the abdomen can help allay pain and cramping.

Myrica cerifera • Wax Myrtle

Myrica root bark is specific for liver disease and hepatic congestion, for biliary insufficiency with nausea, and for a bitter taste in the mouth and halitosis. Wax myrtle, also called bayberry, is also indicated for loss of appetite and long-lasting stomach discomfort after eating and for digestive symptoms that are better with eating acidic agents, such as vinegar or lemon juice on foods to reduce indigestion. Some older texts purport that a craving for such acids or pickled foods are an indication to use *Myrica*. Liver inflammation with jaundice, right upper quadrant pain, a constant sense of fullness, and clay-colored stool may respond to *Myrica*.

Paeonia lactiflora, P. suffruticosa • Peony

Red peony is the whole root of *P. lactiflora*, white peony is the peeled and processed root, and both are used medicinally. Peony is considered to be a blood tonic in TCM, meaning that it supports body fluids, builds the blood, and may promote circulation in the intestine. Peony is traditionally used for pelvic pain, especially when due to blood stasis. It is also specifically indicated for blood deficiency that includes the symptoms of anemia such as pallor, weakness, and dizziness, and for some liver disorders associated with fever, restlessness, excessive sweating, and pain. In traditional Chinese formulas peony is often combined with *Salvia miltiorrhiza*, *Angelica*, or other herbs to treat liver disease. The roots are typically decocted in TCM.

Peumus boldo • Boldo

The bitter leaves are a staple in Latin American herbalism to support the liver and have a general detoxifying, alterative, and cholagogue action.

Podophyllum peltatum • Mayapple

Podophyllum is a strong and potentially caustic herb, used in small amounts only, typically as a tincture. *Podophyllum* is specifically indicated for chronic digestive colic, jaundice and liver disease, and "bilious vomiting." *Podophyllum* is recommended for enlargement of the liver; portal congestion and tendency to hemorrhoids; upper abdominal pain; and heartburn, gagging, and retching due to poor digestion and biliary insufficiency.

Quercus alba • White Oak

Quercus is specifically indicated for fluid stasis in the tissues and vasculature secondary to liver disease or alcoholism, including situations of portal congestion.

Raphanus sativus var. *niger* • Black Spanish Radish

Black Spanish radish preparations made from the fresh fleshy tubers have the unique ability to relax the musculature within the bile duct and gallbladder in cases of biliary colic; include *Raphanus* for acute biliary colic. *Raphanus* can also be included in formulas for chronic biliary insufficiency where there is right upper quadrant pain and chronic constipation.

Rheum palmatum • Chinese Rhubarb

Also called turkey rhubarb, *Rheum* is specific for a sour smell to the body, diarrhea, the sensation of hunger but easily becoming over full, colicky pain about the umbilicus, and sour-smelling stool passed with cramping and straining.

Rumex crispus • Yellow Dock

Yellow dock (also called dock or crispy dock) is indicated for hypochlorhydria, malabsorption, constipation, and digestive insufficiency. *Rumex* is specific for skin eruptions secondary to digestive insufficiency, biliary insufficiency, and poor elimination with toxicity. *Rumex* is also specific for a sore, coated tongue, heartburn, hiccups, chronic gastritis, nausea and anorexia, flatulence and abdominal pain, morning diarrhea, and pruritis related to liver and digestive disturbances.

Schisandra chinensis • Magnolia Vine

The flavorful fruits of this vining shrub, also called five flavor fruit, are used in Asia as a liver and digestive tonic. Tinctures and teas are available to include in formulas for hepatitis recovery, chronic liver disease, and for digestive and liver infections.

Scutellaria baicalensis • Huang Qin

The dried roots of *Scutellaria baicalensis*, which also goes by the common names baical and scute, have broad anti-inflammatory activity and can be included in formulas for liver disease, organomegaly, and hepatic and splenic pain.

Silybum marianum • Milk Thistle

Silybum seeds are very helpful for all matter of hepatic and biliary disorders and associated jaundice and digestive difficulty. The flavonoids in the ripe seeds are called silymarin, which is one of the most well studied of all the hepatoprotective agents. *Silybum* is specifically indicated for infectious or autoinflammatory ailments of the digestive organs—liver, gallbladder, pancreas, spleen, and kidneys. *Silybum* can also improve altered blood composition due to faulty digestion such as hyperlipidemia and hyperglycemia. Milk thistle has been credited with hepatoprotective properties in chronic liver disease since as early as the fourth century BC. Milk thistle leaves are a traditional food, and the seeds have been roasted and ground to use in medicinal beverages. As an herbal medicine, milk thistle seeds are prepared into a variety of medicines to treat gallbladder and biliary disease, jaundice, peritonitis, malarial fevers, bronchitis, insufficient lactation, and varicose veins and to reduce insulin resistance. *Silybum* is specific for liver pain, jaundice, portal congestion, liver problems due to chronic alcohol abuse, gastric ulcers, varicosities due to portal congestion, rectal prolapse, fluid retention and dependent edema due to liver disease, yellowish or clay-colored stools passed with difficulty, nausea and bitter taste in the mouth, a coated tongue, nausea, retching, vomiting of green acid fluid, and an enlarged gallbladder.

Stillingia sylvatica • Queen's Root

Stillingia is a poorly researched plant but mentioned in the folkloric herbal tradition for liver congestion with jaundice and constipation.

Taraxacum officinale • Dandelion

Both dandelion leaves and roots are used medicinally, but the roots are most useful for jaundice, gallbladder colic, biliary insufficiency, pain in the right upper quadrant, headaches due to liver and digestive disturbances, flatulence, enlarged indurated liver, sharp stitching pains, difficult bowel movements, anorexia, a coated tongue, and night sweats due to liver disease.

Terminalia spp. • Sea Almond

Terminalia are tropical trees and include at least 100 different species, several of which have been used in Ayruvedic and other medical traditions of the world. The roots, bark, and leaves are all used medicinally for cardiovascular disease and high cholesterol, as well as liver and digestive disorders, including jaundice with nausea, vomiting, and diarrhea.

Tinospora cordifolia • Guduchi

Tinospora is a traditional Ayurvedic herb used to treat liver and biliary disorders, and tissue culture and animal studies have shown hepatoprotective effects.[85] *Tinospora* is indicated for general liver support as well as for severe hepatobiliary diseases including life-threatening jaundice due to malignant obstruction. It is also used to improve survival rates following biliary drainage surgeries.

Zingiber officinale • Ginger

The roots of ginger can be used as a complementary herb in formulas to treat liver dysfunction, especially to allay dyspepsia and flatulent colic, and for alcohol-induced liver and gastric complaints. The fresh roots can be decocted to prepare the day's drinking water or prepared into a variety of teas, tinctures, and encapsulations.

— CHAPTER FOUR —

Creating Herbal Formulas for Renal and Urinary Conditions

The urinary system consists of the kidneys, ureters, urinary bladder, and urethra, all involved in eliminating wastes, controlling blood volume, and regulating blood electrolytes, pH balance, and red blood cell production. The kidneys are a major organ of elimination; they filter water, electrolytes, nitrogenous wastes from protein metabolism, and water-soluble wastes from the bloodstream. The renal filtering system recycles minerals and electrolytes that help maintain osmotic balance in the tissue fluids and regulate blood pressure. The kidneys are also important in the synthesis of vitamin D and the synthesis of new red blood cells via their production of the bone marrow regulator, erythropoietin.

Of the fluid that passes through the renal glomeruli and tubules, only a small amount becomes urine as minerals, ions, sugars, and amino acids are reabsorbed. An antidiuretic hormone from the posterior pituitary directs the kidneys to hold onto water by increasing the permeability of the distal tubules and collecting duct to water, increasing water reabsorption. The kidneys help judge when to retain more water by sensing sodium concentration. When the concentration is low, the kidneys release renin, which promotes the formation of angiotensin I, and ultimately conversion to angiotensin II via angiotensin-converting enzyme (ACE) stimulation. Aldosterone is also released in response to angiotensin II to help the kidneys hold onto sodium and chloride. Many antihypertensive drugs target ACE, attempting to block its release to reduce blood pressure via enhancing sodium excretion. However, when perfusion through the kidneys is low, for example, due to heart failure, the amount of sodium reaching sensing cells will also be low, triggering the kidneys to retain sodium.

The kidneys respond to sympathetic nervous system innervation by reducing blood flow. Therefore, stress can raise the blood pressure through renal mechanisms. Poor diets that are low in minerals can challenge the kidneys' ability to excrete wastes and maintain osmotic gradients and stabilize blood pH. Minerals are required to enable the kidneys to excrete uric acid and other acidic wastes, and when dietary intake of minerals is poor, more acidic compounds are retained, contributing to disease. Inadequate water intake also challenges the kidneys and leads to supersaturated fluids in the body, making the kidneys more prone to renal calculi.

Protein Intake and Renal Disease

The kidneys are responsible for processing and eliminating the nitrogenous breakdown products of protein metabolism. Excessive protein intake, especially animal protein, may burden the kidneys. Burns, muscle trauma, and muscle-wasting diseases can also liberate a large amount of protein and similarly stress the kidneys. Protein in the urine, proteinuria, is one of the first signs of failing kidney function and occurs with most kidney diseases. Monitor protein intake and optimize the quality and quantity when treating nephropathies.

Screening Urinalysis

A simple urinalysis can be valuable in assessing the basic health and function of the urinary system. For example, a large volume of pale urine can mean that kidneys are not concentrating urine well, while dark and scanty urine can signal dehydration. A strong or foul odor signals an infection; obvious blood in the urine may signal a stone, cancer, or significant infection. Urine with a large amount of sediment or heavy cellular debris can signal toxicity. A large amount of crystalline debris (visible in a microscopic examination) may help diagnose renal stones, or when red blood cell casts are present, acute glomerulonephritis. Using a dipstick test will identify proteinuria, chyluria, trace amounts of blood, and other valuable diagnostic signs.

Teas may be the best vehicles for botanical medicines in many urinary complaints, improving the surface contact of herbs with the urinary mucosal and other tissues. When treating infections, irritations, and mucosal membrane issues, teas are often the most effective form of medicine. Diuretics may be most effective administered as herbal teas because the water alone supports diuresis. On the other hand, herbs in a variety of forms can help in situations of heart failure, aimed at enhancing perfusion, or in cases where high amounts of flavonoids or immune-modulating compounds may be desirable. Therefore, the use of teas is featured prominently throughout this chapter, and tinctures and encapsulations help offer aggressive protocols for many of the conditions covered.

Common Urinary and Renal Conditions

Among the situations most encountered in general practices are simple urinary tract infections, interstitial cystitis, prostate enlargement, renal calculi, and possibly chronic renal failure due to glomerulonephropathy in diabetics. In cases of toxicity in the body, as exemplified by frequent infections, muscle stiffness, low energy, a coated tongue, or skin lesions, there is an old saying, "Open the emunctories," meaning to open the channels of elimination in the body. The urinary system along with the liver and bowels are the major organs of elimination, and for many vague or chronic disorders, simply helping the body to better absorb nutrients and better eliminate wastes can have great benefits to the entire body and is supportive to overall health and vitality. Furthermore, a healthy intestinal ecosystem is less likely to seed the urinary system with pathogenic *Escherichia coli* (*E. coli*) strains that can cause urinary tract infection.[1] Thus, for many common urinary complaints, dietary measures are appropriate to reduce nitrogen loads on the kidneys and optimize intestinal transit time and flora.

Botanical approaches to urinary complaints may include the use of herbal antimicrobial agents that have an affinity for the kidney or bladder such as *Arctostaphylos uva ursi*, *Agathosma betulina* (formerly *Barosma betulina*), and *Juniperus communis*, discussed further in this chapter. In other cases, soothing irritated renal passages can help allay the pain of a bladder infection or the passing of a renal stone and include *Althaea officinalis* tea and *Asparagus* root brews. In the case of poor cellular functioning, as with renal failure, *Pueraria* species, *Silybum marianum*, and *Salvia miltiorrhiza* may be helpful. If the kidneys are poorly perfused due to heart disease, the cardiac glycosides, such as *Convallaria majalis*, may help get more blood flowing; for prostatic enlargement, *Serenoa repens* and *Pygeum africanum* may improve symptoms.

In other cases, renal disorders may involve organic pathologies of the kidney cells and tissues themselves, commonly referred to as nephropathies. The more severe of these can lead to renal failure and may require long-term dialysis, or in some cases they can be fatal. In other cases, the kidneys may become irreparably damaged by systemic diseases including autoimmune disorders, hypertension, drug toxicity, and diabetes. See "Formulas for Acute and Chronic Renal Failure" on page 152 for formulas for nephropathy.

Urinary allergies may also occur in atopic individuals. Allergenic foods, benzoates, food additives, and environmental chemicals can all trigger bladder and urinary symptoms that may stump some physicians when microbes cannot be demonstrated. Interstitial cystitis, chronic prostatitis, and lower urinary epithelial dysfunction (LUED) are other such inflammatory disorders that may mimic the symptoms of bacterial cystitis or infectious prostatitis, but prove aseptic and do not respond to antibiotics. See "Formulas for Interstitial Cystitis and Lower Urinary Epithelial Dysfunction" on page 121 and "Formulas for Prostatitis" on page 149 for more information.

Herbal Diuretics

It's important to understand how and why herbs can enhance urinary output in a variety of clinical situations, and thus I'll focus here on herbal diuretics and how to make sense of their use in formulas for renal and urinary conditions.

Diuretics are primarily indicated for renal insufficiency, when the kidneys are failing to filter the blood, or for congestive heart failure, when the heart muscle is incapable of pushing fluids to the kidneys for processing and the kidneys are poorly perfused as a result. (Formulas for both of these conditions are further discussed in "Formulas for Acute and Chronic Renal Failure" on page 152.) In such situations, diuretics would be only one small tool in a broader protocol addressing renal failure or treating heart disease. The management of heart disease often involves numerous pharmaceuticals from antihypertensive agents and anticoagulants to cardiac glycosides and diuretics. Renal failure is often managed with dialysis when severe and possibly with corticosteroids for nephritic edema. Diuretics, especially the more nourishing mineral-rich herbal diuretics, may also help rid the body of inflammatory wastes in situations of edema, or ascites in liver disease. Diuretics may also help to reduce joint swelling in arthritic disorders and may be helpful to reduce swelling in the genitourinary tissues themselves in cases of chronic irritation, infection, lithiasis, or swelling. Folkloric herbal formulas for gout and rheumatism commonly feature diuretic ingredients, and tissue fluid accumulation due to lymphatic stasis or liver disease may also benefit from the use of diuretics.

Types of Herbal Diuretics

Herbal diuretics can stimulate glomerular filtration via several mechanisms.

Irritant diuretics. Herbs that are high in volatile oils may have direct irritating effects on renal cells, stimulating glomerular filtration as a means of eliminating the offending substances. *Juniperus communis* and *Agathosma betulina* may act as irritant diuretics.

Perfusion-enhancing diuretics. Cardiac glycosides such as *Digitalis purpureum* and *Convallaria majalis* may enhance perfusion to the kidneys in situations of heart failure, and *Salvia miltiorrhiza* and *Silybum marianum* may enhance cellular function and microcirculation in the kidneys to help the glomeruli and tubules function more effectively.

Stimulant diuretics. Coffee, caffeine, and other methylxanthines decrease retention of sodium and chloride by the tubules, retaining more water and increasing urine volume. Flavonoids are also capable of increasing sodium excretion and are certainly a safer and gentler form of "stimulant." Elderberries, for example, have been shown to increase the excretion of sodium.[2]

Osmotic diuretics. Demulcent compounds in *Althaea officinalis*, *Zea mays*, and other plants are large, gummy, mucilaginous molecules that hold a lot of water. Such molecules are large enough not to be reabsorbed by the tubules and cause water to be osmotically retained as well, increasing urine volume, and offering demulcent effects in the urinary passage.

Mineral tonic diuretics. The kidneys must "spend" some minerals in order to retain and recycle desired nutrients and compounds; the minerals are used to create osmotic gradients. When the blood is rich in minerals and calcium, silica, potassium, and magnesium are abundant, excretion of metabolic wastes is enhanced and more water is pulled along in the process. *Medicago sativa, Urtica* species, *Petroselinum* species, and *Equisetum* species may work in this way due to abundant nourishing bioavailable mineral content, and all could be considered renal depurants. Related to the word "purify," a renal depurant is an agent capable of enhancing the clearance of solid wastes through the urine, not only water. Minerals are spent to help excrete molecular wastes such as uric and oxalic acids.

General Diuretic Tea

Parsley and fennel seeds contain both minerals and volatile oils with diuretic effects. *Equisetum* and *Taraxacum* leaves are also very high in minerals and among the most reliable of the folkloric diuretics. Unlike pharmaceutical diuretics, herbs are not noted to deplete potassium or magnesium; herbs actually supply these minerals. Peppermint and licorice are added to this formula primarily for flavor.

Petroselinum crispum seeds or seed powder
Foeniculum vulgare seeds or seed powder
Equisetum spp.
Taraxacum officinale leaf
Mentha piperita
Glycyrrhiza glabra

Combine equal parts of the herbs. Steep 1 tablespoon of the herb mixture per cup of hot water and then strain. Drink freely, 3 or more cups per day.

The Trouble with Thiazide

Thiazide diuretics are one of the most widely recommended first-line pharmaceutical options in the treatment of hypertension, but they have many side effects. While there is no doubt that thiazides enable a rapid hypotensive effect, the drugs may actually cause metabolic abnormalities that promote cardiovascular disease, including hyperlipidemia, insulin resistance and new-onset diabetes, hypokalemia, hyperuricemia, and stimulation of the renin-angiotensin-aldosterone system. Thiazide diuretics are well-known to "waste" potassium and other minerals including magnesium. The thiazides interfere with sodium resorption in the renal tubule and promote its excretion in the urine, helping to reduce hypertension, however, pulling water and other minerals along with it.

Because of this issue, many clinicians will monitor blood potassium and offer a potassium supplement when there is evidence of a deficit, but many practitioners seem unaware that the thiazides also deplete magnesium and other important minerals not routinely monitored with blood tests. Furthermore, the use of thiazide diuretics may increase diabetes,[3] as low magnesium promotes insulin resistance. Although lowering serum sodium is the goal of the thiazide diuretic drugs, sodium and electrolytes may become so deranged as to induce hyponatremia severe enough to necessitate hospitalization due to confusion, imbalance leading to falls, even seizures, and occasionally fatalities.[4] Even minor electrolyte imbalances may increase the risk of mortality.[5] Because such side effects are unpredictable, those using thiazide diuretics should be monitored closely when the drugs are initiated. Thiazides may exacerbate metabolic syndrome, and despite initially lowering blood pressure, many are associated with a decline of renal function.[6] Thiazides are associated with an increase in serum triglycerides and in fasting glucose levels and insulin resistance, which may be related with potassium and magnesium depletion.

Because high mineral herbs can support sodium excretion and there are numerous therapies to help manage hypertension, many herbalists and naturopathic physicians avoid thiazides, with the opinion that thiazides interfere with nutrition and mineral status and may actually cause imbalances to worsen over time. Thiazide diuretics may also alter bone mineral composition[7] and cause excessive calcium retention, particularly in patients with undetected mild hyperparathyroidism.[8]

Simple Herbal Osmotic Diuretic Tea

Mucilaginous herbs such as *Althaea* can offer pain relief in infectious situations, whereas a stimulating diuretic can worsen the discomfort. *Althaea* is best prepared using a long, even overnight soak in cold water (for at least 8 hours) and then bringing the tea briefly to a gentle simmer. The tea may become so thick as to need to drip through a strainer for 15 minutes or more. These thick, slimy teas can be consumed as is, for welcome pain relief in acute situations such a bladder infection, renal calculi, or allergic reaction. They may also be diluted with a separately brewed tea of other flavor-enhancing or therapeutic ingredients.

Althaea officinalis root

Soak 3 tablespoons of *Althaea* root in 3 cups of water overnight. In the morning, bring to brief boil for 5 minutes, turn off the heat, let stand for 20 minutes, and then strain. Drink for a soothing, pain-relieving diuretic.

Diuretic Tincture for Circulatory Insufficiency

When edema occurs due to cardiomyopathy, herbs that increase perfusion to the kidneys and increase glomerular filtration would be more helpful than simple high-mineral herbs such as those in the Herbal Osmotic Diuretic Tea, above. *Angelica*, *Salvia*, and *Ginkgo* have the effect of moving more blood through the kidneys to help diurese. *Juniperus* is an irritant diuretic that is used in a small amount in this formula to stimulate the kidneys and increase glomerular filtration.

Juniperus communis 5 ml
Petroselinum crispus 10 ml
Angelica sinensis 15 ml
Ginkgo biloba 15 ml
Salvia miltiorrhiza 15 ml

Combine the tinctures in a 2-ounce bottle. Take ½ to 1 teaspoon of the combined tincture 3 to 6 times daily. This tincture should be complemented with additional cardiotonic therapies from CoQ_{10} to *Convallaria* to a *Digitalis* prescription.

Diuretic Tea for Renal Insufficiency

Herbs in the form of a tea may be more effective than tinctures. *Solidago* and *Pueraria* may both increase renal cellular function in cases of oliguria or anuria associated with nephropathy. *Equisetum* is added as a mineral diuretic and *Juniperus* is added in a small amount as an irritant diuretic to complement the more nourishing and tonifying base herbs. *Glycyrrhiza* is added for flavor.

Solidago canadensis 3 ounces (90 g)
Pueraria montana 3 ounces (90 g)
Equisetum arvense 3 ounces (90 g)
Glycyrrhiza glabra 3 ounces (90 g)
Juniperus communis 1 ounce (30 g)

Hydrogogues versus Depurants

The use of renal depurants is a concept from traditional herbalism: Herbs are used to help the kidneys excrete solid wastes via the urine. *Solidago*, *Petroselinum*, *Equisetum*, and other herbs were not regarded simply as diuretics, but rather as agents capable of "opening the emunctory" function of the kidneys and supporting the removal of metabolic wastes via the kidneys. In contrast to the simpler concept of a diuretic, a hydrogogue is an agent capable of increasing water excretion, while a renal depurant increases the amount of solute present in the urine. Renal depurants are indicated when there are frequent infections, a tendency to stones, poor glomerular filtration, or symptoms of systemic toxicity (skin eruptions, a coated tongue, joint stiffness, frequent dull headaches, and vague malaise or fatigue).

At times, high-mineral herbs were included in traditional formulas for musculoskeletal complaints, chronic infections, toxicity symptoms in the body, hypertension, and skin diseases in order to help the kidneys to purify the blood as a means of treating wider systemic issues. Thus, a renal depurant is a purifying agent. In practice, herbs that are hydrogogues are usually renal depurants and vice versa. But in philosophy, in the treatment of various systemic diseases, many herbal practitioners include alterative agents to support liver and intestinal function and include renal depurants for an analgous support of the renal elimination pathways.

Steep 1 tablespoon of the herb mixture per cup of water and then strain. Drink 3 cups per day.

Herbal Irritant Diuretic Tincture

The irritant diuretics or caffeine and stimulants are best used for the short term only because they can irritate tissues and occasionally lead to mineral depletion. Urinary irritants such as *Juniperus* and *Petroselinum* are only appropriate in some chronic conditions with a cold, damp, edematous ("stuck" or "stagnant"), and atonic presentation. An example may be an elderly person with poor circulation and rather slow urine production, with the urine appearing dark in color and cloudy and containing much debris. Irritant diuretics are best used in a small proportion in urinary formulas, in a base of more tonifying and nourishing herbs as specifically indicated by the case. Diluting irritants such as *Juniperus* by putting the dose in a glass of water helps prevent excessive stimulation and irritation.

Solidago canadensis 20 ml
Angelica sinensis 20 ml
Petroselinum crispum 10 ml
Juniperus communis 10 ml

Add 1 dropperful of the combined tincture to a glass of water. Drink at least 3 such glasses of tincture water daily.

Diuretic Herbs

Among the following useful diuretics, some contain cardiac glycosides that assist the heart in perfusing the kidneys; others are mineral-rich herbs that help the kidneys create osmotic gradients to best produce urine and process nitrogenous wastes. Note how these diuretic herbs are chosen for specific purposes and indications in the formulas throughout this chapter.

Achillea millefolium
Agropyron repens
Apium graveolens
Arctostaphylos uva ursi
Convallaria majalis
Cytisus scoparius
Equisetum spp.
Galium aparine
Levisticum officinale
Petroselinum crispum, P. sativum (particularly seeds)
Sambucus nigra
Taraxacum officinale
Urtica spp.
Zea mays

Formulas for Cystitis

Simple bacterial cystitis, also called a urinary tract infection (UTI), is a common urinary pathology in general clinical practice, and one that will often respond to mannose powder and herbal teas and tinctures. The most common pathogen is *Escherichia coli* of intestinal origin, which, when introduced into the urethra, may colonize and cause bladder infections. For women prone to cystitis following sexual activity,[9] drinking a large glass of water before and after may deter chronic UTIs. In other cases, diabetes and excessive sugar consumption may create an hospitable environment for urinary pathogens, and therapy will involve managing blood sugar and usually making dietary changes. Cystitis is rare in males and necessitates a diagnostic workup to rule out *Chlamydia* or other pathogen or underlying pathology. The urinary mucosa can be affected by allergic reactivity as can the respiratory mucous membranes, but this is often overlooked as a cause of cystitis-like symptoms in hypersensitive individuals. When all the signs of a UTI are present, but the urine is sterile, consider interstitial cystitis or a urinary allergy.

Herbal teas are recommended for cystitis because the herbs can reach the bladder most effectively, but such teas may be complemented by mannose powder, tinctures, and encapsulated products. Such protocols are usually effective, but if there is no improvement in 48 hours, antibiotics and urine cultures should be considered. Urinary antimicrobials such as *Arctostaphylos uva ursi*, *Agathosma* (formerly *Barosma*), or *Chimaphila umbellata* may be base herbs in UTI formulas, and systemic immune supportive herbs such as *Echinacea* or berberine capsules or vitamin C would be complementary. While keeping a constant trickle of antimicrobial tea going through the bladder is therapeutic, some people are under the impression that they should drink copious amounts of water to "flush the system." Flushing, however, may actually irritate already inflamed tissues, and resting the system may be more effective. The use of demulcent herbs in tea formulas can allay pain and soothe irritated passages, with *Althaea* being a personal favorite.

While *E. coli* is the most common pathogen, a variety of bacteria can infect the urinary tissues, and various

Equisetum for Repair and Regeneration

Horsetail (*Equisetum arvense, E. hymenale,* and related species), although not extensively researched in the modern era, are certainly featured prominently in folkloric herbalism for strengthening connective tissue and bone. *Equisetum* has been emphasized in virtually all the folkloric texts for strengthening the teeth and for healing wounds and broken bones as well as for treating joint, skin, and other connective tissue issues—from arthritis to gum disease to urinary complaints. More than a simple pain reliever or wound-healing agent, *Equisetum* is believed to help remineralize weak tissues. The plant is naturally high in minerals, giving it a strong, rough, even sandpaper-like quality. Another common name for some plants in this genus is scouring rush, which is derived from the use of the gritty stalks to scour pots and pans and sand wood. Many traditional recipes suggest prolonged boiling of the plant to achieve the best mineral extraction, a process that may be enhanced with the addition of a small amount of vinegar or lemon juice.

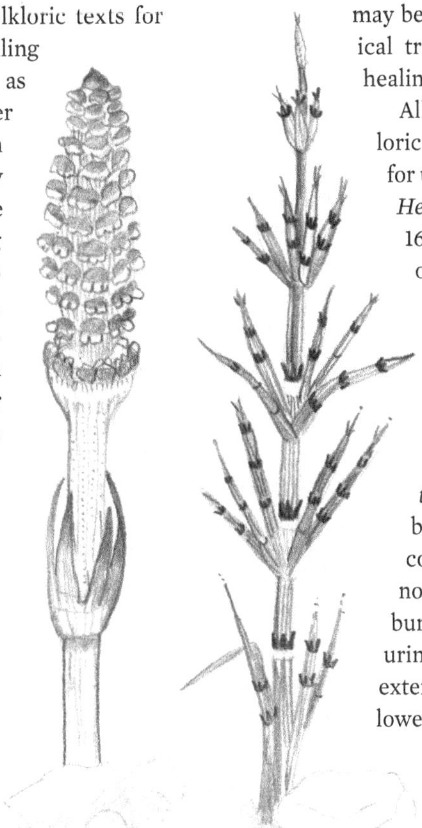

Equisetum arvense, horsetail

Equisetum has also been regarded as having many benefits to the urinary system, acting as a nourishing diuretic in cases of edema, improving urinary ulcers, and tonifying the urinary tissues. *Equisetum* exerts a diuretic effect without wasting minerals[10] as the thiazide diuretics do. The abundant silica salts may chelate heavy metals and reduce inflammation locally in the urinary system. They may also reduce systemic inflammation, which could be one mechanism of its folkloric reputation for reducing joint pain and arthritis. Rheumatoid arthritis is an autoimmune-driven inflammatory process, and heavy metal toxicity and leaky barrier functions are possible etiologic triggers. *Equisetum* may reduce antigen-induced arthritis[11] and offer immunomodulatory effects, and similar mechanisms may benefit interstitial cystitis. Clinical trials report enhanced wound healing following episiotomy.[12]

All the most well-known folkloric herbals suggest *Equisetum* for urinary issues. In his *Complete Herbal,* originally published in 1653, English physician Nicholas Culpeper recommends decocting *Equisetum* in wine to treat urinary stones and "stranguary"—spastic colicky pains in the bladder. William Boericke's *Materia Medica* of 1927 says *Equisetum* works principally on the bladder and is specific for a constant severe urinary urging not relieved by voiding, for sharp burning and cutting pain with urination, deep renal pain that extends into the abdomen and lower back, for urine that passes painfully drop by drop, and for involuntary passing of urine. The folklore also emphasizes *Equisetum* for incontinence in children and the elderly, urinary retention, dysuria during pregnancy and the postpartum period, mucous in the urine, albuminuria, urinary "gravel," an old term referring to urinary calculi (or simply much crystalline debris making a person prone to calculi), urinary ulceration, and hematuria. *Equisetum* may also be pain-relieving and tissue-restorative when prepared into a sitz bath or tub soak and is inexpensive enough to do so daily as part of a protocol in treating cystitis.

urinary pathogens may correlate to the age of the patient and the source of urinary infections. Occasional UTIs in reproductive age women and younger girls are due to *E. coli* of digestive tract origin around 80 percent of the time. *Staphylococcus saprophyticus* is another cause of UTIs in around 10 percent of cases and may occur in otherwise healthy people and be transmitted from casual contact with biofilms in swimming pools, uncooked meat, or other environmental exposures. *Klebsiella* and *Proteus* infections are less common causes of common UTIs. *Enterococcus, Enterobacter, Streptococcus* strains, and *Staphococcus aureus* are less common urinary pathogens, usually found only in susceptible individuals such as diabetics or immune-compromised individuals. Prostatic hypertrophy causes urinary retention and congestion and may also foster atypical pathogens in some cases. The external genitalia can become infected with sexually transmitted pathogens including *Chlamydia trachomatis, Neisseria gonorrhoeae*, Herpes simplex, and *Trichomonas vaginalis*. Common vaginal yeast and other fungi may cause dysuria due to irritation of the external tissues at the urethral outlet. *Lactobacillus* can also overgrow and be the cause of vaginitis and urethritis, but do not typically ascend to infect the bladder. Hot sitz baths may be comforting and therapeutic in such situations, and *Calendula* and *Achillea* teas are good herb choices, along with a few drops of *Melaleuca* oil, to create an antiseptic soak.

Ericaceous plants such as *Arctostaphylos uva ursi* and *Vaccinium* species, including cranberries, are notably high in disinfecting phenolic compounds[13] and mannose. *Arctostaphylos* contains arbutin, which is metabolized into hydroquinone[14] and is readily taken in by *E. coli.* Another polyphenol in *Arctostaphylos*, corilagin, is also active against antibiotic-resistant strains of *Staphylococcus aureus* including[15] methicillin-resistant *Staph aureus* (MRSA) and may significantly increase the antimicrobial effects of β-lactam antibiotics such as oxacillin and methicillin.[16] Although there is a persistent rumor in the herbal community that uva ursi only works when the urine is alkaline, the research on bacterial deconjugation by arbutin has not shown this to be true.[17] (Alkalinizing the urine, however, may be pain-relieving on its own, helping to perpetuate the rumor.) Because *Arctostaphylos uva ursi* has also demonstrated some antiallergy and anti-inflammatory activity, it would also be appropriate in urinary symptoms that are of an inflammatory as well as an infectious nature.[18] Other herbs shown to deter common urinary pathogens include *Rosmarinus officinalis*,[19] *Andrographis paniculata*,[20] and *Berberis* species.[21] *Armoracia rusticana* (horseradish) and *Tropaeolum* (nasturtium) may reduce chronic UTIs.[22] Thujone can act as a urinary irritant, but also contributes to the antimicrobial effects on the bladder by *Thuja, Achillea*, and *Artemisia* species.[23] *Agathosma* and *Eucalyptus* are other antimicrobial plants high in volatile oils that have also been used folklorically for cystitis.

Basic Tea for Simple Cystitis

In this formula, *Rosmarinus, Mahonia* (also known as *Berberis*), and *Arctostaphylos* serve as antimicrobial agents, supported by *Althaea* to soothe the mucosa and *Equisetum* and *Calendula* to help strengthen the uroepithelial barriers.

Arctostaphylos uva ursi
Mahonia aquifolium roots
Equisetum arvense
Rosmarinus officinalis
Althaea officinalis
Calendula officinalis

Combine equal parts of the herbs. Steep 1 tablespoon of the herbal mixture per cup of hot water and then strain. Drink freely, a minimum of 4 cups daily.

Tincture for Severe Urinary Tract Infections

To create a more aggressive treatment for UTIs, a tincture such as this would complement the Basic Tea for Simple Cystitis. The use of both the tea, preferably with mannose in it, and the tincture, all together, can match the efficacy of antibiotics.

Mahonia aquifolium
Echinacea angustifolia, E. purpurea
Andrographis paniculata
Thymus vulgaris
Agathosma betulina

Combine in equal parts. Take ½ teaspoon of the combined tincture every half hour, reducing as symptoms improve over several days.

Urinary Demulcent Tea for Pain Management and Mucosal Repair

Althaea roots are high in mucilage that can be easily extracted by soaking the roots in water overnight or for at least 2 or 3 hours. The resulting tea should be thick and slimy to best soothe an irritated bladder. *Althaea* contains asparagine, an amino compound that may

Herbs for Hematuria

These herbs are emphasized folklorically for hematuria—blood in the urine—whether macroscopic or microscopic. Some older books use the term "brick dust urine," which refers to urine where a red or rusty-colored sediment accumulates at the bottom of the urine specimen cup, indicating the presence of red blood cells.

Achillea millefolium
Agropyron repens
Alnus rubra
Arnica montana
Chimaphila umbellata
Equisetum arvense
Eupatorium purpureum
Hamamelis virginiana
Hydrangea spp.
Mahonia aquifolium
Piper cubeba

Herbs for Pyuria

These herbs are specifically indicated for pyuria, meaning pus in the urine. This can be detected with a microscopic examination of urine or at times, by the naked eye when white sediment accumulates at the bottom of the urine specimen vessel.

Althaea offiicinalis
Arctostaphylos uva ursi
Agathosma betulina
Berberis vulgaris
Chimaphila umbellata
Equisetum arvense and other spp.
Eupatorium purpureum
Hamamelis virginiana
Hydrangea macrophylla and other spp.
Mahonia aquifolium
Piper cubeba
Zingiber officinale

help repair glycosaminoglycans in the urinary mucosal membranes. *Asparagus* species also contain asparagine, giving rise to the compound's name. A species of *Asparagus* native to India, known as shatavari, may be used in a similar manner. To help the tissues recover from a severe infection or to allay pain while taking an antibiotic, a tea such as this would be helpful. This tea does not have strong antibacterial activity on its own.

Althaea 4 ounces (120 g)
Asparagus 4 ounces (120 g)

Soak 1 tablespoon of the herb mixture in 4 cups of water overnight. Bring to a gentle simmer on the lowest possible setting for 5 minutes, and remove from the heat and let stand for a half hour. Strain through a wire mesh or muslin bag and sip constantly throughout the day. More flavorful herbs may be added as desired to the finished brew, or the tea may be combined with a more complex UTI blend such as the Basic Tea for Simple Cystitis on page 114, if treating an infection.

Urinary Trophorestorative Tea

A trophorestorative is a folkloric concept of a medicine that can restore tone and function to organs and tissues, whether excessive or deficient. Terms from TCM and other energetic models refer to tissue status and constitutional tendencies such as atrophic versus hypertrophic, hot versus cold, and excessive secretions versus deficient secretions. When urinary function is sluggish in the elderly, when circulation to the kidneys is deficient in diabetics, or when complaints are chronic with poor recovery, urinary trophorestoratives may be beneficial. Herbs considered to be urinary trophorestoratives include *Solidago, Eupatorium purpureum, Silybum marianum*, and *Pueraria*. When circulation or renal perfusion are impaired, *Angelica, Salvia miltiorrhiza, Ginkgo*, and *Crataegus* may be appropriate additions to formulas for diabetics and those with poor circulation to the kidneys. This formula could be simplified, selecting the best herbs for any one individual, as appropriate.

Ginkgo biloba
Pueraria montana
Salvia miltiorrhiza
Eupatorium purpureum
Agrimonia eupatoria
Solidago canadensis
Equisetum arvense
Glycyrrhiza glabra

Combine equal parts of the dry herbs, such as 2 ounces of each herb, to yield a pound of the trophorestorative tea blend, intended for long-term use. Steep 1 tablespoon of the herb mixture per cup, and then strain. Drink freely, 3 or more cups per day.

Tincture for Cloudy, Dark Urine with Infection Tendency

A tincture such as this may complement the Urinary Trophorestorative Tea for those with chronic urinary issues or elevated blood urea nitrogen (BUN) or creatinine and for those in early stages of renal insufficiency. This formula is better for prevention and maintenance, switching to a stronger antimicrobial formula when needed. *Juniperus* is used here in a small amount as a

Herbs for Thick and Cloudy Urine

These herbs are historically emphasized for urine that is thick and cloudy and contains obvious particulates, precipitates, and debris. These urine conditions are all signs of dehydration, sluggish flow through the kidneys, and various situations of renal insufficiency and renal failure.

Agrimonia eupatoria
Chimaphila umbellata
Collinsonia canadensis
Equisetum arvense
Eucalyptus globulus
Eupatorium purpureum
Hydrangea spp.

Herbs for Cystitis and Pyelonephritis

Infections of the urinary system, including acute cystitis and pyelonephritis, require antimicrobial agents to help deter pathogens. The following herbs help provide antimicrobial effects on urinary tissues.

Achillea millefolium
Allium sativum
Althaea officinalis
Andrographis paniculata
Arctostaphylos uva ursi
Armoracia rusticana
Artemisia absinthium
Agathosma betulina
Chimaphila umbellata
Equisetum arvense
Eucalyptus globulus
Galium aparine
Gaultheria procumbens
Mahonia aquifolium
Piper cubeba
Rosmarinus officinalis
Tropaeolum majus

Folkloric Herbs Specific for Dysuria

Dysuria (painful urination) can be a symptom of urinary infection as well as urinary calculi and possibly allergic or other irritation. These herbs are mentioned in folkloric traditions as being helpful for dysuria.

Agropyron repens
Althaea officinalis
Arctium lappa seed
Arctostaphylos uva ursi
Arnica montana
Cantharis vesicatoria
Eucalyptus globulus
Eupatorium purpureum
Gaultheria procumbens
Hedeoma pulegioides
Hydrangea spp.
Mahonia aquifolium
Petasites hybridus
Petroselinum crispum
Piper cubeba
Zingiber officinale

renal irritant to stimulate slow glomerular filtration due to poor circulation or impaired cellular metabolism.

Pueraria mirifica 20 ml
Silybum marianum 20 ml
Mahonia aquifolium 15 ml
Juniperus communis 5 ml

Take ½ to 1 teaspoon of the combined tincture 3 to 6 times per day.

Tea for Diabetics Prone to Urinary Infection

Diabetics are prone to numerous infections because the high blood sugar creates a hospitable ecosystem for many pathogens. A tea such as this is gently antimicrobial, while providing additional blood sugar regulation with *Mahonia* (also known as *Berberis*) and cinnamon.

Calendula officinalis
Sambucus nigra finely chopped or powdered berries
Mahonia aquifolium
Rosmarinus officinalis
Cinnamomum spp.
Arctostaphylos uva ursi

Combine equal parts of the herbs. Steep 1 tablespoon of the herb mixture per cup of hot water, and then strain. Drink freely, 3 or more cups per day.

Mannose Powder for Acute or Chronic UTIs

Mannose is a sugar found in cranberries and blueberries that is able to interfere with the ability of *E. coli* bacteria to bind to the urinary mucosa and create a biofilm. It is pleasant-tasting and best consumed in water to help it reach the urinary mucosa quickly. It is somewhat expensive, but the effective dose is small and works so well that it is worth the expense.

Mannose powder

Dissolve ¼ teaspoon in ¼ cup to a full cup of water and take as often as hourly for acute infections. A dose once or twice a day can serve as a preventive in those with chronic UTIs.

Tea for UTI in Pregnancy

Bladder infections in pregnancy may induce preterm delivery, or when asymptomatic and persistent, low infant birth weight. Cranberry juice may prevent infections and reduce the risk of acute pyelonephritis.[24] Due to concerns of teratogenic effects, many herbs are avoided during pregnancy, and antibiotics are of concern due to possible harm to intestinal flora. However, ascending urinary infection can trigger premature labor,[25] so intervention is justified. A gentle antimicrobial tea with mannose powder may be an effective alternative to antibiotics, and these herbs are considered safe during pregnancy.

Calendula officinalis 4 ounces (120 g)
Rosmarinus officinalis 4 ounces (120 g)
Cranberry powder 2 ounces (60 g)
Mannose powder

Combine the herbs and cranberry powder. Steep 1 tablespoon of the herb mixture per cup of water and then strain. Add 1/4 teaspoon of mannose powder per cup. Drink continually throughout the day.

Tea Formula for UTIs Due to Indwelling Catheter

Those with an indwelling catheter are prone to UTIs, yet the repetitive use of antibiotics may promote the development of antibiotic-resistant strains of bacteria and necessitate larger doses and more aggressive drugs.[26] Adding herbal tea, cranberry juice or powder, and mannose powder to the daily routine may deter opportunistic infections.

Calendula officinalis
Mahonia aquifolium
Rosmarinus officinalis
Glycyrrhiza glabra
Cranberry powder
Mannose powder

Combine equal parts of the herbs and cranberry powder. Brew 1 heaping tablespoon of the herb mixture per cup of hot water. Add 1/4 teaspoon of mannose powder to at least 1 cup of the tea each day and consume a total of at least 3 cups per day. Large batches of the tea can be prepared and refrigerated to consume for several days, as desired.

Boric Acid Suppositories for Bacterial Vaginosis

Because vaginal infections can lead to bladder infections or simply to dysuria as the urine passes by inflamed tissues, treating bacterial vaginitis is sometimes important

Mannose and Cranberries for UTIs

Cranberry juice is well known by the general public as a treatment for UTIs, but people must be educated to avoid products sweetened with sugar or corn syrup. Cranberry (*Vaccinium macrocarpon*) contains mannose, which deters *E. coli* from adhering to the urinary mucosa.[27] Consuming cranberry teas or mannose in water may reduce the frequency of chronic UTIs.[28] Phenolic compounds abundant in all *Vaccinium* species are credited with at least some of the activity against UTIs. Purified mannose powder is a useful therapy in the management of both acute and chronic UTIs and will often provide some symptom relief in a matter of hours. Use approximately 1/4 teaspoon of mannose in a cup of herbal tea, as often as hourly, reducing to 3 times per day, in the treatment of acute UTIs. It is practical to put mannose powder in an herbal tea for treating acute cystitis.

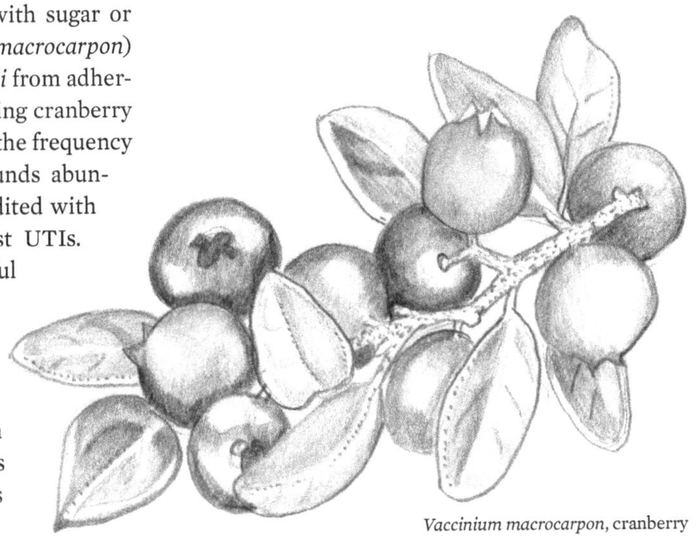

Vaccinium macrocarpon, cranberry

Vaginal Estrogen and UTIs

Estrogen supports healthy urinary epithelia and also maintains vaginal epithelia.[29] A postmenopausal decline in hormones is associated with changes in the urinary mucosa that may increase susceptibility to bladder infections. Postmenopausal women with chronic UTIs may respond to estriol suppositories.[30] Oral hormone replacement therapy may not extend the same benefits, and due to concerns over possible side effects with the systemic hormones, local hormones are preferable for chronic UTIs.[31] Local estrogen creams or suppositories help maintain an acidic pH desirable for deterring microbes, and their use may be complemented by the use of boric acid suppositories.

in the management of chronic UTIs. Treatment of bacterial vaginosis (BV) with antibiotics, such as metronidazole, results in high recurrence rates and secondary *Candida* infections. Intravaginal boric acid has been used for more than a hundred years to treat vaginal infections such as BV. Boric acid is inexpensive, but not many encapsulated products are on the market, such that practitioners may need to make boric acid suppositories for themselves. There is much empirical and anecdotal evidence that boric acid is effective for many types of vaginal infections, but few studies. However, one study reported boric acid suppositories to be as effective as metronidazole in treating BV.[32]

Boric acid suppositories
(food-grade boric acid in 00 gel caps)

Insert 1 or 2 capsules vaginally each night before bed for 5 nights. Take 2 days off and repeat for another 5 days if necessary.

Tincture Formula for Postmenopausal UTIs

Actaea offers gentle hormonal support, and *Angelica* enhances blood circulation in the pelvis. These may help reduce UTI occurrence in chronic sufferers and may be complemented by boric acid capsules. These herbs are not antimicrobial, but rather may improve the integrity of the urinary mucosa and help prevent infections, as part of a broader protocol. Some women may benefit from estrogen creams applied to the external genitals, once or twice a week, to provide hormonal support to the urothelium as well as to the vaginal mucosa.

Actaea racemosa (formerly *Cimicifuga racemosa*)
Angelica sinensis

Combine in equal parts, and take 1 teaspoon of the combined tincture daily.

Formulas for Pyelonephritis

Nephritis is an acute infection of the kidneys that typically ascends from a bladder infection, even if the presentation does not appear to follow a UTI. The onset is often acute with malaise, fever, pain, and sometimes nausea. Pyelonephritis so often involves a mild fever that its absence may suggest pelvic inflammatory disease (PID) or diverticulitis as an alternate cause of the symptoms.[33] Those affected often appear ill or toxic and may have a flushed face and even a foul odor. Urinary symptoms usually accompany, but may be absent in some cases. The abdomen may be tender or rigid with severe infections, and in most cases, the kidneys will be tender to firm palpation, and there will be sharp pain with pressure at the costovertebral angles. White blood cells occur in the urine, along with high numbers of sloughed, football-shaped renal cells and amorphic debris, depending on the type of pyelonephritis. In some cases, tubular casts may be seen in a microscopic exam, such as the red blood cell casts pathognomonic of post-streptococcal glomerulonephritis.

Pyelonephritis is most common in younger adults, especially women, and less common in elders, and acute infections are often bilateral. Renal or urinary obstruction may predispose, including urinary calculi, strictures, and prostatic enlargement, as may neurogenic bladder. As with cystitis, *E. coli* is the most common pathogen responsible for acute pyelonephritis, but *Klebsiella, Proteus, Enterobacter, Pseudomonas, Staphylococcus,* and *Streptococcus* can all be causative agents; urine cultures should be initiated in case herbal formulas are ineffective and antibiotics are required. Fungal infections are less common pathogens associated with pyelonephritis, but can occur in immune-compromised individuals, such as HIV patients or those on long-term steroids or immunosuppressive drugs.

Herbal therapies must be aggressive, which usually means hourly dosing with one or more medicines—a tea, a tincture, and an herbal capsule. Patients who do not improve in 24 to 48 hours should be switched to antibiotics to prevent kidney damage or scarring. Herbal therapy for pyelonephritis is virtually identical to that for cystitis, only with greater vigilance and follow-up to ensure efficacy. The dose should be aggressive, such as by sipping antimicrobial tea constantly through the day, and by the use of additional immune-supportive capsules, antimicrobial tinctures, and possibly hot-tub soaks or hydrotherapy. Traditional folkloric approaches to treating kidney infections suggest that promotion of sweating helps eliminate wastes via the skin, helping relieve the kidneys of some of the workload. A light diet, low in protein, or a period of fasting should be encouraged to ease the kidney's workload as well. Heat packs over the kidneys may provide pain relief as well as possible further antimicrobial support. Avoid harsh antimicrobials, such as *Juniperis communis,* which can irritate already inflamed renal tissues. Demulcent herbs such as *Althaea, Ulmus,* or *Zea mays* (maize, corn) may help allay discomfort but do not provide antimicrobial effects on their own. Demulcents such as *Agropyron repens* (couch grass) and maize may interfere with bacterial adhesion to urinary mucosa,[34] likely due to mannose or related sugars in the mucilage. *Allium sativum* (garlic) capsules may complement herbal teas or may be included in tincture formulas due to broad-spectrum antimicrobial effects, including activity against antibiotic-resistant strains of microbes.[35] *Urtica urens, U. dioica,* or other species may also deter bacterial adherence. Bromelain, a proteolytic enzyme in pineapples, may complement herbal teas and tinctures for treating pyelonephritis, as it has antimicrobial effects and boosts immune response via cytokine and interleukin support.[36] Bromelain may boost the efficacy of antibiotics and increase their levels in the blood and urine when used in combination in treating pyelonephritis.[37]

Urinary Parasites

Parasites can infect the kidneys, particularly in tropical regions of the world. A fairly common vaginal parasite, *Trichomonas,* may infect the kidneys and renal passages in both men and women. *Schistosoma,* a parasite that can be contracted while swimming in tropical rivers or lakes, can infect the kidneys, liver, and other organs and may cause liver and bladder carcinoma. When left untreated, it may also lead to renal obstruction and disease.[38] *Ceratonia siliqua* (carob) may reduce fibrotic changes due to schistosomiasis.[39] Sesquiterpene lactones in *Artemisia absinthium* and *Tanacetum parthenium* have schistosomicidal activity, with parthenolide shown to reduce motor activity and reproductive cycles in adult worms.[40] Schistosomiasis causes significant changes in the bladder wall and may induce the formation of large masses, calcifications, and hydronephrosis;[41] albuminuria and microscopic hematuria are common.[42] *Entamoeba histolytica,* an amoeba, may also invade the renal system and should be considered in those living in, or returning from, tropical locales.

Tea for Acute Pyelonephritis

This tea should be combined with the Tincture for Acute Pyelonephritis, mannose powder, and encapsulations to create an aggressive treatment plan.

Calendula officinalis
Arctostaphylos uva ursi
Althaea officinalis
Urtica urens, U. dioica
Achillea millefolia
Mahonia aquifolium
Mannose powder

Combine equal parts of the dry herbs. Steep 1 tablespoon of the herb mixture per cup of hot water, and then strain. Place ½ to 1 teaspoon of mannose powder in each cup. Drink 5 or 6 cups a day for several days, reducing as symptoms improve.

Pyelonephritis

Herbal Specifics for Pyelonephritis from the Eclectic Literature

The Eclectic physicians were a group of alternative-minded MDs who practiced in the late 1800s and early 1900s and were adept at using herbal medicines in highly specific ways. Many of the following botanicals are available by prescription only due to their strength and potential toxicity and are to be used by experienced clinicians only.

HERB	CONDITION
Aconitum napellus	Chills with a high temperature and with symptoms that come on suddenly with great intensity, with fear and restlessness
Actaea (Cimicifuga) racemosa	Aching back and muscles
Apocynum cannabinum	Edema with renal infections
Gelsemium sempervirens	Flushed, restless, with contracted pupils; possibly paralyzed by the illness

Tincture for Acute Pyelonephritis

This tincture is to complement the Tea for Acute Pyelonephritis and the Encapsulations for Acute Pyelonephritis to best treat the condition.

Agathosma betulina
Chimaphila umbellata
Andrographis paniculata
Rosmarinus officinalis

Combine in equal parts and take 1 dropperful of the combined tincture every 30 minutes for the first day or two, reducing as symptoms improve.

Encapsulations for Acute Pyelonephritis

Garlic and bromelain are options when treating urinary infections, including pyelonephritis. Garlic offers additional antimicrobial support and complements urinary teas and tinctures. Bromelain may enhance the efficacy of antibiotics in cases of lingering and poorly responsive infections.

Bromelain capsules
Allium capsules

Take 2,500-milligram bromelain capsules 3 or 4 times daily for a week. Take 2 *Allium* capsules 3 or 4 times daily for a week. Note that *Allium* capsules are available in different strengths and preparations. Those who experience digestive gas from garlic might take the lower doses.

Formula for "Kidney Congestion and Inflammation" from 1900

Physiomedicalist texts of the early 1900s discuss "kidney congestion," defined as a condition of scanty urination and pain in the small of the back and associated with digestive difficulty, liver burden, and chronic constipation. Many such texts and herb books of the era emphasize liver and alterative herbs in the treatment of chronic urinary pain or infection. *Podophyllum* was used in pyelonephritis formulas when the liver was congested, along with a light diet and daily hot or vapor baths.[43] *Podophyllum* is extremely caustic and cathartic and is to be used only by skilled clinicians and in small doses. This tincture is sweetened and mellowed with a substantial amount of ginger syrup, prepared from *Zingiber officinale*, which is commercially available or can be prepared for one's self. The syrup also helps to temper any potentially harsh effects of the *Podophyllum*.

Syrup of *Zingiber officinale* 150 ml
Eupatorium purpureum 30 ml
Podophyllum peltatum 15 ml
Hydrastis canadensis 15 ml
Althaea officinalis 15 ml
Agathosma betulina 15 ml

This 8-ounce formula is to be taken by the teaspoon, 3 times a day.

Formulas for Interstitial Cystitis and Lower Urinary Epithelial Dysfunction

Interstitial cystitis (IC) is a chronic inflammatory disorder of the bladder mucosa, occurring mostly in young to middle-aged women. In some sufferers, the dysfunction is restricted to the urethra, causing urinary frequency and dysuria, and is referred to as urethral pain syndrome.[44] Vulvodynia, dyspareunia, prostatitis, and proctitis may be variations of the pathology.

IC affects more than 1 million people in the United States alone. The causes of IC are multifactorial and the condition is analogous to leaky gut, in that the urinary epithelial cells lose integrity, allowing inflammatory molecules to pass to the underlying muscle and interstitial tissues, provoking pain and further inflammation. IC is therefore part autoimmune, part hypersensitivity, and part autointoxication, all contributing to the breakdown of urothelium. Potassium levels are abnormally elevated in IC and may induce heightened nervous and electrical sensitivity,[45] and potassium sensitivity tests are emerging as a possible diagnostic marker. Anxiety and depressive disorders often occur in tandem with IC,[46] and herbal formulas for IC may benefit from the inclusion of nervine herbs when indicated.

IC may mimic acute cystitis, yet the urine is sterile. There is often an unremitting desire to urinate and constant tight or burning sensation in the pelvis. In some cases cystoscopy may be normal, but in 90 percent of the cases, small ulcers or pinpoint bleeding, pus and exudates, gross scarring, glomerulations, and fibrosis may be visualized with cystoscopy.

Caffeine and acids may aggravate the symptoms, and a trial elimination of coffee, artificial sweeteners, benzoate preservatives, and other allergens or chemicals should be implemented, especially in allergic patients. Physical therapy is reportedly successful for IC and

Interstitial Cystitis in Men

When IC symptoms occur in men, the diagnosis is more commonly noninfectious prostatitis, lower urinary epithelial dysfunction (LUED), or chronic pelvic pain syndrome (CPPS).[47] Statistics on men with IC (or a variation thereof) report that the average age at diagnosis is 51 years, that the majority also suffer from painful ejaculation, and 40 percent have microscopic hematuria.[48] The most commonly reported symptoms in men are "bladder" pain (subjective suprapubic discomfort as well as tenderness to suprapubic palpation), anterior rectal tenderness with digital exam, urinary frequency, nocturia, and dysuria. With a cystoscopic exam, most men with such symptoms may be shown to have glomerulations in the bladder mucosa.

Support GAGs Molecules to Treat IC

Because IC involves a breakdown in the connective tissue matrix necessary to regenerate the urothelial cells from the basement membrane,[49] agents that support the integrity of glycosaminoglycans (GAGs) are helpful but are a slow and gradual therapy. Breakdown of the GAGs barrier is likely complex and multifactorial, with possible genetic, dietary, emotional, and nutritional factors contributing, leading to autoimmune reactions that trigger mast cell degranulation, the release of substance P, and nerve hypersensitivity.[50,51] Therefore agents that support GAGs, reduce mast cell degranulation, and reduce inflammatory cascades may all be helpful. There are cannabinoid receptors in the bladder, and cannabinoid receptor agonists may reduce urinary pain and relax excessive bladder contractility.[52]

Sulfur for the Bladder

Sulfur is an important nutrient for the bladder, crucial to the integrity of collagen and the connective tissue. Just as glucosamine and dimethyl sulfoxide (DMSO) may help treat musculoskeletal weakness, these sulfur-containing compounds may also help treat interstitial cystitis, bladder ulcers, and possibly bladder cancer. Sulfur plays a role in detoxification processes, and glutathione and metabolic detoxification pathways are sulfur-dependent.

Sulfur-containing compounds are also among the most promising medicines for interstitial cystitis (IC). Glycosaminoglycans (GAGs) molecules of the connective tissue are sulfur-rich, and the only pharmaceutical approved for IC so far is penzosan polysulfate sodium, aka Elmiron.[53] Elmiron is a synthetic analog of heparin, itself considered to be a glucosamine compound, and Elmiron may improve IC slowly over time by providing sulfur and helping to repair the mucosal barrier. Sulfated polysaccharides may reduce excessive vascularity via vascular endothelial growth factor (VEGF) inhibition[54] and stabilize mast cells, reducing histamine release,[55] all of which weaken bladder cell adhesion and contribute to loss of barrier integrity.

DMSO has long been used to help treat musculoskeletal pain with topical application. DMSO is often instilled into the bladder and noted empirically to improve IC symptoms, improving cell adhesion, and supporting connective repair. DMSO may help provide methyl groups and sulfur and reduce inflammation. Glucosamine sulfate, chondroitin, and other sulfur-containing substances may also help support connective tissue integrity in the bladder, as they do for the joints and connective tissue, and can be taken as capsules and liquid supplements to complement the teas and tinctures featured in this chapter. Other natural compounds that may provide bioavailable sulfur include methionine, s-adenosylmethionine (SAMe), N-acetylcysteine (N-AC), choline, phosphatidylcholine, lecithin, and methylsulfonylmethane (MSM).

Allium sativum has long been noted to have numerous antimicrobial and immune enhancing effects and exerts anticancer effects via a variety of mechanisms, including detoxification of carcinogens and promotion of immune modulators.[56] Natural sulfur compounds in garlic help the bladder protect itself from toxins and irritants such as cyclophosphamide,[57] nicotine,[58] and a variety of carcinogens.[59]*Allium sativum* contains sulfides, such as diallyl sulfide and disulfide (known popularly as allin and allicin), that protect the urinary epithelium from protamine toxicity,[60] cyclophosphamide, and nicotine-induced oxidative stress and damage.[61] Garlic sulfur compounds demonstrate dramatic protective effects on tissues when examined histologically, compared to animals not pretreated with garlic.[62] Garlic sulfides help to activate detoxifying enzyme systems and maintain glutathione content in the tissues, which are otherwise depleted by toxins, protecting the urinary mucosal tissues from oncogenic changes.[63] Diallyl disulfide inhibits acetyltransferase enzymes that may be overexpressed in bladder cancer cells,[64] and garlic may induce apoptosis in human bladder cancer cell lines,[65] including transitional cell carcinoma.[66] Some studies have shown garlic to be as effective as bacillus Calmette-Guérin (BCG), a leading therapy for bladder cancer, without the side effects.[67]

Allium sativum, garlic

related conditions, referred to by physical therapists as "high-tone dysfunction of the pelvic floor."[68] Transvaginal manual therapies and trigger point techniques have proven successful in clinical trials.[69]

IC has no known cure and only one FDA-approved drug is available: pentosan polysulfate sodium (Elmiron). The condition is notoriously unresponsive, and urologists may instill medicines such as DMSO (dimethyl sulfoxide), or inject botulinum toxins (Botox) directly into the bladder. Physically distending the bladder (hydrodistension) to break apart adhesions when extensive scar tissue has formed may offer some temporary relief.[70] And when all such methods fail, extreme measures such as partial removal of the urinary bladder or urinary diversion procedures have been employed.[71]

Herbal therapies for IC may be aimed at reducing allergy-induced damage to barrier function and restoring and regenerating urothelial connective tissue, while alleviating discomfort as fast as possible. *Piper methysticum* (kava) may calm increased potassium channel activity locally in the bladder.[72] Kava lactones have a prompt anesthetic effect on the oral mucosa[73] and may exert similar effects on the urinary mucosa, along with spasmolytic and anxiolytic effects.[74] See "*Piper methysticum* for Urinary Pain" on page 125. Because histamine contributes to "leaky" barrier functions, herbal mast cell stabilizers having antihistamine effects may be helpful and include *Ammi visnaga* (khella), which has been used as an antispasmodic for bronchial and urinary passages for centuries. Other mast cell stabilizers include *Glycyrrhiza glabra*, *Tanacetum parthenium*, *Ginkgo biloba*, *Allium cepa*, and *A. sativum*.

Centella asiatica (gotu kola) is emphasized in folklore for healing ulcers, broken bones, skin, and joint and connective tissues and may be included in teas for IC. (See the *Centella* entry on page 240.) *Centella* may prevent excessive scar tissue and fibrosis and promote formation of new connective tissue, hyaluronic acid, and chondroitin sulfate.[75] *Allium sativum* (garlic) protects against bladder injury and degeneration,[76] preventing inflammatory response and promoting protective antioxidant levels. *Allium* is notably high in sulfur compounds, which may be important for urothelium integrity and for regulation of mast cell histamine responses. *Eupatorium purpureum* is a classic botanical in Western folklore for chronic bladder, kidney, and urinary complaints. Research suggests that *Eupatorium* may affect cell-to-cell adhesion via integrin molecules,[77] improving the integrity of the uroepithelium and reducing leaky bladder. Patients with IC express elevated quantities of immunoreactive nerve fibers just below the urothelia and in the detrusor muscle, and some sufferers may experience pain relief from cannabinoids (that is, medical marijuana).[78]

Tea for Interstitial Cystitis

This tea offers herbs traditionally reported to heal urinary ulcers and allay chronic urinary pain. A tea such as this should be complemented with the glucosamine and supplements discussed in this section. All the herbs in this tea are emphasized in the folklore for urinary irritation, bladder sensitivity and pain, and wound healing.

Centella asiatica
Calendula officinalis
Eupatorium purpureum
Glycyrrhiza glabra
Hypericum perforatum
Equisetum spp.
Althaea officinalis

Combine equal parts of the dry herbs, or use more *Glycyrrhiza* to taste. Steep 1 tablespoon of the herb mixture per cup of hot water and then strain. Drink freely, a minimum of 3 cups per day.

Natural Therapies for Bladder Mucosal Lesions

These wound-healing herbs and supplements can help heal urothelial lesions and may be used in teas, tinctures, and encapsulations to promote healing in cases of interstitial cystitis and other urinary epithelial lesions.

HERBS

Allium sativum
Althaea officinalis
Calendula officinalis
Centella asiatica
Glycyrrhiza glabra
Hypericum perforatum
Symphytum officinale
Ulmus fulva
Verbascum thapsis
Zea mays

SUPPLEMENTS

Bromelain
Glutamine
Mannose
Glucosamine sulfate
DMSO
Isothiocyanates
Sulfur

Tea for Relief of Acute Urinary Pain

Piper methysticum (kava) is one of the most effective urinary pain relievers and may allay pain more quickly than the Tea for Interstitial Cystitis. However, it may not be as deep-acting and restorative to urothelial integrity over time. When urinary discomfort is acute, this tea may offer some palliative relief, but it should be complemented with glucosamine capsules and other herbal therapies for the greatest effect over time. This tea blends kava with the urothelial tonic herb *Althaea*. Licorice adds flavor and offers its own anti-inflammatory effects.

Althaea officinalis
Piper methysticum
Glycyrrhiza glabra

To obtain the greatest demulcent effects, combine the kava and licorice together in one bag and the *Althaea* in its own bag. Steep 6 heaping tablespoons of *Althaea* roots overnight in cold water. In the morning add 3 tablespoons of the kava-licorice blend and 10 cups of water, bring to a brief simmer for 5 minutes, remove from the heat, and let stand for 15 minutes. Strain the resultant thick tea, and drink over the course of the day for pain relief, as part of a broader protocol for IC.

Tea for Cramping Pains in the Bladder

Older herbals refer to cramping pain in the bladder as tenesmus. When the muscular layer and detrusor muscle are affected by IC and patients describe their discomfort as tight, squeezing, or cramping, these herbs may be most specific.

Piper methysticum
Lobelia inflata
Foeniculum seeds
Matricaria chamomilla
Glycyrrhiza glabra

Combine equal parts of the herbs. Steep 1 tablespoon of the herb mixture per cup of hot water, and then strain. Drink freely, consuming 3 cups within an hour's time for acute spastic pains of the bladder.

Herbs for Urinary Tenesmus

Cramping in the bladder can occur due to allergic reactivity or neurologic dysregulation. The following herbs are emphasized in the historic literature for tight, cramping, and squeezing sensations.

Apis mellifica
Arctostaphylos uva ursi
Arnica montana
Atropa belladonna
Avena sativa
Conium maculatum
Corydalis yanhusuo
Dioscorea villosa
Equisetum arvense
Hyoscyamus niger
Levisticum officinale
Lobelia inflata
Piper methysticum

Tincture for Interstitial Cystitis

Tea formulas probably "touch" the bladder mucosa to a greater degree than tinctures would, but a formula such as this might complement a tea and help create an aggressive protocol.

Piper methysticum
Apium graveolens
Equisetum arvense
Eupatorium purpureum

Combine in equal parts. Take 1 to 2 dropperfuls of the combined tincture 3 to 6 times a day.

Tincture for Interstitial Cystitis with Stress, Anxiety, and Insomnia

These herbs have nervine effects, and *Piper*, *Hypericum*, and *Apium* are also emphasized for urinary discomfort.

Piper methysticum
Withania somnifera
Hypericum perforatum
Apium graveolens

Combine in equal parts. Take ½ teaspoon of the combined tincture 3 or 4 times a day, reducing as symptoms improve.

Tincture for Severe Symptoms, Tenesmus, and Pain

This formula exemplifies how to use *Atropa belladonna* in a formula for acute urinary spasm. Belladonna is considered a prescription-only botanical available to licensed clinicians only. *Lobelia* is also useful for urinary spasm.

Piper methysticum 20 ml
Lobelia inflata 15 ml
Corydalis yanhusuo 15 ml
Atropa belladonna 10 ml

Take 1 dropperful of the combined tincture every 15 to 30 minutes, reducing as symptoms improve. Omit or reduce the belladonna if dry mouth or visual disturbance occurs.

Piper methysticum for Urinary Pain

Piper methysticum has been used by South Pacific Island communities as an intoxicating beverage and as a sedative, muscle relaxant, anesthetic, and anticonvulsant as well as for relieving urinary pain associated with venereal diseases. Kava has an anesthetic[79] effect on the oral cavity, and the Lloyd Brothers, well-respected manufacturing pharmacists of the early 1900s, claimed kava to be specific for soothing irritated genitourinary passages. Kava teas and tinctures may help allay renal colic and support emptying of the bladder in cases of urinary obstruction and retention. Modern research shows kava to exert an anticancer effect on the bladder.[80] Liberal consumption of kava beverages, as seen with men in Fiji, may offer some protection against prostate and urinary cancer. Cancer in general occurs only a third to a quarter as often in Fiji as it does in non-kava-drinking countries.[81]

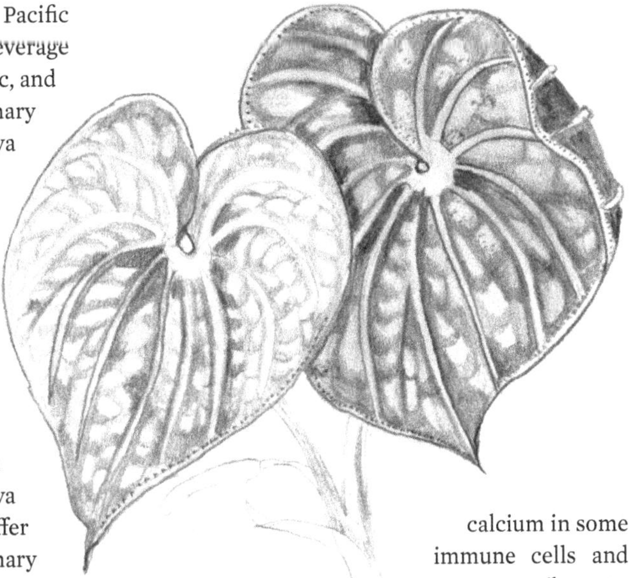

Piper methysticum, kava

Many of kava's physiologic effects are credited to the kavalactones, including kavain, yangonin, and methysticin. Kavalactones may down regulate androgen receptors and reduce excessive stimulation of hormone-sensitive urinary tissues.[82] Kavain inhibits numerous inflammatory pathways including lipopolysaccharides-induced tumor necrosis factor (TNF-α) production.[83] Kava lactones may uncouple sodium and potassium channels from neuronal membranes and thereby reduce nerve impulses to urinary muscles and exert a relaxing effect.[84] Kava leads to significant elevations of intracellular free calcium in some immune cells and may contribute to anti-inflammatory and immune-modulating effects. Kava lactones are also well established to promote *gamma*-aminobutyric acid (GABA) activity and activate the associated ion channels,[85] but less well-known is that there are GABA receptors on mast cells and basophils, which may contribute to kava's anti-inflammatory and immune-modulating effects.[86] Furthermore, kava may also act as a cannabinoid receptor ligand and affect transient receptor potential (TRP) channel pharmacology, all contributing to pain-relieving effects.[87] The kavalactone yangonin, and possibly other kava compounds, bind cannabinoid receptors.[88]

Tincture for Urinary Epithelial Dysfunction in Allergic Individuals

These herbs reduce mast cell activation, histamine, and other inflammatory mediators, and a formula such as this could be the primary tincture in a urinary pain protocol for those with underlying allergic constitutions. I have had good results using such an approach to treat urinary symptoms believed to be of allergic origin.

Tanacetum parthenium
Angelica sinensis
Ammi visnaga
Scutellaria baicalensis

Combine in equal parts. Take 1 dropperful of the combined tincture every several hours for acute symptoms and 3 times per day as a maintenance therapy.

Protocol for Interstitial Cystitis

Following the various herbal and alternative medicine therapy options is useful to treat IC. Some patients may require nearly all of them to recover, while others may be improved by selecting several of the most indicated options.

- Teas and tinctures as exemplified in this section, choosing those most indicated for the patient
- *Althaea officinalis* tea, prepared as a thick, cold infusion, consumed 3 or more cups per day
- *Allium sativum* capsules, 2 capsules, 3 times a day
- Mannose powder, ¼ teaspoon, 3 times a day
- Glutamine powder, 1 teaspoon, 3 times a day
- Glucosamine sulfate, chondroitin, MSM, 2 capsules, 3 times a day. Often available as a combination. They may also be taken individually.
- Ground *Equisetum* powder, in capsules or stirred into soups, applesauce, or smoothies
- *Asparagus* root powder (shatavari), prepared into medicinal food, teas, and smoothies

Formulas for Neurogenic Bladder

Neurogenic bladder is a loss of bladder function and tone due to impaired neurologic input to the bladder and is most often caused by a spinal cord injury, but it may also result from multiple sclerosis,[89] cerebral palsy, parkinsonism, and Guillain-Barré syndrome. Loss of neurologic control of the bladder necessitates the use of a catheter to empty the bladder. Injury to the spinal cord due to lumbar disc herniation or brain lesions from cerebrovascular disorders may also impair neurologic input to the bladder. Detrusor hypotonia sometimes follows childbirth[90] or less commonly, follows an obstructive disorder including severe constipation and megacolon.[91] Amphetamine use, particularly long-term methamphetamine addiction, may promote neurogenic bladder, and there are case reports of transient episodes of neurogenic bladder occurring acutely with the use of the street amphetamine known as Ecstasy.[92] I have seen cases of apparently permanent bladder dysfunction in methamphetamine addicts requiring lifelong self-catheterization.

Allergic reactivity in the urinary mucosa can allow overdistension and damage to the urinary sphincters. Electromyograms of the bladder muscle can help identify neurologic and muscle abnormalities,[93] and cystometrograms may help assess bladder awareness and sensation via instillation of a specific volume of fluid into the bladder. Patients with spinal cord damage have no bladder sensation and will not experience the sensation of a full bladder, even when overdistended, while patients with interstitial cystitis or detrusor overactivity experience a full sensation with only a few drops of fluid. Virtually all patients with spinal cord injuries suffer from neurogenic urinary dysfunction and are subject to ascending infections and septicemia.[94]

Other than catheterization, there are few treatments for neurogenic bladder. Researchers are exploring the use of electrical stimulation of detrusor innervation to improve bladder tone,[95] and surgical therapies attempting to reconnect spinal cord injuries are being developed experimentally with moderate success.[96] *Hypericum* is used for spinal cord and nerve trauma and may be worth a try in teas to improve bladder tone. Urinary trophorestoratives such as *Centella asiatica* (gotu kola) may help preserve what bladder muscle tone exists following neurologic or obstructive injury. Poor circulation to the bladder can cause contractile dysfunction of the bladder in obstructive conditions, and animal studies suggest that *Pygeum africanum* prevents ischemia-induced contraction.[97] Antimicrobial herbs may be needed when urinary retention or indwelling catheters cause chronic UTIs. The use of antimicrobial herbal teas and mannose powder may be effective, as described in "Formulas for Interstitial Cystitis and Lower Urinary Epithelial Dysfunction" on page 121. Encapsulated cranberry deters frequent infections in neurogenic bladder patients,[98] and cranberry juice reduces bacterial adherence at a dose of 3 glasses per day.[99] All of the common well-studied antioxidant herbs and nutraceuticals including green tea catechins, soy genistein, resveratrol, and curcumin may protect muscles from damage in new onset cases or reduce harm in methamphetamine users,[100] but they are unlikely to regain function following long-standing damage or spinal cord injuries. *Salvia miltiorrhiza* may help ameliorate fibrotic changes to the bladder following outlet obstruction.[101] Ligustilide from *Angelica sinensis* is credited with many supportive effects on the vasculature, and animal studies have shown neuroprotective effects following spinal cord injury.[102] If afforded the luxury of time, some of these therapies may help preserve bladder tone and function in neurologic disease, injuries, methamphetamine abuse, or obstructive pathologies.

Tea for Neurogenic Bladder

This complex tea may be simplified if desired, but this kitchen sink approach includes both folkloric herbs and those supported by modern research for neurogenic bladder. *Arctostaphylos* is included here as a deterrent to infections for those relying on a catheter to empty the bladder.

Hypericum perforatum
Calendula officinalis
Arctostaphylos uva ursi
Eucalyptus globulus
Centella asiatica
Althaea officinalis
Salvia miltiorrhiza
Angelica sinensis
Mentha piperita

Combine equal parts of the dry herbs, or use a greater proportion of *Mentha* to taste. Steep 1 tablespoon of the herb mixture per cup of hot water and then strain. If chronic bladder infections are present, ¼ teaspoon of mannose powder may be added to each cup. Drink 3 or more cups a day to support bladder tone and as a preventive against chronic UTIs.

Herbs for Neurogenic Bladder

These herbs may help protect the bladder muscle in situations of trauma, poor circulation, oxidative stress, or other disease. See also "Specific Indications," beginning on page 156, to better understand when to select these herbs for inclusion in formulas to treat neurogenic bladder.

Alpinia galanga
Angelica sinensis
Centella asiatica
Eupatorium purpureum
Ginkgo biloba
Hypericum perforatum
Piper methysticum
Pygeum africanum
Salvia miltiorrhiza
Serenoa repens

Tincture for Neurogenic Bladder

This formula uses the three herbs most heralded in the folkloric literature for loss of urinary tone.

Serenoa repens
Panax ginseng
Eupatorium purpureum

Combine in equal parts, and take 1 teaspoon of the combined tincture 3 to 4 times daily, long term.

Formulas for Urinary Tract Irritability and "Overactive Bladder"

Detrusor hypersensitivity, often referred to popularly as overactive bladder, has several underlying causes, but in many cases, no underlying issue can be found. Overactive bladder involves a hypersensitivity to distention of the bladder leading to discomfort and frequent urination. Around one-third of all cases involve elderly women with cystitis symptoms but no actual infection present, causing frequency and some degree of incontinence.[103]

Other conditions discussed in this chapter are types of urinary overactivity. Interstitial cystitis is a type of bladder hypersensitivity due to a breakdown in the mucosal barrier and irritation of the underlying nerves and detrusor muscle, which shares some traits in common with overactive bladder. Lower urinary epithelial dysfunction and aseptic prostatitis may also involve frequency, pain, and urinary irritability and is further discussed below. Nocturnal enuresis (bed-wetting) in children may be due to urinary irritability.

The underlying cause of urinary overactivity is not always identifiable, but allergic, toxic, and inflammatory triggers may contribute. A thorough history may reveal other allergic issues, a high level of oxidative stress, or liver or bowel toxicity, such that formulas could be tailored to individual constitutions. In some cases, aging may alter neurologic input to the bladder, and hormonal changes may contribute to atrophy of the urothelium. The urinary mucosal cells release neurotransmitters that play regulatory roles in sensing bladder distension and initiating the micturition reflex.[104] Elevated urinary neurotransmitter levels may contribute to heightened sensation, and an overlap between urinary hypersensitivity and mental-emotional symptoms exists. Nervine herbs may be added to teas, and behavioral or cognitive therapies may be helpful. Nitric oxide acts as both a neurotransmitter and a vasodilator and is released from blood vessel endothelial cells. Nitric oxide may be elevated in association with bladder irritability,[105] and *Angelica*, *Leonurus*, and *Ginkgo* may normalize nitric oxide synthesis, explaining the traditional reputation of these herbs as blood movers and antiallergy agents. As with fibromyalgia, anxious personalities, headaches, insomnia, muscle pain, and social and relationship issues may accompany bladder sensitivity. Improving pelvic floor tone with fitness and Kegel exercises may help mechanically and enhance body awareness in those with many mental-emotional contributors.

The main pharmaceutical therapy for overactive bladder is anticholinergic agents, discussed in "Tropane Alkaloids for Urinary Retention and Bladder Spasm," below. Pharmaceutical and natural agents that block muscarinic receptors of acetylcholine can relax the bladder and improve urinary capacity. Anticholinergic drugs may be instilled directly in the bladder. *Atropa belladonna* is a natural anticholinergic and may be considered in small amounts in formulas for overactive bladder but is contraindicated in the situation of neurogenic bladder. Botulinum neurotoxin from *Clostridium* bacteria is most well-known for injecting into facial muscles to temporarily reduce wrinkles, popularly known as "Botox" therapy. Urologists may also inject botulinum toxin into the detrusor muscle with some efficacy,[106] although the

Tropane Alkaloids for Urinary Retention and Bladder Spasm

Cholinergic innervation is the primary regulator of bladder contractility. Muscarinic receptors that receive cholinergic signaling occur in the urothelia and in the nerve fibers and detrusor layer directly underlying the urothelia. The neurotransmitter acetylcholine can decline with age and contribute to increased bladder spasticity, at times severe enough to obstruct the urinary outlet and cause urinary retention.[107] Tropane alkaloids in Solanaceae family plants, such as *Atropa belladonna*, are well-studied anticholinergic agents capable of blocking muscarinic receptors, reducing detrusor contractility, and calming the muscarinic contraction. While muscarinic innervation is the primary regulator, ATP mechanisms also play a role in bladder contractility and are said to be atropine-resistant. Because purines, particularly caffeine, promote this ATP mechanism, many IC and LUED patients report that drinking coffee exacerbates detrusor overactivity, urgency, and frequency, and all purines should be eliminated for several months as an experiment for patients with overactive bladders.

Herbal tropane alkaloids, including atropine, scopolamine, and hyoscyamine, may be used to relax excessive bladder contraction, but the dose must be limited due to the many systemic side effects. Pharmaceutical tropane alkaloid analogs, tolterodine and oxybutynin, may be preferable in some cases, because they have specificity to urinary muscarinic receptor subtypes and have fewer systemic side effects. Nonetheless, the inclusion of small amounts of *Atropa belladonna* or *Hyoscyamus niger* tinctures in formulas for overactive bladder and acute renal colic may yield symptomatic relief at a dosage small enough to avoid undesired systemic effects. Clinical studies have reported that the use of rectal, vaginal, and urethral suppositories have the fewest side effects compared to various oral antimuscarinics. I have made belladonna suppositories for individual patients, with good results. Of the various Solanaceae family (nightshade) tropane alkaloids, *Hyoscyamus niger* has the greatest specificity for the urinary passage, according to folkloric literature, with *Atropa belladonna* emphasized more for digestive and skeletal muscle spasms. *Hyoscyamus* is an extremely strong herb with great toxic potential, but it is especially indicated for severe spasms of the urinary tract, and of value to experienced herbal clinicians. See also "Specific Indications," beginning on page 156, for further comments regarding the use of *Atropa belladonna* and *Hyoscyamus niger*.

results are not permanent, and research is underway to develop a longer lasting protocol. Botox may also improve urinary irritability due to prostate disorders and sphincter anomalies.[108]

Botanical agents used to treat overactive bladder may include nervines, as mentioned above, demulcents, and muscle-relaxing herbs such as *Piper methysticum* (kava). Kava affects ion currents in neurons and has been emphasized in the folkloric tradition for urinary pain and hypersensitivity. Modern research suggests that kava may relax urinary muscle due to direct effects on potassium channels and ion currents,[109] explaining some of its traditional reputation for muscle spasms. Kava pyrenes are noted to reach the urine and may contribute to therapeutic effects on urinary passages.[110] *Hypericum* is a folkloric remedy for bed-wetting[111] and may reduce nerve hyperexcitability. Herbs known as urinary trophorestoratives may be included in formulas for overactive bladder due to effects on genitourinary tone and include *Serenoa, Agrimonia,* and *Solidago.* Herbal muscle antispasmodics may relax the bladder, reduce urinary retention, and improve micturition reflexes. *Valeriana, Dioscorea, Ammi visnaga* (khella), and kava may be included in formulas, especially for patients with concomitant anxiety and insomnia or those with obvious spastic symptoms, micturition inhibition, and urinary retention. For those with systemic allergies, *Angelica, Ginkgo,* and flavonoid supplements may be included in the broader protocol.

Tincture for Overactive Bladder with a Spastic Quality

Piper methysticum is one of the more powerful urinary antispasmodics and is complemented here with the anticholinergic agent *Atropa belladonna. Atropa belladonna* is available only to licensed health care professionals who are skilled in its use as a low-dose botanical. This formula lacks nourishing and restorative ingredients and is intended for short-term use in acute cases, switching to a convalescent formula as symptoms abate. It would be best combined with one of the nourishing teas described in this section.

Piper methysticum 30 ml
Ammi visnaga 20 ml
Atropa belladonna 10 ml

This formula can be dosed 1 to 2 teaspoons 3 to 6 times daily, depending on severity of symptoms. Reduce the quantity of belladonna in the formula if dry mouth or visual disturbances occur.

Botanical Stimulants to Avoid in Overactive Bladder

Purine-based neurotransmitters control purinergic neurotransmission, and these and other neurotransmitters may be involved in detrusor overactivity.[112] Caffeine and other natural amphetamines are classified as purine compounds, and coffee drinking is often particularly aggravating to individuals with overactive bladders and urinary hypersensitivity. Strict avoidance of coffee, chocolate, green and black tea, and any other caffeinated beverages or substances may improve symptoms in some sensitive individuals and is always worth an earnest effort.

Coffee drinking and caffeine consumption also stimulates adrenaline and adrenal activity. The smooth muscle of the bladder is regulated in part by adrenaline,[113] which leads to contraction of smooth muscle cells within the human lower urinary tract. Research suggests that there are many adrenoceptor subtypes, and investigation is underway to identify agonists and antagonists that may play a role in loss of bladder tonicity, excessive tonicity and spasticity, prostatic hypertrophy, and detrusor overactivity.[114] Furthermore, since caffeine inhibits phosphodiesterase[115] and causes cAMP to remain active in muscle cells for a prolonged period of time, caffeine favors muscle contraction. Capsaicin, from *Capsicum* hot peppers, may bind vanilloid receptors found on urinary epithelium and contribute to bladder overactivity.[116] In addition to avoiding coffee and caffeine, some susceptible individuals might try strict avoidance of stimulating spices such as cayenne pepper as well.

Tincture for Overactive Bladder in Allergic Individuals

Scutellaria, Ginkgo, and *Angelica* reduce allergic hypersensitivity via numerous mechanisms. They are combined here with *Hypericum,* noted for its ability to reduce bladder pain and sensitivity.

Scutellaria baicalensis
Ginkgo biloba
Angelica sinensis
Hypericum perforatum

Combine in equal parts. Take 1 to 2 teaspoons of the combined tincture 3 to 6 times daily depending on severity of symptoms.

Tincture for Overactive Bladder in the Elderly

This formula combines the traditional trophorestorative *Solidago* with circulatory-enhancing anti-inflammatories to support uroepithelial reactivity emerging due to old age and oxidative stress.

Panax ginseng
Solidago spp.
Ginkgo biloba
Angelica sinensis

Combine in equal parts. Take 1 to 2 teaspoons of the combined tincture 3 to 6 times daily depending on severity of symptoms.

All-Purpose Tea for Overactive Bladder

This formula combines *Centella* and *Equisetum* to support tissue regeneration, *Althaea* to soothe hypersensitive mucosal surfaces, and *Hypericum* as a classic herb for bladder hyperreactivity. *Lobelia* and *Foeniculum* may have relaxing effects on bladder spasms.

Foeniculum vulgare
Althaea officinalis
Hypericum perforatum
Lobelia inflata
Centella asiatica
Equisetum spp.
Mentha piperita
Glycyrrhiza glabra

Combine equal parts of the dry herbs, or add more fennel or licorice for a more pleasing taste, if desired. Steep 1 tablespoon of the herb mixture per cup of hot water and then strain. Drink 3 or more cups per day.

Tea for Urge Incontinence

Urge incontinence is generally classified as urine leakage with the urge to urinate, in contrast to stress incontinence, which is classified as urine leakage when straining, coughing, or exercising.

Equisetum spp.
Hypericum perforatum
Lepidium meyenii
Medicago sativa
Turnera diffusa
Mentha piperita

Combine equal parts of the dry herbs, or add more mint for taste if desired. Steep 1 tablespoon of the herb mixture per cup of hot water, strain, and take 3 cups per day. I often suggest that patients complete 200 Kegel exercises (conscious contracting of the pubococcygeal muscles) while drinking each cup and not to expect too much from the tea alone.

Formulas for Enuresis and Incontinence

Enuresis, or bed-wetting, can be due to unsuspected infections when the problem occurs in children who have previously gained control over the bladder. Children who are late in gaining bladder control sometimes simply sleep so deeply that they fail to awaken at the sensation of a full bladder. Pinworms, roundworms, and parasites can infect the bladder or genitals and be an overlooked cause of bedwetting. In other cases, bladder allergies can cause intermittent bladder irritability following ingestion of offensive foods,[117] and one study found soy and hazelnuts to trigger urinary reactivity.[118] Some researchers have reported a correlation between bed-wetting and childhood migraines and have found that an allergen-free diet may remedy both complaints.[119]

Any food may be the culprit, but in adults, caffeine and benzoate preservatives may trigger urinary irritation, including lower urinary and prostatic irritation in men. Remove all irritants, colors, artificial flavors, and preservatives from the diet for several months and evaluate for any improvement. Incontinence can occur due to extreme irritability and reactivity or an otherwise asymptomatic infection, but it most often occurs in older women with poor pelvic floor pubococcygeal tone. Kegel exercises and weight loss may help support the urethral outlet and sphincters,[120] but real results require hundreds of Kegels per day and a month's persistence. Therapy for enuresis and incontinence can be tailored to any suspected underlying contributor.

Limiting water intake in the evening may reduce childhood bed-wetting but could hardly be considered a cure or a deep-acting therapy. Some parents resort to setting an alarm to go off at 1 or 2 am and helping affected children get to the bathroom, but this interrupts the sleep of the entire family. Having children sleep in the family bed may make this approach easier. Homeopathic *Thuja* and Kreosotum are remedies for bed-wetting. Urinary tonics such as *Serenoa* or *Collinsonia*, and especially *Hypericum*, are mentioned in herbal folklore to improve bed-wetting, and demulcent herbs such as *Althaea* may allay allergic reactivity. Bladder allergies may be treated in a manner similar to that for hayfever or dermatitis, with *Ammi visnaga*, *Angelica*, *Ginkgo*, *Petasites*, and/or *Tanacetum*, all noted to reduce histamine, mast cells, and platelet activation.

Tea for Enuresis

Drinking a tea may seem illogical for bed-wetting, but if taken in the morning and during the day, it may help to reduce urinary irritability. Here *Hypericum* is aimed at being a tonic for urinary innervation, *Lobelia* as a bladder muscle antispasmodic, and *Althaea* as an agent to soothe the urothelium.

Hypericum perforatum
Lobelia inflata
Althaea officinalis
Glycyrrhiza glabra

Combine equal parts of the dry herbs. Steep 1 tablespoon of the herb mixture per cup of hot water and then strain. Drink freely, 3 or more cups during the day, restricting fluids after 5 pm.

Tea for Bed-Wetting in Allergic Individuals

Allergies may increase bladder spasticity and micturition reflexes in allergic children. These herbs may calm nerve hypersensitivity and reduce histamine, mast cell activation, and other inflammatory mediators. Such a formula would be especially ideal when bed-wetting is noticed to be worse from the ingestion of certain foods or chemicals or occurs in an atonic individual.

Hypericum perforatum
Angelica sinensis
Petasites hybridus

Combine equal parts of the three herbs. Add a fourth equal part of a flavoring herb such as *Matricaria*, *Glychyrrhiza*, or *Mentha* if the remedy is being prepared for a child or anyone who prefers more flavor. Steep 1 tablespoon of the herb mixture per cup of hot water for 10 minutes, then strain. Drink several cups each afternoon or early evening.

Formulas for Bladder Cancer

The most common type of bladder cancer is carcinoma of transitional epithelium. Squamous cell carcinoma is the second most common bladder cancer. It is highly invasive, carries a poor prognosis, and is associated with exposure to irritants and carcinogens. Dyes and environmental chemicals such as benzenes, tobacco, and parasites are known urinary carcinogens, creating inflammation and oxidative stress and, ultimately, DNA damage. Men are more affected by bladder cancer, especially in the elder decades, as long-term irritation creates cellular changes. Bladder cancer often has few symptoms and is therefore frequently diagnosed in fairly advanced stages when pelvic masses and urinary symptoms emerge. Microscopic hematuria may be one of the only signs in early stages, which makes the case for periodic routine urinalysis. Bladder cancer does not respond readily to conventional chemotherapy, and radiation is possible only on large tumors, because the bladder does not withstand radiation to the entire organ in cases of widely disseminated cancers. Thus, prevention is especially important. Bacillus Calmette-Guérin (BCG) is a type of bacterial-derived immunotherapy used to reduce the recurrence and progression of bladder cancer; BCG is instilled directly into the bladder. BCG is most effective when caught early enough to have not invaded the muscle of the bladder or for cancer types confined to the mucosal tissues. Radical cystectomy is performed when this immunotherapy fails and in cases of more aggressive muscle-invasive bladder cancer.

Various toxins and irritants are associated with bladder cancer including nitrosamines used to cure bacon, hot dogs, and lunch meat as well as aniline dyes, hair dyes,[121] and especially cigarette smoke.[122] Some plants are shown to exert anticancer effects in animal studies by protecting the tissues from such mutagens. For example, *Camellia sinensis* (green tea) may protect against nitrosamine-provoked urinary cancers,[123] and carotenoids and retinoids may offer chemoprevention in bladder cancer, especially in smokers,[124] and may help treat bladder cancer as well.[125] *Piper methysticum* (kava) may have activity against bladder cancer as well as deter cancer in general and is further discussed in "*Piper methysticum* for Urinary Pain" on page 125. The mushroom *Grifola umbellata* has been noted to reduce the occurrence of bladder cancer following exposure to carcinogenic nitrosamines as well as to reduce recurrence of urinary cancers in human patients following surgical therapy.[126] *Andrographis* has a protective effect against cyclophosphamide-induced urothelial damage[127] as has *Withania somnifera*.[128] Isothiocyanates, a group of sulfur-containing compounds naturally occurring in the Brassicaceae or Crucifer family (which includes broccoli, cauliflower, cabbage, and horseradish), have also been found to prevent cyclophosphamide-induced urotoxicity in mice.[129] Some bladder cancer strains are hormonally sensitive, and as with reproductive cancers, isoflavones may both prevent and treat bladder cancer.[130] Isoflavones, including genistein, genistin, daidzein, and biochanin, inhibit growth of bladder cancer, reducing angiogenesis and inducing apoptosis.[131] Animal models of bladder cancer show soy to significantly reduce tumor weight when included in animal feed.[132] *Grifola* was investigated as a follow-up therapy for bladder cancer patients having undergone bladder surgery as a means of preventing tumor recurrence and was reported to significantly reduce recurrence rate compared to controls, without toxicity or side effects.[133] *Viscum album* (mistletoe) contains a variety of immunoactive compounds and is used to treat leukemia, Hodgkin's disease, and other cancers. Due to cytotoxic activity credited to mistletoe lectins and other compounds, *Viscum album* is noted to promote tumor regression in many clinical and research scenarios, often in high doses that induce a febrile state. The induction of fever may be part of the cytotoxic and immune modulation mechanisms, as hyperthermia alone has immune benefits. *Viscum* may complement intravesical therapies such as BCG or mitomycin C and promote tumor regression. One series of cases studies at a German clinic reports high doses of *Viscum album* to reduce tumor recurrence in bladder cancer patients with recurrent disease.

Herbs Shown to Deter Bladder Cancer

The following herbs have been found to exert anticancer effects in animal, molecular, or human models of bladder cancer.

Agrimonia pilosa[134]	*Grifola umbellata*[139]
Allium sativum	*Hypericum perforatum*[140]
Andrographis paniculata[135]	*Ligustrum lucidum*[141]
Astragalus membranaceus[136]	*Cantharis vesicatoria*[142]
	Medicago sativa
Brassica spp. (especially broccoli sprouts)[137]	*Piper methysticum*[143]
	Piper nigrum
Camellia sinensis	*Rubia tinctorum*
Ganoderma lucidum[138]	*Serenoa repens*[144]
Glycine max (soy isoflavones)	*Vaccinium macrocarpon*[145]
	Viscum album[146]

Hypericum (St. Johnswort) contains hypericin, the most potent photosensitizer known. Research is underway to use hypericin topically on skin, throat, and accessible cancers, followed by light exposure, as a means of inducing apoptosis. The bladder is one such organ that is somewhat accessible to St. Johnswort and fiber optic therapy. *Hypericum perforatum* may have antiviral and anticancer activity via a unique mechanism of photosensitivity, whereby the flavonoids absorb UV light. The use of such compounds to absorb UV light in the body for medical effects is referred to as photodynamic therapy.[147] Photodynamic therapies are presently showing promise for bladder cancer.[148]

Supportive Tincture for Bladder Cancer

Serenoa repens has shown inhibitory effects to all hormone-sensitive cell lines tested, including breast, prostate, and bladder cancers; it inhibits cell growth and induces apoptosis.[149]

Ganoderma lucidum 15 ml
Serenoa repens 15 ml
Piper methysticum 10 ml
Viscum album 10 ml
Allium sativum 10 ml
Cantharis vesicatoria 1 ml

Take 1 or 2 dropperfuls of the combined tincture 3 to 6 times a day.

Photodynamic Therapy for Bladder Cancer

Plants are the base of the entire food chain due to the ability of chlorophyll to absorb photons of certain wavelengths of sunlight and use this energy to fuse molecules of carbon dioxide and water to form glucose and oxygen. This is the phenomenon of photosynthesis. Flavonoid molecules in plants act as secondary photosynthesizers, absorbing electromagnetic radiation and distributing it about the multiringed molecular structures without being damaged themselves. Flavonoids have anti-inflammatory and antioxidant effects, in part, by similarly being able to absorb free electrons and prevent them from damaging other cell membranes and molecules and boosting tissue antioxidant status.

Hypericin in *Hypericum* is one such flavonoid and is reported to be the most potent natural photosensitizer known. When hypericin is rubbed into the skin, the skin's absorption of UV light may increase so dramatically as to induce third-degree burns. The skin and eyes can also become more sensitive to light if *Hypericum* flavonoids are consumed orally in quantity. This photophenomenon may also lend antiviral activity to *Hypericum*, because viruses can be subdued due to interactions with hypericin and viral proteins as they pass through cutaneous blood vessels close enough to the skin's surface to be affected by UV light. Furthermore, hypericin generates large quantities of singlet oxygen, superoxide anions, and other reactive oxygen species when exposed to UV light, which may be the mechanism by which *Hypericum* photodynamic therapy damages cells and induces apoptosis.[150]

Hypericin, one well-studied flavonoid in *Hypericum*, is a potential anticancer therapy due to its potent photodynamic effects.[151] Early research suggests that hypericin can be instilled into the bladder and spontaneously establish itself in transitional epithelial carcinoma cells,[152] concentrating particularly in bladder cancer cells.[153] Researchers report hypericin to be concentrated in cancer cells some 25 to 30 times above surrounding noncancerous cells, and combining hypericin with fluorescence techniques may be a tool for diagnosing bladder cancer.[154] The profound ability of hypericin to concentrate in bladder cancer cells is being investigated as a means of treating bladder cancer using visible light rather than ionizing radiation, due to the unique capacity of hypericin to take up electromagnetic energy. Research is underway to determine what wavelength and intensity will best induce apoptosis without damaging the bladder muscle.[155] Other researchers are showing efficacy against nasopharyngeal cancers[156] as well as against nonmelanoma skin cancer cells, including actinic keratoses, basal cell carcinoma, and Bowen's disease carcinoma in situ.[157]

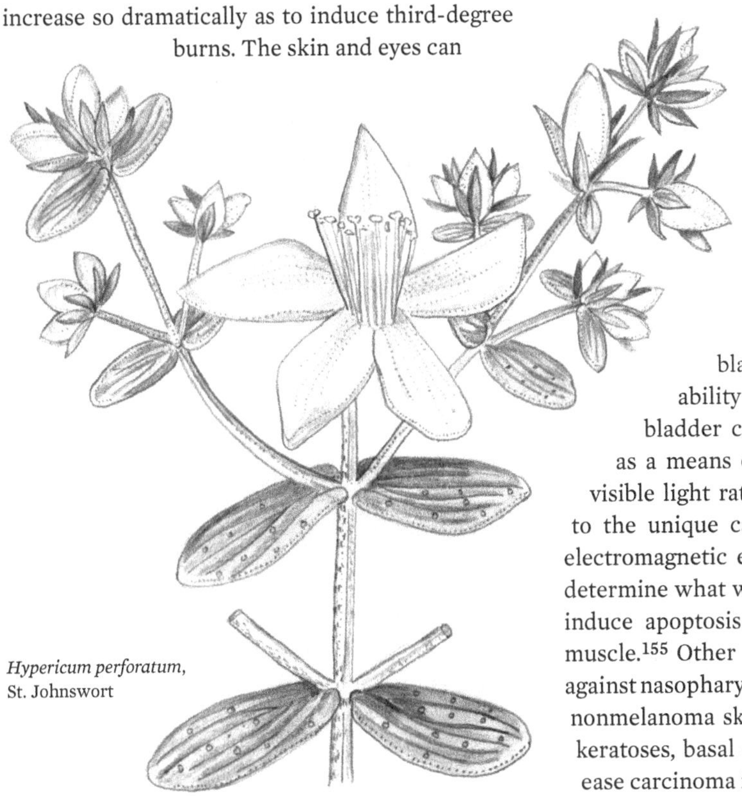

Hypericum perforatum, St. Johnswort

Natural Compounds That May Promote Bladder Cancer

Four naturally occurring compounds that may promote bladder cancer are madder root, coffee, aniline dyes, and tobacco. *Rubia tinctorum* (madder root) induces bladder neoplasms and may damage DNA via adduct formation.[158] Coffee may increase the risk of bladder cancer,[159] but studies are conflicting. Aniline dyes (used to color fabric and wood) may increase the risk of bladder cancer. This was noted as early as the late 1800s when dye workers were noted to have 10 to 50 times the cancer rate, especially bladder cancer, than the rest of the community. Some related hair dyes were also noted to increase urinary cancer and were removed from the market in the 1980s. Tobacco is highly associated with bladder cancer as well as lung cancer; tobacco smoke naturally contains aniline compounds.

BCG for Bladder Cancer

Local therapies, such as instillations of therapeutic agents directly into the bladder, along with partial cystectomy (surgical excision of cancerous regions) are mainstays of therapeutic approaches to bladder cancer. Bacillus Calmette-Guérin (BCG) is a bacterium that is prepared by a pharmaceutical process to use as a bladder cancer vaccine and therapy. BCG is becoming established as a suitable therapy following surgery to prevent tumor recurrence and for the primary treatment of bladder carcinoma in situ caught in the early stages.[160] However, the results are not promising in cases of metastatic disease.

Supportive Anticancer Tea for Bladder Cancer Patients

Agrimonia is a traditional herb for bladder inflammation. It is used to astringe urinary passages when blood or mucous is present in the urine. *Agrimonia* has antioxidant activity[161] due to its flavonoid content,[162] including catechin and epicatechin[163], which are also considered some of the active constituents in *Camellia* (green tea), and which would lend broad anti-inflammatory activity to this brew.

Camellia sinensis
Hypericum perforatum
Calendula officinalis
Agrimonia eupatoria, A. pilosa
Astragalus membranaceus
Glycyrrhiza glabra

Combine equal parts of the dry herbs. Steep 1 tablespoon of the herb mixture per cup of hot water and then strain. Drink 3 or more cups per day.

Aggressive Herbal Protocol for Bladder Cancer

This protocol of herbal encapsulations would be complementary to any herbal teas or tinctures chosen. As with the treatment of all types of cancer, many forms of medicine are typically used at once, as the situation is often dire. Such a protocol would not interfere with other medical interventions or surgical procedures.

- Green tea capsules: 2 or 3 pills at a time, 3 times a day
- *Grifola, Ganoderma* medicinal mushroom blend: 2 or 3 pills at a time, 3 times a day
- Diindolylmethane (DIM) supplement: 1 pill, 3 times a day
- *Andrographis paniculata*: 2 or 3 pills, 3 times a day
- Garlic capsules: 2 or 3 pills, 3 times a day
- St. Johnswort capsules: 2 or 3 pills, 3 times a day
- Saw palmetto capsules: 2 or 3 pills, 3 times a day
- Kava capsules: 1 pill, 3 times a day

Dietary Support Protocol for Bladder Cancer

- Isoflavones in legumes
- Medicinal mushroom preparations
- High-crucifer food diet, broccoli sprouts
- Flaxseeds

Avoid: Cured meats that contain nitrates, food chemicals, and dyes, and possibly coffee in excess of 1 cup per day.

Formulas for Urinary Colic and Lithiasis

Renal calculi, also referred to as urinary lithiasis, result from supersaturation of the urine, allowing solids to precipitate out of solution and form into stones. Miniscule granules in the ureters create inflammation, edema, and obstruction and are moderately painful to excruciating, with patients often resorting to an emergency room visit because of intolerable pain. Dehydration predisposes to renal calculi as the urine becomes more concentrated; the peak incidence of emergency room visits for acute renal colic is in summer months, when sweating and inadequate hydration predispose. Renal calculi may also result from poor circulation or kidney diseases where urine forms very slowly, allowing dissolved calcium oxalates, phosphates, and urates to precipitate out. Precipitates form more readily when the urine is supersaturated or with abnormal pH due to dietary or metabolic deficiencies. A high intake of alcohol, salt, sugar, refined carbohydrates, animal protein, and excess vitamin D can all contribute to calcium oxalate stones. Excessive use of calcium supplements or calcium

Urinary Antispasmodics

Among our most powerful renal antispasmodics are *Piper methysticum*, *Lobelia inflata*, *Viburnum opulus*, *Ammi visnaga*, *Dioscorea villosa*, *Gelsimium sempervirens*, *Atropa belladonna*, and *Hyoscyamus niger*. *Piper methysticum* (kava) is my "ace in the hole" herb because it is powerful, works quickly, and has an affinity for urinary smooth muscle. A *Piper methysticum* alkaloid, kawain (aka kavain) is in part responsible for numbing sensations in oral and other mucous membranes due to effects on potassium ion flow, and it contributes to urinary anodyne effects.

Many Apiaceae family herbs may have the effect of relaxing urinary smooth muscles via inhibiting calcium ions channels. Such plants include *Apium graveolens*, *Petroselinum crispum*, *Foeniculum vulgare*, and *Conium maculatum*. These herbs are emphasized in folkloric formulas for renal pain, bronchospasm, and hypertension due to muscle-relaxing effects on urinary, respiratory, and vascular muscle fibers. *Conium*, however, is potentially toxic and is for use only in small doses and by skilled clinicians. *Ammi visnaga* (khella) is also in the Apiaceae family and is useful to relax spasms in the ureters. (See "*Amni visnaga* for Renal Colic" on page 139.) Use khella in teas or tinctures or apply hot compresses to the back to facilitate the passing of a stone.

Angelica sinensis (angelica or dong quai) is another Apiaceae family plant with numerous circulatory and antiallergy effects. It may help individuals with urinary symptoms due to food or chemical sensitivity as well as those with renal pathologies involving poor circulation such as diabetic nephropathy. *Angelica* has been shown to normalize the vasodilating substance nitric oxide, which when elevated may trigger detrusor and other urinary smooth muscle overactivity. *Angelica* may be best for patients with an overactive bladder, having a less powerful anodyne action than kava or khella, but may be included in recovery formulas or broad protocols, for patients with indwelling stents, or to support recovery following lithotripsy. *Salvia miltiorrhiza* and *Pueraria montana* may also be used in convalescent formulas to improve circulation to the kidneys and urinary tissues.[164] *Aesculus hippocastanum* may be complementary in formulas for acute colic, both internally and topically, helping to reduce edema in the ureters.

Atropa belladonna and *Hyoscyamus niger* are both Solanaceae family plants that contain tropane alkaloids that have antispasmodic effects on the gut and urinary passages, but the oral dosage must be limited due to toxic and possibly hallucinogenic side effects. Tropane alkaloids in these plants, including atropine and hyoscyamine, are muscarinic antagonists that block the stimulating effects of cholinergic activity, thereby relaxing urinary tone.

Therapeutic Agents for Renal and Ureteral Colic

The α1-adrenoceptor antagonist tamsulosin (Flomax) is often prescribed allopathically for those with acute renal calculi. *Serenoa repens* (saw palmetto) is used to reduce the obstructive symptoms of benign prostatic hypertrophy (BPH) and may also be considered as a supportive, although not particularly anodyne, ingredient in formulas for renal colic. *Serenoa* is both an α1-adrenoceptor and muscarinic cholinoreceptor antagonist,[165] but research has concentrated on its effect on prostatic hypertrophy symptoms. No research as yet has investigated whether it might reduce hydronephrosis due to ureteral obstruction or otherwise help to treat related situations of urinary obstruction.

Herbal β-agonists may offer spasmolytic effects on the ureters. Pharmaceutical β-adrenoceptor agonists include albuterol, well established as a bronchial muscle spasmolytic in the treatment of asthma, especially in nebulized form. Albuterol and other β-adrenoreceptor agonists may also improve the passage of renal calculi when used systemically, and possibly when used topically.[166] Many stimulants, the sympathomimetic amines, such as caffeine, aminophylline, and ephedrine are β-adrenoreceptor agonists but appear to have a greater efficacy on bronchial smooth muscle than ureteral muscles; however, small amounts may be complementary in herbal tea formulas.

Tropane alkaloids are found in Solanaceae family plants and include atropine, hyoscyamine, and scopolamine. All have spasmolytic effects on urinary smooth muscle due to antimuscarinic (also commonly referred to as anticholinergic) effects.[167] Tropane alkaloids are a combination of a piperidine and a pyrrolidine ring, and piperidine alkaloids may also have antispasmodic effects on urinary muscles. Piperidine alkaloids also reduce urinary contractility via muscarinic blockade[168] and nicotinic agonism, and such compounds are notably high in *Lobelia*. *Lobelia inflata* is most well-known for treating wheezing and to relax respiratory smooth muscles; however, *Lobelia* may also help to relax cardiac muscle, skeletal muscle, the uterine muscle, and urinary smooth muscle fibers. Much of the antispasmodic effect of *Lobelia* is credited to piperidine alkaloids in the leaves and seedpods, and include lobeline, lobelanine, and lobelanidine. *Lobelia* may act as both an agonist and antagonist to β nicotinic receptors, depending on the receptor subtype and situation.[169] Tropane alkaloids can be used in small doses only as muscarinic antagonists[170] due to side effects such as bradycardia, hypotension, tremors, dry mouth, and visual disturbance, or even central nervous symptoms in large dosage, but they can be a valuable part of an herbal therapy in low and moderate dosages.

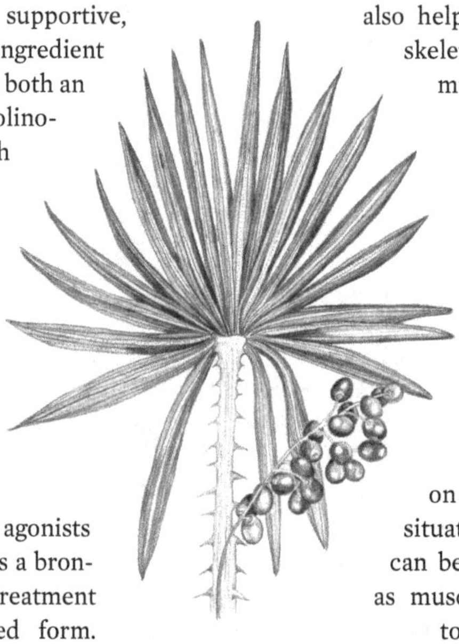

Serenoa repens, saw palmetto

Mucilaginous herbs can offer significant relief of urinary pain when properly prepared. In order to glean the most powerful demulcent effect, such mucilaginous herbs should be macerated many hours and even overnight to best extract the mucilaginous components. Aim for a tea that is thick and slimy. The tea can always be thinned down with mint or licorice teas if the thick nature of such teas is distasteful to an individual patient. *Althaea* is my personal favorite for such purposes; it can be macerated using 1 to 2 teaspoons per cup of cold water. Let stand for many hours before bringing to a brief simmer. Remove from the heat after 5 to 10 minutes, and let stand until cool enough to strain.

Neurologic Control of Urinary Smooth Muscle

Urinary muscle tone is controlled by both cholinergic and adrenergic signaling, along with numerous other mechanisms affecting these pathways.[171] Although the detrusor muscle contracts during micturition under the control of parasympathetic cholinergic mechanisms, the adrenergic system plays a significant role in the resting tone of the bladder. For example, stress incontinence may involve deficient urinary tone where increasing adrenergic input may be beneficial, while in overactive bladder and urinary pain and spasm, blockage of adrenergic signaling may be beneficial. Alpha (α)-adrenergic receptor leads to urinary smooth muscle contraction, while beta (β)-adrenergic receptors promote relaxation of the detrusor and urinary smooth muscle in general.[172] Therefore, blockade of α-adrenergic receptors and promotion of β-adrenergic signaling may reduce ureteral colic pain and improve flow around calculi, reducing obstruction and clearing hydronephrosis. *Lobelia inflata* may relax bladder spasms by promoting β-adrenergic signaling, while *Atropa belladonna* and *Hyoscyamus niger* may relax the bladder through blockage of cholinergic signaling at muscarinic sites. Calcium is essential for the maintenance of action potentials controlling ureteral smooth muscle contraction. In addition to adrenergic agents, calcium channel blockers may inhibit the influx of calcium and therefore have a spasmolytic effect on the activity of the ureter. Nifedipine, a calcium channel blocker inspired by *Ammi visnaga* constituents, is commonly used in the treatment of hypertension and angina and may reduce ureteral contraction without impairing the resting ureteral tone and peristaltic rhythm.[173]

carbonate antacids can increase calcium oxalate in the urine and contribute to stones. *Taraxacum*, *Equisetum*, and *Arctium* may help reduce the formation of acidic mineral crystals such as uric acid, calcium, and phosphorus-based stones[174] and are included in the tea formulas in this section. However, they are preventive and not helpful for the acute pain of passing a kidney stone. The ingestion of large quantities of lemon juice, around six or more lemons per day, acidifies the urine and can prevent, or possible dissolve, urate calculi, the least common type of stone.[175] Ascorbic acid may maintain urinary acidity, and the regular consumption of citrus juices may reduce the occurrence of renal caluli. The formulas in this section would be complemented by drinking ample water with lemon or other fresh squeezed citrus fruit and eating a light diet high in vegetables and low in protein and animal products.

Renal colic pain is not due to the stone itself, but rather is due to obstruction of the urinary flow by a kidney stone, causing increased pressure on the urinary tract wall, smooth muscle spasms of the ureter, edema and inflammation near the stone, and increased ureteral peristalsis. I have seen several patients the day *after* they have been seen in the ER, still in pain, but with an opiate prescription in hand (as well as a hospital bill of a thousand dollars or more). Although opiates take the edge off the pain, the pain returns when the drugs wear off. Opiates also cause lethargy and constipation and do little to actually move or treat the stone.[176] Opiates deaden the pain *sensation*, but they are not renal antispasmodics and do nothing to improve urine flow or resolve obstruction. Herbal medicines can, in fact, relax spastic ureters, and demulcent agents can help lubricate and soothe inflamed urinary mucosal tissues. Herbal protocols for renal colic should be implemented as soon as possible, because the longer a renal calculus sits in the ureter, the more edematous and swollen the ureter may become, contributing to the obstructive symptoms and to the body's inability to dislodge the stone—as the tissues swell, the stone becomes increasingly embedded. Urinary antispasmodic herbs, along with mucilaginous herbs, may reduce ureteral inflammation and spasm, prevent edematous swelling of the tissues, and assist the natural ability of the ureter to move a stone along toward the bladder.

The use of strong irritants and diuretics is contraindicated because they would stimulate already inflamed and irritated tissues and worsen the problem. When a stone is too large to pass, or diagnostic imaging reveals

Herbs for Acute Urinary Pain

Some of these herbs are the most powerful antispasmodics, such as the tropane alkaloid and calcium channels herbs discussed in detail in this section. Others are demulcent agents with anti-inflammatory effects and other mechanisms of action. These herbs are among the most commonly cited in folklore for acute flank pain, renal colic, and lower urinary spasm.

CATEGORY AND CONDITION	HERBS
Tropane alkaloids for intense spastic pain	*Atropa belladonna* *Hyoscyamus niger*
Opiate-, GABA-, and adrenergic agonists for intense spastic pain	*Corydalis yanhusuo* *Dioscorea villosa* *Gelsemium sempervirens* *Hypericum perforatum* *Lobelia inflata* *Mentha piperita* *Piper methysticum*
Demulcent herbs to reduce edema and promote circulation	*Aesculus hippocastanum* *Collinsonia* spp.
Mucilaginous and mucosal anti-inflammatory herbs	*Agropyron cristatum*, *A.* spp. *Althaea officinalis*
Calcium channel blocker, antiallergy/anti-inflammatory agents	*Ammi visnaga* *Angelica sinensis* *Apium graveolens* *Conium maculatum* *Foeniculum vulgare* *Levisticum officinale* *Petroselinum crispum*

numerous stones in the renal pelvis or proximal ureter, extracorporeal lithotripsy may be employed to rumble the stones apart so that they may pass more readily, or direct laser therapy may sublime the stone. Large stones that are damaging the kidney may be surgically removed. Stents are often placed after such procedures to prevent ureteral spasms, but a stent causes moderate pain and discomfort itself. *Andrographis paniculata* prevents kidney infections following lithotripsy as effectively as antibiotics, as demonstrated by leukocytes, red blood cells, and protein content of the urine.[177] *Equisetum* is a traditional herb emphasized for treating and preventing urinary lithiasis, and it may be included in teas for acute and convalescent recovery formulas. (See "Equisetum for Repair and Regeneration" on page 113.) *Equisetum* may improve urinary pH in a manner that reduces the tendency for minerals to precipitate out of solution and contribute to stone formation.[178]

Tea for Acute Renal Colic

This is an example of using *Althaea* as a demulcent in a tea. For the most powerful pain-relieving effects, cold macerate for many hours before simmering, by soaking the herbs in cold water, overnight or at least 4 hours. However, when such a tea is required immediately, prepare a batch without the long maceration, but start macerating a second batch at the same time to be ready to use later in the day. This formula combines the demulcent effects of *Althaea* with the antispasmodic and pain-relieving effects of *Piper methysticum*. *Glycyrrhiza* (licorice) offers its own demulcent and anti-inflammatory effects and improves the flavor of the tea. Avoid inundating the irritated renal passages with a copious amount of fluid, but rather make a large batch of the brew, and sip constantly through the day. Complement this tea with the tinctures and topical applications offered in this section.

Piper methysticum	1 ounce (30 g)
Glycyrrhiza glabra	1 ounce (30 g)
Althaea officinalis	4 ounces (120 g)

Briefly simmer 3 tablespoons of the herb mixture in 10 cups of hot water and then strain. Slowly drink, taking a sip every 5 to 10 minutes, drinking as much as possible in this way over the course of the day. Start macerating a second batch while the first is brewing by soaking 3 tablespoons of the herb mixture in 3 cups of water overnight. In the morning add 6 more cups of water and bring to a simmer.

Tincture for Acute Renal Colic

This tincture can complement the Tea for Acute Renal Colic. Aggressive doses are best, not only to help allay pain, but also to help prevent the ureter from becoming edematous and swelling around the stone. This formula employs *Hyoscyamus*, which is potentially toxic, but is used safely here by including just 4 milliliters in the formula. *Hyoscyamus* is a restricted botanical to be used by experienced clinicians only. When taken every 10 minutes, this formula may provide significant pain relief and also relax ureteral spasms and help a stone to pass.

Ammi Visnaga for Renal Colic

Khellin, a flavone in *Ammi visnaga* (khella), is used traditionally for both asthma and renal colic.[179] Although the bulk of the research on *Ammi visnaga* has focused on mast cell stabilizing effects, inhibiting the release of inflammatory mediators in allergic situations,[180] *Ammi visnaga* has also given rise to several important calcium channel blocking pharmaceuticals including amiodarone, a hypotensive agent; cromolyn, a bronchial antispasmodic; and nifedipine, a urinary relaxant.[181] *Ammi visnaga* also has a gentle diuretic effect, which may promote the expulsion of stones as well as inhibit the reabsorption of citrate and decrease calcium oxalate crystal formation in the kidney.[182] For acute renal colic, *Ammi visnaga* may be used in tinctures, teas, and tea-soaked compresses topically over the back.

Piper methysticum 20 ml
Lobelia inflata 14 ml
Ammi visnaga 12 ml
Aesculus hippocastanum 10 ml
Hyoscyamus niger 4 ml (optional, for licensed practitioners only)

Take 20 to 30 drops of the combined tincture every 10 to 30 minutes for acute renal colic, reducing the dose and frequency as symptoms abate.

Topical Heat Pack for Renal Colic

Lobeline in *Lobelia*, like many alkaloids, is better extracted in an acidic menstruum than in a simple aqueous menstruum. The use of *Hyoscyamus* tincture may boost the antispasmodic effects; if it is available as an oil, that would be suitable as well. *Atropa belladonna* is used here, as it is more readily available, perhaps even as an oil for topical use. To make this heat pack, you must prepare both an oil for application to the patient's back and an herbal tea for the compress.

Astringent Herbs for Cloudy Urine or Stone Tendency

These herbs would be suitable adjuvant herbs in formulas for urinary pain, complementing the specifics listed in "Herbs for Acute Urinary Pain" on page 138. These herbs can help astringe swelling in the tissue, reduce edema, help stones pass, and alleviate ureteral blockade that contribute to the pain of acute renal colic.

Agrimonia eupatoria
Armoracia rusticana
Berberis aquifolium
Berberis vulgaris
Chimaphila umbellata
Equisetum spp.
Eupatorium purpureum
Hedeoma pulegioides
Herniaria glabra
Hydrangea spp.
Lespedeza spp.
Phyllanthus amarus
Rubia tinctorium

Castor oil 1 tablespoon
Mentha essential oil 30 drops
Apium essential oil 30 drops
Lobelia 2 tablespoons dry leaves
Apple cider vinegar as needed to saturate the *Lobelia*, roughly several tablespoons.
Aesculus ½ cup dried nut slices
Atropa belladonna tincture 1 teaspoon

First prepare the oil by combining the castor oil and essential oils. Then prepare the tea by saturating the *Lobelia* leaves with apple cider vinegar and letting it stand for 15 minutes or longer. Add 2 cups of water and the *Aesculus* pieces, and bring to a gentle simmer for 10 minutes. Remove from heat, let stand 10 minutes more, and strain.

Apply the oil to the back and flank. Soak a thick washcloth or dish towel in the hot tea, add the belladonna tincture (or *Hyoscyamus* if available) to the moist compress, and position the compress over the oiled area. The oil may help wick the aescin, lobeline, and atropine in the compress deeper into the body, and the essential oils blended into the castor oil may also offer immediate pain relief. Leave the compress in place for 20 to 30 minutes, and cover with a heating pad or other heat pack for the best anodyne effects. Begin preparing an additional compress while the first compress is in place, so that you can replace with a fresh compress every 30 to 45 minutes. **Caution:** Omit the belladonna should dry mouth or visual disturbances occur.

Convalescent Tea for Urinary Calculi

A formula such as this may help traumatized tissue recover, and for those with chronic stones, drinking this tea may help prevent a recurrence. This large recipe will yield a pound of tea, which can be stored in a glass jar and brewed daily in the convalescent period, and at least 3 or 4 times a week to prevent stones when recurrent.

Salvia miltiorrhiza root 3 ounces (90 g)
Pueraria montana root 3 ounces (90 g)
Angelica sinensis root 3 ounces (90 g)
Urtica spp. root 3 ounces (90 g)
Glycyrrhiza glabra root 2 ounces (60 g)
Apium graveolens seeds 1 ounce (30 g)
Zingiber officinale root 1 ounce (30 g)

Gently simmer 1 heaping tablespoon of the herb mixture in 4 cups of water for 10 minutes. Remove from the heat, let stand covered for 10 minutes more, and then strain. Drink the entire batch over the course of a day. It can be chilled if preferred or mixed into the day's drinking water.

Aesculus: A Synergist in Renal Colic Formulas

Aesculus hippocastanum (horse chestnut) is emphasized in folklore for varicosities and circulation, but it can also be a supportive herb in renal colic formulas, both topical and internal. Eclectic physicians recommended *Aesculus* to reduce ureteral edema and allow a stone to pass,[183] and also included the herb in convalescent formulas for those prone to stones due to poor circulation; contemporary research supports the use of *Aesculus* for venous congestion. *Aesculus* contains triterpenoid saponins, collectively referred to as β-aescin, which are shown to increase plasma membrane permeability by opening up membrane pores, explaining the herb's historical reputation for edema and fluid stasis. β-aescin forms pores in the plasma membrane via effects on the cytoskeleton and cellular adhesion molecules,[184] an action that may also improve the uptake of other molecules into cells and help activate other herbs in a formula. At the same time, β-aescin improves microcirculation and increases venous tone and venous return, other mechanisms that may reduce edema. Although β-aescin increases agonist-induced contraction in urinary smooth muscle,[185] it also appears to boost the antiallergy and anti-inflammatory actions of various compounds,[186] with the net effect being significant antiedema effects.[187] Thus it may serve well as a complementary herb in formulas for renal colic, improving the delivery of other herbs. *Aesculus* may also enhance absorption and provide its own anodyne effects when included in topical preparations for renal colic.[188]

Tea to Prevent Stones

Folkloric herbs for preventing renal calculi include *Eupatorium*, *Equisetum*, and *Collinsonia*. *Collinsonia* is purported to improve circulation and lymphatic flow, reducing any fluid stasis that allows for urinary precipitates to form. *Collinsonia canadensis* contains anti-inflammatory flavonoids,[189] and along with improving any underlying contributors, it may reduce formation of new stones.

Collinsonia canadensis
Eupatorium purpureum
Urtica spp.
Equisetum arvense, E. hymenale
Glycyrrhiza glabra

Combine equal parts of the dry herbs. Gently simmer 1 tablespoon of the herb mixture per cup of hot water and then strain. Drink several cups per day, long term.

Urethral Suppositories

Suppositories such as this formula are not available on the market, and even the ingredients take effort to make or find. Thus, these suppositories are useful only for acute situations when the herbal oils are already prepared and handy. Urethral suppositories to treat spasm could feature *Atropa belladonna* or *Hyoscyamus*, or such oils could be used topically over the back, bladder, flank, and genitals. The tiny molds for making urethral suppositories are available from some pharmaceutical suppliers. When suppositories are not available or too difficult to insert on one's own, a physician might place a single suppository and send the patient home with oral

Atropa and *Hyoscyamus* for Urinary Pain

The tropane alkaloids most commonly found in Solanaceae (nightshade) family plants can be used to treat a variety of painful muscle spasms. *Atropa belladonna* (belladonna) has been historically used to treat acute febrile nephritis, renal engorgement, and nocturnal incontinence. The Eclectic physicians reported the medicine to stimulate the kidneys as well as to relieve irritability of the kidneys and bladder. Belladonna is particularly susceptible to biphasic effects, having one effect at a low dose and differing effects at higher dosages. For example, *Atropa belladonna* may help bed-wetting in children due to hyperexcitability of the bladder, yet it can also improve urinary flow in cases of urinary retention. *Atropa* can be included in formulas for renal colic, due to antispasmodic effects on urinary muscle, and for relieving renal congestion and associated renal pain and fullness. *Atropa* can be used in tinctures as well as topical oils to help reduce the pain of acute renal colic, and it can be crafted into urethral suppositories for treating some acute urinary spastic situations.

Individual tropane alkaloids hyoscine and hyoscyamine have been employed intranasally and intravenously[190] for relief of urinary pain. *Hyoscyamus niger* (henbane), the source of these tropane alkaloids, is another Solanaceae family plant used in traditional medicine for centuries and is often claimed to be the most powerful urinary antispasmodic in that plant family. Include *Hyoscyamus* in formulas for renal colic as well as for urinary retention and distension of the bladder,[191] due to its reliable anti-inflammatory, analgesic, and bladder-relaxant properties. *Hyoscyamus* may reduce neural hyperexcitability[192] as well as provide calcium channel blocking and anticholinergic effects in digestive and urinary smooth muscle. One investigation showed *Hyoscyamus* to have a greater effect on urinary muscle than on digestive muscle, requiring 10 to 25 times the dose to elicit the same antispasmodic action in the gut.[193] Whole plant extracts of *Hyoscyamus* are likely superior to isolated alkaloids or other compounds. Hyoscine N-butylbromide (Buscopan) is a semisynthetic medication based on combining the hyoscine with bromide. The medication has been evaluated in human clinical trials for patients with large distal ureteral stones, measuring 5 to 10 millimeters, but was inferior to α blockers for promoting spontaneous passage of the calculi.[194] Other studies, however, suggest that the combination of α-blockers and antimuscarinics may be more effective against ureteral pain than either therapy alone.[195] While small doses of *Hyoscyamus* and *Atropa* may enhance the expulsion of stones, large dosages may lead to urinary retention or can even cause renal colic. The formulas in this section exemplify appropriate dosing strategies.

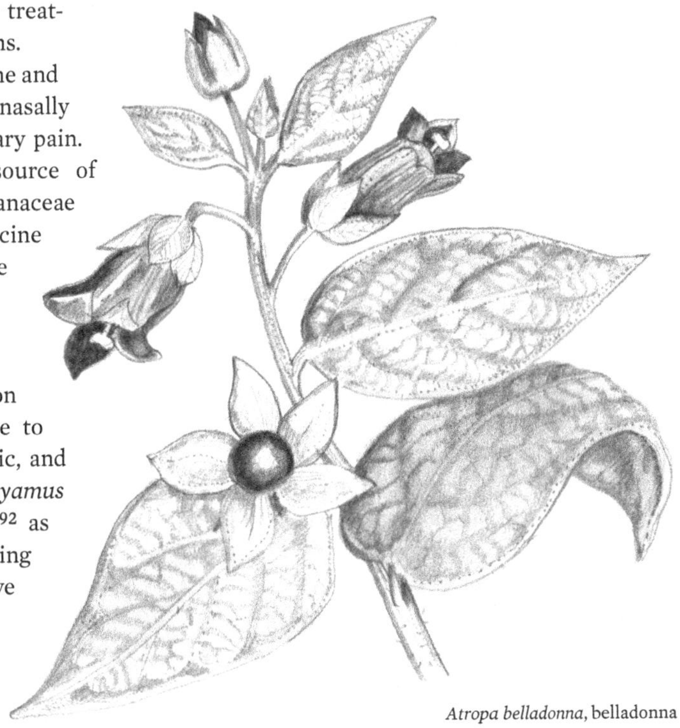

Atropa belladonna, belladonna

Herbs for Urinary Retention

The following herbs do not have a rapid anodyne action as do the herbs listed in the "Herbs for Acute Urinary Pain" on page 138, but rather they may reduce urinary retention over time. These herbs may also be synergists in formulas for urinary pain, especially for the elderly and in cases of urinary obstruction or when prostatic disease impeding urine flow contributes to the discomfort.

Apocynum cannabinum
Curcubita pepo (seed oil)
Serenoa repens
Solidago canadensis
Urtica dioica, U. urens

medicines and a topical oil. Tropane alkaloids are readily absorbed when rubbed repeatedly over the external urethra and genitals.

Apium graveolens seed powder oil 1 teaspoon (5 ml)
Aesculus hippocastanum oil 1 teaspoon (5 ml)
Lobelia inflata oil 1 teaspoon (5 ml)
Atropa belladonna oil 1 teaspoon (5 ml)
Cocoa butter, melted ½ cup (235 ml)
Suppository mold, smallest size, suitable for the urethra

To prepare the oils needed for this formula, plan ahead by macerating 8 ounces of dry herbs in oil. The powdered or finely cut dry herbs are covered with olive, sesame, or apricot oil such that they are completely covered by the oil and nothing is exposed to the air lest they mold. Shake or stir vigorously every day for 2 months, and then press out and filter out all particulates.

When suppositorities are needed, combine the oils in melted cocoa butter using a double boiler, adding the pressed oil to the double boiler once the cocoa butter is melted. Use a small pipette to transfer the oil into the waiting molds, and carefully move the molds to the refrigerator. Once set, take the suppositories out of the refrigerator, wrap them individually in foil or wax paper, and return them to the refrigerator.

A Formula from 1900 for Bladder Stones

A fluid extract is a 1:1 tincture and was commonly produced by the Lloyd Brothers Pharmacy, an herbal medicine manufacturer who produced products for the Eclectic physicians of the era.

Hydrangea fluid extract 2 ounces (60 ml)
Gum arabic 3 ounces (90 g)
Triticum repens fluid extract 1 ounce (30 ml)
Alcea rosea 1 ounce (30 ml)
Glycerine 1 ounce (30 ml)

Add 1 teaspoon of the combined ingredients to a glass of water and drink a dose every 4 hours for 3 days.

Formulas for Prostatic Disease

Prostatic diseases include benign prostatic hypertrophy (BPH), prostatitis, and prostatic cancer, and all are known to have hormonal contributors. Nearly all men develop some degree of prostatic hypertrophy in the latter decades of life, which can be asymptomatic when minimal. However, symptoms can be significant and progressive, interfering with urinary, erectile, and ejaculatory functions,[196] and the condition carries an increased risk of developing prostate cancer, such that BPH is really not all that "benign." Early urinary symptoms include difficulty voiding, weakness of the urinary stream, and difficulty starting urination (commonly referred to as "hesitancy") as well as nocturia and incontinence.

Androgens are presently the main targets of pharmaceutical therapies for BPH because adrenoceptors in the lower urinary tract support tissue contraction and contribute to prostatic congestion and urinary retention.[197] Blockade of α-adrenoceptors may improve urine flow; however, blocking all the adrenoreceptors throughout the body may have feminizing effects in men, thus the need for selective androgen blockers. Prazosin and tamsulosin are among the most widely used pharmaceutical agents for this purpose, and research is ongoing to identify natural adrenoreceptor antagonists with urinary specificity.

Both α and β estrogen receptors are also found on the prostate gland and are involved with fetal development of reproductive organs,[198] tissue proliferation in prostatic hypertrophy, and cancer initiation. Although testosterone contributes to prostatic proliferation and is the main target of pharmaceutical approaches to treating prostatic disease, excessive stimulation by estrogen is also contributory,[199] but sometimes overlooked in medical protocols for BPH. Prostate health is impaired when the balance between α and β estrogen receptors is disrupted, and the same is true when the balance between androgens and estrogens is disrupted.[200] When fetal and neonatal exposure to exogenous estrogens alters genetic expression and regulation of hormone receptors, a variety of diseases and increased risk of hormonal cancers is evident, a phenomenon referred to as endocrine disruption. Early exposure to hormonally active compounds such as diethylstilbestrol (DES)—a profound endocrine disruptor—heavy metals, PCBs, and pesticides,[201] alters steroid regulation and reception and contributes to reproductive disease and cancer, including prostatic disease.

Hormonally active compounds from plants include isoflavones, coumestans, and coumarins. They are commonly referred to as phytosterols and are particularly high in the legume family. *Glycine max* (soy) is one of the most studied sources of phytoestrogens, but *Trifolium* (red clover), *Medicago* (alfalfa), *Glycyrrhiza* (licorice), *Pueraria* (kudzu), and other legumes may help reduce excessive hormonal stimulation of the prostate gland. The inclusion of isoflavones and other phytoestrogens in the daily diet is thought to offer many health benefits, including protection against prostate cancer.[202] Genistein, for example, the most common and most studied isoflavone, may offer a protective effect against prostate cancer when consumed throughout a lifetime, due to protection from the hyperproliferative effects of endogenous and exogenous estrogens.[203]

Aromatase inhibitors are also appropriate for treating BPH. Aromatase enzymes convert androgens into estrogen, and inhibiting this conversion can help reduce excessive estrogenic stimulation.[204] Another target for treating BPH is 5α-reductase, the enzyme that converts testosterone into the more active dihydrotestosterone, with a greater stimulatory and proliferative effect on the prostate gland compared to testosterone. Inhibition of this enzyme may improve prostatic enlargement, and finasteride (Proscar) is a synthetic pharmaceutical used for this purpose. *Serenoa repens*, *Pygeum africanum*, *Urtica*, and *Camellia sinensis* are plants that inhibit

Selective Estrogen Response Modifiers (SERMs)

In tissues possessing both α and β estrogen receptors (ERs), the two subtypes seem to counteract and balance one another. In general, the α subtype directs cellular proliferation; the β subtype directs differentiation and apoptosis.[205] Intensive research is underway to develop therapies that target a specific subtype of receptor, specific to individual tissue types. These therapeutic agents are called selective estrogen response modifiers, or SERMs. For example, genistein, an isoflavone type of phytoestrogen common in legumes, is reported to be an ER agonist, particularly to the β subtype, and thus could be considered a natural SERM. Although plant phytoestrogens exert effects on the endocrine system, research suggests they are protective and of positive benefit when consumed in normal dietary levels, rather than disruptive to endocrine balance as are so many synthetic chemicals.[206] One study of mice compared the results of a high phytoestrogen soy feed to a low phytoestrogen soy-free feed on hormonal parameters in offspring. The low phytoestrogen feed resulted in higher estradiol levels in offspring and induced early puberty in the females and larger prostates and smaller testes in the males, while the high phytoestrogen offspring displayed healthier reproductive status.[207] Furthermore, the group receiving the low phytoestrogen diet displayed a tendency to obesity and altered glucose regulation later in life. Researchers have suggested that these effects may become more pronounced over several generations of similar dietary influences. Genistein is one of the most active and most studied of the isoflavone phytoestrogens. Genistein may affect gene expression due to direct activity at estrogen receptors. Animal studies show that genistein consumed in the maternal diet is noted to pass in breast milk to nursing infants. Prolonged exposure to genistein in early life is reported to reduce prostate size and may reduce the risk of BPH and prostate cancer later in life.[208]

5α-reductase activity,[209] with *Serenoa* shown to reduce circulating levels of active testosterone in humans.[210] *Pygeum africanum* has been the subject of several dozen clinical trials and shows promise in treating BPH symptoms.[211] Genes that control 5α-reductase in the prostate are induced by high-fat diets and inhibited by genistein.[212] Glucosinolates in *Lepidium meyenii* (maca) and other crucifers such as broccoli, cauliflower, brussels sprouts, and cabbage have antiestrogenic effects and may exert antiproliferative and apoptotic actions on prostate cancer cells,[213] without altering blood testosterone levels.[214] Researchers reported that lycopene, zinc, and supplemental vitamin D weekly reduce the progression of BPH, while a diet low in fat and red meat and high in protein and vegetables slowly reduces hypertrophy and the emergence of symptoms overall.[215]

Trifolium Simple Tea for BPH

Trifolium (red clover) isoflavones consumed on a daily, long-term basis may reduce elevated PSA levels by around 30 percent without significantly altering other hormone levels in the body.[216]

Trifolium flowers

Steep 3 tablespoons of dried flowers in 4 cups of water, let stand for 15 minutes, and then strain. Drink the quart of this tea each day.

Purple Grape Seltzer for BPH

Polyphenols in *Vitis vinifera* (grapes)—including procyanidin flavonoids and the well-studied stilbene, resveratrol—inhibit aromatase enzymes.[217] *Vitis vinifera* flavonoids in an aqueous vehicle have been demonstrated

Nettle and Saw Palmetto for BPH

A variety of *Urtica* (stinging nettle) species have demonstrated benefits for prostatic diseases, including BPH and prostatic cancer. *Urtica* reduces excessive hormonal stimulation of the prostate gland via mechanisms that include 5α-reductase inhibition,[218] and inhibition of sex hormone binding globulins[219] and aromatase enzymes.[220] *Urtica* may also reduce adenosine deaminase enzymes, which may be overexpressed in prostate cancer.[221] The phytoestrogens stigmasterol and campesterol in *Urtica* have been shown to inhibit transmembrane sodium and potassium ATPase enzyme activity, reducing prostate cellular growth.[222] Lectins found in *Urtica* roots may also possess antimicrobial and immune-modulating actions. The polysaccharide fraction of aqueous extracts of *Urtica* are noted to have anti-inflammatory effects on the prostate and to significantly reduce both proliferation of the prostate gland epithelia and fibrosis,[223] in part due to activity on epidermal growth factor and prostate steroid membrane receptors. Numerous clinical trials have shown *Urtica* to improve lower urinary symptoms in men with BPH, improving urine flow, urine retention, and prostate size, compared to controls,[224] without significant side effects.

Urtica dioica, stinging nettle

Urtica is commonly combined with *Serenoa* (saw palmetto) in European products for prostatic disease. Several clinical trials have shown efficacy of lower urinary tract symptoms compared to placebo.[225] One study showed the herbal duo to have equal efficacy to an androgen-inhibiting drug in the treatment of BPH.[226] Another double-blinded placebo-controlled clinical trial investigated the effects of a *Serenoa/Urtica* combination on elderly men with symptomatic BPH. The 257 men in this trial were given either the herbal combination or placebo and evaluated for a reduction in symptoms via both subjective reporting and objective testing (urine flow and sonography). The group receiving the herbal combination demonstrated improved urine flow and a reduction of symptoms compared to the group receiving the placebo.[227]

to improve perfusion in cases of bladder obstruction, lending credence to the idea that ischemia contributes to symptoms in men with BPH.[228]

1 cup (60 g) purple grapes (organic, pesticide free), chilled
1 cup (240 ml) legume tea (such as *Astragalus, Medicago, Trifolium*), chilled
1 cup (240 ml) pomegranate juice, chilled
1–2 cups (240–480 ml) seltzer water

Place the chilled grapes in a blender with the chilled tea and pomegranate juice, and puree. Transfer to a large drinking glass and top off with a cup or more of seltzer water. Drink each day. The type of flavonoid-containing fruit and flavor of the tea and juice can be varied to help maintain motivation to drink this beverage daily over the long term.

5α-Reductase Inhibitors

The 5α-reductase enzymes inhibit the local conversion of testosterone into the more potent androgen, 5α-dihydrotestosterone (5α-DHT). The 5α-reductase inhibiting agents, such as the prescription drug finasteride, are a standard treatment for lower urinary symptoms due to prostate enlargement. The following plant compounds and plants are natural 5α-reductase inhibitors.

Astaxanthin	*Piper nigrum*[230]
Genistein	*Piper cubeba*[231]
Piperine	*Pygeum africanum*[232]
Camellia sinensis	*Serenoa repens*[233]
Ganoderma lucidum[229]	*Urtica* spp.[234]

Pygeum africanum for BPH

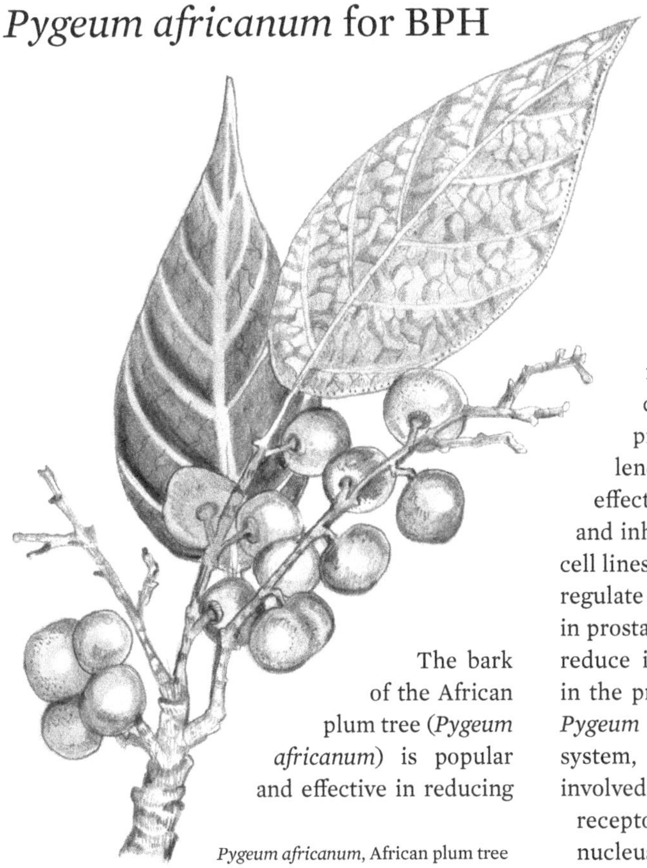

Pygeum africanum, African plum tree

The bark of the African plum tree (*Pygeum africanum*) is popular and effective in reducing lower urinary tract symptoms in cases of BPH.[235] Numerous mechanisms of action have been reported. *Pygeum* contains the constituent atraric acid, which has been shown to bind androgen receptors and act as an antagonist,[236] displaying an ability to prevent stimulation of the gland by testosterone.[237] Part of the antagonism of the androgen receptors may also be due to disruption of the intracellular signal when androgens bind the prostate androgen receptors. These actions lend *Pygeum africanum* an antiproliferative effect on the prostate gland, growth factors,[238] and inhibit mitosis in general in prostate cancer cell lines.[239] *Pygeum* has also been shown to down regulate estrogen receptors and induce apoptosis in prostate cancer cells.[240] *Pygeum* also is noted to reduce inflammatory proliferation of fibroblasts in the prostate gland.[241] Other research suggests *Pygeum* may also affect the adenylyl cyclase system, which, together with protein kinase, is involved with mediating the signal of the cellular receptors to the inside of the cell and to the nucleus of the cell.[242]

Flavonoid Aromatase Inhibitors

Numerous flavonoid molecules have been shown to be aromatase inhibitors, in addition to being potent antioxidant, anti-inflammatory, and anticancer agents. High-flavonoid foods should be used liberally in the diet, and the following herbs may be used in teas, tinctures, and medicinal foods for prostatic and other estrogen-dependent cancers. Among the noted flavonoid aromatase inhibitors are apigenin, hesperitin, chrysin, quercetin, kaemphferol, and resveratrol.[243] While all flavonoids have significant health benefits, specific research indicates that the following plants and fungi contain aromatase inhibitors.

Agaricus bisporus (white button mushrooms)[244]
Alcea rosea (hollyhock)[245]
Camellia sinensis (green tea)[246]
Cicer arietinum (garbanzo bean)
Coix lacryma-jobi (Job's tears)[247]
Euonymus alatus (burning bush)[248]
Hibiscus sabdariffa (hibiscus)
Phaseolus vulgaris (kidney, navy, and other beans)
Punica granatum (pomegranate)[249]
Scutellaria barbata (barbed skullcap)[250]
Isoflavones[251]

Dehydrogenase Inhibitors

The 17β-hydroxysteroid dehydrogenase enzymes are a large family of hormone-processing enzymes occurring in the placenta, endometrium, breast, prostate, small intestine, liver, and bone. These enzymes convert active hormones in the most active form of testosterone, dihydrotestosterone (DHT). There are two principal isozymes of 17β-HSD, Type 2 and Type 5. Type 2 converts testosterone and estradiol into DHT, and Type 5 converts androstenedione into DHT. The 17β-HSD enzymes occur in high levels in the human prostate, and inhibiting these enzyme systems can reduce hormone-driven proliferation of the prostate. The following plants and plant compounds are among the most studied dehydrogenase inhibitors.

Phytoestrogens:
isoflavones, coumarins, and coumestans[252]
Flavonoids[253]
Liposterols found in *Serenoa repens*
Glycyrrhiza glabra[254]

Pumpkin Seed Oil for BPH

Many seeds and nuts are high in lignans metabolized in the body into compounds with hormonal effects, along with sterol in their oils credited with beneficial effects on cholesterol and hormones. Pumpkin (*Curcubita pepo*) seeds and seed oil are a folkloric remedy for BPH and may reduce testosterone-induced hyperplasia of the prostate gland[255] in as little as 3 months' time, without side effects or toxicity.[256]

Pumpkin seed oil

Consume 1 teaspoon to 1 tablespoon daily, off the spoon or in salad dressings and blended drinks.

Phytoestrogen Tea for BPH

This formula features the better-tasting herbs noted to reduce hormonal stimulation of the prostate. Such a tea needs to be consumed regularly for months, if not years, to reduce hormonal stimulation of the prostate.

Trifolium pratense
Medicago sativa
Urtica spp. (especially root)
Camellia sinensis
Glycyrrhiza glabra

Combine equal parts of the herbs. Steep 1 tablespoon of the herb mixture per cup of hot water, and then strain. Drink 3 or more cups per day, long term.

Tincture for BPH

This formula features *Serenoa*, one of the most studied herbs for reducing hormone stimulation of the gland and the basis of the leading pharmaceuticals for BPH. *Serenoa* is not very palatable, so it's best prepared as a tincture rather than a tea. *Serenoa* is also readily available in capsules to complement other tinctures and teas.

Serenoa repens
Pygeum africanum
Lepidium meyenii

Combine in equal parts and take 1 to 2 dropperfuls of the combined tincture 3 or 4 times daily, long term.

Tincture for Infectious Prostatitis Symptoms

Prostate infections can cause pain with urination or ejaculation, abdominal tenderness, pus and blood in the urine or ejaculate, and general infection symptoms such as fever and malaise. The following herbs are emphasized in herbal folklore for such symptoms. Such a formula may be followed with a general prostate tonic and hormonal-regulating therapies, once the infection has resolved. It would also be important to address any underlying tissue congestion and fluid stagnation that predisposes to infection, such as BPH, prolonged sitting, or lymphatic and circulatory stasis or insufficiency.

Arctostaphylos uva ursi 20 ml
Allium sativum 20 ml
Agathosma betulina 10 ml
Thuja plicata 10 ml

Take 2 dropperfuls of the combined tincture every several hours, reducing as symptoms improve. This formula may be combined with antibiotics, mannose powder, or one of the teas described in "Formulas for Cystitis" on page 112.

Herbs for BPH Symptoms in Men

While infectious prostatitis is characterized by pain and infection symptoms, prostatic enlargement is typified by the sensation of a ball in the perineum and difficulty with stopping or starting the flow of urine, such as dribbling urine instead of having a strong urine stream. Because the enlarged prostate impedes the flow of urine, such symptoms are considered to be obstructive rather than infectious. The following herbs are emphasized for "dribbling of urine" and other obstructive symptoms.

Agrimonia pilosa, A. eupatoria
Agropyron repens
Conium maculatum
Curcubita pepo—seed oil
Equisetum arvense, E. hymenale
Galium aparine
Gaultheria procumbens
Lepidium meyenii
Medicago sativa
Serenoa repens
Turnera diffusa
Urtica urens, U. dioica leaf, root, and seed

Formulas for Prostate Cancer

Prostate cancer often emerges following long-standing prostatic hypertrophy and inflammation and is associated with hormonal stimulation, diabetes, smoking,[257] high-fat diets, and pesticide exposure.[258] Diets high in fruits and vegetables offer protective effects against prostate cancer, due to the flavonoid and isoflavones they provide.[259] Men with symptoms of prostatic enlargement and those with elevated PSA levels should be monitored long term to note rising PSA levels, the development of palpable nodules on the prostate, or hematuria—all worrisome findings that are associated with prostate and bladder cancers. When these signs are present, a bone scan may be warranted because the bones are the most common site for prostate cancer metastasis.

Testosterone and estrogen have proliferative effects on the prostate and, as discussed in "Formulas for Prostatic Disease" on page 142, working with hormonal expression and balance may both treat and prevent prostate cancer. Testosterone-lowering drugs, such as finasteride, reduce the risk of prostate cancer by 25 percent but can have feminizing effects systemically. Aromatase enzymes may be overexpressed in prostate cancer and contribute to cancer initiation and progression[260] and are a target in the treatment of prostate cancer. The 17β dehydrogenase enzymes are another target, due to their role in oxidizing steroids and reducing hormonal activity.[261]

Because prostate cancer is often very slow-growing, herbal and chemotherapeutic measures are appropriate to reduce hormonal stimulation. *Pygeum* reduces androgenic stimulation[262] and may inhibit prostate cancer cell lines,[263] induce apoptosis, and down regulate estrogen receptors.[264] *Pygeum africanum* contains atraric acid, which has direct antiandrogen activity, binding to androgen receptors and acting as an antagonist.[265] *Serenoa repens* (saw palmetto) may induce apoptosis and reduce dihydrotestosterone-induced proliferation in prostate cancer cell lines.[266] A simple *Serenoa* tincture has also been shown to exert an antiproliferative effect in hormone-sensitive prostate, breast, renal, urinary, and other cancer cell lines.[267] The liposterols in *Serenoa* are credited with much of these positive effects in enlargement and cancer of the prostate and are reported to reduce 5α-dihydrotestosterone stimulation of the prostate.[268] *Urtica* (nettle roots) has antiproliferative effects on prostate cancer,[269] curcumin of *Curcuma longa* exerts chemopreventive effects against prostate cancer,[270] and *Rhodiola rosea* has demonstrated activity against prostate cancer cell lines.[271] *Vitex* has exhibited anticancer

Phytosterols for Prostate Cancer

Plant phytosterols, such as isoflavones, bind to estrogen receptors and mitigate excessive hormonal stimulation. The most prevalent dietary isoflavone, genistein, is common in legume family plants. Genistein improves the response to radiation in estrogen-positive cancers,[272] has inhibitory effects on prostate cancer cell lines,[273] and influences the expression of steroid-producing and tumor-supressing genes.[274] Supplementing with isoflavones may help slow the progression of prostate cancer without side effects or toxicity.[275] Even soy milk, if consumed 3 times a day for one year, may slow the progression of prostate cancer, stabilizing PSA levels and increasing the level of equol, an isoflavone metabolite, in the serum.[276]

Pollen Extracts and Prostate Cancer

Many types of pollen may be hormonally active and help to improve BPH symptoms. Various human clinical trials have demonstrated improvements in urodynamic measurements.[284] Pollen from *Brassica campetris* contains steroids and inhibits prostate cancer cell lines.[285] *Brassica napus* (rapeseed) contains brassinolide, which is shown to be active against prostate cancer cell lines.[286] *Brassica* pollen is rich in flavonoids including naringenin, luteolin, and kaemphferol, all shown to reduce elevated PSA levels,[287] and in vitro studies show inhibitory effects on prostatic epithelium and fibroblast proliferation.[288] Such pollen extracts also appear active against hormone-insensitive types of prostate cancer.[289]

effects on prostate cancer cell lines.[277] These herbs are included in the following teas and tinctures for prostate cancer and in an aggressive protocol using such formulas, along with isoflavone, resveratrol, lycopenes, and other supplements as warranted. Resveratrol, a stilbene polyphenol in grapes and red wine, may inhibit prostate cancer via modulation of androgen and estrogen stimulation.[278] Resveratrol and other high-flavonoid foods and supplements may be part of the broader treatment protocol for prostatic disease.

Lycopene is another carotenoid type of flavonoid with inhibitory effects against prostate and other cancer. Lycopene is the most abundant carotenoid in tomatoes, also responsible for the pink to red color of rosehips, watermelon, papaya, pink grapefruit, and guava.[279] Lycopene endures tomato processing and is present, and physiologically active, in dried tomatoes, pastes, and powders.[280] Like all carotenoids, lycopene is credited with powerful antioxidant effects. Lycopene has antimetastatic and antiproliferative effects for a variety of cancers, is shown to reduce BPH symptoms,[281] and exerts antiproliferative effects on the prostate,[282] in part via local androgen blockade.[283]

Tincture for Prostate Cancer

Serenoa repens is one of the most studied herbs for the prostate, and molecular research on the plant evolved into the development of pharmaceutical agents aimed at treating BPH and reducing proliferative effects of hormones on the gland. Further animal research shows *Serenoa* to inhibit tumorigenesis and induce apoptosis in prostate cancer models.[290] Astaxanthin may synergize the anticancer effects of *Serenoa*.[291]

Serenoa repens
Pygeum africanum
Lepidium meyenii
Vitex agnus-castus
Rhodiola rosea

Combine in equal parts. Take 1 or 2 dropperfuls of the combined tincture 3 to 6 times per day.

Dietary Recommendations for Prostate Disease and Cancer

Increasing the following foods and spices in the diet and taking antioxidant supplements can be helpful for BPH and also reduce cancer risk. Aim to eat these foods on a daily basis to prevent prostate disease or slow its progression when present. In cases of prostate cancer, the consumption of these foods as well as the use of concentrated doses of these nutrients—a larger dose than what the daily diet could reasonably provide—would be warranted.

- Phytoestrogen-rich foods, such as soy foods
- Blue- and purple-pigmented foods such as cranberries, blueberries, pomegranates, grapes
- Cruciferous vegetables such as broccoli, cauliflower, brussels sprouts, cabbage
- Tomatoes, tomato juice, tomato paste, tomato soups (avoid products that contain added sugar)
- Mushrooms, such as *Ganoderma, Grifola, Agaricus, Shiitake*
- Spices such as black pepper, turmeric, ginger, garlic
- Pumpkin seeds and seed oils
- Zinc, selenium,[292] and other antioxidants

Supportive Tea for Prostate Cancer

Various species of *Solidago* (goldenrod) have been used historically as a urinary tonic, especially for the elderly with a tendency to urinary retention. There has been very little research on *Solidago* but limited study has suggested cytotoxic effects, including against prostate cancer.[293]

Trifolium pratense
Medicago sativa
Urtica urens, U. dioica
Camellia sinensis
Astragalus membranaceus
Solidago canadensis
Glycyrrhiza glabra

Combine equal parts of the dry herbs. If desired, add *Mentha* or increase the amount of *Glycyrrhiza* for better flavor. Steep 1 tablespoon of the herb mixture per cup of hot water and then strain. Drink 3 or more cups per day.

Formulas for Prostatitis

Prostatitis is the most common urological disorder in men. Both infectious (bacterial) and noninfectious types of prostatitis occur, the latter being akin to interstitial cystitis, with inflammatory and possibly autoimmune contributors. Infectious prostatitis may sometimes present acutely with pain, urinary symptoms, and fever, but more often is low grade and insidious. Chronic pelvic pain, diagnosed as prostadynia, is similar to interstitial cystitis in that no pathogen or obvious pathology can be found. It may involve local immune hyperactivity, loss of mucosal and epithelial barriers, and complex autoimmune phenomena. It is sometimes referred to as chronic pelvic pain syndrome. Few standard lab indices are remarkable for this condition, and pain may correlate to elevated cytokines or other markers of autoinflammation.[294] Neurotrophin nerve growth factor, for example, may be elevated and associated with a heightened urinary sensation, as demonstrated experimentally, but this is not yet available as a standard lab test.

As with interstitial cystitis, a variety of toxicity, leaky bowel, anxiety, and other triggers may contribute to tissue and biochemical changes. In some cases of noninfectious prostatitis, a simple shift in normal prostatic flora may trigger inflammation and autoinflammatory responses. Alkalinizing the urine may improve some cases of idiopathic urinary pain, when unrelated to bacterial infections, and may be accomplished with the ingestion of

Zinc for Prostate Health

Semen and the prostate gland are high in zinc, and the mineral appears so essential in preventing prostate infections that some researchers have referred to zinc as the "prostatic antibacterial factor."[295] Chronic prostatitis and chronic pelvic pain syndrome[296] and *Ureaplasma* infections[297] are associated with low zinc levels. Zinc supplementation improves prostatitis symptoms in animal models of chronic bacterial prostatitis,[298] and human clinical studies show oral zinc supplementation to improve chronic bacterial prostatitis. Frequent ejaculation may deplete zinc stores in the prostate gland and increase the susceptibility to infection.

Diagnosis of IC and LUED and the Potassium Sensitivity Test

The potassium sensitivity test is emerging as one diagnostic tool for lower urinary epithelial dysfunction (LUED). Men with chronic prostatitis are more sensitive to potassium in the bladder than men without prostatitis symptoms, suggesting that some so-called prostate symptoms actually have origins in the bladder.[299] Instillation of potassium in the bladder induces variable sensation in men, including pain in the perineum, lower pelvis, lower back, penis, testes, scrotum, or rectum.[300] Potassium channels in urinary epithelia may be excessively stimulated in relation to tissue damage and barrier damage. As in leaky gut syndrome and interstitial cystitis (IC), hyperreactivity of potassium-regulated cellular functions may occur in various states of "leaky" bladder epithelia,[301] contributing to noninfectious prostatitis. The urothelia is damaged, potassium "leaks" into the submucosa and interstitium, resulting in stimulation of stretch receptors and the symptoms of pain, urgency, frequency, and possibly incontinence. Urologists and researchers now use lower urinary dysfunctional epithelium (LUDE) as a more accurate term that includes the diagnoses of IC, chronic prostatitis, overactive bladder, chronic pelvic pain, urethral syndrome, and vulvodynia.[302]

Goldenrod for the Prostate

Solidago (goldenrod) is specific for dysuria due to spasm and tenesmus.[303] Pollen extracts of *Solidago* species are used in Germany to treat chronic prostatitis. Pollen is high in phytosterols and flavonoids, both of which may contribute to the therapeutic effects on the prostate. The high bioflavonoid content may provide anti-inflammatory and vascular-stabilizing effects. Recent research suggests that *Solidago* antagonizes muscarinic acetylcholine receptors[304] and thereby promotes detrusor relaxation, helping to treat frequency, urgency, and overactive bladder. *Solidago* flowers produce abundant amounts of pollen, and therefore, tinctures made from dried flowers contain large amounts of pollen constituents. Pollen itself is reported to have therapeutic effects on the prostate gland,[305] and one clinical trial found pollen extracts to significantly improve the symptoms of chronic prostatitis and chronic pelvic pain.[306] Antineoplastic effects against prostatic tumors have also been reported from *Solidago*.[307]

Solidago, goldenrod

potassium or sodium citrate. Although citrate is an acidic compound, the use of potassium and sodium citrate will alkalinize the urine by helping the kidney to excrete carbonate (carbonic acid). Several clinical studies have noted an improvement in bladder and urethral pain with the ingestion of 4 grams of sodium citrate every 4 hours.[308]

In cases of infectious prostatitis, the usual urinary pathogens, such as *E coli*, are most commonly involved. Other infectious organisms include *Mycoplasm genitalium*, *Chlamydia trachomatis*, *Ureaplasma urealyticum*, and *Trichomonas vaginalis*.[309] In all cases, urinary cultures should be run. HIV-positive men are more susceptible to less common pathogens such as *Staphylococcus aureus* and *S. saporphyticus*, *Haemophilus influenzae*, and *Streptococcus pneumoniae*, *Enterococcus*, *Klebsiella pneumoniae*, *Pseudomonas aeruginosa*, and *Enterobacter cloacae*,[310] and rarely viruses such as *Cytomegalovirus*.[311] Nanobacteria may also be to blame in cases of prostadynia or chronic pelvic pain syndrome and often elude all standard culture and microbiological diagnostic tests.[312]

Although pharmaceutical antimicrobials, such as flouroquinolone, may be initiated immediately,[313] even before cultures are processed, the majority of prostatitis cases are noninfectious. Because of this, herbalists and some clinicians prefer urinary tonics and more thoughtful approaches over antibiotic therapy, in the absence of fever and acute symptoms. Tamsulosin and levofloxacin are other pharmaceuticals prescribed for chronic pelvic pain syndrome and chronic nonbacterial prostatitis.[314] Noninfectious prostatitis can be treated in much the same way as interstitial cystitis (see "Formulas for Interstitial Cystitis and Lower Urinary Epithelial Dysfunction" on page 121). Zinc supplementation is appropriate in all cases of prostatitis because the mineral plays many important roles in prostate health, maintaining semen quality and supporting sperm motility.

Research on botanical therapies for prostatitis is less plentiful than for BPH and prostate cancer. The genitourinary trophorestoratives emphasized throughout this chapter, including *Serenoa*, *Urtica*, *Solidago*, and *Pygeum*, can serve as foundational tonic ingredients in teas and tinctures, adding in antimicrobial herbs such as needed. Berberine, *Allium*, *Rosmarinus*, *Arctostaphylos*, and other herbs can be added for acute or chronic infectious prostatitis. Demulcent herbs such as *Althaea* and *Asparagus* roots may improve barrier functions and allay pain in some cases. Immune modulators such as the reishi mushroom may improve autoinflammatory reactions in other cases.[315] Herbs that are mentioned in the folkloric texts for genitourinary pain in men include many herbs noted to have broad anti-inflammatory effects. *Piper methysticum* has long been used for urinary pain and has potent cyclooxygenase-inhibiting activity.[316] Many umbel family plants are emphasized for urinary pain and spasm historically, and *Daucus carota*[317] and *Apium graveolens*[318] have both been shown to inhibit cyclooxygenase. *Ocimum sanctum* (holy basil) is also mentioned for genitourinary pain and is a cyclooxygenase inhibitor.[319] *Petasites hybridus* (butterbur) has been used for centuries for all manner of painful and inflammatory conditions and inhibits cyclooxygenase and inflammatory prostaglandins.[320] Lycopene, found in tomatoes and other red fruits and vegetables, has an anti-inflammatory effect on the prostate gland and is shown to boost the efficacy of antibiotics against bacterial prostatitis in animal studies.[321]

Supportive Tea for Prostatitis

This tea supplies legume family isoflavones, noted to improve hormonal regulation in the prostate gland, via the leguminous herbs *Trifolium*, *Medicago*, *Astragalus*, and *Glycyrrhiza*. *Urtica* roots improve hormonal regulating enzymes, and *Solidago* is a traditional herb for genitourinary pain, infection, and congestion.

Trifolium pratense
Medicago sativa
Urtica urens, *U. dioica*
Foeniculum vulgare
Astragalus membranaceus
Solidago odora, *S. canadensis*
Glycyrrhiza glabra

Combine equal parts of the dry herbs. Add *Mentha* or *Matricaria* or increase the amount of *Glycyrrhiza* to improve flavor if desired. Steep 1 tablespoon of the herb mixture per cup of hot water and then strain. Drink 3 or more cups per day.

General Tincture for Prostatitis

These herbs are traditional for prostatitis symptoms, some with modern research suggesting hormonal and anti-inflammatory mechanisms of action.

Serenoa repens
Pygeum africanum
Lepidium meyenii
Curcuma longa
Rhodiola rosea

Combine in equal parts, and take 1 to 2 dropperfuls of the combined tincture 3 to 6 times per day.

Tincture for Bacterial Prostatitis

This formula mixes the genitourinary tonic herb *Serenoa*, with antimicrobial synergist herbs to make a formula to treat infectious prostatitis. Only small amounts of *Thuja* (or *Juniperus*) are appropriate, because more could be irritating to inflamed tissues. Once the infection is adequately treated, remove *Thuja* from the formula, switching back to a convalescent or tonifying formula.

Serenoa repens 14 ml
Curcuma longa 12 ml
Allium sativum 12 ml
Andrographis paniculata 12 ml
Thuja plicata 10 ml

Take 1 to 2 dropperfuls of the combined tincture 3 to 6 times per day, reducing the frequency as symptoms abate.

Formulas for Acute and Chronic Renal Failure

Renal failure occurs when kidney function is so impaired as to be ineffectual in clearing nitrogenous wastes from the blood stream or in maintaining fluid and electrolyte balance. Blood urea nitrogen elevates markedly, which is referred to as azotemia, and survival may ultimately require dialysis. Damage to renal cells may occur slowly due to vascular disease and impaired circulation to the kidneys, as in the case of diabetic glomeruloenephropathy, or may occur acutely with rapidly progressive autoimmune nephropathy or acute heavy metal, drug, or other toxicity. "Pre-renal" causes of renal failure include severe burns, dehydration, and liver disease that so derange electrolytes and protein levels as to challenge the kidneys. Pathologies of the kidneys themselves involve cellular, tubular, and glomerular destruction due primarily to vascular disease. Postrenal causes of renal failure include obstructive disorder where stones, tumors, or prostatic enlargement block urine flow and lead to permanent renal damage.

Acute renal failure is diagnosed when urine output decreases and nitrogenous wastes accumulate in the blood. Oliguria is defined by an output of less than 500 millilters of urine per day and is determined by having a person collect all urine for 24 hours. Anuria is the total lack of urine production. Without treatment, the oliguric phase of acute renal failure may last for only a matter of days in severe cases before progressing to total anuria or may progress to anuria over 6 to 8 weeks' time. Blood urea nitrogen (BUN) and creatinine become elevated and tend to climb steadily day by day as renal function declines. Acidosis may occur as the kidneys fail to remove acidic wastes and electrolyte imbalances, particularly potassium elevations and sodium depletion, ensue. When renal failure is due to catabolic situations such as acute infections, trauma, and burns, azotemia may be particularly marked. Anemia may also occur as the kidneys fail to produce erythropoietin.

Chronic renal failure involves the same changes as acute renal failure, but occurring very gradually, over many months' to many years' time. Urine output may be normal, but very dilute and of low specific gravity due to failure of the kidneys to concentrate the urine, even in states of dehydration. Fatigue and digestive upset may be initial symptoms, and many people with chronic renal failure develop pruritis, as the elevated urea remains in the tissue and skin and may precipitate out in the sweat and crystallize on the skin, a phenomenon referred to as "uremic frost." Blood pressure elevates when the body attempts to move more blood to the failing kidneys, which is particularly pronounced in those with underlying vascular disease. Hypertension and altered electrolyte, calcium, and potassium levels burden the heart and can contribute to congestive heart failure. The pharmaceutical management of hypertension in renal failure is challenging because dropping blood pressure can further impair renal perfusion, prompt the release of angiotensin from the kidneys, and create a vicious cycle. Diuretics may help minimally, but dialysis is the main therapy to bring down the blood pressure in such situations. When acidosis is severe, alkalinizing therapies such as sodium bicarbonate or sodium citrate may be helpful. There is little research on botanical agents for renal failure, but when due to underlying vascular disease and diabetic glomerular nephropathy, blood-moving herbs such as *Ginkgo*, *Salvia miltiorrhiza*, and *Angelica* may be helpful. In cases of acute drug or environmental toxicity, *Astragalus*, *Salvia miltiorrhiza*, *Rheum*, and *Cordyceps*[322] are shown to reduce damage. *Cordyceps* may both prevent and ameliorate aminoglycoside-induced[323] and cyclosporine-induced renal damage.[324] *Allium sativum* (garlic) may also protect the kidneys from cyclosporine toxicity,[325] and *Astragalus membranaceus* polysaccharides are shown to protect from streptozocin toxicity.[326] *Silybum marianum* and vitamin E boost renal antioxidant levels and prevent gentamicin-induced nephrotoxicity.[327] *Camellia sinensis* (green tea) attenuates zymosan-induced renal damage,[328] and *Cassia auriculata* protects against gentamicin, cisplatin, and oxidative stress.[329] None of these agents can restore renal function after serious damage has occurred, but they may be helpful to use in cases of early stages of chronic renal failure or as a preventive in those with vascular disease or diabetes.

The herbal therapies with the most research and showing the most promise for retarding decline of renal function are *Salvia miltiorrhiza*, *Silybum*, *Pueraria*, and possibly *Ginkgo*. *Salvia miltiorrhiza* has been shown to reduce oxidative stress in people undergoing hemodialysis and may be an adjuvant therapy.[330] *Pueraria* may support ion channel activity on renal cell membranes,[331] credited to the isoflavone puerarin, and enhanced electrical conduction may improve cellular function and perfusion in renal diseases. Puerarin may also enhance perfusion through gentle β-blocking effects involving direct antagonism of β-adrenoreceptors.[332] *Pueraria* isoflavones also reduce oxidative stress in the kidneys.[333] *Silybum marianum* (milk thistle) has been studied for many decades for its hepatoprotective effects, and research now suggest that *Silybum* may protect renal cells as well.[334] Milk thistle may reduce renal inflammatory markers and reduce elevated albuminuria in diabetic nephropathy.[335] When renal disease is due to clotting or coagulation disorders, *Gingko biloba* may improve coagulation profiles without increasing bleeding risk.[336] *Rehmannia glutinosa* enhances creatinine clearance and sodium excretion in ischemia-induced acute renal failure[337] and may be included in teas and formulas for long-term use. *Astragalus membranaceus* and *Angelica sinensis* are a traditional duo for renal failure and vascular disease in China, and animal research suggests the combination may improve renal circulation and reduce fibrosis of the renal tubules.[338] Animal studies show improvements in glomerular filtration, renal plasma flow, and sodium excretion, evidencing enhanced renal perfusion and cellular function.[339] The duo of *Astragalus* and *Angelica* reduces lipid accumulation in animal models of nephritic syndrome.[340] *Ligusticum* and *Rheum* are also reported to reduce the progression of renal injury,[341] and studies suggest an ability to reduce azotemia and improve cellular function.[342] *Cordyceps sinensis* may also improve cellular function,[343] and a scant amount of research on *Epimedium* has indicated the same.[344]

Outside of diabetic and ischemic glomerular nephropathy, autoimmune disease can impair renal function and lead to failure, often related to progressive fibrosis, as occurs with nephritic and nephrotic syndromes. In these cases, herbs that target autoinflammatory reactivity may be appropriate in formulas for renal failure. *Bupleurum* is widely used in TCM for inflammation of the abdominal organs, and studies suggest renoprotection against immune-complex nephritis credited to saikosaponins.[345] *Cordyceps* may protect against mesangial cell hypertrophy and glomerulosclerosis,[346] common in autoimmune nephritis,[347] including lupus[348] and Sjögren's disease. *Cordyceps* also supports tubule regeneration and sodium pump activity[349] and decreasing elevated BUN and creatinine.[350] *Panax* species may reduce fibrotic processes in the kidneys[351] and also protect against a variety of nephrotoxins known to cause renal failure, such as cisplatin and streptozocin.[352] *Panax* may preserve renal function in cases of diabetic nephropathy, reducing oxidative stress,[353] protecting the renal tubules, down regulating genes involved with renal inflammation,[354] and improving nitrogen balance,[355] renal protein synthesis,[356] and renal function. The Chinese herb *Tripterygium wilfordii*, traditional for autoimmune disorders, may improve renal function in cases of systemic lupus and nephritic syndrome.[357] *Ganoderma* reduces proteinuria in animal models of lupus, decreasing autoantibodies and other inflammatory processes.[358]

Rheum palmatum for Chronic Renal Failure

Animal research has suggested that *Rheum palmatum* may improve cellular function in cases of renal failure.[359] One human clinical trial employed *Rheum* in tandem with dialysis in renal failure patients and reported the herb to slow the progression of the disease compared to placebo, improving creatinine, blood urea nitrogen, urates, and phosphates.[360] These findings have been duplicated in several studies,[361] reducing proteinuria in glomerulosclerosis patients,[362] reducing elevated fibronectin in proliferative glomerulonephritis patients,[363] improving creatinine clearance in chronic renal failure patients,[364] and improving general urinary metabolism of nitrogenous wastes.[365] Emodin in *Rheum* is reported to cause renal damage as well as to have a nephroprotective effect, slowing the sclerotic process in renal disease, so the dosage and formulation of *Rheum* appears important. (See also "Formulas for Acute Pancreatitis" in chapter 2.)

Basic Tea to Support Renal Function

These herbs all have anti-inflammatory, antifibrotic, and immune-modulating effects and would be one possible all-purpose tea for renal failure.

Astragalus membranaceus
Centella asiatica
Camellia sinensis
Glycyrrhiza glabra
Salvia miltiorrhiza

Combine equal parts of the dry herbs. Steep 1 tablespoon of the herb mixture per cup of hot water and then strain. Drink 3 or more cups per day.

Herbs for Renal Insufficiency

The folkloric writings on herbs for kidney disease often mention the following herbs, emphasizing some for basic renal failure leading to systemic fluid accumulation and dependent edema, others for renal failure due to poor circulation, and others still for specific symptoms of renal failure including oliguria and proteinuria. See "Specific Indications," starting on page 156, for further guidance in choosing herbs to include in formulas for renal insufficiency.

CONDITION	HERBS
Systemic edema due to renal insufficiency	*Apium graveolens* *Apocynum cannabinum* *Arctium lappa* *Armoracia rusticana* *Juniperus communis* *Petroselinum crispum* *Panax ginseng* *Pueraria montana* *Solidago canadensis*
Renal failure associated with poor circulation in kidneys	*Aesculus hippocastanum* *Angelica sinensis* *Apocynum cannabinum* *Capsicum annuum, C. frutescens* *Collinsonia canadensis* *Convallaria majalis* *Crataegus* spp. *Ginkgo biloba* *Juniperus communis* *Salvia miltiorrhiza* *Solidago canadensis* *Zingiber officinale*
Progressive pliguria	*Convallaria majalis* *Digitalis purpurea* *Galium aparine* *Lespedeza* spp. *Petroselinum crispum* *Serenoa repens* *Solidago canadensis, S.* spp.
Albuminuria or proteinuria	*Chimaphilla umbellata* *Silybum marianum* *Solidago* spp.

Tincture for Autoimmune Nephritis

Autoimmune nephritis involves vasculitis and fibrosis, and these herbs may deter fibrotic processes and complement an *Astragalus, Salvia miltiorrhiza,* or all-purpose tea. The herbs in this formula are emphasized in folklore as well as modern research for vasoprotective effects. This formula could also be prepared as a tea. In addition, *Ginkgo* and reishi mushroom capsules may be taken separately, roughly 2 of each, 3 times daily.

Bupleurum falcatum
Angelica sinensis
Rheum palmatum
Pueraria montana
Cassia auriculata

Combine equal parts of the herbs, and take 1 teaspoon of the combined tincture 4 times per day.

Tincture for Autoimmune-Related Renal Damage

Bupleurum has particularly been shown to help inflammatory damage to abdominal organs, and *Cordyceps* has been shown to protect from a variety of nephrotoxins that may initiate autoimmune reactivity. *Artemisia annua* contains the sesquiterpene lactone artesunate, credited with anticancer and immune-modulating activity,[366] and the plant is shown to induce apoptosis in renal cancer cells.[367] Therefore, the plant may offer protective effects to renal cells.

Bupleurum falcatum
Cordyceps sinensis
Artemisia annua
Rheum palmatum

Combine equal parts of the herbs, and take 1 teaspoon of the combined tincture 4 times a day.

Tincture for Renal Failure with Heart Failure

When weak heart action fails to perfuse the kidneys adequately, the use of cardiotonics such as *Convallaria* and *Cactus* may complement the basic perfusion-enhancing herbs *Angelica* and *Salvia miltiorrhiza.*

Convallaria majalis 15 ml
Cactus grandiflorus 15 ml
Angelica sinensis 15 ml
Salvia miltiorrhiza 15 ml

Take 1 to 2 dropperfuls of the combined tincture 3 or more times per day.

Astragalus and *Angelica* Tea for Chronic Renal Failure

Many renal diseases involve gradual fibrosis of the kidneys, replacing functional cells with fibrotic tissue. Herbs that attenuate fibrotic processes are appropriate for patients with chronic renal failure. *Astragalus* and *Angelica* may attenuate injury to renal tubules,[368] speed the recovery following ischemic injury,[369] and limit fibrosis.[370]

Astragalus membranaceus
Angelica sinensis

Combine equal parts of the dry herbs. Simmer 1 teaspoon of the herb mixture per cup of hot water, and then strain. Drink 3 or 4 cups per day, long term.

Tincture for Renal Damage Associated with Exposure to Nephrotoxins

Silybum marianum (milk thistle) is most well-known for its hepatoprotective effects, but continued research has shown renoprotective effects as well. Milk thistle is combined here with other herbs noted to have renoprotective effects.

Silybum marianum
Pueraria montana
Cordyceps sinensis
Salvia miltiorrhiza

Combine equal parts of the herbs, and take 1 teaspoon of the combined tincture 4 times per day.

Tincture for Stagnant Renal Function in the Elderly

When elderly patients have scant, cloudy urine, associated with general frailty, *Panax ginseng* and *Solidago* may support basic renal function. *Juniperus* is a renal irritant and not to be used in anything but very small dosages, lest it further inflame congested kidneys. A perfusion-enhancing tea may complement this tincture.

Panax ginseng 30 ml
Solidago canadensis 25 ml
Juniperus communis 5 ml

Take 1 to 2 dropperfuls of the combined tincture 3 or more times per day.

Tincture for Acute Renal Inflammation Associated with Infectious Processes

Streptococcal infections may initiate inflammatory immune complexes that can cause pyelonephritis, or obstructive infections may damage the kidneys. *Allium* has broad antimicrobial activity, as well renoprotective effects, and is included as synergist to the immune-modulating, anti-inflammatory herbs.

Allium sativum
Silybum marianum
Cordyceps sinensis
Curcuma longa

Combine equal parts of the herbs, and take 1 teaspoon of the combined tincture 4 times per day.

Tincture for Renal Failure Associated with Progressive Anuria

Traditional writings have asserted *Solidago* to help in cases of rapidly progressive anuria. *Rheum* and *Pueraria* may increase perfusion and provide additional protective effects. Because this formula includes a small amount of the potentially irritating *Juniperus*, it would be best employed for situations of renal failure and renal insufficiency and not for acute infectious problems.

Solidago canadensis 20 ml
Rheum palmatum 20 ml
Pueraria montana 15 ml
Juniperus communis 5 ml

Take 1 teaspoon of the combined tincture 3 to 6 times daily.

Tincture for Polycystic Kidneys

Polycystic disease is a multifactorial autoimmune disorder characterized by large cysts in the kidneys. These can be a minor issue or so extensive and destructive as to interfere with renal function. There is no published research on herbal medicines to treat polycystic kidneys, so these herbs are chosen based on folkloric indications and my best guess.

Solidago canadensis, S. odora 20 ml
Eupatorium purpureum 20 ml
Phytolacca decandra 10 ml
Iris versicolor 10 ml

Take 1 teaspoon of the combined tincture 3 to 6 times daily.

Specific Indications: Herbs for Urinary Conditions

Achillea millefolium • Yarrow

Achillea flowering tops treat hematuria, astringe capillaries, and offer antimicrobial activity in cases of cystitis. Use *Achillea* as an alternative for people prone to chronic urinary infection and for urological complaints associated with chronic digestive and vascular congestion. *Achillea* is classified by the Eclectic physician Harvey Wickes Felter, MD, in *The Eclectic Materia Medica*, published in 1922 and a favorite reference text of the author, as a urinary sedative, stopping hematuria, and having a soothing effect on the bladder.

Aconitum napellus • Aconite

Caution: Aconite is an extremely toxic botanical for experienced clinicians only. Aconite roots are used in extremely small dosages as a synergist in formulas where symptoms begin suddenly, accompanied by chills and vascular congestion, such as occurs with the initial onset of acute renal infections.

Aesculus hippocastanum • Horse Chestnut

Medications made from horse chestnuts relax spasms of the ureters by reducing inflammation and congestion and may be employed both internally and topically for acute renal colic. *Aesculus* may reduce edema in the ureter and facilitate the passage of renal calculi. *Aesculus* may also improve circulation in the pelvis, reduce vascular congestion, and treat renal disease associated with vascular congestion, liver stasis, and hemorrhoids.

Agathosma betulina • Buchu

Also known as *Barosma betulina*, the leaves of this plant are high in volatile oils with urinary disinfectant effects, and buchu has been recommended for bacterial cystitis and prostatic hypertrophy or inflammation.[371] The phenolic and terpene volatile oils in buchu are stimulating and disinfecting to the urinary tissues, and one of the volatile oils, pulegone, can cause GI, urinary, and uterine irritation. Pulegone is also an emmenagogue and abortifacient and is found in *Hedeoma* (pennyroyal), as well; both plants are noted to be stimulating disinfectants. Buchu is specifically indicated for abnormally acidic urine and mucous in the urine and recommended homeopathically for prostatic disease, mucopurulent discharges, and urinary gravel (a term referring to a large amount of particulate and debris in the urine with a tendancy to form renal calculi).

Agrimonia eupatoria • Agrimony

Agrimonia leaves are used by numerous cultures worldwide, especially as a mucosal astringent and normalizer, and it is specifically indicated for deep, colicky lumbar pain, with foul, muddy urine.[372] *Agrimonia* is also specific for elderly men with dribbling of urine. The aerial parts of *Agrimonia* contain flavonoids with anti-inflammatory effects, and modern research suggests anticancer activity, including induction of apoptosis and cytokine responses.[373]

Agropyron • See *Elymus*

Allium cepa, A. sativum • Onions and Garlic

Onions (*A. cepa*) and garlic (*A. sativum*) can be included in formulas for urinary infections including cystitis and prostatitis, particularly in those with cold constitutions, digestive disorders, and hypertension. Because *Allium* bulbs have anticancer properties, they might also be included in prostate and bladder cancer formulas and protocols. *Allium sativum* has been shown to protect the bladder epithelium from cyclophosphamide-induced damage. Diallyl sulfide and disulfide are credited with the ability to activate detoxifying enzyme systems, protect the digestive mucosal tissues from a variety of carcinogens, and inhibit acetyltransferase enzymes, which may be overexpressed in bladder cancer cells. One of the present pharmaceutical therapies for superficial bladder cancer, BCG (bacillus Calmette-Guérin), has many side effects such that it cannot be used for all patients. *Allium* therapy may have similar efficacy with virtually no such side effects or toxicity.[374]

Althaea officinalis • Marshmallow

The demulcents in *Malvaceae* (mallow) family plants such as *Althaea* are presumed to be the primary constituents responsible for the urinary anodyne effects. The roots of *Althaea* are cold-infused to produce thick, mucilaginous teas offering soothing effects, both direct and reflexive, on the urinary mucosa.

Alcea rosea • Hollyhock

Flavonoids in hollyhock petals inhibit aromatase activity in testicular cells and antagonize estrogen β receptors.[375]

Ammi visnaga • Khella

Ammi visnaga has numerous anti-inflammatory and smooth muscle–relaxing effects, useful to relax spasms in the ureters. *Ammi* tinctures and teas prepared from the ripe seeds may be consumed orally, or hot compresses of *Ammi visnaga* infusions may be applied to flank and back to facilitate the passing of a stone.

Andrographis paniculata • King of Bitters

Both the leaves and roots of *Andrographis* are used medicinally, and this herb has been used for all manner of infections, from simple colds to more serious illnesses. *Andrographis* has antimicrobial effects for simple bladder infections and may be an effective alternative to antibiotic therapy following extracorporeal lithotripsy procedures. *Andrographis* may protect against cyclophosphamide-induced urothelial damage[376] and optimizes anti-inflammatory cytokines and immune compounds such as tumor necrosis factor, interleukins, and interferons, which are suppressed by cyclophosphamide. Antiangiogenic effects have also been demonstrated by *Andrographis*. The plant is contraindicated in pregnancy.

Angelica sinensis • Angelica

Angelica species, also called dong quai, have numerous circulatory and antiallergy effects, improving urinary symptoms due to food or chemical sensitivity. The roots of this herb may also improve renal pathology involving poor circulation such as diabetic nephropathy. *Angelica* normalizes the vasodilating substance nitric oxide when low; however, when nitric oxide is elevated and triggering detrusor overactivity, *Angelica* may reduce nitric oxide and benefit those with overactive bladder.

Apis mellifica • Honey Bee

Although this is an insect, not a plant, bee venom has long been prepared into tinctures and listed in folkloric herbals. *Apis* is traditional to relieve constant tenesmic pain in the bladder and urethra and to help burning, stinging urination. *Apis* is specific for acute and sudden onset of violent burning urinary pains. Small material doses of *Apis* mother tincture (a tincture prepared purposefully diluted at a 1:10 ratio) are suggested for edema and renal engorgement and acute urinary retention; 3X dilutions (where a mother tincture is further serially diluted) are suggested for bladder irritation. A "material" dose is a term from homeopathy, signifying that a bit of the bee venom is materially present in the tincture, unlike a homeopathic dilution, where the mother tincture is so diluted as to no longer contain a material dose. A traditional remedy for burning dysuria is to place 2 milliliters of *Apis* mother tincture in a half glass of water and take by the teaspoonful every hour.

Apium graveolens • Celery

Celery stalk juice, root, and seed extracts are used as nerve, bronchial, and urinary relaxants. *Apium* is high in the volatile oil called apiol. *Apium* is specifically recommended for urinary retention and edema.

Apocynum cannabinum • Dogbane

Apocynum roots contain cardiac glycosides and *Apocynum* medications are specific for urinary ailments related to circulatory weakness and systemic fluid retention. *Apocynum* is also indicated for bladder distension, mucous in the urine, and burning urination and is specific for the patient who develops edema with urinary infections. Due to potential cardiac toxicity, *Apocynum* is to be used only by skilled clinicians who cautiously employ small dosages.

Arctium lappa • Burdock

The roots of *Arctium*, also called gobo root, are valuable alterative agents and useful in cases of dysuria and dropsy due to renal obstruction. Scudder suggests *Arctium* for chronic urinary problems in which tissues and organs are debilitated. *Arctium* enhances absorption, metabolism, and elimination through the bowels, supporting optimal repair and regeneration of bowel and urinary tissues. In addition to the roots, *Arctium* seed preparations can help treat the urinary system.

Arctostaphylos uva ursi • Uva Ursi

The leaves of uva ursi, also called kinnickinick, contain the phenolic glycoside arbutin, and they are commonly used for acute cystitis, pyelonephritis, and urethritis. However, uva ursi teas and tinctures are also appropriate for diabetic polyuria and some cases of lithiasis. Uva ursi is appropriate for pelvic pain that is cramping, heavy, and dragging in character and is particularly indicated where complaints are chronic, though it should not be used long term. Uva ursi is indicated for chronic urinary irritation and atony with a tendency to tenesmus and catarrh. Felter stated that uva ursi was especially indicated for persons with feeble circulation, and numerous sources report *Arctostaphylos* to be a general diuretic.[377] Infusing the dried leaves is a common practice, but the urinary antiseptic arbutin is best extracted by gently simmering the leaves for 20 to 30 minutes. Modern research notes *Arctostaphylos* to have activity against antibiotic resistant infections such as MRSA as well as to potentiate β-lactam antibiotics.[378] *Arctostaphylos* also has some anti-inflammatory activity, making it appropriate for inflammatory conditions and allergic phenomena of the renal passages. Avoid *Arctostaphylos* during pregnancy, or use it with caution, due to possible oxytocic activity that may induce abortion.

Armoracia rusticana • Horseradish

Spicy horseradish root preparations may be used as synergists in formulas for frequent, painful urination (edema) to help renal calculi pass, and to reduce the likelihood of recurrence. Older texts recommend *Armoracia* for dependent fluid accumulation and edema, formerly referred to as "dropsy", where improving renal processing of fluid could be beneficial. *Armoracia* may deter urinary pathogens and provide sulfur to help repair connective tissue in cases of leaky bladder and uroepithelial barrier weakness. Horseradish was formerly identified as *Cochlearia armoracia*.

Arnica montana • Leopard's Bane

Arnica may be included as a synergist in formulas for urinary pain, cramping sensations in the bladder, dysuria, and red sediment in the urine. The entire plant is prepared into a tincture, and homeopathic preparations are classic medicines for trauma to the kidneys or for blood in the urine following vigorous exertion. *Arnica* can be a valuable agent to treat neurogenic bladder due to spinal cord injury and parapalegia.

Asparagus officinalis • Asparagus

All parts of various asparagus species are high in mucilage, including the stalks, but the roots are most used medicinally and contain asparagin, which has a soothing restorative effect on urinary mucosal membranes. *Asparagus* contains sulfuric acids that are metabolized by the kidneys and excreted via the urine. *Asparagus racemosus* (shatavari) is used medicinally in Ayruvedic medicine as a mucous membrane tonic for the genitourinary system.

Atropa belladonna • Belladonna

All parts of the plant contain tropane alkaloids and medicines are most often prepared from the fresh or dried leaves. **Caution:** Potentially toxic, *Atropa* is used by skilled clinicians to treat renal colic, acute febrile nephritis, renal engorgement, and nocturnal incontinence. Eclectic physicians reported belladonna to stimulate the kidneys as well as to relieve irritability in the kidneys and bladder. Like many physiologically strong herbs, *Atropa* has a biphasic effect, where opposite actions are seen at differing dosages. Belladonna may improve bed-wetting in children when due to poor circulation in the pelvis or hyperexcitability of the bladder. Belladonna has a noteworthy antispasmodic effect on urinary smooth muscle, relaxing urinary spasm and associated urinary retention. Belladonna is also specific for renal congestion and associated pain and fullness. *Atropa belladonna* may be crafted into urethral suppositories for some acute urinary spastic situations. Include belladonna as a synergist in formulas for the polyuria of diabetes insipidus. Herbal tropane alkaloids, including atropine, scopolamine, and hyoscyamine, may be used to relax excessive bladder contraction, but the dose must be limited due to the many systemic side effects.

Avena sativa • Oats

Medicines prepared from "milky" oats, meaning the fresh flowering tops, have a restorative effect on the nervous system and can be a base trophorestorative herb when urinary issues are associated with stress or long-standing nervous exhaustion. Folkloric herbals have recommended milky oats for bladder spasms, impotence, and genitourinary weakness, especially when due to alcoholism, drugs, insomnia, or overwork.

Barosma • See *Agathosma*

Berberis vulgaris • Barberry

Barberry is indicated for urinary symptoms with renal pain, back pain, and pain in the thighs and for urine that is thick, red, and mealy in appearance. *Berberis* species contain antimicrobial isoquinoline alkaloids including berberine in the yellow inner bark, which is used to make tinctures or dried to use in teas.

Calendula officinalis • Pot Marigold

Calendula flower petals or fresh, whole plant succus medications may be added to tea and tincture formulas respectively for ulcerative lesions of the bladder or for uroepithelium disorders, such as occur with interstitial cystitis. Dried *Calendula* petals are useful in tea formulas for simple bladder infections.

Camellia sinensis • Green Tea

Green tea leaves contain powerful antioxidants including the catechin found in numerous animal and cell line studies to prevent urinary cancers caused by exposure to nitrosamines[379] and other carcinogens. *Camellia sinensis* is also an aromatase inhibitor.[380]

Cantharis vesicatoria • Spanish Fly

Cantharidin, a potent irritant and vesicant found in the Spanish fly (a type of beetle), is frequently used in homeopathic medicine or in highly diluted form for relief of bladder pain and the symptoms of cystitis. While the scientific binomial for *Cantharis* has changed to *Lytta*, most products are still marketed under the name *Cantharis*. Cantharidin when consumed in a substantial material dose is a profound irritant and induces cytotoxicity in the bladder. However, cantharidin is also reported to have cytotoxicity against bladder cancer cell lines, inducing apoptosis and stimulating immune-modulating cytokines.[381] Therefore, *Cantharis* mother tincture might be used as a synergist in more nourishing and soothing formulas for bladder cancer. *Cantharis* is specifically indicated for bladder afflictions accompanied by a sense of erotic excitement or sexual stimulation and for fairly superficial sensations of burning, tickling, teasing, and irritation of the distal urethra with a constant desire to urinate. *Cantharis* is a potent irritant with many systemic toxic effects and is used only in a diluted form, a drop or two at a time. *Cantharis* is purported to have specific activity on urinary sphincters, and the Eclectic authors recommend small, cautious dosages for nephritis. Felter stated of this substance: "but when recklessly employed it is a dangerous medicine, producing or aggravating the very conditions sought to be relieved by it."

Capsicum annuum • Cayenne

Capsicum fruit tincture prepared from various cultivars may be used as a synergist in formulas for cold, weak, sluggish constitutions and poor digestion. *Capsicum* is specific for cramping in the bladder and for urinary urging without producing much urine. *Capsicum frutescens* is also referred to as cayenne and is considered a cultivar of *Capsicum annuum*.

Centella asiatica • Gotu Kola

Centella is indicated for bladder ulcers, mucosal inflammation, and autoimmune and connective tissue disorders. The entire plant may be prepared into tinctures, dried for tea, and encapsulated. In Asia, the fresh plants are juiced and sold as a beverage mixed with sugarcane juice. *Centella* may be useful for uroepithelial disorders such as interstitial cystitis, because it supports the synthesis of hyaluronic acid and basement membrane glycosaminoglycans.

Chimaphila umbellata • Prince's Pine

Whole *Chimaphila* plant extracts are reported by Eclectic authors to improve recuperative powers of the urinary mucosa[382] and to reduce albuminuria[383] and "brick dust" urine. Like *Arctostaphylos uva ursi*, *Chimaphila* contains arbutin, an effective urinary antiseptic. *Chimaphila* has affinity for both the renal and lymphatic systems and is specific for scanty urine production where the urine is thick and ropy. *Chimaphila* may be included in formulas for prostatic enlargement, urinary retention, and the sensation of a ball in the perineum.

Collinsonia canadensis • Stoneroot

Collinsonia roots are specific for pelvic complaints associated with vascular, lymphatic, and mucous membrane congestion. *Collinsonia* is recommended for urinary complaints where there is a sense of weight, engorgement, and constriction and is specific for urinary problems and gravel secondary to venous atony. *Collinsonia* may improve hemorrhoids and other varicosities, especially when associated with a heavy, tense sensation. *Collinsonia* is appropriate for individuals with a history of crystals and much debris in the urine.

Conium maculatum • Poison Hemlock

Caution: *Conium* species are potentially toxic and to be used by skilled clinicians only. All plant parts contain the central nervous system depressant piperidine alkaloid coniine; the roots or fresh seeds can be prepared into medicines to be used in very small doses. *Conium* alkaloids have nicotinic effects in autonomic ganglia and at high dosage act as paralyzing neurotoxins at neuromuscular junctions. *Conium* has an antispasmodic effect on urinary muscle when taken orally and may be used topically or crafted into urethral suppositories for acute urinary pain. *Conium* is specific for tight sensations, urinary spasm, difficulty starting and stopping urinary flow because of cramping, and dribbling of urine in the elderly. *Conium* may be helpful when included in formulas for the prostate gland and swelling of the testes, capable of improving urinary retention due to detrusor muscle overactivity. Small or homeopathic dosages of *Conium* show promise in treating breast and cervical cancer. Toxicity symptoms include bradycardia, burning sensation in mouth, hypersalivation, tremor in hands, ataxia, and pupillary dilation. *Conium* should be removed from a formula at once should any such symptoms occur. Toxic dosages lead to cardiac and respiratory depression that may lead to coma and death from respiratory failure. The plant is teratogenic and must be avoided during pregnancy.

Convallaria majalis • Lily of the Valley

Convallaria roots are primarily a cardiovascular medicine, but they can be included in formulas for urinary complaints related to vascular insufficiency, where the kidneys are poorly perfused and there is venous stasis and edema. *Convallaria* is specific for aching in the pelvis and bladder that is associated with urinary retention or frequent, scant urination.

Corydalis yanhusuo, C. cava • Corydalis

Corydalis, also called turkey corn, is a nervine herb used for general pain, tension, and muscle spasms. Medicines prepared from the roots may have an antispasmodic effect on urinary muscles, although the plant is less specific and powerful for this than other herbs emphasized in this chapter. It is mentioned here as an agent to help provide weak, opiate-like pain relief when appropriate and to provide a gentle calmative anodyne effect in formulas to treat urinary pain.

Crataegus oxyacantha, C. monogyna • Hawthorne

Berries and young leaves and flowers are all used medicinally for strengthening the heart and blood vessels via various anti-inflammatory and vascular stabilizing effects. Use hawthorn as a synergist where circulatory issues, diabetes, or vascular injury and inflammation underlie urinary complaints. *C. oxyacantha* is also known as *C. rhipidophylla.*

Cucurbita pepo, C. maxima • Pumpkin

Oil from pumpkin seeds can be consumed to treat benign prostatic hypertrophy (BPH). Pumpkin seeds contain lignans, including secoisolariciresinol and lariciresinol[384] and are high in sterols.[385] Squalene, a steroidlike compound, is also abundant in pumpkin seeds and may contribute to the hormonal effects of *Cucurbita,* as do the α, β, and gamma tocopherols.[386] Regular consumption of whole pumpkin seeds or pumpkin seed oil may improve overactive bladder and reduce testerone-driven hypertrophy of the prostate.

Dioscorea villosa • Wild Yam

Dioscorea tubers contain anti-inflammatory steroidal saponins and have antispasmodic effects in smooth muscles of hollow organs including the stomach, biliary tree, intestines, bronchi, uterus, and ureters. Taken as a tea or tincture, *Dioscorea* may be helpful in formulas for spastic pains in the bladder and renal colic.

Elymus repens • Couch Grass

Elymus is a weak diuretic capable of improving urinary irritation and dysuria. The juicy rhizomes are indicated for incontinence due to urinary irritation, purulent cystitis, hematuria, and strangury[387] as well as pyelitis and kidney irritation. Eclectic authors such as Felter and John Milton Scudder, MD, wrote about *Elymus* using its previous name, *Agropyron,* and also recommended it for prostatitis, prostatic adenoma,[388] prostatic enlargement, and gonorrhea.[389] *Agropyron* is specifically indicated for an intense burning sensation and a constant desire to urinate.[390] *Agropyron* is high in mannose noted to prevent bacterial adherence to bladder epithelium.[391] *Agropryron* contains around 3 percent inositol and mannitol[392] as well as agropyrene, a terpenoid credited with antimicrobial action. *Elymus* was formerly classified as *Agropyron,* and many sources may still list it that way.

Equisetum arvense, E. hymenale • Horsetail

The mineral-rich aerial parts of *Equisetum,* commonly called horsetail or scouring rush, are indicated when there is blood in the urine, and for chronic bacteria or frequent cloudiness and debris. Felter recommended *Equisetum* for "Cystic irritation, tenesmic urging to urinate; nocturnal urinary incontinence; and renal calculi,"[393] and used *Equisetum* for acute prostatitis and enuresis related to bladder irritation. *Equisetum* is notably high in silica compounds, which may lend strength and flexibility to connective tissue as well as help the kidneys excrete heavy metals and reduce toxicity, hypersensitivity, and autoimmune reactivity. *Equisetum* is a connective tissue tonic, supporting the structural integrity of bones, joints, bladder mucosa, skin, hair, and finger nails. Sitz baths of *Equisetum* may allay pain in cases of pelvic spasms, pelvic inflammatory disease, and musculoskeletal tightness. The most potent *Equisetum* teas are prepared by soaking the herb in cold water overnight and then simmering briefly in the morning. To leach the greatest amount of silica and minerals from the tea, a long soak at low heat, such as all day in a slow cooker, is recommended.

Eschscholzia californica • California Poppy

The fresh leaves and flowers contain low levels of papaverine alkaloids, which can bind GABA and possibly opiate receptors to offer gentle nervine and anodyne effects. *Eschscholzia* may have an antispasmodic effect on urinary muscles. Use it in formulas for renal colic or to help wean from opiate medications.

Eucalyptus globulus • Eucalyptus

Eucalyptus leaves are high in volatile oils and are used as a stimulating, cooling, and astringing antimicrobial agent for irritation of the bladder, nephritis, and chronic urinary catarrh. *Eucalyptus* is specific for cramping and tightness in the bladder and for catarrhal states of the urinary mucosa. *Eucalyptus* may also help alleviate pain via antinociceptive activity.

Eupatorium purpureum • Gravel Root

Eupatorium purpureum root preparations are recommended for hematuria, chyluria, increased urinary sediment, urates, and urine that appears to contain solids and is dark and cloudy. *Eupatorium* is specifically indicated for dysuria, for irritable bladder where there is a deep aching sensation in the pelvis due to pathologies of the bladder or prostate, and for frequency, and shooting pains in the urethra. Another Eclectic physician, Finley Ellingwood, MD, claimed *Eupatorium* could promote elimination of crystals and reduce tendency to form calculi, and Felter recommended it for bladder or urethral irritation and inflammation, with blood and gravel present in the urine.

Galium aparine • Cleavers

Galium is a cooling "refrigerant diuretic" indicated for suppressed urine, dysuria and bladder irritation that accompanies uterine disease or prostatic irritation. The long thin scratchy aerial parts are used to prepare medicines including tinctures, teas, and fresh poultices. *Galium* has a historical reputation in lymphatic disorders and cancers, used both topically over lymphatic growths and internally.

Ganoderma lucidum • Reishi Mushroom

The woody reishi mushroom is inedible but it is prepared into teas, tinctures, glycerites, and powders as an immune-supportive herb to include in formulas for bladder and prostate cancer. It has also been shown to have 5α-reductase inhibiting effects.[394] *Ganoderma* may speed recovery for toxin-induced renal damage and urinary inflammation.

Gaultheria procumbens • Wintergreen

Wintergreen leaves and berries are high in camphorous volatile oils and salicylates. Tinctures and concentrated essential oil products may be used as synergists in formulas for bladder and prostatic irritation. Wintergreen essential oil may be included in topical formulas for renal infections, such as in hot packs over the kidneys for pyelonephritis.

Gelsemium sempervirens • Yellow Jessamine

Caution: The roots and rhizomes of yellow jessamine, also called yellow jasmine, are used to prepare herbal anodynes, but it can be a highly toxic botanical if misused. Due to powerful spinal nerve depressant effects, overdoses of *Gelsemium* can be fatal due to suppression of the respiratory centers of the brain stem. *Gelsemium* can be included as a low-dose synergist in formulas to treat renal pain and neuralgia. *Gelsemium* is specific for profuse, clear, watery urine with sense of chill and tremors as well as for urinary retention and dysuria.

Ginkgo biloba • Ginkgo

Ginkgo leaves have numerous circulatory and antiallergy effects that may help individuals with urinary symptoms due to food or chemical sensitivity as well as those with renal pathologies involving poor circulation such as diabetic nephropathy. *Ginkgo* normalizes the vasodilating substance nitric oxide, which, when low, leads to blood stasis and fluid congestion and, when elevated, is associated with detrusor overactivity. Use *Ginkgo* as a synergist in formulas for patients with allergy, hyperreactivity, atopy, poor circulation, and vascular stasis.

Glycyrrhiza glabra • Licorice

Licorice roots are a widely used herbal medicine due to broad anti-inflammatory and immune-modulating properties. *Glycyrrhiza* enhances mucous production by mucosal goblet cells in the gut, contributing to ulcer-healing effects, and it may have similar effects on the urinary mucosa. *Glycyrrhiza glabra* is reported to inhibit the steroid dehydrogenases, due in part to glycyrrhetinic acid content,[395] which may lend hormone-balancing effects.

Grifola frondosa, G. umbellata • Maitake

Grifola is a delicious edible mushroom. Like many medicinal mushrooms, maitake (also called hen of the woods) has been found to have immune-enhancing and anticancer effects. *Grifola frondosa* and *G. umbellata* have shown activity against prostate and bladder cancer cell lines, inducing apoptosis.[396] *Grifola* may protect from bladder carcinogens such as nitrosamines and may prevent bladder cancer recurrence.

Hamamelis virginiana • Witch Hazel

Stems and young branches can be pruned from *Hamamelis* shrubs and decocted to yield tannins that have anti-inflammatory effects on swollen tissues. It is a traditional remedy for varicosities, used both topically and internally. *Hamamelis* is specific for venous congestion, hemorrhoids, varicocele, and heavy aching pain in the genitals as well as for testicular complaints and orchitis.

Hamamelis is specific for blood and mucous in the urine and is said to be a urinary astringent.

Hedeoma pulegioides • Pennyroyal

Hedeoma leaves are high in volatile oils and can be prepared into teas and tinctures to include in formulas for renal colic, urinary frequency, dysuria, renal pain, and ureteral pain. Pennyroyal essential oil is used topically as a flea and tick repellant and is not commonly consumed orally due to abortifacient and toxicity concerns.

Herniaria glabra • Rupturewort

The aerial parts of the creeping *Herniaria* are used medicinally as an antispasmodic for the renal musculature, and both the common and the Latin names refer to the traditional use of the plant for hernias and to strengthen the pelvic musculature.

Hydrangea arborescens • Hydrangea

The root and rhizome of *Hydrangea* are considered a renal tonic to include in formulas for renal calculi, renal colic, blood in the urine, pain at the costal vertebral angle, and prostatic disease. *Hydrangea* is specific for burning in the urethra, difficulty starting urination, and white sediment in the urine. *Hydrangea* is recommended for both acute and chronic renal problems, especially where there is an increase in urinary phosphates and alkaline urine. Felter recommended *Hydrangea* for "Vesical and urethral irritation, with gravel; difficult urination; deep seated renal pain; bloody urine, irritation of the bronchial membranes." *Hydrangea* is one of the few botanicals thought to be safe during an acute infection or episode of renal colic. *Hydrangea* is said to be most effective when taken as hot tea, and it was combined with *Piper methysticum*, *Actaea racemosa*, and *Gelsemium* in Eclectic formulas of the early 1900s.[397]

Hyoscyamus nigra • Henbane

Of the various Solanaceae family (nightshade) plants that contain tropane alkaloids, early American authors reported *Hyoscyamus* to have the greatest specificity for the urinary passages. All the aerial parts are used—leaves, flowering tops, and seeds—to prepare a potent and potentially toxic medicine. *Hyoscyamus* is specifically indicated for severe spasms of the urinary tract, such as renal colic.

Hypericum perforatum • St. Johnswort

Hypericum buds contain a blood red fluid high in flavonoids credited with many of the plant's medicinal effects. *Hypericum* has an affinity for the nerves and may reduce bladder irritability in cases of bed-wetting, nervous bladder, and hypersensitivity in the bladder mucosa. *Hypericum* may improve detrusor overactivity, interstitial cystitis symptoms, and bladder tone in postmenopausal women. Some of the flavonoids in *Hypericum*, such as hypericin and hyperforin, are potent photosensitizers that are being explored as a possible photodymic therapy against some cancers, including bladder cancers. More general anticancer effects, such as antiviral, preventive and apoptosis-enhancing effects, are also reported for *Hypericum*.

Juniperus communis • Juniper

Juniperus berries are high in volatile oils, and juniper tincture is specific for atonic and noninflammatory conditions of the kidneys and bladder. *Juniperus* may reduce mucous discharges when of an atonic, atrophic origin, but it is too irritating and stimulating for use in acute inflammatory or infectious conditions. *Juniperus* may have antioxidant properties[398] and inhibit the synthesis of inflammatory eicosanoids.[399] *Juniperus* is best reserved for chronic renal insufficiency or atonic situations; in small doses, it may stimulate circulation and glomerular filtration. *Juniperus* is also specific for chronic nephritis and pyelitis associated for vascular congestion related to stasis and poor circulation, but not acute hyperemia due to active inflammation. Use juniper cautiously, in small dosages only, where a warming, stimulating urinary antiseptic is needed.

Lepidium meyenii • Maca

Maca roots are claimed to improve erectile function as well as improve urinary function in those with BPH. In the high Andes where the turniplike roots are indigenous, maca is also used for general weakness and atony. Modern histological studies have shown *Lepidium* to reduce the proliferative effects of testosterone on prostatic cells[400] and to reduce prostate size in mice,[401] without reducing the testosterone level itself.[402] This suggests that *Lepidium* inhibits the proliferative effects of testosterone on the prostate gland by a mechanism other that affecting active testosterone levels.[403] Both tinctures and aqueous extracts of maca have proven effective in reducing prostate size in rats with induced prostatic hyperplasia.[404] Glucosinolates are present in *Lepidium* and credited with some of the physiologic effects, and glucosinolates from other Brassicaceae (crucifer family) plants such as broccoli, cauliflower, and cabbage have demonstrated hormonal activity, including antiestrogenic effects. Glucosinolates in maca are claimed to have antiproliferative and apoptotic actions on the prostate. Alkaloids, steroids, tannins, saponins, and other constituents of maca may contribute to the medicinal effects.

Lespedeza capitata • Roundhead Bushclover

Various species of this legume are used medicinally. The entire aerial portion is dried or prepared into extracts, and in some cases the roots are included in medicines as well. *Lespedeza* is specific for chronic renal disease and insufficiency with reports that it can help the kidneys clear blood urea nitrogen (BUN) and creatinine from the blood, normalizing these values. *Lespedeza* may slow the progression of chronic renal disease and inflammation into sclerosis and failure.

Levisticum officinale • Lovage

Levisticum roots, leaves, and seeds are all used medicinally and may be included in diuretic tea mixtures. This Apiaceae family plant has numerous circulatory-enhancing and calcium channel blocking effects, and it is anti-inflammatory and antispasmodic to urinary passages.

Lobelia inflata • Pukeweed

Lobelia flowering tops are most noted for their antispasmodic effects on bronchial smooth muscle, but they also have antispasmodic effects on the urinary organs and passages. *Lobelia* can be included in formulas for tenesmus of the bladder and renal colic.

Lycopus virginicus • Bugleweed

Lycopus aerial parts are most used as an endocrine herb, especially in the treatment of hyperthyroidism, but they may be included in formulas for profuse watery urine related to hypermetabolic states and diabetes.

Mahonia aquifolium • Oregon Grape

Medicines prepared from the root bark of *Mahonia aquifolium* are indicated for dysuria, burning sensations, and urinary calculi. *Mahonia aquifolium* is specific for stitching cramping pains in the bladder, thighs, and groin, urine with thick mucous, and bright red, mealy sediment. *Mahonia* is also indicated when urinary pain is accompanied by chronic infections, skin eruptions, and liver congestion. This plant and many others in this genus are also known as the genus *Berberis*.

Medicago sativa • Alfalfa

The young aerial parts of *Medicago*, also referred to as lucerne, are highly nourishing, containing vitamins D and K, calcium and other minerals, as well as isoflavones offering hormonal-balancing effects in cases of BPH or in cases of urinary atony due to postmenopausal urinary atrophy. *Medicago* is specific for urinary frequency; its nourishing minerals help the kidneys move wastes and toxins across osmotic concentration gradients, supporting elimination via the urine.

Panax ginseng • Ginseng

Panax roots are one of the most prized herbal medicines in China, where they are used as vitality and longevity tonics. Ginseng roots are widely researched for improving a variety of cell stress pathways and offering antioxidant, immune-modulating, and hormone-regulating effects. *Panax* may reduce oxidative stress in the kidneys and can be used in cases of renal insufficiency, especially when weakness and frailty accompany.

Petasites hybridus • Butterbur

Both the roots and leaves of *Petasites* are used medicinally for inflammation of the airways, vasculature, and urinary passages. *Petasites* roots are specific for urinary cramping, crawling sensations in the urethra, and urethral discharges. Petasin and isopetasin are credited with many positive effects on inflammatory markers, and although most research focuses on migraines and lung reactivity, similar mechanisms may benefit urinary passages, given the folkloric use of the plant for urinary pain. *Petasites* contains small amounts of pyrrolizidine alkaloids, which when consumed in large or continuous quantity, as with cattle who graze on plants containing these compounds, are associated with liver damage. *Petasites* should not be consumed by pregnant or lactating women and should not be taken for months on end.

Petroselinum crispum • Parsley

Parsley seeds are an excellent diuretic, even in cases of anuria associated with renal failure and heavy metal toxicity. They are specific for urinary pain, burning and tingling, perineal pain and sensation, "voluptuous" itching of genitals, and hemorrhoids. Concentrated and isolated apiol found in the essential oil of parsley may be toxic and cause nerve damage, paralysis, and paraesthesis in the hands, feet, and lower legs; however, whole parsley extracts and foods are not considered dangerous. Large doses of parsley tincture or tea may also be abortifacient in some rare cases and should be avoided during pregnancy. Also known as *P. sativum*.

Piper cubeba • Cubebs

The peppery fruits of *Piper cubeba*, a relative of black pepper (*Piper nigrum*) and kava (*Piper methysticum*) are specific for urinary pain, blood and mucous in the urine, and cystitis. Cubebs has an affinity for urinary mucous membranes and is specific for burning sensations and profuse mucous discharges from the urethra, as occurs with gonorrhea or other sexually transmitted diseases. Cubebs may improve the activity of other ingredients in a formula. Cubebin, a flavonoid found in many types of pepper, is credited with some of the physiologic activity, and piperine, a major alkaloid of black peppercorns, has been shown to be antiandrogenic via 5α-reductase when tested in isolation.[405] *Piper cubeba* has also been found to down regulate androgen receptors in general.[406] As male pattern baldness is also associated with the activity of this enzyme, *Piper* species are also being explored as potential oral and topical therapy for baldness.

Piper methysticum • Kava Kava

Sometimes called the "intoxicating pepper," kava is in the same family as black pepper, and the large roots have powerful urinary anodyne and antispasmodic effects. Kava also possesses antimicrobial activity for urinary pathogens and sexually transmitted diseases (STDs). Although it is not certain whether kava has significant antimicrobial action on the urinary tract for cystitis, there are numerous historical reports of kava's use in syphilis, venereal diseases, and urinary symptoms. Kava is numbing to the tongue and mouth and also to the ureters and bladder as it passes through. Kava may be used as a sedative, antispasmodic anodyne for pain in urinary structures, its anesthetic effect relieving irritation of the bladder, improving dysuria and the ability to completely empty the bladder. Kava may even help the passage of kidney stones due to quieting spasms in the ureters. Due to peripheral relaxation, kava is valuable for muscle pain, tension, and cramps[407] that are seen with anxiety, nervousness, and fibromyalgia. In addition to urinary anodyne effects, Eclectic physicians and authors listed kava for tooth, nerve, stomach, and intestinal pain.

Piper nigrum • Black Pepper

Black pepper fruits, commonly referred to as peppercorns, are used mainly as a synergist in formulas, improving the absorption and activity of other ingredients. However, *Piper* has many medicinal effects in small doses, including anticancer activity against prostate and bladder cancer. *Piper nigrum* has been found to be a 5α-reductase inhibitor.[408]

Podophyllum peltatum • Mayapple

The small applelike fruits of the mayapple are used to produce a caustic botanical medicine that has occasionally been recommended for chronic urinary infections that are accompanied by liver congestion, constipation, and poor, sluggish digestion. Resins from the fruits include podophyllin, having anticancer and direct caustic and ablative effects on tissues. "Kidney congestion," defined by the symptoms of scanty urination and pain in the small of the back, was said by the Eclectic authors to be a specific indication for *Podophyllum* when liver and digestive issues accompany renal insufficiency.

Populus tremuloides • Quaking Aspen

Aspen bark is noted to contain salicylates and have anti-inflammatory effects on connective tissue in cases of arthritis and traumatic injury. However, anti-inflammatory and antioxidant properties may lend *Populus* a deeper connective tissue stabilizing effect apart from its salicylate content.[409] Folkloric traditions report aspen to have an affinity for the digestive and urinary systems, especially when associated with heat, night sweats, and indigestion.

Pueraria spp. • Kudzu

Pueraria roots contain isoflavones credited with many hormonal-balancing effects. *Pueraria* roots are also a renal trophorestorative, indicated for all manner of renal cell damage, inflammation and autoimmune conditions, renal failure, and poorly functioning kidneys. *Pueraria* is used as an herb of rejuvenation in many traditional cultures, from reversing graying of the hair to maintaining skin and sexual function to improving circulation to body organs. Circulatory-enhancing effects may contribute to *Pueraria's* ability to improve renal cellular function in cases of renal failure and insufficiency. Both *P. mirifica* and *P. lobata* are used medicinally. Some taxonomists prefer to classify *P. lobata* as a variety or subspecies of *P. montana*.

Punica granatum • Pomegranate

Pomegranate fruit flavonoids may prevent and improve vascular weakness in urinary complaints associated with vascular issues. The tiny seeds in pomegranate fruits contain estrone, which may improve hormonally related urinary complaints.

Pygeum africanum • African Plum Tree

Pygeum is specifically indicated for urinary symptoms due to prostatic enlargement and may possibly have effects against prostate and bladder cancer. The plant has antiandrogenic activity that may reduce proliferative effects of testosterone on the prostate and cancer cells.

Rhus aromatica • Sweet Sumac

The fruits of *Rhus aromatica* are edible and the leaves and bark are used medicinally. *Rhus aromatica* is related to poison oak and ivy but is less allergenic, and it is indicated for chronic irritability of the urinary passages leading to loss of urinary tone. *Rhus* is used for enuresis and bed-wetting in homeopathic and botanical traditions and is specific for urinary complaints of diabetic and the aged. *Rhus* is specifically indicated for an inability to concentrate urine, resulting in pale urine with elevated albumin levels. *Rhus* is also indicated for loss of bladder tone leading to enuresis and constant dribbling of urine.

Rhus toxicodendron • Poison Ivy

The allergenic leaves are prepared into a homeopathic medicine, commonly referred to as "Rhux tox," especially indicated for tearing pains that are made better by moving the affected part, causing people to be restless, and to toss and turn. Homeopathics or diluted 1:10 extracts, referred to as mother tinctures, of *Rhus* are specific for dark scant urine with blood and sediment, burning pains with stiffness in the body, and a tendency to sepsis.

Rosmarinus officinalis • Rosemary

Rosmarinus has broad antimicrobial effects that make both teas and tinctures of the plant appropriate for simple urinary tract infections.

Rubia tinctorum • Common Madder

Rubia is so named because the roots have long been used to make a red dye. *Rubia* is a traditional herbal remedy for kidney stones, and some Ayruvedic remedies for renal pain and calculi include *Rubia* as an ingredient. *Rubia* is also reported to have possible efficacy against bladder cancer. The use of the plant may produce a bright red urine. The tea is not palatable, and a tincture may be preferred.

Scutellaria lateriflora • Skullcap

Scutellaria leaves and young sprouts have nervine effects and may be included in formulas for patients with chronic pelvic pain and interstitial cystitis when an anxious or nervous temperament underlies.

Serenoa repens • Saw Palmetto

Serenoa fruits are prepared into tinctures, capsules, and proprietary products, especially to treat BPH, and are historically recommended for atony or weakening of urinary organs such as occurs in the elderly or postmenopausal women. *Serenoa* will promote the flow of urine and has been recommended for kidney stones and problems with urination, especially when due to old age and loss of genitourinary tone. In Germany, *Serenoa* has been referred to as "the plant catheter," owing to its ability to diurese when the kidneys are functioning poorly. The German physician and author R. F. Weiss recommends *Serenoa* for bladder pathology of nervous origin, including abnormal detrusor muscle and sphincter tone, and to improve autonomic function of urinary organs. Traditional writings recommend *Serenoa* to be specific for dysuria due to spasm and tenesmus, suggesting utility in acute renal colic. Pollen extracts of *Serenoa* are also used for chronic prostatitis, and the high bioflavonoid content may provide anti-inflammatory and vascular-stabilizing effects. *Serenoa* inhibits the formation of dihydrotestosterone due to its liposterol content and is highly regarded in herbal medicine for the treatment of prostatic enlargement due to both adenoma and benign hyperplasia.[410] *Serenoa repens* is noted to inhibit tumorogenesis and induce apoptosis in animal models of prostate cancer.[411] Human investigations have supported the ability of *Serenoa* to inhibit 5α-reductase and reduce circulating levels of active testosterone.[412] By the same mechanisms, *Serenoa* may be useful in women with hirsutism and androgen excess, such as with polycystic ovaries.

Solidago odora, S. virgaurea. S. canadensis • Goldenrod

The flowering tops of *Solidago* are used as renal tonics, specific for atonic and chronic conditions. *Solidago* can be stimulating to atonic organs but can be overly stimulating to inflamed urinary structures. *Solidago* is best used following acute situations to improve renal function and prevent a recurrence of stones or infections. Due to the abundant pollen it contains, fresh and dried *Solidago* can also be allergenic to respiratory passages for those with a tendency to hayfever. *Solidago* has been reported to reduce albuminuria in various types of nephritic syndromes and hemorrhagic nephritis.[413] It may promote diuresis via a direct action on the renal function. *Solidago* is specific for increasing urine production in cases of anuria and oliguria, and it is indicated for scanty urine and for thick, dark red or brown sediment in the urine.

Taraxacum officinale • Dandelion

The leaf makes an excellent nontoxic mineral-rich and nourishing diuretic, and the roots have useful alterative effects, indicated with liver congestion or intestinal dysbiosis occurring in tandem with urinary complaints. The diuretic effects of *Taraxacum* leaves are powerful and yet gentle on the body, are nourishing, and do not cause wasting of potassium or other minerals. Because tinctures do not extract the minerals as well as teas, water extracts are preferred when the minerals are desirable to support renal function. Adding a bit of vinegar or lemon juice to the macerate helps extract minerals out of *Taraxacum* leaves.

Thuja occidentalis, T. plicata • Cedar

T. occidentalis is northern white cedar; *T. plicata* is western red cedar. *Thuja* leaf preparations are specifically indicated in cases of incontinence occurring in children or in elderly men with prostatic enlargement and are used in small amounts in formulas as a synergist or specific. *Thuja* leaves are high in volatile oils, which can reach renal passages following oral ingestion, and have both antimicrobial and stimulating effects. However, in larger dosages, *Thuja* can be irritating, and *Thuja* tincture is to be avoided in acute infections or acute inflammation of the kidneys. Folkloric herbals emphasize *Thuja* preparations for the genitourinary symptoms of sexually transmitted diseases. *Thuja* has antiviral properties and is used homeopathically for venereal warts and for a split stream of urine due to warty growths in the urethra. *Thuja* is also specifically indicated for a sensation of trickling in the urethra after urination, urinary frequency, dysuria, and leakage of urine due to infection and irritation, symptoms of chronic prostatic hypertrophy.

Turnera diffusa • Damiana

Turnera leaves have a folkloric reputation as being an aphrodisiac. However, such effects are not sudden or powerful; instead *Turnera* is most specific for impotence and diminished libido, urethral discharge associated with prostatic disease, mucous in the urine, and incontinence in the elderly. *Turnera* is a reproductive tonic and has not yet been the subject of much scientific research.

Ulmus fulva • Slippery Elm

This elm is "slippery" due to the presence of mucilage in the tree bark, which along with various mucosal nutrients may have an anodyne effect for bladder and pelvic pain. Slippery elm bark may be included in teas to promote healing of ulcerations and inflammations of the bladder mucosa.

Uncaria tomentosa • Uña de Gato

This tropical vine, which is also called cat's claw, is used in South America for wound healing and immune support. Bark from *Uncaria* vines is a reported antiulcer remedy used for gastritis, colitis, and leaky bowel syndrome. It is used as a cure-all in Amazonian herbalism. Immune-stimulating alkaloids pteropodine and isopteropodine have been isolated[414] as well as anti-inflammatory quinovic glycosides.[415] *Uncaria* may relieve pain via the "curare-like" ability of hirsutine, an indole alkaloid, to inhibit neuromuscular transmission.[416] This same alkaloid is credited with providing antispasmodic and anesthetic actions in cases of bladder pain and spasm.[417]

Urtica urens, U. dioica • Nettles

Nettle (also called stinging nettle) leaves and roots may serve as nourishing renal tonics and nonirritating diuretics, with the root being emphasized for beneficial hormone-balancing activity. *Urtica* preparations have been the subject of much modern research confirming benefits to renal lithiasis, prostatic enlargement, and hormonally stimulated prostate cancer and disease. Extensive anti-inflammatory and hormone-balancing mechanisms have been elucidated, and teas, tinctures, and encapsulated products have all been identified.

Vaccinium macrocarpon • Cranberry

Cranberry interferes with the ability of *E. coli* to adhere to the urinary mucosa and is therefore appropriate in the management of both acute infections and as a preventive for those who suffer chronic bladder infections. Cranberries contain mannose and phenolic compounds, both credited with antimicrobial effects on the bladder. Dried cranberry powder is available and can be included in teas or medicinal drinks.

Verbascum thapsus • Mullein

Verbascum leaves and roots contain mucilage and are most commonly used for respiratory irritation. The roots of this plant may also improve nervous tone in the bladder, specifically the trigone muscle.[418] *Verbascum* is recommended for pressure in the bladder, enuresis, burning with urination, and dribbling of urine. In

American folklore, *Verbascum* has been simmered with milk and administered to enfeebled patients suffering from wasting and consumptive diseases. It has also been employed to irrigate the bladder in cases of chronic cystic irritation.

Viscum album • Mistletoe

Viscum lectins have been researched for anticancer effects for several decades due to emphasis of the plant for malignancy in folkloric traditions of ancient Europe. The young leaves and sprouts of mistletoe can be included in formulas for bladder, prostate, and other cancers. One recent clinical trial used weekly intravesicle *Viscum* instillations for bladder cancer patients having undergone bladder resection surgery as a treatment to prevent cancer recurrence. *Viscum* was reportedly associated with a positive outcome without local or systemic side effects.[419]

Zea mays • Corn

The fresh "silk" of the corn or maize plant has long been considered a demulcent diuretic, and early American Eclectic authors have recommended *Zea* for increased phosphates and urates in the urine.[420] Chinese researchers have also noted *Zea* to reduce the tendency to urinary lithiasis.[421] Like *Aloe vera* and *Symphytum officinale* (comfrey), *Zea* silk and fruit contain allantoin, a mucosal restorative molecule,[422] and like *Equisetum*, corn silk contains silica, another tissue-restorative agent.

Zingiber officinale • Ginger

Ginger root is a well-studied anti-inflammatory herb that can be included as a supportive ingredient in a variety of urinary formulas. Older herbal texts suggest that ginger is specific for urinary frequency, mucous and sediment in the urine, urethral discharge, and a sense of burning and stinging during urination.

— CHAPTER FIVE —

Creating Herbal Formulas for Dermatologic Conditions

One of the greatest philosophical differences in herbal versus allopathic approaches to the treatment of skin disorders is that most herbalists and alternative medical practitioners emphasize the importance of gastrointestinal health on the health of the dermis. Accordingly, many herbal formulas for acne, psoriasis, boils, and dermatitis include the use of alterative and bowel-supportive herbs such as *Arctium lappa*, *Mahonia aquifolium*, or *Taraxacum officinale* root. To this same end, many herbalists and naturopathic physicians also work with people on their diet to support intestinal health and optimize nutrition as a way of treating chronic skin conditions.

Many skin conditions may also be related to underlying allergic and atopic phenomena, and herbs that affect mast cells, histamine, and inflammatory mediators are other important ingredients in dermatologic formulas. *Tanacetum parthenium*, *Angelica sinensis*, *Curcuma longa*, and many other herbs may reduce inflammation and atopic phenomena in the skin. Notice how many formulas in this chapter include alteratives and antiallergy herbs as foundational ingredients.

In other cases, vulnerary agents such as *Calendula officinalis, Equisetum arvense*, or *Centella asiatica* are included to support wound healing and connective tissue regeneration. *Centella asiatica* contains the triterpenoid wound-healing compounds asiaticoside, asiatic acid, and madecassic acid, which are found to reduce excessive fibrosis in situations of scleroderma, extensive scar formation, and keloids.[1] Both epithelial and vascular regeneration are promoted by *Centella asiatica*, and research shows *Centella asiatica* to be an angiogenic agent,[2] promoting growth factors and extracellular hyaluronic acid-binding proteins.[3]

Both poor circulation and diabetes can contribute to chronic fungus, skin infections, and poor wound healing. Some formulas in this chapter include *Ginkgo biloba* or other circulatory-enhancing agents that aid when poor perfusion contributes to skin complaints. Hypoglycemic agents such as *Cinnamomum* spp., *Allium sativum*, or *Opuntia ficus-indica* are also featured in formulas for cases when high blood sugar and diabetes contribute to nail, foot, or other skin fungus.

All these botanical therapies are aimed at improving the entire "ecosystem" of the skin, making it less hospitable to acne bacteria and fungi and less susceptible to inflammatory processes, poor wound healing, or excessive scar tissue formation. Herbalists contend that supporting optimal microbiota in the skin is a superior therapy for treating common acne and fungal infections of the skin than is the use of antibiotics such as tetracycline for acne or steroids for eczema. In nearly all cases, acne and eczema will recur when such antibiotic and steroid medications are discontinued, which is evidence that they are superficial therapies that do nothing to change the underlying pathology, and are detrimental to overall health when used long term. The following formulas explain and exemplify how to work at a deeper level, using herbs to treat common skin complaints and diseases.

Formulas for Burns, Bites, Stings, Wounds, and Trauma

Traumatic injuries to the skin affect nearly everyone over the course of a lifetime. Herbal therapies can promote healing and help alleviate pain and discomfort. Such herbs may be effective both topically and internally. Homeopathic medicines are often helpful for bites and stings and are specific for various presentations and qualities of pain. Home remedies can also be useful for minor kitchen burns and cuts, and many households have common remedies on hand such as onions (for poultices), an *Aloe* plant, or simple ice packs.

Following are homeopathic, topical, and internal remedies to help allay acute pain and itching due to venomous insects. While snakebites require expert medical care, and antivenom where available, *Echinacea* is a traditional North American herbal option, and several homeopathics detailed below may also help while en route to the hospital.

Field Poultices for Venomous Bites and Stings

The following herbs are classic remedies for bee stings and other insect venom. These herbs or mud and ashes are commonly available "in the field," where such stings typically occur. Herbalists may employ a so-called "spit poultice" where *Plantago* or *Stellaria* leaves are simply chewed, and the masticated pulp is applied topically to the lesion. The pulp can be covered with a piece of torn leaf to help keep it moist and active for 15 to 30 minutes. Mud and clay can have natural drawing effects. Where available, vinegar or Epsom salt compresses may also be helpful.

Plantago ovata or *P. lanceolata*: Plantain leaves employed as a spit poultice
Stellaria media: Chickweed leaves employed as a spit poulice
Charcoal or ashes: Activated charcoal is ideal, but in a pinch, can be prepared from campfire ash and moistened with water
Mud: Obtained from a stream bank or made out of dirt and water

Any of the above may be applied topically and left in place for 15 to 30 minutes. A single application may suffice for minor stings, while wasp or other more painful stings may require repeat application as the poultice dries.

Essential Oils to Allay Itching

Mint, lavender, and tea tree essential oils can relieve stinging and itching sensations when topically applied. Small bottles of essential oils are light and easy to include

Some Herbal Specifics for Skin Presentations

This chapter covers a wide range of skin conditions. Here is a listing of some appropriate herbs for formulas for several common skin complaints you may need to treat. Rather than addressing specific diagnoses, such as acne, scabies, or dermatitis, these herbs are organized as being specific for common *presentations*, regardless of the formal diagnosis. Folkloric herb books commonly employ herbs for specific *symptoms*, and the following herbs can be chosen as specifics or synergists to include in formulas to complement base herbs that may be more specific to the diagnosis.

CONDITION	HERBS
Hot red stinging skin	*Aloe vera* *Apis mellifica* *Sanguinaria canadensis*
Ulcerated skin with discharges	*Apium graveolens* *Betula pendula* *Calendula officinalis* *Geranium maculatum* *Phytolacca decandra* *Quercus alba*
Sense of crawling in the skin	*Apium graveolens* *Hypericum perforatum* *Scrophularia nodosa*
Dry, scaling dermatitis	*Avena sativa* (baths) *Centella asiatica* *Fucus vesiculosus* *Phytolacca decandra*
Pustular skin lesions	*Alnus rubra* *Echinacea angustifolia* *Iris versicolor* *Juglans nigra* *Melaleuca alternifolia* *Phytolacca decandra* *Quercus alba*

Homeopathic Therapy for Venomous Bites and Stings

These homeopathic remedies are emphasized for treating itching from simple mosquito bites to helping resolve the pain of a bee sting as quickly as possible to allaying the serious pain and tissue damage following snakebites.

REMEDY	CONDITION
Apis mellifica	Pink puffy skin; intense itching; burning, stinging, smarting sensations
Cantharis vesicatoria	Swollen, blistering, and burning stings; chemical burns
Carbolicum acidum	Severe allergy; pallor; systemic illness from allergic reactivty
Crotalus horridus	Discolored, mottled skin; vascular congestion; rapid development of serious systemic symptoms
Lachesis mutus	Purple blue discoloration; sense of tightness or constriction; snakebites
Ledum palustre	Small puncture wound with local soreness and little or no redness, purple-blue discoloration
Rhus toxicodendron	Burning pain with muscle stiffness; pustular or scabbing rashes, poison oak rash
Tarentula hispanica	Irritability; nervousness; skin sensitive to touch; sense of constriction; skin-crawling sensations
Urtica urens	Intensely itchy, stinging pain; hives, vesicular skin eruptions

in the first aid kit when camping or backpacking or to keep in emergency supplies in trunk of the car.

Mentha piperita
Lavandula augustifolia
Melaleuca alternifolia

Simply apply a few drops of one of the above essential oils to mosquito bites, bee stings, and other insect stings, directly on the skin. Effects are typically immediate, and the essential oil can be reapplied as needed, when the effects wear off over time.

Echinacea Tincture for Spider Bites and Snake Venom

Snake venom spreads so rapidly because it contains hyaluronidase, an enzyme that breaks down the hyaluronic acid "glue" that supports connective tissue scaffolding and structure. Hyaluronidase enzymatically disrupts the integrity of connective tissue, allowing the venom to disseminate rapidly and deeply. *Echinacea angustifolia* is a natural hyaluronidase inhibitor[4] and a Native American snakebite remedy. In this formula *Echinacea angustifolia* is combined with *Viscum album* and *Phytolacca decandra* to promote white blood cell–driven immune modulation and support cell-to-cell adhesion. *Silybum marianum* will help protect the liver, should the venom get that far.

Echinacea angustifolia 15 ml
Viscum album 15 ml
Phytolacca decandra 15 ml
Silybum marianum 15 ml

Take 1 dropperful of combined tincture every 5 to 10 minutes while enroute to an emergency room for antivenom treatment and appropriate medical care.

Epsom Soak for Snake Venom

Both magnesium and sulfate found in Epsom salts (magnesium sulfate) play roles in ion channels and other gated channels on cellular membranes[5] and may be able to affect the venom load that is taken into the tissues following a snakebite. Epsom salts can affect pain signaling as well as prevent the spread of infections by affecting cell adherence. Epsom salt soaking is a time-honored home remedy for both pain and skin infection. Epsom salt soaking is highy regarded for its ability to treat vascular inflammation when topically applied, and research has shown efficacy in treating thrombophlebitis[6], which may result from serious envenomations.

Epsom salts

Dissolve the entire quart-size package (as affordable) in hot bathwater and soak for 30 minutes at a time, 3 or more times daily on the first day of an acute wasp, snake, jellyfish, or other sting. Alternately 1 to 2 cups of Epsom salts may be dissolved in a quart of hot water and used to soak cloths to use as a compress. Leave in place for 30 minutes, and repeat 3 or 4 times over the course of the day.

Aloe vera Pulp for Burns

There are more than 400 species of *Aloe*, and *Aloe vera*, *Aloe ferox*, and *Aloe arborescens* are the most commonly used medicinal species. Keep live *Aloe vera* plants on hand to treat the occasional first-degree to second-degree small burns. "Fillet" cut an *Aloe vera* leaf and hold the slices upright, allowing the small amount of orange-brown resin to drip away. Use the tip of a spoon to scrape off a mass of pulp on the inside. Use the fresh pulp directly on the skin or add it to herbal teas and tinctures using a blender or preparing by hand.

Topical Burn Spray

Aloe vera is one of the most well-known herbs for burns, and *Aloe vera* polysaccharides reduce inflammation via reducing nitric oxide release and activity.[7] *Hypericum perforatum* is a good complement to *Aloe vera* in helping to heal injured nerve endings. Rosewater and lavender essential oil are both pain-relieving, plus they provide nervous system support via the olfactory nerves into the limbic system to help calm a traumatized patient. This formula blend can pass through a spray nozzle or can be placed in a regular tincture bottle and applied via a cotton ball. The use of a spray is least painful to a patient and recommended. When no such formula is available, the topical use of honey or egg whites is both pain-relieving and healing.

Rosa damascena (rose) water 30 ml
Aloe vera gel 14 ml
Hypericum tincture 12 ml
Lavandula angustifolia essential oil 4 ml

Combine in a 2-ounce spray bottle. Shake prior to each use, and spray on burned skin every 30 minutes.

Internal Formula to Support Healing of Burns

For serious burns, the use of vulnerary herbs may help speed healing and reduce scarring. *Symphytum officinale* is a healing agent of long standing and *Calendula officinalis* is known to promote circulation to the dermis and enhance connective tissue regeneration at the basement membrane. Research is limited, but *Calendula officinalis*

Homeopathics for Burns

REMEDY	CONDITION
Apis mellifica	Minor burns; pink, rosy skin with burning and stinging sensation
Atropa belladonna	Bright red flushed skin with a throbbing sensation
Cantharis vesicatoria	Severe blistering and burning pain
Causticum hahnemanni	Chemical burns; severe dermatitis
Urtica urens	Stinging and itching sensations

is believed to facilitate wound healing via promotion of glycoprotein and collagen synthesis, hastening dermal regeneration.[8] *Hypericum perforatum* helps reduce pain and heal injured nerve endings. *Centella asiatica* (gotu kola) has been shown useful in promoting epithelial regeneration following burns.[9] This formula may be prepared as a tincture or a tea.

Symphytum officinale
Centella asiatica
Hypericum perforatum
Calendula officinalis

Combine in equal amounts. Take 1 teaspoon of the combined tincture 6 times daily for a week or more. To prepare as a tea, use ½ ounce of each herb to make the blend. Steep 1 tablespoon of the herb mixture per cup of hot water and strain. Drink 5 or 6 cups per day if possible, continuing for a week or two.

Simple Spray to Allay Burn Pain

Lavandula angustifolia essential oil is wonderful to heal burns and allay pain. This simple recipe will prepare a 1-ounce bottle of medicine to be used with a spray nozzle. This spray is useful to include in first aid kits for sunburns, kitchen burns, and other minor burns.

Lavandula angustfolia essential oil 5 ml
Water 25 ml

Combine the oil and water in a 1-ounce bottle with a spray nozzle and spray on affected area frequently.

Kitchen Medicines for Burns

These home remedies are found in most households and can be used for grease splatters and other minor kitchen burns.

Egg whites. Since eggs are usually stored in the refrigerator, egg whites are cold and soothing for burns, plus the sterile proteins and other compounds in them help promote healing.

Honey. Honey applied topically is a natural bandage, promotes healing, and smothers any aerobic opportunistic bacteria.[10] Research on using manuka honey reports a significant improvement in healing time in cases of serious burns.

Onion juice. *Allium cepa* (onion) is rich in quercetin and other immune-modulating compounds. Place a sliced onion directly against burned skin or better yet, mince and place the pulp against the skin.

All-Purpose Herbal Wash Powder

This powder-based formula is handy to use as an herbal wash for infected wounds and stasis ulcers and to irrigate purulent ear canals. The powder can also be prepared as a sitz bath for infants with infected diaper rash lesions and as an herbal douche for vaginal discharges and inflammation.

Calendula officinalis powdered flowers
Hamamelis virginiana powdered bark
Symphytum officinale powdered root
Geranium maculatum powdered root
Mahonia aquifolium powdered root

Combine equal parts of these powders and store in an airtight jar, in a dark cupboard if possible. Prepare a wash for topical use by infusing 1 teaspoon per ½ cup of hot water. Apply to the affected part by soaking a cloth or piece of gauze in the liquid and placing against the area to be treated. Leave in place for 10 to 30 minutes, allowing to air-dry afterward. You can also make large quantities to use as a douche or a sitz bath.

Tincture Formula for Acute Skin Trauma

This formula contains the most well-known vulnerary herbs combined in a tincture to take aggressively following any serious skin wounds to support healing. It is taken orally to complement topical applications.

Hypericum perforatum
Symphytum officinale
Calendula officinalis
Centella asiatica

Combine in equal parts. Take 1 teaspoon of the combined tincture hourly for several days, reducing to every 2, then every 3, then every 4 hours as pain and symptoms abate.

Calendula Succus Simple for Lacerations and Abrasions

Calendula officinalis (calendula) is a first aid kit staple. A succus has lower alcohol content than a tincture and is made from fresh juice. *Calendula officinalis* succus is unlikely to sting very much when topically applied. Use a dropperful of succus on a sterile gauze pad and apply topically, repeating each hour until the wound has scabbed over.

Calendula-Comfrey Poultice for Lacerations and Abrasions

Calendula officinalis (pot marigold or calendula) and *Symphytum officinale* (comfrey) are a classic duo for trauma and wounds. *Calendula* enhances circulation and connective tissue regeneration in the dermis, and comfrey contains the cell-proliferating agent allantoin.

Calendula officinalis succus
Symphytum officinale roots

When fresh comfrey roots are available, thoroughly wash several inches of a supple root, then mince or grate, and cover in a small amount of water. Add *Calendula officinalis* succus and allow it to stand for 10 minutes to yield a gummy mass for topical application. Dry comfrey roots can be macerated in hot water until mucilaginous. *Symphytum officinale* root tincture would also do in a pinch, but fresh roots, or long-macerated dry root preparations are superior. If using fresh root pulp, place the gummy mass on a thin gauze pad, place the fabric side against the skin, saturate with *Calendula officinalis* succus and cover with plastic or more gauze and tape to hold in place. Apply 3 to 5 times daily for acute injuries, allowing the injury to air-dry between applications.

You can also make this poultice by combining equal parts of *Calendula* and *Symphytum* tinctures, applying to a sterile gauze pad, and using topically, or by making a paste out of the combined dry powders and water.

Tincture to Control Profuse Bleeding in Skin Lacerations

Achillea millefolium and *Hamamelis virginiana* are folklorically classic herbs to control bleeding, both topically and internally. *Achillea millefolium* is said to be named after Achilles, a hero of Greek mythology who used the plant to staunch bleeding wounds in soldiers. These styptic herbs are combined with *Calendula officinalis* to support wound healing.

Achillea millefolium
Hamamelis virginiana
Calendula officinalis

Combine in equal parts. Several milliliters of the tincture may be placed on a sterile gauze pad and used topically to staunch bleeding. This formula may also be taken internally by the teaspoon every 5 to 10 minutes for hemorrhagic situations.

Formula for Trauma with Extensive Bruising

Following motor vehicle accidents, bad falls, and tissue trauma, the use of herbs high in flavonoids can help speed the resolution of bruising and abatement of pain. *Hypericum perforatum* is specific for bruising, strains, sprains, and injuries to highly innervated areas such as the spine, hands, and feet. This formula may also be prepared as a tea when the dry herbs are on hand. Bromelain at a dose of 500 to 1,000 milligrams 3 to 6 times daily is also appropriate and may be taken along with the tea or tincture.

Hypericum perforatum
Aesculus hippocastanum
Hamamelis virginiana
Vaccinium species

Combine in equal parts. Take 1 teaspoon of the combined tincture every hour, reducing the frequency as pain and symptoms improve.

To prepare as a tea, steep 1 teaspoon of the dry herb mixture per cup of hot water and then strain. Drink as much as possible, at least 3 cups daily, and up to 8 to 10. Continue for 3 or 4 days, then cut the dosage in half, and continue for another 3 to 7 days, depending on the severity of the trauma.

Hypericum-Hippophae Oil for Ecchymoses and Nerve Trauma

Hypericum perforatum (St. Johnswort) and *Hippophae rhamnoides* (sea buckthorn) are some of our best topical remedies for nerve, blood vessel, skin, and tissue trauma. Prepare *Hypericum* and *Hippophae* into a salve or simply combine into a skin oil to use topically. Essential oils may be added for pain and medicinal effects, such as *Mentha piperita* essential oil for pain relief, and *Helichrysum italicum* for additional vulnerary action. *Hypericum* promotes collagen production by fibroblasts[11] and is especially specific in formulas for fingertip, spine, and other nerve trauma. *Centella asiatica* may also be specific for injury to nerves and highly innervated areas by promoting axonal regeneration.[12] *Centella asiatica* tincture, but not the tea, has been shown to promote a marked increase in neurite growth with oral consumption. This formula uses *Centella asiatica* powder, only because *Centella asiatica* oil is not readily available. Consider making your own oil by macerating the powder in olive, sesame, or other oil. Blend these ingredients together when the need arises.

Hypericum perforatum oil 30 ml
Hippophae rhamnoides oil 15 ml
Centella asiatica powder or oil, if available
7.5 ml oil or ¼ ounce (7 g) powder
Helichrysum italicum essential oil 7.5 ml

Combine the oils in a shallow bowl and stir in the *Centella asiatica* powder and *Helichrysum italicum* essential oil. Transfer to a 2-ounce bottle and use topically for bruises and nerve injuries. If you have pre-prepared *Centella asiatica* oil, combine it in equal parts with the other fixed oils, adding a small amount of *Helichrysum italicum* oil. Apply as often as possible in the acute phase following trauma, such as 5 times daily in the first 48 hours, reducing frequency until pain is resolved and healing is well established.

Formulas for Skin Infections

Most skin infections, such as a simple cut that has become infected, respond readily to topical approaches and herbal remedies. Some serious infections, however, such as tetanus or toxoplasmosis, may enter through the skin and lead to systemic complications that warrant antibiotics or aggressive topical treatment. This section includes skin washes and topical disinfectants for common boils and minor infections as well as tinctures and other herbal support to use orally to support the immune system, have direct antimicrobial effects, or enhance the efficacy of antibiotics when needed. It may go without saying, but care must be taken to avoid spreading microbes to other body parts or other people when treating skin infections. Compresses, bandaging material, and any material contacting infected lesions should be considered to be infectious waste and disposed of properly. Patients must be reminded that clothing, bedsheets, and other fabric can easily become contaminated and lead to the spread of infection or lead to transmission to other people. Scrupulous hand washing, bandaging of lesions, and frequent laundering of bedding and clothing is essential. It *is* possible to treat skin infections with herbal medicines, but for best results soaks, compresses, and both topical and internal therapies should be combined, and clinicians should be prepared to consider antibiotic therapy for any infection that does not readily respond over the period of 2 to 3 days. Fever and malaise are worrisome signs of disseminated infections and warrant antibiotic therapy. New papules around a primary skin lesion are an indication that the infection is spreading and that new lesions may erupt. Diffuse redness, pain, fever, swelling, and tenderness to touch are also worrisome signs of cellular spread or that actual cellulitis is emerging and reasons to initiate antibiotic therapies at once. Herbal therapy can still be used in tandem with antibiotics.

When skin infections are chronic, from cystic acne to repetitive crops of boils to minor wounds that tend to become infected easily, the use of alterative herbs for digestive and liver support may reduce the chronic infections. Along with improving the diet, optimizing the liver's ability to clean the blood and the bowels to eliminate wastes can deter opportunistic infections. Many of the older herbal textbooks recommend "opening the emunctories" in cases of chronic infections, referring to the notion that the organs of elimination—the emunctories—are vital to keeping the blood and tissues cleared of metabolites that may invite and support pathogenic bacteria. Many older textbooks speak of "morbid matter," meaning simply normal waste products that have accumulated, creating a fertile breeding ground for opportunistic microbes. The complexion is a reflection as to the health and vitality of the entire body, and when the skin is chronically infected, improving the health of the entire body—starting with the liver and bowels—may be especially important.

Topical Application for Infected Wounds

For wounds that become swollen, tender, red, hot, or purulent, topical antimicrobials are indicated. Animal bites or cat scratches may introduce rabies, toxoplasmosis, or other serious pathogens, however, so all such wounds should be evaluated by a physician, who may prescribe oral antibiotics and other therapeutic measures. The herbs in this formula are astringent and anti-inflammatory as well as strongly antimicrobial, and will quickly improve most simple skin infections. This formula may be complemented by taking an oral herbal antimicrobial, vitamin A, and zinc.

Calendula officinalis powder 1 ounce (30 g)
Achillea millefolium powder 1 ounce (30 g)
Hydrastis canadensis powder 1 ounce (30 g)
Iodine 20 drops into each 2 cups (480 ml) of prepared wash
Melaleuca alternifolia (tea tree) essential oil 20 drops into each 2 cups (480 ml) of prepared wash

If possible, soak the affected lesion in a solution of Epsom salts in hot water while this formula is being prepared. Dissolve 1 cup of Epsom salts in a large pan of water and soak the hand or foot, or soak a cloth in the water to prepare a compress to use topically. Blend the three herb powders and place 1 heaping tablespoon of the combined powders in 2 cups of hot water. Stir well and allow it to cool to a comfortable temperature; allow the powder to sink to sediment. Add 20 drops each of iodine and tea tree oil. Soak a gauze pad or piece of soft, clean fabric in the liquid and use it to wash the affected part. Soak another pad in the solution and apply to the wound. Leave in place for 15 or 20 minutes, then allow to air-dry. Repeat as often as possible throughout the day, at least 3 times. This formula may also be prepared using tinctures of the three herbs. Add 1 to 2 dropperfuls of each herb to a cup of hot water, and add the iodine and tea tree oil as above.

Skin Wash for Skin Infections

Although it can stain the skin and fabrics, iodine is a wonderful disinfectant that has been in use for hundreds of years and is included in hospital disinfectants to this day. This skin wash based on a traditional Eclectic formula that combines iodine with the anti-infective herbs *Echinacea angustifolia* (coneflower) and *Commiphora myrrha* (myrrh). Betadine is a commercially available iodine solution that may be used as the iodine in this formula.

Echinacea angustifolia powder ½ ounce (14 g)
Commiphora myrrha powder ½ ounce (14 g)
Iodine 15 drops
Water 4 ounces (120 ml)

Pour 4 ounces of boiling water over the dry herb powders in a saucepan. Stir well and put a lid on the pan, allow the liquid to reach a comfortable temperature. Strain out the powders and add 15 drops of iodine to the solution. Use to wash infected wounds. To avoid contaminating the solution, pour it onto a washcloth, rather than re-immersing the washcloth in the solution. Discard the solution after 24 hours and make anew. Wash the afflicted skin as often as hourly, allowing full air-drying between washes.

Herbs for Skin Infections

The following herbs are listed as specifics in Eclectic, homeopathic, and folkloric reference texts for chronic boils and pustules.

Achillea millefolium
Baptisia tinctoria
Echinacea angustifolia
Juglans nigra
Smilax ornata

These herbs are listed as specifics in Eclectic, homeopathic, and folkloric reference texts for acute skin abscesses.

Calendula officinalis
Echinacea angustifolia
Equisetum arvense
Hydrastis canadensis
Juglans nigra
Melaleuca alternifolia
Origanum vulgare
Salix alba
Veratrum album (small doses internally of only a few drops of tincture at a time; may be applied topically with less toxicity concerns)

Simple for Staph Infections of the Skin

Mahonia aquifolium (Oregon grape), also known as *Berberis aquifolium*, contains the powerful and broad-spectrum antimicrobial berberine. Berberine in *Mahonia aquifolium* and other berberine-containing herbs is reportedly active against more than 40 strains of *Staphylococcus aureus*,[13] including antibiotic-resistant strains.[14] Berberine inhibits *Staph* from binding to mucosal fibroblasts[15] and deters bacterial adhesion to fibronectin via enzyme inhibition.[16] *Mahonia aquifolium* is especially specific for purulent discharges and infected mucous membranes and is helpful both topically and internally.

Mahonia aquifolium 2 to 4 ounces chopped root bark, enough to last for 3 or 4 days

Gently simmer 1 teaspoon per cup of hot water for 10 minutes in a covered vessel. Strain and use as a skin wash, soak, or tea for oral consumption. Consume 2 or 3 cups each day; use another cupful to prepare several compresses each day. If there are no significant improvements in 24 hours, consider other therapies.

Tincture for Staph Infections of the Skin

This formula is based on modern bacterial culture and sensitivity research on plants shown to have significant activity against staphylococcal infections and would be helpful for antibiotic resistant infections and staphylococcal scalded skin syndrome. Unlike antibiotics, whole plant extracts appear unlikely to cause bacterial resistance because they work by many different mechanisms of action at once. Bacteria often mutate to work around a single mechanism, but when using multiple herbs that have multiple mechanisms of action working in tandem, it becomes close to impossible for bacteria to evolve quickly enough to overcome them all at once. As with the above tea, this tincture must be combined with topical antimicrobial skin washes to treat staphylococcal infections adequately.

Curcuma longa 15 ml
Achillea millefolium 15 ml
Origanum vulgare 15 ml
Allium sativum 15 ml

This formula will fill a 2-ounce bottle. When taken at a dosage of 1 teaspoon per hour, it will be gone in several days, the time at which a follow-up visit should be scheduled to reevaluate serious staphylococcal infections. Other treatments should be chosen if there are not significant improvements in 48 hours.

Disinfectant Liniment Based on Jethro Kloss

Jethro Kloss was a "nature cure" advocate who published what was to become a classic home health book in 1939 entitled *Back to Eden*. Many of his formulas featured *Capsicum annuum* (cayenne) powder. This liniment based on Kloss's writing will most certainly sting broken skin, so I recommend using this only on unbroken skin following trauma or for abscesses and other infections with unbroken skin. It is also important to note that this formula uses isopropyl rather than ethyl alcohol, so it can never be taken orally. Label well!

Hydrastis canadensis powder 1 ounce (30 g)
Echinacea angustifolia powder 1 ounce (30 g)
Capsicum annuum powder ¼ ounce (7 g)
Isopropyl alcohol 1 pint (470 ml)

Macerate the powdered herbs for 4 weeks in the alcohol, shaking daily. Apply with a cloth or cotton ball to infected and inflamed areas. Label the liniment container as NOT for oral consumption—for topical use only.

Topical Compress for Skin Infections

Here is another formula based on research on herbs shown to have activity against a broad spectrum of gram-positive and gram-negative bacteria. *Calendula officinalis* is added to help strengthen connective tissue and prevent bacterial dissemination into underlying tissue while promoting wound healing. The formula calls for small quantities because only several teaspoons are needed for each compress. This formula is enough to produce eight or nine compresses, a reasonable dispensing quantity for 3 or 4 days' use.

Mahonia aquifolium powder 1 ounce (30 g)
Calendula officinalis powder 1 ounce (30 g)
Achillea millefolium powder 1 ounce (30 g)
Origanum marjorana (marjoram) essential oil 10 drops per cup of strained infusion
Melaleuca alternifolia (tea tree) essential oil 10 drops per cup strained infusion

Infuse 2 teaspoons of the mixed powder per cup of hot water, steep, strain, and add the marjoram and tea tree essential oils. Soak a clean cloth in the fluid and apply to open lesions for 20 minutes every hour with air-drying in between. Use the cupful of liquid over 24 hours, using a new clean cloth for each compress. Prepare the formula anew when it's gone or on the following day. Continue for 2 days, reevaluating therapy if there is no significant improvement by that time.

Activated Charcoal Compress for Staph Infections

7Song, an impressive herbalist who specializes in first aid in challenging situations, locations, and populations, uses activated charcoal topically for staph infections. Charcoal in a water slurry binds pathogens and toxins, offering an absorptive effect. The compress can be combined with a skin wash and oral antimicrobial therapy for best results.

1 tablespoon or more activated charcoal, depending on the size of the area to be treated
Water
2 sterile gauze pads, slightly larger than the area to be treated

Place the charcoal in a small bowl and mix with a small amount of water—just a teaspoon at a time to achieve a thick charcoal paste. Use a spoon to spread the charcoal paste on a gauze pad, and spread the charcoal all the way to the edges of the pad. Top with a second gauze pad and moisten with just a few drops of water to ensure the charcoal will act as an effective absorbant when placed against the skin. Put the moist side down, directly atop the staph lesions, and tape into place. Leave in place for an hour, and then take care to properly dispose of the contaminated compress. Repeat every several hours throughout the day. Consider an antibiotic or other therapy if there are no signs of improvement in 3 days.

Oral Tincture for Drug-Resistant *Staph*

This formula is based on modern bacterial culture and sensitivity research regarding herbs able to inhibit drug resistant *Staph aureus*, often referred to in the medical community as MRSA—Methicillin Resistant *Staphylococcus aureus*. Propolis is not an herb, but rather a natural product made by bees from natural plant resins; propolis is available as a tincture. As with the above oral formulas, this formula must be complemented with topical disinfectant skin washes or compresses.

Glycyrrhiza glabra 15 ml
Allium sativum 15 ml
Andrographis paniculata 15 ml
Propolis 15 ml

Dosing will depend on the situation, but a teaspoon of the combined tincture by mouth hourly or every several hours is a strong useful dose. Reduce dose as symptoms improve.

Tea Formula for Staph Infections of the Skin

A tastier tea than the above *Mahonia* simple, here is another recipe for internal use as a supportive measure against staphylococcal skin infections. For serious infections, use teas, tinctures, and topical agents all together for an aggressive approach that may match the power of antibiotics; this tea would not be a powerful therapy all on its own.

Mahonia aquifolium 1 ounce (30 g)
Zingiber officinale 1 ounce (30 g)
Glycyrrhiza glabra 2 ounces (60 g)

Gently simmer 1 heaping teaspoon of the herb mixture per cup of hot water for 5 minutes in a covered saucepan. Let stand 10 minutes more and strain. Drink as much as possible, at least 3 or 4 cups each day, hourly if possible.

Topical Compress for Drug-Resistant *Staph*

Bees collect plant resins to make propolis, which is available as a tincture. This formula calls for a full ounce of the tincture but less-expensive substitutes are an option. *Populus* species (cottonwood) buds are particularly high in resins and can be soaked in vodka or even rubbing alcohol as a substitute for propolis tincture. Cottonwood buds can be gathered for oneself in early spring or obtained from a dry herb supplier. Herbs prepared in isopropyl (rubbing) alcohol must not be consumed orally and are only appropriate for topical use. The plants in this formula have all been shown to have activity against drug-resistant staph strains,[17] and this compress can be prepared at home by herbalists to dispense to patients.

Allium sativum, minced 2 cloves
Mahonia aquifolium, powder ½ ounce (15 g)
Hot water 1 cup (240 ml)
Propolis tincture or cottonwood tincture 30 ml
Eucalyptus globulus essential oil 10 drops
Melaleuca alternifolia essential oil 10 drops

Making a cottonwood bud tincture for oneself requires macerating the whole buds in vodka or when intended for topical use only, rubbing alcohol. Place the buds in a glass canning jar, just cover with alcohol, and shake vigorously everyday for 6 weeks. Strain, store in a glass bottle, and label appropriately.

To prepare the topical compress, place the minced garlic and the *Mahonia aquifolium* powder in the bottom of a blender and cover with 1 cup boiling water. Keep the lid on the blender and allow it to steep for 10 minutes. Puree on a high setting. Strain through a wire mesh kitchen strainer into a cup or small bowl. Allow to cool to room temperature and add the propolis tincture (or cottonwood tincture/isopropyl alcohol) and 10 drops of each essential oil. Saturate a cotton ball or gauze pad and apply to affected part, leaving in place for 15 minutes. Prepared a fresh compress and reapply every several hours. Make fresh each day.

Echinacea Simple for Chronic Boils

Echinacea angustifolia is very specific for chronic boils and will often work as a simple. For best results, address any underlying causes such as hyperglycemia, poor circulation, or poor digestion and intestinal health. *Echinacea angustifolia* can be used for acute boils or for the onset of abscess formation, but it is especially effective for resolving a chronic tendency to boils.

Echinacea angustifolia

Take 1 teaspoon of the tincture 3 to 4 times a day for 3 to 6 months.

Astringent Wash for Impetigo

This formula works well as a skin wash to use prior to the application of the propolis skin shellac formula below. This formula is antimicrobial, anti-inflammatory, and above all drying, which is important because impetigo lesions typically involve thick, moist, honey-colored crusts. If lesions can be encouraged to scab over more quickly, they will resolve faster.

Quercus alba or *Q. rubra* powder
Calendula officinalis powder
Achillea millefolium powder
Melaleuca alternifolia (tea tree) essential oil
Origanum vulgare (oregano) essential oil
Iodine liquid preparation

Combine equal amounts of the three powders. Steep 2 teaspoons of the combined powders in a cup of hot water. Add 10 drops of tea tree and oregano essential oils and 5 drops of iodine. Soak a sterile gauze pad or soft cloth in the liquid and apply topically. Repeat every 3 or 4 hours, if possible, allowing to air dry in between applications. This formula may also be followed with the application of the skin shellac, below.

Herbs for Impetigo

Impetigo is due to streptococcal and staphylococcal infections of the skin that cause ulcers, papules, and pustules that form moist, thick honey-colored crusts. These herbs are listed as specifics in Eclectic, homeopathic, and folkloric reference texts.

Achillea millefolium
Coptis chinensis
Curcuma longa
Hamamelis virginiana
Hydrastis canadensis
Melaleuca alternifolia
Zingiber officinale

Propolis Skin Shellac for Impetigo

A skin "shellac" relies on the sticky and antimicrobial qualities of plant resins, which in addition to treating skin infections, will cover a lesion and act as a protective Band-Aid-like, yet breathable covering. Impetigo is a highly contagious streptococcal or staphylococcal infection of the skin. Propolis is a resinous compound made by bees and is active against many microbes including staph and strep. Propolis tincture is commercially available but may be hard to find. Both myrrh and propolis are sticky and can help form a gummy seal over impetigo crusts, and this formula works to both disinfect and protect. Children with impetigo must be encouraged to avoid touching the skin lesions.

Propolis tincture 14 ml
Commiphora myrrha tincture 14 ml
Melaleuca alternifolia essential oil 2 ml

Combine the herbs and shake well to incorporate the essential oil. Saturate a cotton ball with the liquid and apply topically to the affected lesion. For smaller lesions, a moist cotton ball can be taped in place to maintain long-term contact such as overnight when possible.

Formula for Impetigo and Streptococcal Infections

Impetigo may involve both streptococcal and staphylococcal organisms. This formula is based on modern bacterial culture and sensitivity research on herbs with strong activity against streptococcal infections. Some of the most readily available herbs are included in this formula.

Cordyceps sinensis
Allium sativum
Mahonia aquifolium
Achillea millefolium

Combine in equal parts, and take ½ to 1 teaspoon of the combined tincture as often as hourly, reducing as symptoms improve.

Herbs for Streptococcal Infections

These herbs are listed as specifics in Eclectic, homeopathic, and folkloric reference texts for presentations typical of streptococcal skin infections with hot, red, dry skin.

Atropa belladonna (small doses only, due to potential toxicity, or a homeopathic or drop dose)
Sanguinaria canadensis (can be irritating orally; dilute with water or milder herbs)

These herbs are listed as specifics in Eclectic, homeopathic and folkloric reference texts for streptococcal infections with coldness and systemic toxicity. Because strep bacteria can release endotoxins, and the inflammatory response itself can liberate cytokines, extensive streptococcal infections of the skin such as scalded skin syndrome, may cause systemic symptoms including malaise, loss of appetite, achiness, nausea, and headaches. Some patients with extensive streptococcal infections, including those of the skin, may appear flushed, hot, and obviously "sick" looking, and these herbs are noted in herbal folklore to be indicated in such situations.

Astragalus membranaceus
Capsicum annuum
Cinnamomum spp.
Commiphora myrrha

Other Supportive Measures for Acute Erysipelas

Homeopathic *Atropa belladonna* is indicated for bright pink skin, with a sense of heat, swelling, and possibly a throbbing sensation. Topical compressess, as discussed in this chapter, or simply heat applications, may speed the resolution of acute erysipelas. Simple hot compresses can also be prepared by placing several drops of tea tree, oregano, lavender, or other essential oil on a hot wet cloth.

Topical Compress for Streptococcal Infections of the Skin

This is a topical version of the above tincture, using herbs noted to have activity against streptococcal infections. Treating acute strep with herbs is done in the same manner as treating acute staph. Thus, formulas such as the All-Purpose Herbal Wash Powder on page 173 are convenient.

Mahonia aquifolium
Achillea millefolium

Combine equal parts of the dry herbs or powders, and gently decoct 1 tablespoon of the herb mixture per 2 cups of hot water. Strain the decoction and use the liquid to saturate a compress. A cotton ball serves well for a small compress; use a wash rag for a compress to cover a larger area. The solution may also be used to wash infected wounds, followed with the application of a compress. To avoid contaminating the solution, pour it onto a washcloth, rather than immersing the washcloth in the solution. Discard the solution after 24 hours and make anew. Wash the afflicted skin as often as hourly, allowing full air-drying between washes.

Internal Formula for Erysipelas

Erysipelas is a superficial infection of the skin due to *Streptococcus*. This formula is based on the recommendation of folkloric and Eclectic physician authors. Erysipelas may provoke a fever and systemic symptoms of nausea and vomiting but may be treated separately if needed.

Echinacea angustifolia 15 ml
Achillea millefolium 15 ml
Mahonia aquifolium 15 ml
Phytolacca decandra 8 ml
Allium sativum 7 ml

Take 1 teaspoon of the combined tincture once every hour to every several hours, reducing as symptoms improve.

Formulas for Poor Wound Healing, Keloids, and Scars

Poor wound healing may occur due to malnutrition or protein deficiency, and in the elderly, it may be due to poor circulation or weak digestion and assimilation. The elderly may have both poor circulation and malnutrition as the aging digestive system fails to adequately absorb the nutrients ingested. In other cases, minor injuries may induce excessive inflammatory responses and lead to excessive scar tissue and keloid formation. Consider circulatory support when appropriate, such as with *Ginkgo biloba, Capsicum annuum*, and *Zingiber officinale*. Assess the diet and offer nutritional education and supplements where needed. When extensive scarring and fibrosis are apparent, choose herbs noted to reduce excessive fibrin deposition, including *Astragalus membranaceus, Centella asiatica*, and *Curcuma longa*.

Tincture for Scarring or Keloid Formation

Those whose minor wounds heal poorly may be helped with the herbs in this formula, aimed at preventing keloid formation or extensive scarring. Keloids are an excessive deposition of fibrin. They can occur in anyone but are more common in people of color. *Curcuma longa* has been shown to reduce excessive scarring and fibrous changes and promote dermal blood vessel growth. *Scutellaria baicalensis* (scute or huang qin) may help reduce excessive inflammatory responses in wounds, and *Centella asiatica* and *Calendula officinalis* are some of our best general healing and vulnerary agents. *Centella* improves microcirculation in the skin, and both *Hypericum perforatum*[18] and *Centella*[19] have positive effects on fibroblast regeneration.[20] *Centella* is appropriate to promote connective tissue formation, even for those with excessive connective tissue response.[21] *Centella* supports appropriate collagen synthesis.[22] Eotaxin is a cytokine released from fibroblasts that promotes collagen and eosinphilic response in allergic inflammation. *Scutellaria* contains flavonoids, baicalein, and baicalin that inhibit eotaxin production by fibroblasts[23] and thereby may also reduce excessive scarring and keloids.

Centella asiatica
Calendula officinalis
Curcuma longa
Scutellaria baicalensis

Combine in equal parts. Those with a history of keloid formation can take the combined tincture by the teaspoon 4 or 5 times a day immediately following any wound, trauma, or injury.

Topical Application for Tendency to Keloids

Those with a tendency to keloids may be able to prevent them by treating all wounds promptly with a topical therapy such as this one. *Curcuma longa* and *Centella asiatica* have stabilizing and anti-inflammatory effects on collagen and connective tissue, and garlic and onions can supply sulfur to promote healthy connective tissue regeneration. Onion poultices may be helpful when tinctures or powders are unavailable.

Curcuma longa
Allium sativum
Centella asiatica

Combine the tinctures or dry powders in equal amounts to make a topical application. Use the combined tincture to saturate a cotton ball or gauze pad to tape into place over the wound. Mix a combined powder with water to make a thin paste. Spread the paste between two thin pieces of muslin or cotton fabric and apply to the affected area.

Onion Poultice for Keloid Prevention or Therapy

Allium cepa (onion) is high in quercetin and may help prevent keloids in those with a prior tendency. Onions in a topical gel preparation are noted to reduce keloids

Herbs for Skin Infections That Tend to Scarring or Keloids

These herbs are emphasized in historical literature as specific for skin infections that scar and lead to keloid formation. They may be considered for both topical and internal use.

Allium cepa
Avena sativa
Calendula officinalis
Centella asiatica
Curcuma longa
Equisetum arvense
Thuja occidentalis

Centella for the Skin

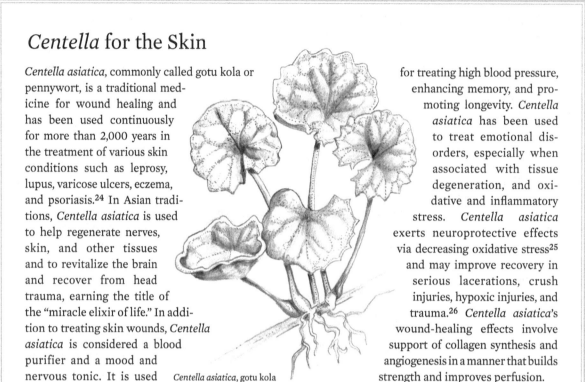

Centella asiatica, commonly called gotu kola or pennywort, is a traditional medicine for wound healing and has been used continuously for more than 2,000 years in the treatment of various skin conditions such as leprosy, lupus, varicose ulcers, eczema, and psoriasis.[24] In Asian traditions, *Centella asiatica* is used to help regenerate nerves, skin, and other tissues and to revitalize the brain and recover from head trauma, earning the title of the "miracle elixir of life." In addition to treating skin wounds, *Centella asiatica* is considered a blood purifier and a mood and nervous tonic. It is used for treating high blood pressure, enhancing memory, and promoting longevity. *Centella asiatica* has been used to treat emotional disorders, especially when associated with tissue degeneration, and oxidative and inflammatory stress. *Centella asiatica* exerts neuroprotective effects via decreasing oxidative stress[25] and may improve recovery in serious lacerations, crush injuries, hypoxic injuries, and trauma.[26] *Centella asiatica*'s wound-healing effects involve support of collagen synthesis and angiogenesis in a manner that builds strength and improves perfusion.

Centella asiatica, gotu kola

and hypertrophic scars in human trials.[27] Mince a fresh onion or puree in the blender. Onions slices or minced pulp can be applied directly to the skin for superficial abrasions, or cloths may be saturated with the moist pulp and placed directly against the skin for deep wounds and raw abraded skin. Cover with another cloth or piece of plastic wrap and leave in place for 20 minutes. Repeat at least 3 times per day, using *Calendula officinalis* salve between applications.

Tincture for Poor Healing with Tendency to Infection

Those with poor circulation, diabetes, excessively inflammatory reactivity, and general systemic toxicity may heal poorly, and each such circumstance can be addressed directly, where appropriate, with a long-term therapy. This formula, however, can be used acutely in all such circumstances to promote wound healing. Compare this all-purpose formula to the one that follows, which is aimed more specifically at poor circulation as an underlying cause of poor wound healing.

Echinacea angustifolia
Calendula officinalis
Achillea millefolium

Combine in equal parts and take 1 teaspoon of the combined tincture 4 or 5 times a day, beginning immediately following any wound, trauma, or injury.

Tincture for Poor Healing Due to Poor Circulation

The elderly, heavy smokers, and those with circulatory insufficiency may heal poorly when inadequate oxygen and nutrients in the injured tissue impairs healing and worsens inflammatory response. *Ginkgo biloba* promotes circulation, *Curcuma longa* and *Centella asiatica* act as connective tissue promoting agents, and *Echinacea angustifolia* can help prevent opportunistic infections. See also the list of "Herbs for Skin Infections with Poor Circulation" on page 183 and compare and contrast the information about these herbs in "Specific Indications," beginning on page 237, to help fine-tune herbal formulas for individual patients. *Calendula officinalis* has been shown to promote healing of chronic venous leg ulcers,[28] and thus it might be considered topically for anyone experiencing poor wound healing, as a complement to this internal systemic formula.

Ginkgo biloba
Curcuma longa
Centella asiatica
Echinacea angustifolia

Combine in equal parts. Take 1 teaspoon of the combined tincture hourly, beginning immediately following any wound, trauma, or injury.

Herbs for Skin Infections with Poor Circulation

These herbs are mentioned in folkloric texts as capable of improving circulation in the skin and specific for infections related to circulatory stasis.

Angelica sinensis
Calendula officinalis
Chelidonium majus
Cinnamomum spp.
Syzygium aromaticum
Ginkgo biloba
Phytolacca decandra
Quercus alba
Zingiber officinale

Formulas for Skin Allergies, Dermatitis, and Hives

One goal of herbal formulas for allergic skin disorders is to reduce histamine, mast cell activation, platelet release of cytokines, and other inflammatory mediators. Another goal is to reduce acute reactivity as well as the chronic tendency. Improving intestinal health and liver function can reduce allergic triggers in the body, and thus alterative and liver supportive herbs may be as important in formulas for atopic skin disorders as they are for acne, boils, and skin infections. Many allergic patients will also benefit from supplementing essential fatty acids, avoiding food allergens, and possibly supplementing vitamin A (or beta-carotene), selenium, zinc, or other nutrients. For acute allergic skin eruptions, palliative herbs are often valuable in formulas to relieve itching, burning pains, or other discomfort. In the following formulas, note how anti-inflammatory and antihistamine herbs can be used for acute conditions, and alteratives and general stabilizing herbs can be included in long-term management plans to reduce the underlying state of allergic inflammation. Note also how topical applications and anodynes may complement acute formulas to best address pain and irritation.

Dermatitis and eczema are common skin complaints that occur most often in those with atopic constitutions; these conditions often run in families. As with chronic infections, poor gut health can trigger chronic reactivity in the skin, and therefore, liver and digestive-enhancing herbs can often be useful foundational ingredients in dermatitis formulas. Alterative herbs suit such a purpose and can be used in teas and tinctures, complementing antiallergy, anti-inflammatory, and wound-healing herbs. In many cases, those with eczema have concomitant asthma, hay fever, or other allergies, referred to as atopy, and may respond to herbs that reduce allergic reactivity. Because one role of antioxidant nutrients is to temper inflammatory reactivity, supplementing with these nutrients or consuming nutrient-dense food can be part of treatment protocols to reduce hyperreactivity in the skin.

General Dermatitis Tincture

Dermatitis can occur due to underlying atopic constitutions but may be exacerbated by stress, toxemia, nutritional deficiencies, hypo- and hyperthyroidism, and other factors. Each patient should be evaluated individually so that any such circumstances can be addressed accordingly. The following formula offers general support in all situations. *Glycyrrhiza glabra* and *Angelica sinesis* provide antiallergy support to reduce atopy, and *Centella asiatica* and *Calendula officinalis* are specific for resolving of all types of skin lesions. Modern research reports *Angelica sinensis* to be effective in treating allergic disorders[29] and antihistamine and antiserotonin effects have been demonstrated.[30]

Calendula officinalis
Glycyrrhiza glabra
Angelica sinensis
Centella asiatica

Combine in equal parts and take 1 to 2 teaspoons of the combined tincture 3 or 4 times daily.

Tincture for Chronic Eczema

Eczema is a chronic dermatitis related to underlying atopic hypersensitivity. It may be associated with other atopic conditions such as hay fever and asthma. Eczema tends to run in families and may be aggravated by stress, bowel reactivity, food allergen exposure, dry skin, and essential fatty acid deficiency. The herbs in this formula are aimed at reducing histamine and immune hyperactivity and would be best paired with an alterative and immune-modulating tea. *Astragalus membranaceus,* for example, has been noted to reduce atopic reactivity in the skin.[31] Essential fatty acids, zinc, beta-carotene and vitamin A, and B vitamins will be appropriate in most cases. Seek out possible dietary contributors to address, and consider probiotics, glutamine powder, and other bowel support in cases of leaky gut, which can often contribute to systemic hyperreactivity. The use of liver support and lipotropic formulas would be complementary to this formula.

Tanacetum parthenium
Glycyrrhiza glabra
Centella asiatica
Astragalus membranaceus

Combine in equal parts and take ½ teaspoon of the combined tincture 3 or 4 times a day to reduce the occurrence of chronic eczema. Take 1 teaspoon every hour for new episodes and acute outbreaks.

Tea for Chronic Eczema

Calendula officinalis is best known as a simple wound-healing agent, but it does have significant anti-inflammatory effects as well. *Calendula officinalis* reduces allergic skin reactions associated with chemotherapy[32] and allergen-induced edema.[33] *Calendula officinalis* triterpenoids are credited with anti-inflammatory activity,[34] and the herb is safe to use long term. The Tincture for Chronic Eczema, above, and this tea may be used in tandem for best results.

Arctium lappa root powder 4 ounces (120 g)
Astragalus membranaceus shredded root 2 ounces (60 g)
Foeniculum vulgare seed powder 2 ounces (60 g)
Glycyrrhiza glabra 2 ounces (60 g)
Calendula officinalis flowers 1 ounce (30 g)
Centella asiatica leaves 1 ounce (30 g)

This formula yields 12 ounces. Store in 4-ounce quantities in tightly sealed containers for up to several months, replenishing as necessary. To make the tea, gently simmer 1 heaping tablespoon of the herb mixture in 4 cups of water, then steep, covered, for 10 minutes. Strain and drink as much as possible, at least 3 cups per day.

Glycyrrhiza Compress for Atopic Dermatitis

Glycyrrhiza glabra tea compresses, and especially concentrated glycyrrhizinic acid products, are helpful topically due to the steroidal components and anti-inflammatory properties.[35]

Glycyrrhiza glabra shredded root

Simmer 1 teaspoon in 1 cup of water for 10 minutes, and let stand until cool enough to strain. Soak a soft cloth in the liquid and apply to the affected area, 3 or more times a day.

Bentonite Clay for Atopic Dermatitis

For "wet" lesions such as vesicles or bullae or for weeping or raw lesions, apply a paste of bentonite clay to absorb and dry secretions and offer a moderate anti-inflammatory effect.

Bentonite clay (or French clay, green clay, and so on)

Make a wet paste by diluting 1 to 2 tablespoons of the dry clay powder with warm water, blending to a smooth consistency with a fork. Apply the paste to the affected area and allow to air-dry. Wash off with warm water. It is best to avoid allowing a large amount of clay to go down a household drain because clogs may occur.

Tincture for Chronic Dermatitis and Hives

Alteratives can be added to formulas for hives, eczema, and other types of dermatitis, especially when constipation, digestive difficulty, gas, bloating, coated tongue, liver congestion, and poor elimination of wastes contribute to allergic inflammation. Many other formulas that provide alterative, antiallergy, and skin-supportive effects could be equally effective. Consider this formula in terms of the *actions* provided by the herbs. *Apium graveolens* could be substituted for *Angelica sinensis* for antiallergy effects, *Taraxacum officinale* could be substituted for *Arctium lappa* for alterative effects, and *Equisetum arvense* might be substituted for *Centella asiatica* for connective tissue supportive effects.

Calendula officinalis
Arctium lappa
Angelica sinensis
Centella asiatica

Combine in equal parts and take 1 to 2 teaspoons of the combined tincture 3 or 4 times daily.

Tincture for Chronic Eczema in Atopic Patients

If poor diet, leaky gut, or impaired bowel health do not seem to contribute to a patient's chronic dermatitis, the long-term use of herbs to reduce allergic and inflammatory process is appropriate. This formula is just one example, and substitutions can be made. For instance, if a patient has joint pain and asthma, rather than hay fever and headaches, *Petasites hybridus* may be more appropriate than *Tanacetum parthenium*. *Calendula officinalis* may be more specific than *Centella asiatica* in a patient whose skin eruptions tend to become infected, rather than ulcerated. The goal is to combine herbs that reduce allergic reactivity by a variety of different mechanisms, rather than by the same mechanism of action. In this formula, *Tanacetum parthenium* acts on histamine and inflammatory molecules, *Glycyrrhiza glabra* has cortisol and immune-modulating properties, *Centella asiatica* has direct effects on the skin and connective tissue, and *Astragalus membranaceus* balances cytokines and immune responses.

Tanacetum parthenium 30 ml
Glycyrrhiza glabra 30 ml
Centella asiatica 30 ml
Astragalus membranaceus 30 ml

Take ½ teaspoon of the combined tincture 3 or 4 times a day to reduce the occurrence of chronic eczema. Take a teaspoon every hour for new episodes and acute outbreaks.

Tea for Chronic Eczema, Dermatitis, and Hives

This tea is based on the premise of reducing allergic and inflammatory processes, and the antiallergy herbs in it were selected for their taste. Alteratives or other herbs as specifically indicated can be substituted, but consider flavor when making choices. A separate tincture would be complementary to this tea, especially for acute flare-ups, helping to create a more aggressive therapy.

Foeniculum vulgare seeds 1 heaping tablespoon
Glycyrrhiza glabra 1 tablespoon
Calendula officinalis flowers 1 tablespoon
Centella asiatica leaves 1 tablespoon

Crush the *Foeniculum* seeds in a mortar and pestle or run through a coffee grinder reserved for the purpose. Add the crushed seeds to 4 cups of water and simmer briefly. Add the remaining ingredients, steep in a covered vessel, and then strain. Drink freely, at least 3 cups per day.

Herbs for Itching Skin

These herbs are all mentioned in traditional and Eclectic herbal literature for itchy skin, hives, and red blotchy skin eruptions. Most of these traditional herbs are known mast cell stabilizers and platelet antiaggregators, giving them antihistamine and other antiallergy effects. Consider one or more of these herbs in formulas for hives and allergic skin conditions.

Angelica sinensis
Apium graveolens
Crataegus oxyacantha
Ephedra spp.
Petasites hybridus
Picrorhiza kurroa
Salix alba
Sanguinaria canadensis
Scutellaria baicalensis
Tanacetum parthenium
Urtica urens

Tincture for Acute Hives (Urticaria)

The following herbs have antihistamine and mast cell–stabilizing effects as well as noted effects to stabilize platelets and the release of inflammatory mediators. While this formula may help an acute episode of hives subside as quickly as possible, a long-term approach to reduce the hyperreactivity is usually helpful, such as treating any underlying leaky gut, supplementing with antioxidant nutrients, fish oil, and plant-based essential fatty acids, or treating stress.

Ephedra spp. 15 ml
Tanacetum parthenium 15 ml
Ginkgo biloba 15 ml
Phytolacca decandra 15 ml

Take 1 dropperful of the combined tincture every 10 to 30 minutes, reducing as symptoms improve.

Traditional Chinese Duo for Urticaria

These herbs are a classic duo from Traditional Chinese Medicine for treating hives, due to their ability to clear heat and toxins. The pair is also used to treat infections and is referred to as Yin Qiao San.

Lonicera japonica
Forsythia suspensa

Decoct equal parts of the herbs using 1 tablespoon per cup of water, or find a similar formula in pill form or as granules. The tea may be consumed at a dose of 1 cup per hour, reducing the frequency as itching and inflammation subside. Where available, these herbs may be mixed with the herbs featured in the other formulas for urticaria.

Topical Salve for Eczema

This salve will not "cure" eczema on its own—appropriate systemic and dietary approaches are needed as well. But it can help heal fissures, and prevent secondary infections in chronic skin lesions. *Daucus carota* seed oils and extracts inhibit cyclooxygenase,[36] helping to boost the anti-inflammatory effect of this topical formula.

Calendula officinalis oil ½ cup (120 ml)
Centella asiatica oil ½ cup (120 ml)
Beeswax 1 ounce (30 g)
Daucus carota (carrot seed) oil 1 teaspoon
Vitamin E oil 1 teaspoon

Combine the *Calendula* and *Centella* oils and warm them. Melt the beeswax in a double boiler or microwave oven and add to the warm combined oils. Add the carrot seed and vitamin E oils, stir briefly, and quickly pour the mixture into 1- or 2-ounce containers. Apply frequently throughout the day.

Topical Oil for Dermatitis

Quality *Matricaria chamomilla* oil or any herbal oils using *Hippophae rhamnoides* (sea buckthorn) oil as a base are hard to come by. Because sea buckthorn has excellent healing and anti-inflammatory properties for skin disorders, it makes a good base herb to prepare this formula. *Lavandula angustifolia* essential oil may reduce mast cell–mediated allergic reactivity, including histamine and immunoglobulin E (IgE) release,[37] and helps boost the antiallergy effects of the formula.

Matricaria chamomilla flowers, dry 1 ounce (30 g)
Hypericum perforatum buds, dried 1 ounce (30 g)
Hippophae rhamnoides oil 2 ounces (120 ml)
Symphytum officinale oil to cover as needed
Melaleuca alternifolia (tea tree) essential oil ¼ ounce (7 ml)
Lavandula angustifolia essential oil ¼ ounce (7 ml)

Place the *Matricaria* and *Hypericum* in a blender with the *Hippophae* oil. Pulverize as finely as possible. Transfer to a glass storage container such as a canning jar. Top off with *Symphytum* oil or additional *Hippophae* oil to ensure that the herbs are entirely covered. Store in a dark cupboard and shake at least once a day for 4 to 6 weeks. Filter through a mesh jelly bag or a muslin-lined stainless steel mesh strainer to remove the particulate. Rebottle the finished oil and add the tea tree and lavender essential oils, blending well. Apply to affected areas 3 or more times daily.

Anti-Pruritic Oral Tincture

Intense itching is often due to histamine and other inflammatory mediators. *Panax ginseng*, *Angelica sinensis*, and *Petasites hybridus* can reduce histamine and inflammation, and *Hypericum perforatum* is traditional for uncomfortable itching due to its anti-inflammatory effects on nerve endings. *Petasites* should be avoided in those with liver disease, and due to the content pyrollizidine alkaloids, its use should be limited to short term to be on the safe side. *Petasites* is also contraindicated during pregnancy and lactation. This formula is intended for short-term use; use for only a day or two.

Panax ginseng
Angelica sinensis
Hypericum perforatum
Petasites hybridus

Combine in equal parts and take 1 teaspoon of the combined tincture every 5 to 10 minutes for acute pruritus, reducing the frequency of the dose as symptoms improve.

Simple Spray to Allay Itching

This simple all-purpose spray can help relieve itching due to hives, mosquito bites, dermatitis, and healing wounds and burns. This palliative topical approach would make a good complement to an oral formula for hives.

Mentha piperita essential oil 5 ml
Water 25 ml

Combine in a 1-ounce bottle with a spray nozzle. Spray on the affected area frequently, shaking the bottle well before each use.

Anti-Pruritic Topical Spray

Mentha piperita (peppermint) oil is the most effective thing I have tried on myself and others to alleviate acute itching. The resins of *Grindelia squarrosa* and *Commiphora myrrha* (myrrh), along with the *Aloe vera* gel base, make the perfect complementary vehicle for the mint oil. Resins help dry suppurative, itching lesions such as poison oak and ivy, and the formula also provides antimicrobial effects, which is helpful in cases of children scratching itchy mosquito bites and chicken pox lesions.

Aloe vera gel 22 ml
Grindelia squarrosa tincture 7 ml
Calendula tincture 7 ml
Commiphora myrrha tincture 7 ml
Mahonia aquifolium tincture 7 ml

Herbal Specifics for Intense Itching

The following herbs are recommended for topical application to relieve intense itching.

Avena sativa. Oatmeal baths are excellent for chicken pox, poison oak or ivy, or any other situation of extensive itching over the entire body.

Hypericum perforatum. Teas and tinctures can be used orally for chronic itching, especially when accompanied by painful ulcers or scabs and neuralgic pain.

Mentha piperita. The essential oil may be used topically on mosquito bites, eczema, or any unbroken skin when there is acute itching.

Picrorhiza kurroa. The tincture can be taken orally for chronic itching, especially when associated with digestive disorders, liver disease, or asthma.

Scrophularia nodosa. The tincture may be taken orally for dermatitis, seborrhea behind the ears with a tendency to crusting, and prickling and itching sensations in the skin.

***Syzygium aromaticum* (formerly known as *Eugenia*).** The essential oil may allay itching and tingling in cases of neuralgia or temporarily deaden severe pruritis of eczematous lesions when topically applied.

Herbs for Itchy Skin with Eruptions

These herbs are all mentioned as being specific for itchy skin eruptions in folkloric herbals and traditional literature, regardless of the medical diagnosis. These herbs are taken orally (some can also be applied topically).

Ammi visnaga
Angelica sinensis
Apium graveolens
Chelidonium majus
Echinacea angustifolia
Grindelia squarrosa
Juglans nigra
Mentha piperita
Picrorhiza kurroa
Salix alba
Scutellaria baicalensis
Smilax ornata and related species
Syzygium aromaticum
Urtica urens

Lavandula angustifolia essential oil 5 ml
Mentha piperita essential oil 5 ml

Combine in a 2-ounce bottle and attach a spray nozzle. Shake well and spray on itching lesions, as often as desired. The liquid may also be applied using a cotton ball or gauze pad.

Oatmeal Baths to Allay Itching

Avena sativa is a classic remedy for allaying itchy skin due to dermatitis, poison oak, multiple bug bites, chicken pox, and other common causes. For best results, use the largest quantity of oats possible (they are not that expensive) and don't towel off after the bath. Let the oat water dry on the skin to prolong the antipruritic effects.

Avena sativa (oats) 5 to 10 cups (or even more)
Old pillowcase

If you have time, powder the dry oats in a blender or grinder to yield a fine powder, which will be the most powerful medicine. Place the oats in a large stockpot and add roughly 4 or 5 cups of water per cup of oats. Bring to a gentle simmer and hold until the oats are thick and gluey. Meanwhile, prepare a bathtub with warm water for bathing. Pour the oatmeal through an old pillowcase into the waiting warm bath. Add 3 or 4 additional cups of uncooked oats to the pillowcase and tie a large knot at the open end. Place the pillow in the hot bath like a giant tea bag. The more oats you are able to use, the milkier the water will become, and the more powerful the effects can be. Soak at least 15 to 20 minutes in the water, and air-dry upon getting out. Repeat 2 or 3 times a day for severe pruritus.

Tincture for "Prickly Heat" and Itchy Skin in Hot Weather

Infants and young toddlers may have immature development of the autonomic regulation of sweat glands, and they may quickly develop tiny papular rashes in hot weather, sometimes referred to as "prickly heat." This condition does not appear to be terribly uncomfortable to infants and does not warrant the chore of administering difficult-to-consume tinctures or teas. However, if the problem persists into adolescence or is particularly uncomfortable and bothersome, this tincture can be helpful. The herbs in this formula are recommended in folkloric literature for those who develop itchy skin in hot weather.

Smilax ornata 15 ml
Salvia officinalis 10 ml
Thuja occidentalis 5 ml

Take 1 or 2 dropperfuls of the combined tincture 3 times a day for prevention or hourly for acute itching.

Protocol Options for Eczema and Allergic Dermatitis

Severe skin reactivity usually occurs in those with underlying atopic or hypersensitivity disorders; eczema, asthma, hay fever, hives, and bowel reactivity can all occur in tandem. Treating the underlying reactivity and addressing leaky gut or other contributors may improve, and even cure, the tendency to acute skin rashes. The following list offers nutritional supplements and herbs to use to provide immediate relief in acute situations. For best results, large and frequent dosages are best. Avoid all food allergens immediately and begin probiotic and bowel support as well. High stress and anxiety, adrenal or thyroid imbalance, and nutritional deficiencies can all contribute to chronic skin reactivity and should be investigated and treated accordingly.

Nutrients	Homeopathics	Herbs Internally	Herbs Topically
Linum usitatissimum (flax) seed oil	*Rhus tox*	*Ephedra sinica*	*Avena sativa* (oatmeal) baths
Beta-carotene	Sulfur	*Tanacetum parthenium*	*Matricaria chamomilla* soaks or poultices
Zinc		*Calendula officinalis*	*Mentha piperita* (mint) and *Lavandula angustifolia* essential oils
		Glycyrrhiza glabra	

Compresses for Dermatitis

Apigenin, a component of *Matricaria chamomilla* and *Chrysanthemum leucanthemum* and *C. indicum*, has anti-inflammatory effects and may support barrier functions of the skin.[38]

Matricaria chamomilla
Chrysanthemum leucanthemum or *C. indicum*

Steep the herbs for 10 minutes, using 1 tablespoon per cup hot water, and then strain. Soak a cloth in the liquid and apply to the affected skin. Leave the compress in place for roughly 10 minutes, and repeat several times, and again throughout the day, whenever possible.

Herbs for "Wet" Atopic Dermatitis

Dermatitis may be classified as wet or dry. Wet dermatitis is characterized by weeping skin, vesicular eruptions, and raw, oozing skin lesions. Plants that are astringent and anti-inflammatory may help such lesions heal. These herbs are mentioned in the folkloric literature for dermatitis, itchy skin, and weeping lesions and may be utilized both topically and internally.

Apium graveolens
Betula pendula
Camellia sinensis
Geranium maculatum
Glycyrrhiza glabra
Hamamelis virginiana
Hydrastis canadensis
Hypericum perforatum
Matricaria chamomilla
Quercus alba
Salix alba
Sambucus niger
Smilax ornata
Urtica urens

Tincture to Aid Withdrawal from Steroids

Many patients are managed for decades on steroidal creams (and oral steroids, for more severe cases) to treat eczema and chronic inflammatory disorders of the skin. While steroids can be effective in suppressing inflammation, they do nothing to correct the underlying cause of the problem, and they suppress the production of corticosteroids in the adrenal gland. The use of adaptogenic, adrenal supportive herbs can assist recovery of the adrenal gland. *Centella asiatica* promotes granular tissue, the strength of the granular tissue, and hydroxyproline content in both normal skin and steroid-suppressed, poorly healing skin.[39]

Centella asiatica
Glycyrrhiza glabra
Panax ginseng
Eleutherococcus senticosus

Combine in equal parts and take ½ to 1 teaspoon of the combined tincture 3 or 4 times daily, continuing for 3 to 6 months.

Herb Therapies for Itchy Skin with Dryness

Chronic dermatitis is often "dry" and characterized by lichenification, rough and dry skin, accentuated skin lines, and in some cases, deep fissures that may crack and bleed. Itchy skin can be related to severe dryness, either constitutionally or due to essential fatty acid deficiency. Constitutionally dry skin is often worse in cold weather when the heating of homes furthers dries out the air. Note that *Hypericum perforatum*, *Matricaria chamomilla*, *Glycyrrhiza glabra*, and *Camellia sinensis* are used for both wet and dry dermatitis. *Symphytum officinale* roots and oils are specific for dry cracked skin. The oils on this list are commercially available and good sources of essential fatty acids that can be consumed orally at a dose of 1 tablespoon per day. Quality nut and seed oils can also be taken internally. The oral consumption of essential fatty acids will often help soften and lubricate the skin in 1 to 3 months.

Althaea officinalis teas and tinctures
Avena sativa baths
Camellia sinensis tea or as capsules
Centella asiatica tea or tincture
Chelidonium majus tincture or capsules
Fucus vesiculosus, used as a culinary spice
Glycyrrhiza glabra teas and tinctures orally, or to prepare into compresses
Hippophae rhamnoides oil, to consume orally or use topically
Hypericum perforatum tea, tincture, or encapsulation orally, or to prepare into compresses or macerate in oil to yield a product for topical use
Juglans nigra tincture or as capsules
Linum usitatissimum oil, by the tablespoonful or as capsules
Matricaria chamomilla tea or tincture orally, or to prepare into compresses for topical use
Oenothera biennis oil, by the tablespoonful or as capsules
Olea europaea oil; can be used in the diet and is the most common base oil for extracting other plants and making St. Johnswort, comfrey, or calendula oil.
Phytolacca decandra tincture, in small amounts
Ribes species oil, taken by the tablespoonful or in capsules
Symphytum officinale teas or tincture orally, or to extract into oil for topical use

Dry, Scaly Skin? Check the Thyroid

Chronic dry scaly skin can be a sign of hypothyroidism. The problem can be particularly severe on the shins. Run blood tests to rule out hypothyroidism; thyroid support may improve the situation quickly. Vitamins D and E, zinc, selenium, and fish oils can all help dry skin in general.

Eczema Formula from the Lloyd Brothers

The Lloyd Brothers were a pharmacy and manufacturer of medicinal grade plant products, catering to the so-called Eclectic physicians of the late 1800s and early 1900s. This formula, from a century ago, exemplifies the belief that bowel issues may underlie chronic skin complaints. *Echinacea angustifolia* and *Phytolacca decandra* may help cell-to-cell adhesion and tighten the gap junctions of intestinal mucosa cells in cases of leaky gut and bowel toxemia contributing to skin disease and eczema. *Mahonia* (also known as *Berberis*), *Rhamnus purshiana* (formerly *Cascara sagrada*), and *Iris versicolor* are all mild alteratives, and *Rhamnus* is also a strong irritant laxative. This formula would be most appropriate for those with eczema concomitant with constipation, acne, and/or bowel symptoms.

Echinacea angustifolia 15 ml
Mahonia aquifolium 15 ml
Rhamnus purshiana 10 ml
Iris versicolor 10 ml
Phytolacca decandra 10 ml

Take ½ to 1 teaspoon of the combined tincture 3 or 4 times a day, reducing as the skin lesions clear.

Tincture for Dry Skin Concomitant with Hypothyroidism

Low thyroid function can cause dry scaly skin, sometimes in the extreme. These herbs can improve skin dryness by supporting healthy thyroid function.

Coleus forskohlii
Commiphora mukul

Fucus vesiculosus
Panax ginseng

Combine in equal parts and take 1 to 2 teaspoons of the combined tincture 3 times daily.

Herbs for Seborrhea

These herbs are emphasized in the folkloric literature for seborrheic dermatitis, which is seen most in infants but may persist to young adulthood. Symptoms may include crusts behind the ears, in nasal creases, and on the eyelid margins. It is associated with an atopic tendency, and parents whose children have cradle cap, a type of seborrheic dermatitis, should be cautioned to take great care introducing foods to these babies, because food allergies can develop and further trigger atopic conditions, including eczema and asthma in adulthood. The seed oils included in this list are for oral consumption and can be stirred into baby foods to increase the intake of essential fatty acids. Nursing mothers should be advised to consume these oils as well because they can be passed on in the breast milk.

Juglans nigra
Linum usitatissimum (flax) seed oil
Oenothera biennis (evening primrose) oil
Olea europaea (olive) oil
Ribes nigrum (black currant) oil
Scrophularia nodosa
Smilax ornata
Stillingia sylvatica
Thuja occidentalis
Trifolium pratense
Viola odorata (as poultices)

Herbs for Skin Eruptions of the Hands and Feet

These herbs are emphasized in the Eclectic and folkloric literature as being specific for eczema and other skin rashes that particularly affect the hands and feet. Consider including one or more of these herbs in formulas for dermatitis when the hands or feet are affected.

Angelica sinensis
Centella asiatica
Hypericum perforatum
Sanguinaria canadensis
Smilax ornata or *S. regelii*
Stillingia sylvatica
Thuja occidentalis

Tincture for Dermatitis Related to Drugs or Chemical Sensitivity

Chemically sensitive and allergic individuals may quickly develop dermatitis and itching following ingestion of drugs or exposure to detergents, perfumes, solvents, and other household chemicals. (See also "Formulas for Erythema Multiforme" on page 219). This formula is based on my own clinical experience, and the herbs may reduce chemical sensitivity and allergic phenomena in general and may help those who are reacting to laundry soap, shampoo, or other substances. For the most powerful therapy, use a tincture such as this with alterative herbs and bowel support, if needed, and nutritional supplements such as essential fatty acids, zinc, selenium, and beta-carotene and vitamin A. Omit *Petasites* for those with liver disease or during pregnancy and lactation. Due to its pyrrolizidine alkaloid content, the use of *Petasites* should be limited to short-term usage for acute situations. *Tanacetum* may be a possible alternate herb choice.

Astragalus membranaceus
Petasites hybridus
Ganoderma lucidum
Scutellaria baicalensis

Combine in equal parts and take 1 teaspoon of the combined tincture 3 times a day, continuing for at least 3 months to allow the herbs to take effect.

Tincture for Perioral Dermatitis

Perioral dermatitis can be due to a food allergen triggering contact dermatitis around the mouth. Acidic foods such as citrus are the most common offenders, but any common food allergen may contribute. The condition is most common in those with allergies and an atopic constitution, and in women between 20 and 60. Some sufferers note lipstick and cosmetics as a trigger. B vitamins are sometimes therapeutic, and there is no harm in trying B vitamin supplements in all cases. Essential fatty acids may also be employed; these oils often improve allergic conditions. The topical use of *Calendula officinalis* salve can be helpful to alleviate irritation and protect the skin from food. The following tincture uses *Tanacetum parthenium* as a helpful antiallergy herb, and *Calendula officinalis* and *Centella asiatica* can improve the skin's protective barriers and promote healing.

Calendula officinalis
Centella asiatica
Tanacetum parthenium

Remedies for Hot Red Stinging Skin

These herbs are emphasized in folklore, especially in the discipline of homeopathy, for skin reactivity associated with bright red or pink skin accompanied with a hot or burning sensation.

Aloe vera
Apis mellifica
Atropa belladonna
Cantharis vesicatoria
Sanguinaria canadensis

Combine in equal parts and take ½ to 1 teaspoon of the combined tincture 3 to 6 times daily, reducing as the skin improves.

Anti-Inflammatory Supportive Formula for Dermatitis Herpetiformis

Dermatitis herpetiformis is an eczema-like skin condition characterized by chronic vesicles, with initial onset similar in appearance to a herpes lesion or cold sore. The condition is closely associated with inflammation of the bowels, especially celiac disease and gluten intolerance[40], and occurs most often in those of northern European descent.

As with celiac disease, dermatitis herpetiformis is associated with particular genetic markers. Like eczema, the skin lesions are intensely pruritic vesicles occurring in broad crops, and vigorous scratching will produce oozing lesions, crusts, and scabs. Antipruritic topical compresses such as those described earlier in this chapter can be helpful for symptom relief. The condition responds to gluten avoidance and efforts to improve gut health, including taking glutamine powder, probiotics, and herbal formulas for gastrointestinal conditions as detailed in chapter 2. Dapsone is considered the gold standard pharmaceutical for dermatitis herpetiformis because the condition responds so readily that it is practically diagnostic. Dapsone is an antimicrobial and its mechanism of action in this condition is not yet understood.

Short-term herbal treatment for acute eruptions would be focused on antiallergy and anti-inflammatory herbs. This formula includes *Glycyrrhiza* and *Curcuma* to provide general anti-inflammatory support, while *Tanacetum* provides an antihistamine effect. *Ulmus* is included to support the bowel integrity in the event that leaky gut may contribute to dermatitis.

Glycyrrhiza glabra
Curcuma longa
Tanacetum parthenium
Ulmus fulva

Combine in equal parts and take 1 to 2 teaspoons combined tincture 3 to 6 times daily, reducing as symptoms improve. The *Ulmus* might also be separated out of this tincture and consumed instead as a tea or in capsules.

Dermatitis Herpetiformis and Celiac Disease

Celiac disease is an immune mediated small intestinal enteropathy that is triggered by exposure to dietary gluten in genetically predisposed individuals, recognized at least since the first century AD.[41] Those with celiac disease may present with a diversity of symptoms, particularly malabsorption with diarrhea, steatorrhea, weight loss, or failure to thrive. Additional signs and symptoms include anemia, IBS-like abdominal symptoms, neuropathy, ataxia, depression, short stature, osteomalacia, and osteoporosis, with many such symptoms being related to nutritional deficiencies due to enteropathy or inflammatory and immune activation. Dermatitis herpetiformis is a cutaneous manifestation of gluten sensitivity characterized clinically by herpetiform (blisters or vesicles) clusters of intensely itchy urticated papules and small blisters distributed on the extensor aspects of the elbows and knees, over the buttocks, and on the scalp.

Treating Neurodermatitis

Neurodermatitis, also known as lichen simplex chronicus, is a chronic skin condition that starts with a minor itch or nervous tendency to pick or scratch at the skin. Scratching can trigger further itching, and the creation of a lesion perpetuates the tendency to continue touching and picking at the skin. A thickened leathery region of skin results, usually on the limbs or neck, due to easy access. As cutaneous nerves are irritated, the itching sensation can be intense, making it difficult to resist repetitive scratching, which creates a vicious itch-scratch cycle. Some people must glove their hands at night to keep from scratching or bandage the area to break the cycle. Mint essential oil may be a useful anti-pruritic to help allay the itching. Those with dry skin or eczema may be prone to neurodermatitis, especially during the winter months, and the use of essential fatty acids in the diet may benefit dry skin, eczema, and all types of dermatitis. Those with anxiety may also be prone to pick at tiny skin lesions or bug bites and overstimulate the nerve endings in the skin, heightening itching sensations. The use of stress reduction practices and the inclusion of anxiolytic herbs in such formulas may benefit such individuals.

Tincture for Neurodermatitis

Because neurodermatitis has a nervous origin, the use of nervines along with wound-healing agents are appropriate. Patients must stop touching the affected area, and if they cannot command themselves to do so, covering the lesions with bandages may be useful. *Scutellaria baicalensis* and *Verbena hastata* in this formula are nervines, while *Angelica sinensis* is anti-inflammatory, and *Centella asiatica* offers wound-healing support.

Scutellaria baicalensis
Angelica sinensis
Centella asiatica
Verbena hastata

Combine in equal parts and take ½ to 1 teaspoon of the combined tincture 3 to 6 times per day reducing as the lesion clears.

Formulas for Folliculitis

Folliculitis involves an inflammation of hair follicles. The condition may be minor and superficial or deep into the tissues, causing hair loss, sometimes permanent if scarring is extensive. Hair follicles can be invaded by bacteria, fungi, or viruses that infect the follicle and surrounding skin. The condition is not typically serious, but infections can spread from tiny localized areas around individual hair follicles to extensive crusts on the scalp or other hairy areas. Shaving and bikini waxes can sometimes introduce bacteria into the hair follicles and result in folliculitis of the face or groin. This is sometimes referred to as "barber's itch" when occurring in the beard area (also known as pseudofolliculitis barbae and sycosis barbae), as normal *Staphylococcus* bacteria are introduced into tiny razor cuts.

Soaking in hot tubs that harbor heat-tolerant *Pseudomonas* bacteria can result in folliculitis of leg and arm hair. Such bacterial types of folliculitis will often produce tiny pustules around the hair follicle. Pityrosporum folliculitis is an itchy condition caused by *pityrosporum* yeast. When a superficial infection descends more deeply into the skin, painful boils and carbuncles can result. Small boils will usually drain on their own or with simple heat packs and gentle squeezing, but deep carbuncles may need to be incised and drained and may heal slowly with scarring or keloid formation. Folliculitis will usually respond to natural remedies, as exemplified below. When folliculitis is related to shaving, reduce shaving frequency and employ warm compresses over the affected area once or twice a day. Follow with gentle exfoliation using a soft washcloth, as described below.

Tincture for Chronic Folliculitis

Folliculitis is an inflammation of individual hair follicles, and like a boil, it often responds to *Echinacea angustifolia*. *Mahonia aquifolium* and *Achillea millefolium* act as supportive herbs with both antimicrobial and alterative properties to support liver and digestive health, which may underlie chronic folliculitis. *Phytolacca decandra* has effects on white blood cells and may improve lymph stasis, which can be an underlying invitation to chronic skin infections.

Echinacea angustifolia 15 ml
Mahonia aquifolium 8 ml
Achillea millefolium 4 ml
Phytolacca decandra 3 ml

Take 1 teaspoon of the combined tincture 3 to 6 times a day, reducing as skin improves. This yields a 1-ounce formula that can be doubled or quadrupled or refilled and continued for 2 or 3 months or more.

Applying Topical Herbal Compresses for Folliculitis

Folliculitis is most common on the face, underarms, or groin and may be associated with shaving the area. A regimen of herbal compresses and gentle skin scrubbing should be implemented as soon as any irritation begins in those with a history of folliculitis. Shaving should be avoided until the malady is entirely resolved.

1. Cover affected area with a hot towel compress prepared by simply soaking a small hand towel in hot water. A drop or two of iodine solution may be placed on the compress as well when pustules are present. If possible, cover the compress with a dry towel, and cover this with heat such as from a heating pad, hydroculator, or microwaved-prepared hot pack, leaving in place for 15 minutes to "steam" the area, facilitate circulation, and open the pores.
2. "Debride" the area using a textured towel, cloth, or facial scrubbing pad or brush, by gently scrubbing the area with a soft circular motion, being careful to not further inflame the area. Omit this step for highly inflamed or infected cases or those with concomitant dermatitis and allergic reactivity of the skin.
3. Prepare additional small gauze pads or soft cloth compresses with the Tincture for Chronic Folliculitis and several drops of tea tree oil and/or marjoram essential oils. Apply the pads to the affected area.
4. Repeat the procedure 3 or more times daily, allowing the area to fully air-dry after each treatment.
5. A small amount of *Calendula officinalis* salve may be rubbed into the affected area after air-drying.

Formulas for Childhood Skin Eruptions

This somewhat vague category embraces the observation that many common childhood infections, including measles, chicken pox, and Fifth's disease, may be accompanied by skin eruptions. Herbal folklore lists many herbs that are indicated for these childhood infections—referred to in folkloric textbooks as "viral exanthems"—meaning infectious diseases that involve skin eruptions. Clinicians should alter formulas accordingly if a child is acutely ill with a fever or has nausea and vomiting or if the skin lesions are pruritic or ulcerated and oozing, creating an infection risk. Such situations are covered throughout this chapter; here, I focus on the herbal notions from folkloric traditions.

Tea for Childhood Skin Eruptions and "Viral Exanthems"

An *exanthem* is an old term for skin eruption, and a *viral exanthem* refers to a skin eruption associated with the onset of an infectious illness, particularly in children. This formula is based on recommendations from historical literature. Because this formula is intended for ill children, care has been taken to make a palatable tea, rather than a tincture or encapsulation.

Astragalus membranaceus
Eupatorium perfoliatum
Nepeta cataria
Sambucus nigra
Mentha piperita
Matricaria chamomilla
Glycyrrhiza glabra

Combine equal parts of the dry herbs. Use 1 tablespoon of the herb mixture per cup of hot water, and bring to a brief simmer. Remove from the heat immediately as soon as a gentle simmer has been reached. Steep for 15 minutes in a covered pan and then strain. Drink as much of the tea as possible. This tea can be frozen in ice cube trays and then prepared into ice chips to serve in a cup with a spoon for children suffering from sore throats or simply to encourage greater consumption.

Hand Soak for Paronychia

Infections of the nail bed are best treated by topical applications and soaks. This simple formula uses iodine and *Melaleuca* (tea tree) oil as disinfectant agents and Epsom salts as a drawing agent and anti-inflammatory.

Herbs for Skin Eruptions

These herbs are specific for skin eruptions associated with viruses and infectious illnesses in traditional herbal literature.

Astragalus membranaceus
Cordyceps sinensis
Eupatorium perfoliatum
Euphrasia officinalis
Glycyrrhiza glabra
Lomatium dissectum
Phytolacca decandra
Sambucus nigra berries and flower

For skin eruptions that are particularly severe, extensive, and painful, these herbs are emphasized as specific in the folkloric literature.

Eucalyptus globulus
Glycyrrhiza glabra
Hamamelis virginiana
Hypericum perforatum
Lomatium dissectum
Mentha piperita
Quercus alba
Sanguinaria canadensis
Syzygium aromaticum
Veratrum album (a potentially toxic herb; use a low dose only; see the herb list at end of this chapter for further guidance on dosage)

These herbs are specifically indicated in folkloric and historical literature for skin eruptions associated with colds, congestion, and tender swollen lymph nodes. These herbs have also been classified as lymphatic herbs or "lymphogogues" in traditional herbal philosophy.

Calendula officinalis
Collinsonia canadensis
Galium aparine
Iris versicolor
Phytolacca decandra
Schrophularia nodosa

Systemic antimicrobials can be used as well for severe or unresponsive infections.

Hot water 4 to 5 cups (1 liter)
Epsom salts (magnesium sulfate) 1 cup (240 g)
Iodine solution 20 drops
Melaleuca alternafolia (tea tree) oil 20 drops

Combine the ingredients in a pot and soak the affected hand for 10 to 20 minutes. Allow the hand to fully air-dry, and then apply 1 or 2 additional drops of iodine to the affected nail and bandage for several hours. Allow the affected nail to get adequate air for many hours each day; do not bandage it full time.

Formulas for Fungal Infections of the Skin

The *Tinea versicolor* fungus is a common fungal strain that infects the skin and alters pigmentation causing small circular lesions on the skin, especially on the torso, that may be either darker or paler than unaffected skin. *Tinea* affects especially teens and young adults and may be aggravated by hot humid conditions or a high sugar intake, which can feed fungi. Other species of *Tinea* fungus may infect the scalp and be noticed when patches of hair loss occur, or when inexplicably tender areas occur on the scalp. *Tinea crucis* of the groin, otherwise known as "jock itch," may occur in humid environments or when the undergarments are chronically damp as with heavy exercise or certain work and environmental situations. Vaginal yeast infections can result when women take antibiotics long term, such as taking tetracycline for acne or eating a high-sugar diet, and will often respond to boric acid suppositories to simply acidify the vagina and favor beneficial flora such as *Lactobacillus acidophilus* species.

Athlete's foot (*Tinea pedis*) is a particularly stubborn fungal infection that often begins in moist areas between toes, especially when the feet are habitually moist without air circulation. Athlete's foot can extend to the entire foot and may cause itching or sometimes pain, as the skin becomes inflamed and may crack, blister, or ulcerate. In other cases, only the toenails are affected, a condition that is notoriously resistant to therapy and often persists for years or even decades.

Ringworm is another fungus that can invade the skin, and pets may be a vector of infections. The moldlike dermatophytes that cause ringworm are especially common in kittens and puppies, but ringworm can be contracted by close contact with infected humans or even with contaminated clothing, hair brushes, or linens.

In all cases, improved air circulation is helpful, from wearing loose-fitting pants without undergarments while at home to cleaning athletic shoes and airing the feet whenever possible to using a sun lamp to kill off fungi. Fungal infections may occur in diabetics due to high sugar levels providing a hospitable ecosystem for fungi, or in those with poor digestion or immune function. The following formulas exemplify ways to craft herbal formulas specific to these situations.

Formula for Chronic Fungus with Intestinal Dysbiosis or Digestive Weakness

Predisposition to chronic fungal infections of the nails, feet, or intertriginous zones of the skin may be blamed on a number of contributors, including intestinal dysbiosis, which allows fungus to overgrow in the bowels and from there, seed the rest of the body. When digestive symptoms and chronic antibiotic use accompany skin fungus, the following formula may complement topical antifungals. These herbs are both alterative, to improve digestive function, and antifungal, yet they support rather than deter healthy intestinal flora.

Mahonia aquifolium
Allium sativum
Tabebuia impetiginosa
Achillea millefolium

Combine in equal parts and take 1 teaspoon of the combined tincture 3 times a day or more as needed.

Formula for Fungus with Weak Immunity

In contrast to the previous formula, this formula may improve chronic nail and skin fungus in those with weak immune systems. Patients who suffer from frequent colds, flus, bladder infections, or simple wounds that become infected easily may also have problems with chronic fungal skin infections. If digestive function does not seem to be an underlying cause of the fungal infection, supporting the immune system may reduce the ease with which fungus can thrive in the skin.

Allium sativum
Panax ginseng
Andrographis paniculata
Echinacea angustifolia

Combine in equal parts and take 1 teaspoon of the combined tincture 3 times a day or more as needed.

Formula for Chronic Fungus in Diabetics

High blood sugar is very inviting to microbes, including fungi, and diabetics are prone to all manner of infections, including skin fungus. *Mahonia aquifolium* in this formula is antifungal, and it also enhances digestion and helps regulate blood sugar to support desirable intestinal microbes. *Allium sativum* and *Cinnamomum burmannii* and other *Cinnamomum* species also combat high glucose as well as have antifungal properties.

Allium sativum
Cinnamomum burmannii
Mahonia aquifolium
Foeniculum vulgare

Combine in equal parts and take 1 teaspoon of the combined tincture 3 times a day or more as needed.

Topical Protocol for Fungal Infections of the Skin

The following topical options, along with avoiding sugar and consuming a healthy diet, can be used to complement the oral tinctures and teas for treating fungal infections of the skin and nails.

Vinegar soaks. Dilute apple cider vinegar with a double portion of water or herbal tea, and use as a footbath to soak athlete's foot or as a sitz bath for "jock itch" and genital fungus. Take care to dry body crevices thoroughly when done.

Boric acid soaks. Dissolve 1 heaping teaspoon of pure boric acid powder per cup of hot water and use as a soak, douche, or sitz bath. Another option is to soak a gauze pad in the liquid and apply to the affected part.

Antifungal salves. Use a salve prepared from *Allium sativum*, sulfur, and *Lavandula angustifolia* or *Melaleuca alternifolia* oils topically, frequently throughout the day.

Antifungal herbal footbath. Prepare a strong tea of *Achillea millefolium* and *Mahonia aquifolium*,

Nutritional Support for Chronic Fungal Infections

Vitamin A and beta-carotene. These nutrients are vital to skin health having immune-supportive effects and an ability to deter the entrance of pathogens into the skin. Eat plenty of carrots and green leafy vegetables and use a commercial supplement for 3 months to a year.

Zinc. Zinc is also immune-supportive and deters the entrance of pathogens into the skin. Zinc deficiency manifests as poor wound healing, dermatitis, acne, seborrhea, dry and scaling skin, and a susceptibility to skin infections. Zinc can be nauseating on an empty stomach, so be sure to take supplemnts with foods. Supplement for 3 months to a year.

Caprylic acid. This short-chain fatty acid is taken up by fungi, but impairs their reproductive ability. Although caprylic acid occurs in coconut oil and milk, including human breast milk, the best therapeutic benefit in treating stubborn fungal infections may be had from the daily use of a commercial supplement.

strain, and add a few drops of *Melaleuca alternifolia* and *Origanum vulgare* essential oils. Use as a soak where possible (footbath, sitz baths), or soak a cloth in the mixture and apply to the infected part as often as possible.

UV therapy. Allow as much fresh air and direct sun or UV light as possible to affected area. For cases of athlete's foot, go barefoot as much as possible. Wear a loose nightgown or robe without underclothes for vaginal yeast and jock itch. Expose the affected part to direct sunlight for 10 or 15 minutes, if possible, or shine a sunlamp directly at the infected part.

Astringent herbs. Propolis, *Quercus alba, Agrimonia eupatoria, Hamamelis virginiana, Myrica cerifera,* and *Geranium maculatum* can be used to prepare sitz baths and washes when there is swelling, discharge, and pain.

Essential oil preparations. *Thymus vulgaris, Abies sibirica, Melaleuca alternifolia, Origanum vulgare,* and *Lavandula angustifolia* can be used topically, usually by placing just 10 to 20 drops in an ounce of oil. Toenail fungus requires tenacious application of 2 or 3 times a day for many months, while skin fungus often responds more readily.

Neem oil therapy. Apply *Azadirachta indica* (neem) oil topically to the skin or nails multiple times per day.

Herbs for Fungal Skin Infections

The following herbs are mentioned as specific for fungal infections in Eclectic, homeopathic, and other traditional herbal texts. Consider these both topically and internally for acute and chronic fungal infections such as vaginal infections, athlete's foot, jock itch, and nail fungus.

Achillea millefolium
Allium sativum
Azadirachta indica
Commiphora myrrha
Coptis chinensis
Curcuma longa
Eucalyptus globulus
Syzygium aromaticum
Hydrastis canadensis
Juglans nigra
Lavandula angustifolia
Mahonia aquifolium
Melaleuca alternifolia
Sanguinaria canadensis
Tabebuia impetiginosa

Herbal Preparations for Fungal Infections of the Feet

The following herbs are mentioned as specific for foot fungus in Eclectic, homeopathic, and other traditional herbal texts. If you have the time and inclination, in addition to using the herbs as suggested below, many of these herbs could be macerated in vinegar, traditionally called "acetracta"; the filtered finished product is combined with water, using 1 to 2 cups (240–480 ml) per gallon (3.8 liters) for use in foot soaks. Thoroughly dry the feet afterward and expose to strong sunlight or a sunlamp. Repeat once or twice a day for many months for best results.

Allium sativum. Use minced raw garlic or garlic powder.

Achillea millefolium. Brew a strong yarrow tea.

Azadirachta indica. Rub neem oil into the feet or use in topical formulas.

Syzygium aromaticum. Strong clove tea or clove essential oil may be used in formulas.

Herbs for Fungal Infections of the Feet in Diabetics

The following herbs are mentioned as specific for chronic fungus in diabetics in Eclectic, homeopathic, and other traditional herbal texts. These herbs can be prepared into various formulas to use internally. Most of these herbs are noted to improve glucose metabolism and must be continued for many months to improve blood sugar and gradually deter chronic fungal problems.

Allium sativum
Cinnamomum spp.
Ginkgo biloba
Gymnema sylvestre
Opuntia ficus-indica
Panax ginseng

Herbs for Fungal Infections with Raw Moist Lesions

The following herbs are mentioned in Eclectic, homeopathic, and other traditional herbal texts as specific for fungal infections with raw, moist lesions. Consider these herbs topically to astringe lesions and offer antifungal and anti-inflammatory effects.

Agrimonia eupatoria
Alnus rubra
Calendula officinalis
Coptis chinensis
Echinacea angustifolia
Geranium maculatum
Hamamelis virginiana
Hydrastis canadensis
Quercus alba
Thymus vulgaris

Herbs for Genital Fungus

The following herbs and substances are mentioned in Eclectic, homeopathic, and other traditional herbal texts as specific for jock itch, genital and gluteal crease infections in diabetics or others, and vaginal discharges and infections, commonly called "yeast infections."

Coptis chinensis
Geranium maculatum
Hamamelis virginiana
Hydrastis canadensis
Mahonia aquifolium
Melaleuca alternifolia
Mentha piperita
Vinegar
Boric acid

Juglans nigra. Walnut hull preparations have broad antifungal effects but can stain the feet and fabric.

Melaleuca alternifolia. Tea tree oil can be used directly on the feet or included in herbal footbaths with other herbs.

Help for Fungal Infections with Intense Itching

Several herbs are mentioned in Eclectic, homeopathic, and other traditional herbal texts as specific for itchy fungal infections. *Allium sativum* can be used internally to deter chronic fungus and topically when prepared as a foot soak. *Lavandula angustifolia* and *Mentha piperita* can be used topically to allay itching, the essential oils being the most effective. Use *Quercus alba* topically to astringe new eruptions and moist itching lesions. *Sanguinaria canadensis* should be diluted with other herbs and used topically to combat fungus and help allay itching.

Formulas for Scabies and Lice

An infestation of the *Sarcoptes scabiei* mite in skin is commonly referred to scabies. The mites burrow superficially and lay eggs in the skin, creating severe itching and a bright red rash. Due to the severe pruritic nature of the condition, the skin often becomes scabbed, raw, and prone to secondary bacterial infections. Topical pharmaceuticals akin to pesticides, such as lindane, may be presecribed, but herbs may be just as effective without the toxicity or allergenicity of the drugs.

Lice include some 5,000 species of wingless insects in the order *Phthiraptera*, and three different types of lice exist: head, body, and pubic. Lice feed on blood and skin secretions, causing itching and crawling sensations, and they lay eggs, known as nits, that adhere to hairs and quickly hatch. Lice can be the vectors of other diseases such as typhus. As with scabies, prescription therapies for lice include topical lindane and the antiparasitic oral medication ivermectin, which carries some risk of

Herbs for Scabies

These herbs are noted in modern research to be toxic to scabies organisms without being harmful or too harsh on the skin. These herbs may all be prepared into skin washes, compresses, or ointments to treat scabies.

Allium sativum
Azadirachta indica (neem oil)
Eucalyptus globulus
Hedeoma pulegioides
Origanum vulgare
Phytolacca decandra
Thymus vulgaris

neurotoxicity. Ivermectin is contraindicated in those with liver and kidney disease, women who are breast-feeding, and children under the age of five. Because lice are readily killed by herbal therapies, most herbalists recommend trying the approaches below first.

Both scabies mites and lice can contaminate clothing and bedding, which must be laundered in hot water and dried on the highest heat setting to rid them of the parasites. The parasites will not live long away from the human body, but until the infestation is controlled, great effort and care is usually required to prevent spread among family members, school classmates, and other intimate contacts.

Topical Oil for Scabies

This formula will kill scabies mites and can be used in conjunction with one of the anti-itch sprays in "Formulas for Skin Allergies, Dermatitis and Hives" on page 183. Essential oils may be more effective than tinctures to deter scabies, lice, and other vermin, as this formula exemplifies. A sulfur soap may be complementary and used in the morning shower. Sulfur ointments are also available and can be alternated with the application of the herbal formula.

Allium sativum oil 90 ml
Melaleuca alternifolia (tea tree) essential oil 10 ml
Origanum onites (oregano) essential oil 10 ml
Lavandula angustifolia essential oil 10 ml

Combine the oils in a 2-ounce bottle. Bathe in the morning, then apply the combined oil on every inch of skin from the neck down. Repeat the application midday or after work and again before bed. Continue for several weeks or more.

Herbs for Lice

These aromatic herbs and essential oils are some of the best lice deterrents to use topically on the body and hair. Take care to keep the oils away from the eyes because even contact with the fumes can cause a stinging sensation. However, aim to make the treatment as strong as possible on the scalp to ensure that the lice are killed.

Azadirachta indica (neem oil)
Eucalyptus globulus (eucalyptus)
Hedeoma pulegioides (pennyroyal)
Lavandula spp. essential oil (lavender)
Melaleuca alternifolia essential oil (tea tree)
Thymus vulgaris essential oil (thyme)

Topical Oil for Pediculosis (Lice)

Natural essential oils are a kinder and gentler option than harsh and potentially toxic pharmaceuticals, but care must still be taken to keep them out of the eyes. Even the fumes can sting. The thyme oil in this formula is particularly irritating to the skin and should only be used in combination with the other ingredients to dilute it.

Lavandula angustifolia (lavender) essential oil 3 ml
Melaleuca alternifolia (tea tree) essential oil 3 ml
Thymus vulgaris (thyme) essential oil 1 ml
Origanum onites (oregano) essential oil 1 ml

Combine the oils in a 1-ounce bottle and shake to homogenize. (Omit the thyme oil for those with sensitive skin.) For head lice, apply 10 to 20 drops of the oil mixture to freshly washed hair—use less for very short hair, more for very long, thick hair. Massage in thoroughly, close to the scalp. Be careful not to scratch or injure the skin in any way, which would increase the chance of irritation or inflammation. Cover the hair with a plastic bag to keep the fumes in and cover the bag with a towel twisted into a turban. Leave in place for 15 to 30 minutes. Then remove the coverings and wash the hair several times to help get the greasiness out. Rinsing with diluted vinegar (1 cup per quart water) will help dissolve the glue that attaches nits to the hair and reduce the need to use a special lice comb. For pubic or body lice, add 10 to 30 drops of the oil mixture to 1 tablespoon olive oil, and shake well. Massage into the affected body areas—start with a low dose of essential oil and increase the number of drops as tolerated.

Formulas for Warts

Warts are a type of skin tumor formed in reaction to the presence of a virus in the skin, particularly condyloma and papilloma viruses. There are more than 100 different types of human papillomavirus (HPV), and when these viruses persist, they are associated with various types of skin cancer, especially cervical, anal, and throat cancer. Warts are sometimes classified by where they occur in the body—such as plantar warts on the bottom of the foot or genital warts, which are sexually transmitted. Many children and adolescents develop small round rough-textured warts on the hands and fingers, referred to as *verruca vulgaris*, which often self-limiting as the body's immune system eradicates the virus. But in severe or persistent cases topical measures and immune support may be necessary. Various caustic methods of destroying warts exist, from gentle salicylic acid application to the more aggressive and painful freezing with liquid nitrogen and cauterizing with lasers and electrodes. The following formulas are aimed at creating topical applications to support the body's antiviral response.

Topical Podophyllin Preparation for Warts

Podophyllin is a caustic resin from *Podophyllum peltatum* (mayapple) that has antiviral and immune-modulating properties.[42] The application of podophyllin can both ablate the wart tissue and deter the underlying viruses. The causticity of podophyllin is painless, but care must be taken to dot only a small quantity on the wart itself, because podophyllin will kill healthy tissue with equal ease. Podophyllin is appropriate for verruca warts on the hands and body or condyloma warts on the genitals. Because the HPV virus associated with genital warts may increase the risk of cervical and other cancers, further assessment and follow-up is prudent in addition to eradicating visible warts.

Antiviral Herbs and Topical Treatments for Warts

The following herbs and compounds have been found to exert antiviral properties when taken orally or to have direct caustic and toxic effects on viruses when applied topically. See "Specific Indications," starting on page 237, for further details on selecting these as options in the management of chronic warts. For best results, use both oral antiviral support and topical agents to treat large, numerous, or stubborn warts. Plant latex and topically applied salicylates and chloroacetic acid are all traditional remedies to ablate warts and can be stand-alone therapies or combined with other oral and topical approaches below.

ANTIVIRAL AND IMMUNE-SUPPORT HERBS AND MUSHROOMS

Allium sativum
Astragalus membranaceus
Baptisia tinctoria
Ganoderma lucidum (reishi mushroom)
Glycyrrhiza glabra
Hypericum perforatum
Juglans nigra
Lentinus edodes (shitake mushroom)
Lomatium dissectum
Melissa officinalis
Phytolacca decandra
Podophyllum peltatum (topical use only)
Sanguinaria canadensis
Thuja occidentalis

ESSENTIAL OILS FOR TOPICAL APPLICATION

Lavandula angustifolia
Melaleuca alternafolia
Thymus vulgaris

SOURCES OF PLANT LATEX TO APPLY TOPICALLY

Chelidonium majus
Figs, unripe, sliced
Bananas, latex from a fresh tree or the inner fruit peel
Taraxacum officinale (fresh stems and leaves)

OTHER COMPOUNDS TO APPLY TOPICALLY

Bicholoroacetic acid (BCA)
Salicylic acid (also for use in formulas)

Internal Suppositories and Topical Agents for HPV Infection

Human papillomavirus (HPV) can cause genital warts in both men and woman and is associated with an increased risk of cervical cancer. Naturopathic physicians may employ a procedure known as an escharotic treatment, in which somewhat caustic substances such as *Sanguinaria*, bromelian, and zinc chloride are applied topically. The herbs in this formula are researched to inhibit human papillomavirus and induce apoptosis in infected cells. They are to be used only under a physcian's guidance, because *Podophyllum peltatum* and *Thuja occidentalis* can be irritating, if not caustic, if improperly prepared and utilized. There are only a few types of suppositories on the market that feature *Thuja occidentalis* and vitamin A; when unavailable, such suppositories would have to be made for oneself. *Curcuma longa* suppositories will permanently stain fabric.

Azadirachta indica (neem)
Curcuma longa (turmeric)
Podophyllum peltatum (mayapple)
Thuja occidentalis (cedar)
Vitamin A

Oral Supplements for Support for HPV

Try some of these supplements in rotation or in tandem for a more aggressive approach, as part of an overall treatment plan for HPV.

Vitamin C: 3 to 4 grams/day
Vitamin A: Includes retinol, retinaldehyde, retinoic acid, and provitamin A carotenoids including beta-carotene. Each is dosed differently; consult the product label and manufacturer for the optimal dosage.
Flavinoids, proanthocyanidins, catechins: As much as possible, as per product and label suggestion
B vitamins: Folic acid, 5 to 10 milligrams/day and B_{12}, 1,000 micrograms/day
Indole 4 carbinol/DIM: 400 milligrams/day
Green tea catechins: 1,500 milligrams/day

Herbs for Venereal Warts, Fleshy Growths, and Polyps

These herbs are mentioned in folkloric herbals as herbs that are specific for genital warts and fleshy growths and polyps of the cervix and anus. Both the *condyloma* and *human papillomavirus* can cause such lesions and, although minimally researched, these herbs may have activity against these viruses.

Hydrastis canadensis
Thuja occidentalis
Sanguinaria canadensis

Podophyllin dry resin 30 mg
Tincture of benzoin 30 ml

Mix the two ingredients thoroughly and place in half-dram or 1-dram bottles. Protect the skin surrounding the wart by coating it with *Calendula officinalis* salve. Use a cotton swab to apply the preparation to the wart. Hand warts may be bandaged, but this is not easily accomplished for genital warts. Wash off 4 to 5 hours later. Repeat every 3 to 5 days for larger warts or as needed.

Antiviral Tincture for Warts

Calendula officinalis may improve the skin's protective barrier, while the other herbs in this formula have antiviral and immune-enhancing properties. For those with multiple, extensive, or recurrent warts, a formula such as this would complement the use of a topical wart-ablating therapy, such as Topical Podophyllin Preparation for Warts.

Melissa officinalis 60 ml
Hypericum perforatum 60 ml
Echinacea angustifolia 60 ml
Phytolacca decandra 30 ml
Calendula officinalis 30 ml

Take 1 teaspoon of the combined tincture 3 to 5 times daily for several months. If there is no improvement in 3 months, more aggressive ablative therapies may be necessary.

Basic Protocol for Antiviral Support

This protocol can be helpful for an outbreak of herpes cold sores or any type of virus.

- Limit or restrict simple sugars and carbohydrates.
- Take a daily dose of vitamin C.
- Take a lysine supplement.
- Drink plenty of *Glycyrrhiza glabra* (licorice) and *Melissa officinalis* (lemon balm) tea.

Aspirin and Castor Oil for Warts

Salicylates are gentle caustic agents on the skin and can easily be obtained from aspirin. **Warning:** Do *not* use acetaminophen (Tylenol), ibuprofen (Advil), or other pain relievers.

1 aspirin tablet
1 to 2 teaspoons castor oil

Crush the aspirin in a small mortar and pestle. Combine with the castor oil to create a thick paste. Apply the paste topically to warts each night and cover with a bandage. Repeat nightly for 2 to 4 weeks.

Formulas for Acne

Acne is a well-known and common skin condition that most often afflicts teens and young adults, due to the association with hormonal activation at puberty. Because testosterone can increase sebum production in the skin, adolescent males may be particularly affected, as skin pores and hair follicles become clogged and create an environment where *Propionibacterium* proliferate. Acne can affect both genders, however, and may persist well into adulthood.

There is a strong familial tendency to acne, and most naturopathic physicians and herbalists believe diet and nutrition may also play a role, although some dermatologists deny such association. Alterative herbs are often indicated and folkloric tradition emphasizes such herbs for numerous types of skin lesions and infections. Improving digestion and bowel health typically improves the complexion, and increasing the intake of fresh fruits and vegetables will provide important skin nutrients such as beta-carotene and zinc. Reducing sugar and "junk" food and optimizing hormonal balance will also often improve acne, and such therapies are preferred over antibiotic therapy and tretinoin (Retin A), a prescription usually reserved for severe cystic acne due to its hepatoxicity and teratogenicity. Herbal formulas to improve hormonal balance and reduce bowel toxicity are exemplified below, and along with various topical and skin care routines, they will often improve the complexion in several months' time. Although commercial skin washes containing azelaic acid, benzoyl peroxide, or salicylic acid are readily available to help reduce excessive oil production, many of them are somewhat harsh and can cause inflammation and scaling. Their use may even trigger oil production in response to the drying effects and thus worsen the problem. Use such products with care, such as only every third day, increasing to every other day if tolerated.

Tincture for Acne with a Hormonal Component

Acne may present as obviously connected to hormones, such as premenstrual outbreaks of face blemishes and adolescent acne that begin at puberty. Since all types of acne will benefit from liver and digestive support, *Mahonia aquiafolium* is included here as an alterative agent and has also been shown to deter *Propionibacterium* associated with acne.[43] *Vitex agnus-castus*, commonly thought of as a women's herb, will benefit both young men and women by helping to balance reproductive hormones. *Serenoa repens* may temper testosterone-driven sebum excess that can clog pores and lead to inflammation. *Serenoa repens* has also been shown to reduce sebum production when applied topically.[44]

Vitex agnus-castus 15 ml
Echinacea angustifolia 15 ml
Serenoa repens 15 ml
Mahonia aquifolium 15 ml

Take ½ to 1 teaspoon of the combined tincture 3 to 6 times daily, reducing as skin improves.

Herbal Specifics for Acne

These herbs can be included in tinctures or taken as capsules to best address various presentations of acne.

HERB	CONDITION
Commiphora mukul (guggul)	Dry rough skin, slow metabolism, stagnation
Curcuma longa	Liver burden or toxic state with skin blemishes
Echinacea angustifolia	Pustules; tender, hot, red cysts; boils
Gymnema sylvestre	Acne related to diabetes, poor blood sugar regulation
Vitex agnus-castus	Hormone-related acne

General Protocol for Adolescent or Adult Acne

1. Steam and gently scrub the face each day, but avoid soap and chemical cleansers.
2. Drink ample water or herbal teas and limit coffee, alcohol, and sugary drinks.
3. Consume ample fiber, avoid constipation, and take a probiotic supplement.
4. Consume ample fresh fruits and vegetables to provide beta-carotene and other essential nutrients.
5. Limit fried foods and animal fats, which contribute pro-inflammatory fatty acids to the body.
6. Supplement with essential fatty acids (EFAs) such as flax or fish oil.
7. Beta-carotene, zinc, and B vitamins may be supplemented; this is especially important for adolescents who eat poorly.
8. Exercise to the point of breaking a sweat at least 3 times each week.
9. Expose the face to mild sunlight each day where possible. Consider UV light therapy as an alternative.
10. Adopt stress management techniques where appropriate.

Tincture for Acne with Poor Intestinal Health

When the liver and bowels are unhealthy or inundated with toxins and wastes to metabolize and excrete due to poor diet, the skin may try to take up the slack and act as a secondary organ of elimination. Supporting the liver and optimizing bowel health will benefit all types of skin conditions including acne. These classic alterative and liver herbs are complemented by *Echinacea angustifolia,* which is specific for boils, pimples, and skin that is prone to pustules.

Arctium lappa
Achillea millefolium
Silybum marianum
Echinacea angustifolia

Combine in equal parts and take ½ to 1 teaspoon of the combined tincture 3 to 6 times daily, reducing as skin improves.

Tincture for Rough Greasy Skin with Eruptions

Based on Eclectic literature, these herbs are recommended for those with poor complexions, oily skin with many blackheads, and a tendency to acne, pustules, and skin eruptions. This formula would be complemented by an alterative tea, lipotropic formulas, probiotics, and digestive and dietary support.

Curcuma longa 12 ml
Arctium lappa 12 ml
Gentiana lutea 2 ml
Iris versicolor 2 ml
Juglans nigra 2 ml

Take 1 to 2 dropperfuls of the combined tincture 3 or 4 times a day, reducing as the complexion improves. Double or triple this recipe once it is known to be tolerated, as at this dosage, that 1-ounce bottle will not last very long.

Topical Wash for Acne

Hibiscus sabdariffa has wound-healing effects[45] and is combined here with *Quercus alba* for its astringent effect. *Camellia sinensis* (green tea) polyphenols are anti-inflammatory when topically applied, and *Calendula officinalis* may improve lymphatic circulation in the skin.

Topical Herbs for Poison Oak/Ivy Dermatitis

These herbs are specified in herbal folklore as being remedies to speed the resolution of skin eruptions due to poison ok and poison ivy.

Actaea racemosa
Alnus rubra
Geranium maculatum
Grindelia squarrosa
Lavandula angustifolia essential oil
Lobelia inflata
Sanguinaria canadensis

Hibiscus sabdariffa dry flowers
Quercus alba dry shredded bark
Camellia sinensis dried leaves
Calendula officinalis dried flowers
Melaleuca alternifolia essential oil
Mentha peperita essential oil

Combine the first four herbs in equal parts (1 ounce of each is a good amount to start with). Steep 1 tablespoon of the dry herb mixture in 1 cup of freshly boiled water for 15 minutes and then strain. Place ¼ cup of the strained infusion in a small cup or dish and add 5 drops each of tea tree and mint essential oils. Use promptly with cotton balls or other gauze pad to apply to the freshly washed face. Repeat 3 times per day.

Hepatoprotective Tincture for Patients Taking Tretinoin Medications

Those with severe, scarring, and disfiguring acne are sometimes treated with tretinoin-based oral medications. However, tretinoin is a known hepatotoxic medicine often discouraged by naturopathic physicians and herbalists. This formula may be helpful when patients are using such pharmaceuticals. It contains the liver-protective herbs *Silybum marianum*, *Curcuma longa*, and *Mahonia aquifolium*, all having anti-acne and alterative effects. *Echinacea angustifolia* is specific for pustules and boils and complements the liver-protective herbs.

Silybum marianum
Curcuma longa
Mahonia aquifolium
Echinacea angustifolia

Combine in equal parts and take ½ to 1 teaspoon of the combined tincture 3 to 6 times daily, continuing for the duration of prescription drug use.

Hibiscus Tea for Wound Healing

Hibiscus species are not commonly mentioned for skin applications, but the tea accelerates re-epithelialization in wounds when topically applied.[46] One study showed positive effects in promoting healing in chronic anal fissures.[47] *Hibiscus sabdariffa* tea can be consumed internally as well as used as an ingredient in various skin washes for wounds ranging from simple blemishes to chronic lesions to traumatic injuries.

Hibiscus sabdariffa, hibiscus

Tincture for Low-Grade Chronic Skin Eruptions

This formula combines herbs specifically mentioned in the folkloric literature for chronic skin eruptions and skin blemishes. It would work best in conjunction with a skin care routine, clean diet, and an alterative tea.

Arctium lappa 30 ml
Juglans nigra 15 ml
Coleus forskohlii 5 ml
Iris veriscolor 5 ml
Phytolacca decandra 5 ml

Take ½ to 1 teaspoon of the combined tincture 3 to 5 times a day, reducing as the complexion improves.

Formulas for Acne Rosacea

Acne rosacea is a chronic disease of the microvasculature of the face and pilosebaceous units and is characterized by dilation of superficial capillaries over the cheeks and nose and in some cases by papules, pustules, scaling, and inflammation as well as red, blotchy skin. The condition is most common in people of northern European descent; it has sometimes been referred to as "the curse of the Celts" and may affect 10 percent of those with Celtic heritage. Rosacea is more common in women, with initial onset between the ages of 30 and 50. It is notoriously difficult to cure, and while there are a variety of different triggers, they can often be unnoticed by the patient. Because rosacea can involve psychological, infectious, immune, and vascular triggers,[48] formulas might include nervines, immune-modulators, antimicrobials, anti-inflammatories, and vascular stabilizers. In some cases, the eyes may be affected; this is referred to as ocular rosacea.

Acne vulgaris (common acne) and acne rosacea are very distinct conditions with disparate therapies. Pharmaceutical approaches for acne rosacea include isotretinoin creams (such as Accutane) and various antibiotics, but these are ineffective in some cases. Some cases of acne rosacea are associated with low stomach acid, digestive difficulty, and small intestinal bacterial overgrowth (SIBO), and antibiotics may be most effective in these populations. Some sufferers find that any agent associated with vasodilation, such as strong sun, sweating, coffee, and hot spices, aggravates dilated capillaries. Some association between Demodex skin mites and rosacea has been identified, but eradicating the mites does not necessarily resolve the condition. The mites may trigger an autoinflammatory reaction, and rosacea is sometimes classified as an autoimmune disease for this reason. Use of some medications, especially steroids, may cause temporary rosacea, and general topical treatments for acne vulgaris may worsen the inflammation and aggravate rosacea.

Pharmaceutical approaches to treating acne rosacea include topical ivermectin cream (10 milligrams/gram) to help kill Demodex mites, topical metronidazole gel, topical tretinoin, oral tetracyclines such as doxycycline and minocycline, and oral azithromycin.[49]

Tea for Acne Rosacea

Rosacea has been associated with an increased risk of cardiovascular disease,[50] so therapy might also emphasize herbs with an affinity to blood vessels and the ability to reduce vascular inflammation, such as those

Daily Sunscreen

The underlying mechanisms that cause rosacea involve abnormal cytokines, hormones, and neuropeptides. Sunlight may contribute to cutaneous inflammation and disrupt normal immune modulation in rosacea patients[51] and thus, faithful use of sunscreen on a daily basis is recommended for all rosacea patients.

Avoidance of Flushing in Acne Rosacea

A prolonged and excessive flush response is seen in rosacea, and abnormal substance P release triggers mast cell degranulation and contributes to lower pain and heat tolerance.

Flushing may be better prevented than the results treated. Help educate patients to use mechanical means—such as sun hats, fans, topical compresses, and water spritzers—to avoid facial flushing. Those with rosacea are also advised to take care outdoors, when exercising heavily, or when moving from a colder environment to a warmer one. Alcohol, caffeine, and hot culinary spices may also cause flushing that promotes an inflammatory response in the skin. Aggressive topical skin care routines—such as strong essential oil applications, chemical peels, microdermabrasion, or even vigorous scrubbing—can worsen skin inflammation and should be avoided.[52]

containing polyphenol flavonoids. Hibiscus, berries, and hawthorne may be logical in the diet, and tastier flavonoid-containing herbs are appropriate for teas, such as the following example.

Crataegus oxyacantha berry powder
Centella asiatica
Calendula officinalis
Astragalus membranaceus
Hibiscus sabdariffa
Glycyrrhiza glabra

Combine in equal parts, and gently simmer 1 tablespoon of the herb mixture per cup of hot water for just 1 to 2 minutes. Let steep in a covered pot for 15 minutes and then strain. Drink 3 or more cups per day, long term. As you become familiar with the flavor of this tea, you can adjust the quantities of the ingredients to taste.

Preventive Tincture for Acne Rosacea

The skin of rosacea patients is often highly sensitive, and the emergence of skin lesions may be preceded by a decade of frequent flushing and blushing. Herbs that protect the blood vessels and reduce congestion can be one component of a comprehensive treatment for rosacea. All the herbs in this formula are anti-inflammatory with an affinity for blood cells and blood vessels and were chosen based on the author's experience, as little folklore or research exists.

Crataegus oxyacantha
Vaccinium myrtillus
Angelica sinensis
Hypericum perforatum

Combine in equal parts and take ½ to 1 teaspoon of the combined tincture 3 to 6 times daily, reducing as skin improves.

Seaweed Poultice for Rosacea and Skin Inflammation

Sulfated polysaccharides from various algae species are unique natural products that have an affinity for mammalian glycosaminoglycans and are uncommon in

Azelaic Acid Face Wash and Topical Treatments for Acne Rosacea

Astringent herbal washes may reduce inflammation and tighten dilated capillaries. The topical application of 15 percent azelaic acid preparations exerts both anti-inflammatory and antimicrobial effects and improves acne rosacea, reduces inflammatory papules and pustules, and lessens the severity of the erythema.[53] The topical azelaic acid medications are available as creams, gels, and foams, and clinical studies have shown that topical azelaic acid is as effective or more so than metronidazole in the treatment of acne rosacea.[54] Azelaic acid is produced by *Malassesia furfur* (formerly *Pityrosporum ovale*), a yeast strain that is typically found on healthy skin, and can also be commercially produced from oleic acid found in wheat bran and wheat germ oil, barley, and rye. The fact that beneficial fungal strains are found in healthy skin further supports the idea that long-term bowel health support and gentle skin care routines may be superior to oral antimicrobials and harsh skin disinfectant washes for many skin conditions, including rosacea. Azelaic acid reduces excessive keratin in the skin and thereby discourages acne bacteria that thrive in the high keratin environment. Azelaic acid can also reduce melanin production in hyperpigmentation disorders (See also "Formulas for Pigmentation Disorders" on page 222.) As azelaic acid can be irritating if too strong or too frequently applied, it is recommended to take a break of a day or two should any irritation occur, or dilute the products with *Aloe* gel. In the same way, while *Malassesia furfur's* natural production of azelaic acid is helpful to deter excessive keratin buildup, overgrowth of the fungus is associated with other skin conditions including seborrhea, dandruff, folliculitis, and tinea versicolor.

Sources of Sulfated Polysaccharides

Red algae (Rhodophyceae). Red algae are the largest phyla of algae and include *Chondrus crispus*, *Gigartina*, *Eucheuma cottonii*, and *E. spinosum*. Red algae contain the sulfated polysaccharide carrageenan. *Chondrus crispus* is also known as Irish moss and is used to prepare mucilaginous gels to soothe throat or skin pain or to extract carrageenan to be used as a gelling agent in the food industry.

Brown algae (Phycophyta). *Cladosiphon okamuranus* and *Fucus vesiculosus* are brown algae and sources of non-sulfated alginic acid and laminaran polysaccharides, and sulfated polysaccharides referred to as fucoidans, widely studied for immune modulating effects. Brown algae are often marketed commercially by their common names, including kombu, bladderwrack, wakame, and hijiki.

Green algae (Cyanobacteria). Green algae also contain sulfated polysaccharides such as ulvan, and include *Ulva* and *Codium*. Many green algae are edible and are sold under the name "sea lettuce."[55]

Skin Care Options for Rosacea

Patients with rosacea and seborrhea should avoid washing with soap because the alkaline pH of soap can be damaging to skin barriers. Acidic washes or neutral foaming cleansers are more appropriate. Sulfur-based soaps may be helpful to some patients. Creams containing glycerin can provide valuable moisture, and ceramide ointments can help replace the lipids found naturally in the skin.

Rosacea with Rhinophyma

While women are more often affected by acne rosacea than men, men who are affected are more prone to rhinophyma, an enlargement and thickening of the skin of the nose. The onset is often between the ages of 50 to 70 and can be associated with alcoholism and other causes of telangiectasia and fibrosis besides rosacea. Rhinophyma does not respond to topical or oral medications and can only be corrected via surgical procedures, removing the tissue buildup and various types of laser therapies. Treating the underlying condition, however, may stall the severity and progression of the condition.

higher plants.[56] When topically applied, seaweeds, or isolated sulfated polysaccharides, may reduce inflammation in the skin via positive immune modulation.[57] For this formula, it is logical to use the least expensive and most readily available species that contains the desired sulfated polysaccharides, such as *Chondrus crispus* (Irish moss). Those with iodine allergies should be cautious using seaweed topically.

Chondrus crispus
Matricaria chamomilla

Prepare 1 cup of *Matricaria chamomilla* tea by steeping 1 tablespoon in a cup of boiling water. Strain and place the liquid in a small bowl. Soak 1 cup of *Chondrus crispus* in the tea, stirring occasionally to help it become gummy. Apply the moss to the face and leave in place for 15 to 30 minutes. Rinse off with cool water.

Topical Poultice for Acne Rosacea

Because rosacea involves disrupted epithelial barrier function, excessive water can be lost from the stratum corneum and contribute to inflammation. Therapies, both topical and systemic, that help the skin retain moisture are indicated. Avoid any creams or lotions that contain perfumes or irritants. Those containing *Aloe vera* gel, ceramides, algae, and simple oils, while not curative, may reduce the severity and progression of the disease. *Arctostaphylos uva ursi* is most known to herbalists for treating bladder infections, but the plant contains phenolic compounds, which may help reduce hyperpigmentation when applied topically. While purified hydroquinone, a breakdown product of uva ursi's arbutin, can be too irritating to apply to the face for patients with rosacea, whole uva ursi powder is well tolerated. Uva ursi may help reduce hyperpigmentation while glycerin helps the skin retain moisture.

Arctostaphylos uva ursi powder
Aloe gel
Glycerin

Prepare a paste by blending roughly equal parts of uva ursi powder, aloe, and glycerin, and apply to the clean face. Leave in place for 15 to 30 minutes, and rinse off with warm water. Use roughly every other day.

Eyelid Wash for Rosacea with Ocular Involvement

Ocular rosacea may require that eyelid washing be included in daily hygiene. The pharmaceutical cyclosporine is sometimes prescribed topically to improve immune modulation. Use of warm herbal compresses, eyelid massage, and gentle eyelid scrubs may also help. *Glycyrrhiza* (licorice) has been shown to improve bacterial-induced blepharitis and may be helpful.[58] Research is lacking, but *Foeniculum* (fennel) seeds and rosewater are traditional remedies to include in eyewashes for eyelid inflammation.

Calendula officinalis petals 2 ounces (60 g)
Matricaria chamomilla flowers 2 ounces (60 g)
Foeniculum vulgare seed powder 2 ounces (60 g)
Glycyrrhiza glabra root powder 2 ounces (60 g)
Rosewater

Combine the dry herbs to yield ½ pound of dry herbs. Steep 1 heaping tablespoon of the herb mixture in 1 cup boiling water. Let stand until cool and then strain several times to remove all particulate possible. Combine roughly 2 tablespoons of the strained liquid with 1 tablespoon rosewater. Use this liquid as eye drops and to gently scrub the eyelids. Repeat 3 times throughout the day, or as often as possible. Make fresh each day, and continue for at least 3 months.

Demodex Mites and Rosacea

The presence of Demodex mites on the face can sometimes promote rosacea. The mites trigger inflammatory reactions, and around 80 percent of rosacea sufferers can be found to harbor the mites. Various pesticide-like chemicals, such as permethrin, have been found to deter these mites, but 1 part camphor essential oil diluted with 2 parts glycerin was shown in one study to benefit rosacea patients proven to have Demodex mites present based on biopsy.[59] Clove and tea tree essential oils are also reported to have significant effects against Demodex mites.[60] Again, supporting the overall beneficial microbes in the body may prove better in the long run than relying on any skin disinfectant, herbal or not. Furthermore, while permethrins are generally regarded as safe to humans when used topically and are sometimes tempting options for treating lice or scabies, they are very harmful to bees and to fish. Since less environmentally toxic options are available to deter Demodex mites, lice, and scabies, many herbalists prefer essential oils as safer options for the planet.

Tincture for Rosacea with Digestive Insufficiency

Some people with rosacea have been shown to have hypochlorhydria and poor digestion. This formula uses the vascular protectants *Crataegus* and *Vaccinium* combined with digestion-enhancing herbs. *Rumex* can promote hydrochloric acid if hypochlorhydria is suspected.

Crataegus spp. berries
Vaccinium myrtillus
Rumex spp.
Silybum marianum

Combine in equal parts and take ½ to 1 teaspoon of the combined tinture 3 to 6 times daily, reducing as skin improves.

Tincture for Rosacea with Papules, Pustules, and Fluid Stagnation

Some cases of rosacea involve primarily congested blood vessels and "broken" capillaries (telangiectasias); other cases are acneiform, displaying pustules and papules as well. This formula combines the vascular protectants *Crataegus* and *Vaccinium* with herbs specific for pustules and cysts, *Echinacea* and *Phytolacca*. Individual herbs could be separated out of this formula as appropriate for different individuals.

Crataegus spp.
Vaccinium myrtillus
Echinacea purpurea
Phytolacca decandra

Combine in equal parts and take ½ to 1 teaspoon of the combined tincture 3 to 6 times daily, reducing as skin improves.

Therapies for Telangiectasias

Telangiectasias, also known as spider veins, are small dilated blood vessels, especially common on the face. They may simply be due to normal aging in those genetically predisposed or may occur as part of a broader disease entity affecting the vasculature. For example, CREST syndrome, a condition related to scleroderma, involves telangiectasias as part of the symptom complex. Rosacea is a condition that also involves dilated capillaries and venules in the face. When telangiectasias occur on the lower limbs, such as around the ankles, they are often associated with vascular congestion and varicosities. Liver and circulatory congestion may induce telangiectasias, and alcoholism may predispose as the liver burden and portal congestion backs up blood flow through the entire body. For the same reason, constipation and varicose veins can predispose to telangiectasias of the ankles.

Telangiectasias may be irradiated or injected with sclerosing agents that cause the capillary or venule to collapse as a cosmetic therapy, referred to as sclerotherapy. Laser therapy may be used on the facial telangiectasias. Although sclerotherapy is fairly simple and can be done as an outpatient procedure, it does involve injecting fairly toxic sclerosant medications into the veins and carries the risk of escaping the local area and traveling to the heart, lungs, and even brain and inducing ischemic episodes or stroke.

Natural therapies are aimed at preventing new telangiectasias from forming, but will not resolve existing telangiectasias. Because bioflavonoids are important nutrients for strengthening vein and artery walls, berries such as blueberry and bilberry to provide proanthocyanidin bioflavonoids may be part of a long-term approach to improving blood vessel integrity. Liver or bowel congestion can cause mild portal hypertension, which in turn may increase vascular congestion, especially in the lower limbs. Therefore, digestive support, liver and alterative herbs, and avoidance of liver toxins and alcohol may also be considered therapies where appropriate. Sedentary habits can also worsen venous reflux in those with weak or failed valves in the veins, such that getting more exercise and especially avoiding prolonged sitting can help those with spider veins and varicosities of the legs.

Herbs for Telangiectasias

To help those with telangiectasia, address vascular congestion, support the strength and integrity of blood vessels walls with colorful flavonoid foods and herbs, and support liver function as individually appropriate. Consider teas, tinctures, encapsulation, and medicinal food including any or all of the following:

Calendula officinalis
Crataegus oxyacantha
Curcuma longa
Hypericum perforatum
Punica granatum
Silybum marianum
Vaccinium myrtillus

Formulas for Psoriasis

Psoriasis is a chronic skin condition characterized by skin plaques where skin cells regenerate more quickly than normal, creating patches of reddish lesions covered with fine white scales. The lesions bleed readily if the scales are removed. Psoriasis plaques occur almost anywhere, particularly the torso, scalp, elbows, knees, hands, and feet. Psoriatic lesions may be painful and itchy or may be painless, and the affected skin can be friable, easily injured, and prone to cracking and bleeding. Pitting or crumbling nails are also typical in those with psoriasis, and at times the nails may appear as if dotted with small oil stains.

Psoriasis is classified into various types, including pustular, guttate, and erthyrodermic presentations. At times, psoriasis can be severe; large sheets of epidermis are shed, and the condition requires expert care and possibly hospitalization. Those with psoriasis may also have an increased risk of other chronic inflammatory disorders and metabolic syndromes including obesity, diabetes, renal disease, eye diseases, hypertension and cardiovascular disease, and other autoimmune disorders. Around 10 percent or more of those affected by psoriasis also have a particular type of arthritis known as psoriatic arthritis.

Psoriasis may be classified as an autoimmune disorder and may involve a genetic immune dysfunction. The condition will often wax and wane in relationship to stress, diet, or unidentified triggers and is often highly resistant to treatment. Some pharmaceutical drugs such as ibuprofen, beta blockers, lithium, iodides, or hydroxychloroquines (an antimalarial drug) may aggravate psoriasis, as can skin trauma, injuries, and sunburn. Smoking and viral infections may also aggravate psoriasis.

While no therapy is completely curative, pharmaceutical cortisone and similar steroid creams are prescribed to reduce acute flare-ups and reduce discomfort. Salicylate creams may reduce scaling, but can be drying to the skin and cause hair loss if used on the scalp. Methotrexate and cyclosporine are powerful immune suppressants that help to halt accelerated cell division, and they are often quite effective. However, they have many side effects and require constant monitoring for potential toxicity, so they are usually reserved for the most severe cases. Newer immune-modulating drugs, such as adalimumab (Humira), can also be effective, but the high cost may be prohibitive. Some of the most important nutrients for skin health, including vitamins A and D, may reduce lesions and flare-ups. Synthetic forms of both nutrients, such as calcipotriene and tretinoin, can be prescribed, but herbalists and naturopathic physicians prefer the natural nutrients when possible due to side effects and toxicity issues with the synthetics. Pine tar ointments are traditional and used to slow the rapid cell turnover rates, often in tandem with controlled exposure to ultraviolet light. Pine tar soaps and shampoos are available for this purpose. Psoralen is used to photosensitize the skin to ultraviolet A (UVA) to improve the efficacy of light therapy. This is sometimes referred to as PUVA (psoralen and ultraviolet A) therapy. The following formulas detail how to use photosensitizing herbs for this purpose.

Herbal medicines may also improve immune modulation and help treat psoriasis, and general liver and digestive support may reduce any toxic load in the body that may trigger autoinflammatory reactions in the skin. Epsom salt baths, oatmeal baths, and *Aloe vera* applications can also be comforting to itching, burning, or skin discomfort.

Nutritional Protocol for Psoriasis

These nutrients are very valuable in treating a variety of skin conditions.

Vitamin D_3: 1,000 milligrams or more per day
Vitamin A: 500 IUs/day. Mixed carotenoids are safer but the actual vitamin will often yield a more rapid result when treating skin conditions. Mixed carotenoids may be used to maintain the effects.
Zinc: Because zinc is available in many different forms, the dosage may vary. In general, aim for 30 milligrams of elemental zinc per day, which may require 90, 100, or more milligrams of a zinc chelate such as zinc gluconate. Take zinc with food as it is commonly nauseating on an empty stomach.
Selenium: 200 to 400 micrograms/day

Supportive Tincture for Psoriasis

Psoriasis involves greatly accelerated cell turnover in the dermis. *Coleus* may affect cell function and reduce this excessive epithelial proliferation.[61] *Petroselinum* is included here for possible photosensitizing effect that is known to promote healing in psoriasis—this is also discussed in the formula Photosensitizing Therapy for Psoriasis.[62] *Curcuma* has been noted to promote healing of psoriatic lesions, possibly due to positive effects on connective tissue. *Tanacetum* is anti-inflammatory and may help reduce the severity of the eruptions.

Coleus forskohlii
Tanacetum parthenium
Petroselinum crispum
Curcuma longa

Combine in equal parts and take ½ to 1 teaspoon of the combined tincture 3 to 6 times daily, reducing as skin improves.

Photosensitizing Therapy for Psoriasis

Lomatium, *Petroselinum*, and *Angelica* are all in the Apiaceae family and known to contain photosensitizing furanocoumarin compounds called psoralens[63] Phototherapy was a mainstay of psoriasis therapy prior to the advent of steroids and pharmaceutical drugs. The amount of the photosensitizing compounds in these tinctures is quite small, and thus frequent, large dosages over an extended period of time are necessary to exert a photosensitizing effect. By that same token, this formula is unlikely to be too harsh or induce sunburn. Small dosages of the potentially toxic *Veratrum* can interfere with epithelial division[64] and is included as a synergist in the formula. Veratrum is contraindicated in pregnancy and should be removed from the formula if any digestive upset occurs. Start slowly and ramp up the dose to prevent any unexpected extreme sun reactivity.

Lomatium dissectum 10 ml
Petroselinum crispum 10 ml
Angelica sinensis 10 ml
Veratrum viride 100 drops

Take 1 teaspoon of the combined tincture per day. In addition, strive for 1 to 2 hours' exposure to midday sunlight. Use artificial light such as a sunlamp or tanning booth if real sunshine is impossible. Repeat daily for 2 weeks, increasing the dosage of tincture gradually to 3 to 5 teaspoons taken through the early morning. If effective, this therapy may be continued indefinitely, such as employing for 30 days, every other month, long term.

Apiaceae Juice for Photosensitization

The amount of photosensitizing psoralens in the juice of Apiaceae family plants is quite small, and drinking juice is unlikely to be a powerful therapy all on its own. However, when combined with an oral tincture, a vitamin D supplement, and a topical application, this juice may be a useful adjuvant therapy for psoriasis.

3 or 4 fresh celery stalks
1 or 2 parsnip roots
1 large bunch fresh parsley

Use a juicer to turn the ingredients into a batch of fresh juice. Make the juice daily for 2 weeks and drink it each morning. Also aim to get several hours' worth of sunshine at midday, or artificial light such as in a tanning booth if real sunlight is impossible to access. Evaluate results and repeat if encouraging.

Photosensitizing Topical Application

Caution: Unlike the psoralens consumed orally in foods and foodlike preparations, the topical use of photosensitizing agents can cause significant light sensitivity and result in severe burns. Use this therapy only under the guidance of an experienced clinician. *Hypericum* is one of the most potent photosensitizing agents known,[65] and *Citrus bergamia* (bergamot) is another strong photosensitizer. In this formula, these herbs are combined with parsley and carrot seed oils to be used *cautiously* to induce slight sunburn, which can often promote healing of psoriatic lesions.

Hypericum oil 30 ml
Carrot seed essential oil 60 drops
Citrus bergamia essential oil 20 drops
Petroselinum essential oil 20 drops

Combine the oils in a 1-ounce bottle. Use sparingly, applying just 2 or 3 drops on each affected area, increasing as tolerated. Apply to lesions and expose to sunlight or artificial sunlight. Increase amount applied and duration in the sun as tolerated, but being careful to avoid sunburn. Stop at once if any irritation or inflammation occurs. If it does, wait 1 week and try again at half the irritating dose, or less than half. If no sun sensitivity is induced with 2 or 3 drops, increase the dosage and duration of sun exposure gradually and cautiously.

Psoriasis Ointment Based on TCM

This formula is based on Traditional Chinese Medicine (TCM) applications for topical use.[66] The use of topical *Curcuma* is noted to improve psoriasis,[67] and *Curcuma* may be used as a simple as well as a skin wash, an ointment, or a poultice where available. Temporary yellow staining will result in those with light complexions and in some other cases.

Phellodendron amurense bark powder 2 ounces (60 g)
Scutellaria baicalensis root powder 2 ounces (60 g)
Curcuma longa rhizome powder 2 ounces (60 g)
Rheum palmatum root powder 2 ounces (60 g)
Glycyrrhiza glabra root powder 2 ounces (60 g)
Sulfur powder 2 ounces (60 g)
Olive oil

Combine the powders in a large canning jar and add enough olive oil to cover. Blend thoroughly and shake each day, ensuring that the powders are fully submersed in the oil. After 6 weeks, press out the oil and use topically.

Decoction for Psoriasis Based on TCM

This formula is based on traditional remedies used in China to treat psoriasis,[68] typically in association with phototherapy.

Glycyrrhiza glabra
Rehmannia glutinosa
Angelica sinensis
Smilax ornata and other species
Salvia miltiorrhiza

Combine equal parts of the dry herbs. Decoct 1 teaspoon of the herb mixture per cup of water, simmering gently for 10 minutes. Allow to stand 10 minutes more, and then strain. Drink 3 or more cups per day.

Formulas for Lichen Planus

Lichen planus is a poorly understood autoinflammatory condition of the skin and mucous membranes that may be triggered by viral infections and allergic hypersensitivity reactions. Possible allergens that may trigger lichen planus include heavy metal exposure, diuretics, iodides, antibiotics, and dyes. While the lesions may be uncomfortable, the condition is not typically serious but can cause discomfort and self-consciousness; the condition is associated with a slight increase in skin cancer risk. Roundish pink, red, to purple flat-surfaced plaque may occur on the skin and scalp. White, scarlike lesions may occur in the mouth and on the tongue, and the fingernails may be deformed. The condition is usually self-limiting and may resolve in a few months to a year and a half. In rarer cases, lichen planus may affect the vulva and the oral cavity and cause a burning sensation or pain with swallowing. Immune suppressants are the primary pharmaceutical therapy for severe or painful cases. As the cause is elusive, herbal treatments are aimed at general liver support to help clear drugs and toxins; general systemic anti-inflammatory herbs can also be helpful. Treat any underlying allergies or bowel issues directly where possible and find alternatives for all pharmaceutical drugs possible. Herbal remedies may speed resolution and help to allay pain, and antiviral herbs or mast cell–stabilizing herbs may be indicated when viruses such as hepatitis or allergens are suspected to have triggered the condition. Vitamin A supplements may be beneficial. As with psoriasis, ultraviolet A therapy may speed resolution of the lesions.

Formula for Lichen Planus Skin Lesions

This general formula uses herbs that are both liver-supporting and anti-inflammatory. This recipe may be prepared as a tincture or a tea.

Curcuma longa
Calendula officinalis
Astragalus membranaceus
Glycyrrhiza glabra

Combine in equal parts and take 1 teaspoon of the combined tincture 3 to 6 times a day. Start with 3 times a day and double if no improvements are seen in a month.

As an alternative, take the *Curcuma* as a tincture, and prepare the remaining three herbs as a tea by gently simmering 1 teaspoon per cup of water, and aiming to drink 3 or more cups a day.

Mouthwash for Oral Lichen Planus

About half of those afflicted with lichen planus have oral lesions, and at times the oral lesions may be painful. This formula uses the anti-inflammatory and mucous membrane healing herb *Glycyrrhiza* as a base. *Commiphora myrrha* is specific for infections, inflammations, and lesions of the oral mucosa. *Sanguinaria* is a potentially caustic herb and must be diluted as exemplified in this formula. It is also specific for oral lesions and burning sensations in the mouth.

Glycyrrhiza glabra 15 ml
Commiphora myrrha 10 ml
Sanguinaria canadensis 5 ml

Place 1 teaspoon of the combined tincture in a small sip of water and swish the liquid around the mouth as long as possible, then spit it out. Repeat hourly to every 2 to 3 hours as possible.

Aloe and Licorice for Oral Ulcers

Glycyrrhiza glabra is noted to heal digestive and aphthous ulcers, thus it makes sense to try for lichen planus, canker sores, herpangina, and other ulcerative lesions of the mouth and gums. Licorice is available as a thick, sweet solid extract so it can be both pleasant-tasting and sticky enough to adhere to the lesions for a longer period than a tea or tincture. *Aloe vera* gel may improve oral lesions[69] and can be made reasonably palatable by combining with licorice.

Licorice solid extract 1 ounce (30 g)
Aloe vera gel 1 ounce (30 g)

Combine the licorice solid extract and *Aloe* gel in a small wide-mouthed container, such as a salve vial. Homogenize by shaking or stirring vigorously. Rub ½ teaspoon of the blend on individual lesions or simply place it in the mouth and allow to dissolve in the saliva.

Curcuma Capsules for Lichen Planus

Curcuma has been shown to improve lichen planus, and it can be part of a long-term maintenance protocol.[70]

Curcuma capsules

Take 2 capsules, 2 times per day, with a meal or a fat to enhance absorption. *Curcuma* powder can be blended into nut milk or a smoothie to provide the needed fat, or the capsules can be simply chased with a spoonful of cod liver oil or any other dietary fat, such as a salad with olive oil dressing on it.

Herbal Treatment for Pityriasis

Pityriasis rosea is a benign skin condition of unknown etiology that affects mainly adolescents and young adults. The condition usually begins with a rough, pink, single ovoid "herald patch" on the torso, followed by the eruption of numerous smaller patches. The lesions are typically arranged in a somewhat linear fashion referred to as a Christmas tree pattern. The lesions are not painful but are occasionally pruritic. There are few or no other accompanying symptoms, but occasionally headache, sore throat, and slight fever hint at a viral underlying cause, although no organism has yet been identified. The condition usually resolves on its own over a few weeks to a month. Topical therapies such as *Calendula* salve may help the lesions to resolve more quickly. The only therapies commonly reported to speed resolution are superfatted soaps and oatmeal baths.

Formulas for Alopecia and Hirsutism

Male pattern baldness is referred to medically as androgenetic alopecia (AGA), defined as a hereditary, androgen-dependent, progressive thinning of the scalp hair that follows a defined pattern. The accumulation of 5α-dihydrotestosterone (DHT) in dermal papilla cells is implicated in androgenetic alopecia. DHT decreases cell growth by inducing cell death and increasing the production of reactive oxygen species.[71] Hair loss can involve changes to hair fiber caliber, density per unit area, and/or the duration of anagen and telogen in the hair growth cycle. Scalp dermoscopy is used routinely in patients with androgenetic alopecia, as it facilitates the diagnosis and differential diagnosis with other diseases, allows staging of severity, and helps monitor the progress of the disease and response to treatment.[72] Because the condition progresses over time, the most success in halting hair loss is seen with early intervention.[73] Leading pharmaceutical therapies for AGA include oral finasteride (especially for men) and topical minoxidil (for men and women).[74] The male pattern of hair loss typically involves recession at the temples and vertex baldness, while the female pattern is often diffuse hair loss and miniaturized hairs and hair follicles over the midfrontal scalp. As with men, the most common cause of alopecia in women is androgenetic, and the condition affects approximately one-third of adult Caucasian women.[75] Male pattern baldness may be associated with an increased risk of prostate cancer, as both may involve increased sensitivity to testosterone, resulting oxidative stress in the tissues.[76] However, one study following more than 35,000 men did not find a correlation between baldness and prostate cancer.[77]

In addition to a genetic predisposition to AGA, other factors that may exacerbate the process include poor nutrition, poor circulatory health, and stress. Acute stress causes hair loss in animals and humans. One investigation of serum cortisol level and glucocorticoid receptor expression in patients with severe alopecia areata, a condition where hair follicles are destroyed, showed a lower expression of glucocorticoid receptors compared to controls,[78] and researchers believe this contributes to pathological changes in the scalp and contributes to alopecia. Therefore, adaptogenic and nervine herbs are logical to include in formulas for patients with hair loss where stress and adrenal activation is suspected to play a role, although research to support this is lacking.

Other types of alopecia occur with skin diseases, many mentioned throughout the chapter including tinea capitis (fungal infections of the scalp), severe folliculitis or cellulitis of the scalp, and nervous pulling of the hair, referred to as trichotillomania. Sudden changes in hormones can cause hair loss, and some women lose hair postpartum. High fevers and severe stress may cause episodes of hair loss, while inadequate protein intake and poor nutritional status may halt hair growth. Chemotherapy often causes hair loss as the drugs interfere with cell division and shut down the rapid cell turnover in hair follicles. Autoimmune diseases may cause alopecia areata. Harsh shampoos, hair dyes, perming, bleaching, and daily heat or curling can cause the hair to become brittle and fragile. Hypothyroidism may be associated with hair loss or coarse dry hair, and anemia and iron deficiency are associated with hair loss.

Hirsuitism is a condition of excessive hair growth, and while some men may sport something akin to a pelt, even this is considered normal, and the term is more commonly used to apply to women with facial hair, chest hair, or coarse, dark, and excessive body hair. As discussed above, such women may often be found to have elevated testosterone, and hirsuitism is one symptom of the endocrine imbalances typical of polycystic ovarian syndrome (PCOS). Hirsuitism may also be familial due to inheritable genetic traits. There are no effective herbal treatments to reverse excessive hair growth, but herbal therapies may correct hormonal imbalances that underlie the complaint. The primary therapies to remove unwanted facial or body hair are electrolysis and laser therapies.

The Androgen Paradox

Androgens stimulate beard and body hair growth yet suppress head hair growth. Androgens are associated with receding hairlines and baldness in androgenetic alopecia, and in women, androgens will promote hirsutism while causing thinning of the hair on the scalp. Those with androgenetic alopecia may have increased expression of androgen receptors, making them more susceptible to body hair growth, concomitant with loss of hair from the scalp.

Topical Antifungal Oil for Scalp Fungus

Scalp fungus can cause a spotty loss of scalp hair and can be definitively diagnosed with a biopsy to differentiate the condition from autoimmune-associated alopecia areata. Treat any underlying maladies such as diabetes or immune deficiency that predispose to fungal infections, and try a topical antifungal application such as the following. Almost all essential oils are potent antimicrobial agents, and *Azadirachta indica* (neem) has activity against fungus, lice, and other scalp issues. Neem oil is available commercially to prepare this formula.

Azadirachta indica 15 ml
Melaleuca alternifolia essential oil 5 ml
Origanum vulgare essential oil 5 ml
Mentha spicata essential oil 5 ml

Fill a 1-ounce bottle with the *Azadirachta* (neem carrier oil) and the three essential oils. After washing hair, rub approximately 1 tablespoon of the combined oil into the scalp, cover the scalp with plastic (such as a plastic bag) to keep the fumes in, and wrap the scalp with a towel twisted into a turban. Wait 15 to 30 minutes, remove the towel and bag, and then wash hair again. It will require lathering up several times to remove all the oil. Repeat daily if the scalp is not inflamed or irritated by the essential oils. Or repeat every other to every third day, as necessary to avoid irritation.

Simple for Female Hirsutism

Hirsutism in women may be genetic, but otherwise it is almost always related to elevated androgens or abnormal metabolism of peripheral androgens, as is the case with PCOS. *Serenoa* inhibits the conversion of testosterone into the more active dihydrotestosterone[79] and may thereby reduce some of the hormonal stimulation of the hair follicles. This will especially help to deter further hair follicles from beginning to produce dark coarse hairs, but reversal of facial or chest hair growth is more arduous. Not too much should be expected of *Serenoa* on its own, because hirsutism requires treating the underlying hormonal imbalance, as some of the following formulas exemplify.

Serenoa repens

Take 2 capsules, 3 times per day, long term.

Tincture for Female Hirsutism Associated with PCOS

Polycystic ovarian syndrome (PCOS) may affect 10 percent or more of women and elevated androgens is a hallmark of the syndrome; hirsutism is a common accompanying symptom. *Glycyrrhiza* may reduce faulty pituitary stimulation of excessive androgen production[80] as well as support normal menses in amenorrheic women. *Vitex* promotes progesterone serving to oppose the elevated androgens, and *Serenoa* inhibits the conversion of testosterone into the more active dihydrotestosterone. *Mentha spicata* (spearmint) also exerts antiandrogenic effects.[81]

Glycyrrhiza glabra
Mentha spicata
Serenoa repens
Vitex agnus-castus

Combine in equal parts and take ½ to 1 teaspoon of the combined tincture 3 to 6 times daily, reducing as skin improves. Alternately, *Vitex* and *Serenoa* may be taken alone, as capsules, and complemented with the Tea for Hair Loss and Thinning Menopausally on page 216.

Tea for Female Hirsutism

A little known investigation about *Mentha spicata* (spearmint) suggests that it may decrease elevated androgens in women with hirsutism.[82] Fennel may promote peripheral metabolism of androgens. *Urtica* and *Thea* are better-tasting androgen modulators than *Serenoa*, making them better for a tea. *Glycyrrhiza* helps make the tea sweet and is also hormone-balancing and supportive of the adrenals.

Mentha spicata 4 ounces (120 g)
Glycyrrhiza glabra 4 ounces (120 g)
Foeniculum vulgare 4 ounces (120 g)
Urtica urens 2 ounces (60 g)
Camellia sinensis 2 ounces (60 g)

Combine to make 1 pound of tea. Store the tea in an airtight plastic bag or container and use long term. Steep 1 tablespoon per cup of hot water, strain, and drink freely.

Traditional Chinese Duo for Hair Loss

Laminaria japonica and *Cistanche tubulosa* are a traditional duo used to treat dandruff, improve the health of the scalp, and remedy hair loss. Laminarias are seaweeds commonly known as kelp (other seaweeds also go by the same common name) and are a source of fucoidans, a group of sulfated polysaccharide with numerous inflammatory effects.[83] Studies have reported *Laminaria japonica*, synonymous with *Saccharina japonica*, to have positive effects on hair growth when combined with *Cistanche tubulosa* and taken orally.[84] Other studies suggest that the topical application of *Laminaria* polysaccharides may benefit the skin.[85] Those with an allergy to iodine should use seaweed with caution.

Laminaria japonica
Cistanche tubulosa

If combination products are not available, take 1 capsule of each herb, 2 or 3 times a day, or 2 capsules of the combination 3 times a day. Dry bulk herbs may also be found to prepare into hair rinses or a scalp paste to rinse off. Combine 1 pound of each herb to decoct using 10 teaspoons in 10 cups of water. Simmer for 10 minutes and allow to stand for a half hour. Strain into a pitcher, which can be set in the shower for use after bathing to rinse the freshly washed hair.

Alternately, the powders of each herb may be combined in equal parts and used to prepare a paste by drizzling warm water into the blend. Spread on the scalp, cover with a plastic bag, followed by a towel twisted into a turban. Leave in place for a half hour and rinse off outside by carrying a bucket of warm water to the lawn or garden to prevent burdening the bathtub or sink drains with too much particulate.

Hair Rinse for Alopecia Due to Infections

In addition to the hormone-sensitive causes of male and female pattern baldness, other cases of alopecia can be due to scalp infections and may present with hair loss

Autoimmune Alopecia Areata

Alopecia areata involves loss of hair in circular patches due to an autoimmune attack of the hair follicles and is reported to occur more frequently in allergic individuals than in those unaffected by atopy.[86] The use of an oral allergy or eczema formula, as exemplified in this chapter, may also be worth a try in such patients.

Coffee in Skin and Scalp Products

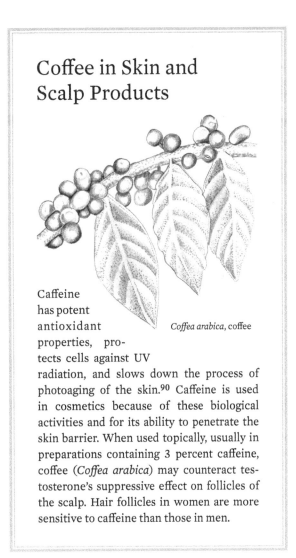

Coffea arabica, coffee

Caffeine has potent antioxidant properties, protects cells against UV radiation, and slows down the process of photoaging of the skin.[90] Caffeine is used in cosmetics because of these biological activities and for its ability to penetrate the skin barrier. When used topically, usually in preparations containing 3 percent caffeine, coffee (*Coffea arabica*) may counteract testosterone's suppressive effect on follicles of the scalp. Hair follicles in women are more sensitive to caffeine than those in men.

in certain locations where fungal infections or other lesions are present. Coffee has been demonstrated to increase the absorption of other ingredients in topical formulas.[87] *Adiantum capillus-veneris*, the maidenhair fern, is mentioned as a hair tonic in older herb books and as a remedy to treat chronic wounds. Research shows decoctions of the root to promote angiogenesis and to protect fibroblasts from oxidative and radiation damage.[88] *Adiantum* may also protect against androgenetic alopecia.[89] *Thuja* provides antifungal and antimicrobial effects and may be omitted in the absence of an infectious lesion.

Adiantum capillus-veneris root 2 teaspoons
Coffea arabica, freshly brewed 1 cup (240 ml)
Thuja plicata, *T. occidentalis* essential oil 20 drops

Prepare a tea with the *Adiantum* by steeping it in 2 cups hot water and then straining. Combine the warm coffee and *Adiantum* tea, and stir in the *Thuja* essential oil. Use as a scalp rinse, or use a sponge or gauze pad to repeatedly apply to the lesion. The preparation should rinse out easily. Employ daily for 2 weeks and evaluate for regression of the lesion.

Scalp Treatment for Alopecia Associated with Scalp Fungus

This is another antifungal recipe based on the antifungal properties of sulfur and bergamot essential oil. Because both these substances are potentially irritating, they are placed in a soothing base of rosewater and glycerin. Some people have allergic reactivity to sulfur products. Before implementing this treatment on the scalp, it's best to test for an allergic reaction. To do so, place a tiny dot of this formula on the wrist and wait an hour or two. Check for redness, swelling, itching, or burning sensations, and wash off with soap and water should any irritation occur.

Sulfur powder 1 tablespoon
Bergamot (*Citrus bergamia*) essential oil 30 drops
Glycerin 30 ml
Rosewater 30 ml

Mix in small bowl by stirring vigorously with a fork. Apply to affected area and leave in place for half an hour, and then shampoo. Do *not* use this treatment for those with known sulfur allergies.

Tincture for Hair Loss and Thinning Menopausally

While androgens are associated with inappropriate facial hair in women, they are also associated with male pattern baldness and menopausal thinning of the hair. Therefore, *Serenoa* is both a treatment for excessive hormone stimulation and for androgen-pattern hair loss. *Foeniculum* also helps reduce elevated androgens by helping to metabolize the hormones peripherally,[91] and *Actaea* can promote estrogen, helping to achieve hormonal balance.

Foeniculum vulgare
Actaea racemosa
Serenoa repens

Combine in equal parts and take ½ to 1 teaspoon of the combined tincture 3 to 6 times daily, reducing as hair thickness improves. Employ for at least 6 and up to 12 months to give the gentle hormonal support a chance to be effective.

Tea for Androgenetic Alopecia

Both men and women are susceptible to hormone-induced thinning of scalp hair, known as androgenetic alopecia. The condition is difficult to remedy, as systemic hormone-regulating therapies are needed, along with therapies that may make hair follicles less easily suppressed by hormones. Long-term therapies are essential to see any improvements. This tea can be prepared to drink as well as to use as a scalp wash.

Urtica urens powdered root
Glycyrrhiza glabra powdered root
Pueraria montana powdered root
Camellia sinensis dried leaves
Trifolium pratense dried flowers
Mentha spicata dried leaves

Combine in equal parts. Steep 1 tablespoon of the herb mixture per cup of hot water and then strain. Drink each day and/or use as a scalp wash.

Tea for Hair Loss and Thinning Menopausally

This nourishing and delicious tea is intended to reduce elevated androgens while supporting healthy estrogen levels. *Mentha spicata* may promote the metabolism of peripheral androgens, and *Urtica* can inhibit the conversion of active testosterone. *Medicago* contains hormonally active isoflavones that can support estrogen levels. *Glycyrrhiza* works via multiple mechanisms to help correct complex hormonal imbalances.

Medicago sativa 2 ounces (60 g)
Glycyrrhiza glabra 2 ounces (60 g)
Mentha spicata 2 ounces (60 g)
Urtica urens leaf and chopped root 2 ounces (60 g)

Antiandrogen Mechanisms of Action

Herbs and pharmaceuticals may help reduce excessive testosterone levels via a variety of mechanisms that reduce androgens. The enzyme 5α-reductase converts testosterone to the more powerful dihydrotesterone (DHT), and therefore 5α-reductase inhibitors can reduce testosterone-driven stimulation of the prostate or hair follicles. Androgen receptors may be blocked by agents capable of binding or blocking activity and preventing DHT from activating the receptors. Aromatase is the enzyme that converts testosterone and progesterone into estrogen in peripheral tissues, and aromatase inhibitors are prescribed to breast cancer patients. Aromatase promoters, on the other hand, may increase clearance of testosterone from the tissue via increased conversion into estrogen, and several herbs may display this antiandrogenic mechanism. Prolactin can inhibit the pituitary's release of Luteinizing Hormone (LH) and Follicle Stimulating Hormone (FSH), which in turn can reduce elevated hormones levels. Prolactin inhibition, such as in *Vitex*, may contribute to the herb's ability to reduce elevated testosterone in women with polycystic ovarian syndrome (PCOS). Androgen blockade helps treat not only PCOS, but also female hirsuitism, prostatic hypertrophy, and androgenetic alopecia.

5α-reductase inhibitors: Finasteride, *Serenoa, Urtica*
Androgen receptor antagonists: Spironolactone and cyproterone acetate
Aromatase promoters: *Paeonia, Epimedium*
Prolactin inhibitors: *Vitex*

Steep 1 tablespoon of the herb mixture per cup of hot water for 10 to 15 minutes and then strain. Drink freely, at least 3 cups per day.

Hair Plaster to Promote Hair Growth

These herbs are traditionally used in various cultures to promote hair growth when applied topically to the scalp. Even though copper peptides are expensive products, they are recommended for a variety of skin lesions. Liquid copper and zinc solutions are also available and less expensive options compared to propriety copper sulfides. Although the products are made for oral consumption, they may also be used in this manner in various recipes intended for topical application.

Polygonum multiflorum powder 2 tablespoons
Pueraria montana powder 2 tablespoons
Glycyrrhiza glabra powder 2 tablespoons
Castor oil 2 tablespoons
Coffee, brewed strong 100 ml
Rosmarinus essential oil 20 drops
Copper and zinc solution (commercial product) 5 ml

Place all the ingredients in a small bowl and blend vigorously with a fork into a thick paste. Apply to freshly washed hair, cover the hair with a plastic bag, and then wrap with a hot towel twisted into a turban. Leave in place for 30 to 60 minutes. Remove the coverings and rinse out the plaster by several washings with a gentle shampoo.

An application such as this may be followed with pharmaceutical options such as finasteride, minoxidil, or progesterone, if chosen. Some physicians prescribe specially compounded products containing such ingredients, prepared by a compounding pharmacist. The herbal hair plaster and the pharmaceutical applications may be rinsed off using one of the herbal hair rinses that follow.

Legume Hair Rinse for Androgenetic Alopecia

Many legume family herbs such as *Trifolium* and *Pueraria* are 5α-reductase inhibitors due to the isoflavones genistein, daidzein, and formononetin they contain. Topical application of *Trifolium pratense* has been found to retard hair loss in human subjects experiencing a receding hairline,[92] and it stimulates the synthesis of extracellular matrix proteins in the vicinity of hair follicles. Formononetin may promote hair regrowth when topically applied, helping hair follicles to recover normal size and stimulating regrowth of lost hair.[93]

Options for Addressing Male and Female Pattern Baldness

Hair loss in older men and women can be due in large part to changes in estrogen and testosterone ratios. Because of this, herbs that support estrogen and reduce testosterone may be helpful for women. Herbs that help maintain the long-standing androgen-to-estrogen balance may be helpful for men. Employ one or more of the treatment approaches that follow for at least a year and evaluate for efficacy.

Soy supplements
Soy foods
Genistein
Bioidentical hormone replacement therapy
Thyroid and adrenal hormonal support
Urtica urens capsules and/or tea
Serenoa repens capsules
Pueraria montana capsules and/or tea
Topical plasters and hair rinses, as described in this section

Medicago sativa 1 pound (455 g)
Trifolium pratense 1 pound (455 g)
Rosmarinus officinalis 1 pound (455 g)
Glycyrrhiza glabra 1 pound (455 g)

Combine the dry herbs and store the resulting large volume of herb mixture in 5 or 6 large resealable plastic bags. Fill a large stockpot three-quarters full of water, bring to a boil, and add 2 heaping cups of the herb mixture. Allow to steep for 15 minutes, and then strain into a big bowl or pitcher. Store the herbal rinse in the bathroom tub or shower. At the close of each shower or bath, use the stored tea to rinse out the hair. Also, use it to rinse hair after treatment with the Hair Plaster to Promote Hair Growth, above. When used frequently for many months, the hormonal effects of the topical herbs may help improve hormonal balance in the scalp.

Capsules for Androgenetic Alopecia

Because herbal products formulated for benign prostatic hypertrophy (BPH) contain 5α-reductase inhibitors, they may also be appropriate for androgenetic alopecia. Look for a product formulated for BPH that contains all of the following, which is not difficult.

Improving Scalp Circulation to Allay Hair Loss

Minoxidil is a vasodilating pharmaceutical that improves circulation to the follicles and may slow or stop hair loss and promote hair regrowth when applied topically. Minoxidil is a potassium channel opener, causing hyperpolarization of cell membranes, promoting vasodilation and allowing more oxygen, blood, and nutrients to reach the follicles. Minoxidil must be applied once or twice daily to achieve results and must be continued to maintain the effects. Allergic contact dermatitis may result from minoxidil, and some patients experience hair loss rather than hair growth.[94]

Natural agents that have vasodilating effects—from topical arginine to counterirritants, such as *Capsicum*, and warming volatile oils—may have therapeutic activity. Scalp massage and hydrotherapy to the head may be other natural means of improving circulation to the hair follicles. The turmeric relative *Curcuma aeruginosa* inhibits 5α-reductase, and may boost the efficacy of minoxidil.

Alpha reductase inhibitors limit testosterone-driven hair loss.[95] Herbal 5α-reductase inhibitors include:

Avicennia marina[96] (white mangrove)
Camellia sinensis (green tea)
Carthamus tinctorius (safflower)
Chrysanthamum spp. (chrysanthemum)
Citrullus colocynthis[97] (bitter apple, bitter cucumber)
Curcuma longa (turmeric)
Cuscuta reflexa[98] (dodder)
Ganoderma lucidum (reishi)
Glycyrrhiza glabra (licorice)
Lepidium meyenii (maca)
Magnolia officinalis (magnolia)
Mangifera indica (mango)
Mentha spicata (spearmint)
Panax ginseng (ginseng)
Phyllanthus emblica, P. niruri (Indian gooseberry, stonebreaker)[99]
Polygonum multiflorum (fo ti, he shou wu)
Pueraria montana (kudzu)
Serenoa repens (saw palmetto)[100]
Camellia sinensis (green tea)
Thuja occidentalis (northern white cedar)
Trifolium pratense (red clover)
Urtica dioica (nettle root)

Serenoa repens
Urtica dioica
Pumpkin seed liposterols
Zinc

Consume 2 or 3 capsules at a time, 2 to 3 times a day.

Hair Rinse for Androgenetic Alopecia

Carnosic acid in *Rosmarinus* inhibits 5α-reductase,[101] and the topical use of *Rosmarinus officinalis* leaf extract promotes hair regrowth in animals, where hair loss has been induced by the administration of testosterone. Using green tea topically on the scalp may protect the hair follicles from normal skin aging that contributes to male pattern baldness.[102] Hair loss is accompanied by apoptosis of keratinocytes, and *Eclipta* and *Angelica* have protective and proliferative effects on keratinocytes. *Angelica* applied topically promotes hair growth and enhances the size of hair follicles.[103] *Eclipta alba* is used traditionally to promote hair growth,[104] and the topical application of *Eclipta* promotes hair growth in mice with genetic hair loss related to abnormal keratinization.[105] The topical application of *Chrysanthemum* flavonoids promotes hair growth in animal studies.[106]

Rosmarinus officinalis 1 pound (455 g)
Eclipta prostrata 1 pound (455 g)
Camellia sinensis 1 pound (455 g)
Angelica sinensis 1 pound (455 g)
Chrysanthemum spp. 1 pound (455 g)

Combine the dry herbs and store the resulting large volume of herb mixture in 5 or 6 large resealable plastic bags. Fill a large stockpot three-quarters full of water, bring to a boil, and add 2 heaping cups of the herb mixture. Allow to steep for 15 minutes and then strain into a big bowl or pitcher. Store the herbal rinse in the

Mr. He's Black Hair—*Polygonum multiflorum*

Polygonum multiflorum root goes by the common names fo ti and he shou wu and is a traditional medicine in China for promoting hair growth and preventing the hair from turning gray. He shou wu can be translated as "Mr. He's black hair," due to the legend that the plant was used to keep the gentleman's hair black. More than 100 chemical compounds have been isolated from the plant, including stilbenes, quinones, and flavonoids. Traditional uses include antiaging, antihyperlipidemia, anticancer, and anti-inflammatory effects as well as promoting immune modulation and neuroprotection. However, at high doses, the quinones, such as emodin and rhein, may lead to hepatotoxicity, nephrotoxicity and embryonic toxicity;[107] topical use, however, has not shown these concerns. The stilbenes and other compounds may promote growth of dermal papilla cells; one study showed this effect to a greater degree than conferred by minoxidil.[108] The traditional use of *Polygonum multiflorum* in Asia to prevent hair from turning gray, or even to reverse graying, may involve effects on melanin, α-melanocyte-stimulating hormone (α-MSH), and melanocyte-regulating pathways.[109] Animal studies suggest hair growth may also be promoted via increased expression of fibroblast growth factor.[110] *Polygonum multiflorum* is a 5α-reductase inhibitor[111] and is shown to promote hair growth by inducing the anagen phase in resting hair follicles by inducing the expression of β-catenin.[112]

Polygonum multiflorum, fo ti

bathroom tub or shower. At the close of each shower or bath, use the stored tea to rinse out the hair. This hair rinse can also be used to rinse hair after treatment with the Hair Plaster to Promote Hair Growth on page 217, or following pharmaceutical applications. When used frequently for many months, the hormonal effects of the topical herbs may help improve hormonal balance in the scalp.

Formulas for Erythema Multiforme

Erythema multiforme occurs due to the deposition of immunoglobulin complexes in the cutaneous microvasculature and is thought to be triggered by exposure to a noxious agent or allergen. From 2 to 3 percent of all children prescribed a pharmaceutical may develop a cutaneous drug eruption.[113] Skin lesions are variable in their appearance, hence the name "multiforme" but most commonly are bright pink plaques occurring on the skin and possibly on the mucous membranes as well. "Target" lesions are also typical, so called because a larger pink lesion appears to have a bull's eye in the center.

The skin lesion may be pruritic, and in chronic cases, skin injuries may result in a new lesion. Erythema multiforme may be quite mild, referred to erythema multiforme minor, or severe and potentially fatal, referred to as erythema multiforme major, also known as Stevens-Johnson syndrome. This more severe type of erythema multiforme may be life-threatening as large regions of epidermis detach, which leads to infection, electrolyte depletion, and shock. Large and small vesicles may emerge and rupture, and the mouth, tongue, conjunctiva, and other tissue may be affected. Stevens-Johnson syndrome is considered a dermatologic emergency.

The long list of etiologic triggers includes infectious organisms, drug reactions, sunlight, and radiation. Erythema multiforme may occur in association with collagen vascular diseases, leukemia, myeloma, cancer, and Hodgkin's disease. Any suspicious symptoms should be fully evaluated to rule out cancer or occult disease. While the major form of erythema multiforme should be treated in the hospital, herbal and other supportive measures are appropriate for the minor form.

Erythema multiforme can affect the oral mucosa as well as the skin. Because the oral lesions are often painful, a mouth rinse such as the Mouthwash for Oral Lichen Planus on page 211 can be helpful.

Topical Compresses for Erythema Multiforme

There is no research on herbs for this condition, and the herbs in this compress formula are selected due to their general anti-inflammatory and healing properties on the skin. *Ganoderma lucidum*, the reishi mushroom, has shown some efficacy for autoimmune-driven inflammatory skin reactions when applied topically.[114]

Calendula officinalis
Achillea millefolium
Centella asiatica
Hamamelis virginiana
Ganoderma lucidum

Combine equal parts of the dry herbs. Steep 1 to 2 teaspoons of the herb mixture per cup of hot water, strain, and use to prepare a cloth compress. Apply topically to skin lesions several times a day.

Tincture for Erythema Multiforme Associated with Drug Reactivity

When erythema multiforme is obviously connected with the initiation of a new pharmaceutical, herbs that reduce histamine and allergic reactivity such as *Tanacetum* may help. The liver herbs *Curcuma* and *Silybum* may help the liver clear the drug as efficiently as possible. *Calendula* is added to promote healing of the skin.

Curcuma longa
Tanacetum parthenium
Silybum marianum
Calendula officinalis

Combine in equal parts and take ½ to 1 teaspoon of the combined tincture 3 to 6 times daily, reducing as skin improves.

Tincture for Erythema Multiforme Associated with Underlying Viruses

When erythema multiforme is obviously connected with the initiation of a viral infection such as herpes, the antivirals in this formula may help.

Glycyrrhiza glabra
Hypericum perforatum
Melissa officinalis
Phytolacca decandra

Combine in equal parts and take ½ to 1 teaspoon of the combined tincture 3 to 6 times daily, reducing as skin improves.

Therapies for Erythema Nodosum

Erythema nodosum is an inflammatory skin eruption of the lower legs, mainly the shins, and is characterized by red tender nodules. Erythema nodosum involves inflammation of subcutaneous fat cells due to a drug reaction or reactivity to a wide variety of other noxious agents. There may be a genetic predisposition to the immune hyperreactivity of erythema nodosum. The condition is most common in younger people and is usually self-limiting, resolving within a month's time. The onset of erythema nodosum is often preceded by sore throat, fever, and flulike symptoms including malaise, joint pain, and weakness. The lesions are tender, red, and about 1 to 10 centimeters in diameter. Lesions resolve in a manner similar to a bruise, turning purple, then yellow brown, and finally green. The lesions are most common on the shins, but may also occur on the thighs, arms, face and neck, and trunk. There are several variations of erythema nodosum, including ulcerating and hemorrhagic forms, and the condition can be triggered by a long list of infections including *Streptococcus*, *Mycoplasma*, *Histoplasma*, and Epstein-Barr. It can also be triggered by pregnancy, autoimmune disease, irritable bowel syndrome, cancer, lymphoma, and valley fever. Drugs including sulfonamides, oral contraceptives, penicillin, and bromides can trigger the skin inflammation. In a full one-third to one-half of all cases, however, no such inciting cause is identified. Because there are so many possible causes, the occurrence of erythema nodosum necessitates a full exam and cessation of all pharmaceuticals possible to determine an underlying cause. Blood panels, throat cultures, chest X-rays, and other diagnostic panels can help rule out occult infections or cancer.

Herbal therapy should address any underlying cause where possible. Herbal treatment would be a combination of anti-inflammatory, liver support and detoxification, and antimicrobial agents. Soothing topical compresses may be used to help relieve any discomfort. Immunosuppressive agents such as steroids or colchicine are sometimes prescribed, but each has significant side effects, so such pharmaceuticals are usually reserved for severe cases.

Tincture for Erythema Nodosum

Some forms of erythema nodosum may become chronic, particularly due to repeated streptococcal infections. In such cases, a formula like this one is especially appropriate, because it aims at reducing hypersensitivity reactions in general. The herbs in this formula provide anti-inflammatory, antimicrobial, and liver-supportive properties.

Echinacea angustifolia
Glycyrrhiza glabra
Hypericum perforatum
Astragalus membranaceus

Combine in equal parts and take 1 to 2 teaspoons of the combined tincture 3 to 6 times a day, reducing as symptoms improve.

Formulas for Pemphigus

Pemphigus is a rare skin disorder associated with autoimmune-induced vesicles and bullae in the skin, often large and extensive. The condition results from autoantibodies to desmoglein, an adherence-promoting compound involved with desmosomal cell-to-cell connections. There are at least three subtypes of pemphigus: pemphigus vulgaris, pemphigus foliaceus, and paraneoplastic pemphigus, as well as numerous rare subtypes and related conditions. The vulgaris type is most common and may be misdiagnosed as eczema; the most severe is the paraneoplastic type and involves painful lesions in the mouth, esophagus, and airways, resulting in irreversible tissue damage. This type of pemphigus can be fatal due to overwhelming infection in the open lesions. High doses of steroids are often prescribed in such cases, but this runs the risk of damaging the intestines, leading to intestinal perforation and sepsis. Patients with pemphigus require expert care and may be treated in a burn ward due to extensive raw lesions. Herbal therapies are not an alternative to standard medical care for this condition, but rather useful as a complementary therapy.

Supportive Tincture for Pemphigus

There is no advanced research on alternative medicine for pemphigus. Natural approaches could include avoiding any and all allergens and nonessential drugs to combat autoimmune reactivity. Herbs that support hepatic and renal clearance of drugs and toxins, such as *Silybum* or *Curcuma*, and herbs that act as immune modulators, such as *Astragalus*, would be supportive. Herbs that reduce allergic hyperreactivity, such as *Scutellaria baicalensis* and *Angelica sinensis*, would also be indicated.[115] With this condition, large amounts of fluid may be lost through the skin, so an effort to replace fluid and electrolytes is also appropriate.

Curcuma longa
Astragalus membranaceus
Scutellaria baicalensis
Angelica sinensis

Combine in equal parts and take 1 to 2 teaspoons of the combined tincture, 3 to 6 times daily.

Supportive Tea for Pemphigus

Because pemphigus is such a serious condition, using a multipronged treatment approach is indicated. This tea offers anti-inflammatory and skin-supportive herbs.

Glycyrrhiza glabra 4 ounces (120 g)
Centella asiatica 2 ounces (60 g)
Mahonia aquifolium 2 ounces (60 g)
Calendula officinalis 1 ounce (30 g)
Achillea millefolium 1 ounce (30 g)

Steep 1 tablespoon of the herb mixture per cup of hot water and then strain. Drink freely throughout the day.

Formulas for Pigmentation Disorders

Both hypo- and hyperpigmentation disorders can occur and may result from destruction of melanocytes (pigment-producing cells) or excessive stimulation of melanocytes. Hyperpigmentation can also result from adrenal and hypothalamic disorders, because the release of corticotropic-releasing hormone (CRH) from the brain causes melanocyte-stimulating hormone (MSH) to be released as well. The excess MSH stimulates melanocytes and results in hyperpigmentation in Cushing's syndrome and related adrenal disorders. Excessive stimulation of melanocytes in facial skin can occur at times of high hormonal loads in the body, and increased facial pigmentation is referred to as melasma. When it occurs during pregnancy, it is sometimes referred to as "the mask of pregnancy." So-called "age spots" occur when melanocytes become damaged by sun and oxidative stress and result in pigmented lesions on sun-exposed areas such as the face, chest, shoulders, or hands. In other cases, autoimmune reactivity to melanocytes may lead to areas of hypopigmentation, such as in vitiligo. Herbal therapies can be tailored to individual cases and underlying cause.

Antimelanin Tincture

Melasma is an acquired skin condition involving increased pigmentation, characterized by symmetrical and confluent gray-brown patches usually on the areas of the face exposed to the sun. One such hyperpigmentation disorder, colloquially referred to as the "mask of pregnancy," is associated with high hormone levels and occurs particularly in women of color. Pigmented striae on the abdomen may occur in Cushing's syndrome due to the tendency of elevated corticotropic-releasing hormone to promote the release of melanocyte-releasing hormone. Nonhormonal causes of hyperpigmentation include silver toxicity and elevated iron in cases of hemochromatosis. Such cases will not respond to this formula, which is aimed at reducing elevated MSH. *Glycyrrhiza uralensis* is both a tyrosinase inhibitor and an endocrine-modulating herb. Glucosamine, procyanidin,

Herbal Constituents for Treating Hyperpigmentation

Arbutin was one of the first herbal constituents studied to inhibit melanogenesis via effects on tyrosinase, the enzyme that metabolizes tyrosine. Arbutin has now been prepared into proprietary micellized products aimed at improving transdermal delivery in the treatment of hyperpigmentation disorders of the skin.[116] Arbutin occurs in *Arctostaphylos uva ursi* and the saxifrage species *Bergenia crassifolia*; skin washes from the crude leaves or plants may be one therapeutic approach in treating hyperpigmentation. Arbutin is metabolized into hydroquinone in the body, and many hydroquinones are reported to have skin-bleaching effects when used topically. Furthermore, normal skin microflora may hydrolyze arbutin into hydroquinone[117] and be less irritating than hydroquinone.

Azelaic acid is another naturally occurring tyrosinase inhibitor produced from fatty acids found in wheat, rye, and barley, but is also produced by yeasts that normally populate the skin. Azelaic acid is particularly used for pigment changes following acne and other skin blemishes due to particular effects on abnormal melanocytes.

Kojic acid is a naturally occurring molecule derived from the fungi *Acetobacter*, *Aspergillus*, and *Penicillium*. Kojic acid also inhibits tyrosinase, but may be more irritating than the hydroquinones or azelaic acid. It is often combined with hydroquinone in commercial products.

Glycolic acid is an alpha hydroxy acid that occurs naturally in fruits and is used to treat acne and skin blemishes and to lighten the skin. Glycolic acid accelerates desquamation of epithelial cells and brings impurities to the surface of the skin quickly. It also directly reduces melanin formation in melanocytes by tyrosinase inhibition.

berry flavonols, and grape seed oil are also safe and supportive tyrosinase inhibitors and would complement an herbal tincture. Sunscreen should be used faithfully.

Salvia miltiorrhiza
Glycyrrhiza glabra or *G. uralensis*
Vaccinium myrtillus

Combine in equal parts and take ½ to 1 teaspoon of the combined tincture 3 to 6 times daily, reducing as skin improves.

Supportive Tincture to Metabolize Hormones

For cases of melasma due to high hormone levels, such as the mask of pregnancy, make certain that the liver is in good health and able to metabolize hormones.

Curcuma longa
Silybum marianum

Combine in equal parts and take 1 to 2 teaspoons of the combined tincture 3 to 6 times daily.

Herbal Skin Bleach for Melasma

Because correcting long-standing hormonal imbalance can be difficult, some patients choose to bleach the skin for cosmetic reasons. Several naturally occurring plant acids can bleach the skin, including azelaic and salicylic acids, and inhibit melanin synthesis in the skin, as does arbutin found in *Arctostaphylos uva ursi*.[118] *Hippophae rhamnoides* (sea buckthorn) contains catechins with many skin benefits, including a skin-bleaching effect of its own.[119] The oil may be applied to the skin following the application of the skin wash.

Arctostaphylos uva ursi leaves
Salix alba bark
Hippophae rhamnoides oil

Combine *Arctostaphylos* and *Salix* in equal amounts. Simmer 1 heaping tablespoon of the herb mixture in 3 cups of water, reducing it to a volume of 2 cups. Let stand covered for 30 to 60 minutes and then strain. Use as a skin wash, twice a day, continuing daily for 3 months. Follow each use of the skin wash with an application of *Hippophae* oil. Take a 24- to 48-hour break from this regimen if any burning or irritation occurs.

High-Phenol Antimelanin Juice

Because phenolic compounds appear to reduce hyperpigmentation, regular consumption of health-promoting anthocyanidin fruits can certainly do no harm and may be complementary to the other formulas listed previously. Choose 100 percent pure juice without any added sugar. Even fruit juice may provide too much sugar, so it may be best to dilute several ounces of juice with tap or bottled water. Bioflavonoids in berries help the skin to absorb and process UV light, which reduces melanocyte stimulation.

Plants for Pigmentation Disorders

Polyphenols in plants absorb UV radiation and help protect the plants themselves from radiation, providing potent antioxidant effects. When consumed or used topically by humans, polyphenols protect the skin and connective tissue from radiation photo damage. Polyphenols include plant flavonols and colorfully pigmented berries are among the richest sources. Pomegranate flavonoids are shown to reduce melanogenesis, in part via absorbing UV radiation that would otherwise stimulate melanocytes to release melanin. Melanogenesis is a complex process involving tyrosinase and tyrosinase-related proteins, and pomegranate flavonoids also inhibit melanin production via tyrosinase activity.[120] Tyrosinase prevents hyperpigmentation by interfering with melanogenesis. *Camellia sinensis* polyphenols also inhibit melanin accumulation and melanin synthesis due to limiting tyrosinase activity in a concentration-dependent manner.[121] *Magnolia grandiflora* flowers have antimelanogenic effects due to similar mechanisms.[122] *Panax ginseng* ginsenosides promote tyrosinase activity in melanocytes and exert a melanogenic effect,[123] however, they have also been shown to inhibit excessive melanogenesis.[124] Silymarin from *Silybum marianum* strongly prevents photocarcinogenesis and prevents melanin production; one human clinical trial showed significant improvement in melasma patients when silymarin was topically applied.[125]

Pomegranate juice
Blueberry juice
Purple grape juice
Black cherry juice

Combine any of the juices in whatever proportion desired. Place several ounces of juice in the bottom of a tall glass, add ice or a lemon slice if desired, and top with water. Drink 2 or more cups per day.

Tincture for Hyperpigmentation on the Shins

Pigmentation on only the lower shins is not related to melanin, but rather to vascular stasis. This formula combines herbs noted to improve venous circulation. Elevating the legs whenever possible to support venous return may also help improve skin discoloration.

Curcuma longa
Ginkgo biloba
Hypericum perforatum
Trifolium pratense

Combine in equal parts and take 1 teaspoon of the combined tincture 4 to 6 times daily.

Tyrosinase Inhibitors for Hyperpigmentation

Tyrosinase is an enzyme involved in melanin synthesis, and agents that inhibit tyrosinase will improve hyperpigmentation disorders. The list of plants and isolated constituents shown to inhibit tyrosine and melanogenesis is extensive, and the most nourishing and readily available botanicals have been used in the hyperpigmentation formulas in this chapter. Isolated molecules and compounds noted to act as tyrosinase inhibitors include hydroquinone, azelaic acid, kojic acid, and certain licorice extracts, when topically applied.[126] Other depigmenting agents include retinoids, ascorbic acid, niacinamide, N-acetyl glucosamine, and soy compounds. Hydroquinone inhibits melanogenesis and may eventually induce necrosis of whole melanocytes—its use in cosmetics is banned in some countries due to possible skin irritation, paradoxical hypermelanosis, or occasional permanant loss of melanocytes. To achieve any results, all such tyrosinase inhibitors must be used several times per day for several months.

Topical Oil for Age Spots

Age spots can be difficult to resolve without bleach, chemical peels, and ablation. This formula may be able to resolve some lesions and certainly help retard the development of additional age-related lesions if used faithfully on a long-term preventive basis. It is also helpful for sun-exposed, freckled skin of the elderly. *Larrea* and *Curcuma* oils are not commonly available on the market but are easy to make for oneself. Use sunscreen faithfully each day, along with attentive use of hats, gloves, and long sleeves when sun-related skin cancers or precancerous lesions are a concern.

Larrea tridentata oil 100 ml
Curcuma longa oil 75 ml
Ricinus communis oil 40 ml
Podophyllum peltatum resin 25 ml

Combine all in a 4-ounce bottle and apply to the chest, hands, and any other problem area every morning and night to treat and retard the development of age spots. When *Podophyllum* resin is not available or too expensive, it may be omitted from the formula.

Simple Resin Treatment for Age Spots

Age spots can be difficult to resolve with any method other than mechanical or chemical methods. Prevention with the use of sunscreen applied three or more times daily and protective clothing is the best medicine. For existing age spots, *Podophyllum peltatum* (mayapple) can be processed to yield a caustic resin that can destroy warts and sclerified age spots with topical application. Apply dots of the resin on age spots, once a day for 7 to 10 days. Take a break for 2 weeks to allow any resulting rawness or lesions to heal, and repeat. The purified resin has become very expensive, but since a small amount is effective, some physicians may wish to invest in a small bottle of purified podophyllin resin.

Podophyllin resin ⅓ ounce (10 g)
Tincture of benzoin 10 ml

Combine the resin and sticky benzoin tincture in a 1-ounce bottle and shake vigorously to homogenize. Dispense to patients by transferring to small 1 or 2 dram (4 or 8 ml) bottles. Patients may use a cotton swab to apply a tiny dot to each lesion. Wait an entire week or more to reapply in the case of any irritation, pain, redness, or ulceration.

Topical Oil for Scaly or Precancerous Moles

Similar to the Topical Oil for Age Spots, this formula is slightly stronger and more caustic, intended to dot onto scaly or precancerous moles rather than rub into large areas of skin. Some of the ingredients needed to prepare this formula, such as the *Sanguinaria* oil, are not commercially available and will need to be prepared for oneself.

Ricinis communis 15 ml
Larrea tridentata oil 15 ml
Phytoclacca decandra oil 15 ml
Podophyllum peltatum resin 5 g
Sanguinaria canadensis oil 5 ml
Thuja plicata essential oil 5 ml

Combine all in a 2-ounce bottle and shake well. Use a screw-on cap rather than a dropper because the *Thuja* essential oil will degrade the rubber dropper over time. Use a cotton swab to dot the formula on moles and lesions. Take care to protect the surrounding skin by applying *Calendula* salve or another greasy ointment. Cover with a small bandage and wash off with soap and water 4 hours later. Apply once a day for 7 to 10 days. Take a break for 2 weeks to allow any resulting rawness or lesions to heal, and repeat.

Tea for Vitiligo

Vitiligo is a lack of melanocytes in patchy regions of the skin and sometimes the scalp. The cause is not well understood but an autoimmune phenomenon is believed to contribute, and some patients may have antibodies to melanocytes. More than 50 percent of the cases begin before age 20. The condition is also loosely related to endocrine disorders including thyroid and adrenal imbalances and diabetes, thus these conditions should be ruled out and treated directly if identified. Genistein, an isoflavone common in legumes, promotes melanin production and prevents UV destruction of melanocytes. Note that *Glycyrrhiza* is included both in formulas for hyperpigmentation and for vitiligo because it can promote both melanin production and metabolism. The immune-modulating effects of *Glycyrrhiza* may also contribute to its beneficial effects in vitiligo, and glycyrrhizin increases tyrosinase gene expression and tyrosinase-related proteins, critical for melanin synthesis.[127]

Medicago sativa
Trifolium pratense
Glycyrrhiza glabra

Combine equal parts of the dry herbs. Steep 1 tablespoon of the herb mixture per cup of hot water. Strain and drink freely, 3 or more cups per day.

Melanin-Stimulating Tincture for Vitiligo

Based on modern research, this combination of Apiaceae family plants with black pepper may help promote melanin synthesis and/or prevent destruction of melanocytes. Note that *Piper* tincture may not be commercially available and may need to be individually prepared.

Herbal Photosensitizers

Hypericum. Although most photosensitizers are in the Apiaceae family, the most potent photosensitizer known is *Hypericum perforatum* (St. Johnswort). The use of *Hypericum* products on the skin can cause third-degree burns when exposure to strong sun follows. Even the oral use of *Hypericum* can cause sun sensitivity in some individuals, and cattle that graze on large amounts of *Hypericum* frequently develop photophobia.

***Apiaceae* family essential oils.** *Daucus carota* (carrot) and *Petroselinum crispum* (parsley) seed oils contain photosensitizing psoralens and related compounds. When these compounds are concentrated in essential oils, they become photosensitizing when used topically.

Ammi visnaga (khella) is also in the Apiaceae family. It is not available as an essential oil, but is available as a tincture that can be taken orally.

Citrus bergamia. Bergamot essential oil contains bergapten, which has a photosensitizing action. Some essential oils are made "bergapten free" to avoid photosensitization. Be sure to choose the type you need for the desired effect.

Nutritional Support Approaches for Vitiligo

Agents that promote tyrosinase, gently stimulate melanocytes, and provide tyrosine may promote and help protect melanin in the skin.

Soy products and foods, genistein supplements
Beta-carotene and vitamin A
Tyrosine
Copper
Phenylalanine

Ammi visnaga
Foeniculum vulgare
Angelica sinensis
Piper longum

Combine in equal parts and take 1 to 2 teaspoons of the combined tincture 3 to 6 times a day. Reduce the proportion of *Piper* should any oral or gastric irritation occur.

Ginkgo Capsules for Vitiligo

Ginkgo biloba, as a monotherapy, has shown efficacy in reducing or resolving vitiligo lesions in human trials.[128] Capsules may be found in 60, 120, or 150 milligrams per pill. The higher dosage ranges are recommended.

Ginkgo pills, 150 milligrams or higher

Take 2 capsules, 2 or 3 times per day.

Topical Oil for Vitiligo

Photosynthesizing agents used in tandem with UV light have been a standard therapy for vitiligo (and psoriasis) for a hundred years or more, but the results vary, and severe sunburns can result if the therapy is not done cautiously. Both oral and topical psoralens, found naturally in many Apiaceae family plants, are photosensitizing. Because the photosensitizing effect of bergamot (*Citrus bergamia*) essential oil is typically regarded as an undesirable side effect, some manufacturers remove bergaptens, the photosensitizing compound present in the oil, and label the bergamot oil as being "bergapten free." For this purpose, the bergaptens are desirable and a product that contains them is ideal.

Hypericum perforatum oil 22 ml
Daucus carota (carrot) seed essential oil 4 ml
Citrus bergamia essential oil 4 ml

Combine the oils in a 1-ounce bottle. Use sparingly, just 2 or 3 drops on a vitiligo lesion, increasing as tolerated. Apply to lesions and expose to sunlight or artificial sunlight. Increase the amount applied and the duration in the sun as tolerated, but being careful to avoid sunburn. Stop at once if any irritation or inflammation occurs. Wait 1 week and try again at half the irritating dose, or less than half. If no sun sensitivity is induced with 2 or 3 drops, increase the dosage and the intensity and duration of sun exposure gradually and cautiously.

Formulas for Hyperhidrosis

Hyperhidrosis is a condition of excessive sweating and can be due to stress and anxiety, genetic tendency, or in the case of menopausal hot flashes, hormonal imbalances. The following formulas address various presentations and causes.

Tincture for Menopausal Hot Flashes

The majority of women will respond to *Salvia officinalis* in the management of menopausal hot flashes, but *Salvia* can be drying in those who already have a dry constitution. If this is a problem, mixing it with cooling, moistening demulcents will usually remedy the problem.

Salvia officinalis
Vitex agnus-castus
Actaea racemosa
Angelica sinensis

Combine in equal parts and take 1 to 2 teaspoons of the combined tincture, 2 to 5 times a day.

Tincture for Hyperhidrosis with Hyperthyroidism

An excessive metabolic rate that occurs with hyperthyroidism can also promote hyperhidrosis and may be addressed with this formula containing *Melissa* and *Leonurus* to support thyroid function. *Salvia* serves as a drying base to complement the herbs noted to reduce elevated thyroxine.

Salvia officinalis
Melissa officinalis
Leonurus cardiaca

Combine in equal parts and take 1 to 2 teaspoons combined tincture 2 to 5 times a day.

Tea for Foul-Smelling Perspiration

Because the skin is a secondary organ of elimination, strong body odor and foul-smelling perspiration is an indication of less-than-optimal elimination and metabolism via the kidneys and especially bowels and support of undesirable skin microbes. This formula is a basic alterative aimed at assisting elimination via the bowels and supporting a healthy intestinal ecosystem, which in turn will often put less of a burden on the skin to eliminate unprocessed metabolic wastes left in the body tissues and fluids.

Arctium lappa
Rheum palmatum
Taraxacum officinale
Cinnamomum spp.
Glycyrrhiza glabra

Combine equal parts of the dry herbs. Simmer 1 teaspoon of the herb mixture per cup of hot water for 10 minutes. Let stand, covered, another 10 minutes more and then strain. Drink 3 or more cups per day.

Oral Oil for Skin Dryness with Deficient Secretions

Patients with Sjögren's syndrome and related mixed connective tissue diseases may have autoimmune destruction of tears, saliva, and intestinal secretions and may respond to *Pilocarpus*, a powerful secretory stimulant. Outside of autoimmune diseases, patients with systemic and constitutional dryness may have insufficient sweat gland functioning and/or essential fatty acid status in the body. Consider formulas using a tiny amount of *Pilocarpus* mixed into a quality seed oil, choosing from the options below.

Pilocarpus jaborandi tincture 1 ounce (30 g)
Linum usitatissimum seed oil 3 ounces (90 g)
Olea europaea oil 3 ounces (90 g)
Oenothera biennis oil 3 ounces (90 g)
Ribes nigrum oil 3 ounces (90 g)

Combine the *Pilocarpus* tincture into one of the above oils in a 4-ounce bottle. Shake vigorously prior to each use, because the tincture will separate out to the top. Take 1 to 2 teaspoons daily.

Herbs for Hyperhidrosis Due to Unknown Causes

These herbs are listed in older herbals for excessive sweating unrelated to menopause, hyperthyroidism, or stress. *Hypericum* was mentioned especially for sweating of the scalp.

Centella asiatica
Crataegus oxyacantha
Hydrastis canadensis
Hypericum perforatum
Rosmarinus officinalis
Salvia officinalis

Tincture for Hyperhidrosis with Emotional Distress

Excessive sweating can be so severe and embarrassing that some people choose to resort to surgical destruction of the nerves to the sweat glands. For less severe cases, this formula will be both drying to perspiration, due to the *Salvia*, and calming to underlying stress that may promote excessive sweating.

Salvia officinalis
Hypericum perforatum
Melissa officinalis

Combine in equal parts and take 1 to 2 teaspoons of the combined tincture 2 to 5 times a day.

Formulas for Skin Creams and Antiaging and UV-Protection Products

The cosmetic industry is a multibillion dollar industry, and just as research on vitamins and specific nutrients transformed the vitamin industry into the "nutraceutical" industry, so, too, the growing research on topical substances to support the skin has transformed the cosmetic industry into the "cosmeceutical" industry. Many people may seek preventive formulas to address the effects of aging; others may wish to use the most effective and well-researched creams or lotions as preventive agents. Some people have small areas of sun damage and wish to prevent a transformation into skin cancer, while others may be undergoing radiation therapy for cancer and have the need to protect the skin. The following formulas offer protection and therapy for all such situations.

Tincture to Increase Circulation in the Skin

Poor wound healing and extensive scarring may occur in the elderly as a result of impaired circulation to the skin. *Curcuma* can improve blood congestion in the liver as well promote angiogenesis in the skin. *Centella*, *Achillea*, *Angelica*, and *Crataegus* may improve circulation.

Achillea millefolium
Curcuma longa
Centella asiatica
Angelica sinensis
Crataegus oxyacantha

Combine in equal parts and take 2 dropperfuls of the combined tincture 3 or more times daily.

Herbs for Masks, Poultices, Lotions, and Other Topical Skin Products

These herbs and other substances work by a variety of protective mechanisms including antioxidant, protectant, and regenerating. Oils, toners, spritzers, lotions, masks, and face washes can all be crafted to feature these ingredients.

Aloe vera gel	*Hypericum perforatum*
Calendula officinalis	*Symphytum officinale*
Camellia sinensis	Caffeine
Hamamelis virginiana	Glutamine
Hippophae rhamnoides oil	Retinoic acid

Versatile Skin Care Oil

Although it is too oily for many to use directly on the face, this oil is excellent to apply to the body immediately after bathing to help retain water in the skin, while providing the medicinal effects of the herbs. This is an example only; other herbs can be substituted and the proportions altered in a myriad of ways. In this formula, note that *Centella* oil is not commercially available, but is not difficult to make.

Hippophae rhamnoides oil 60 ml
Calendula officinalis oil 15 ml
Symphytum officinale oil 15 ml
Castor oil 15 ml
Centella asiatica oil 10 ml
Helichrysum spp. essential oil 3 ml
Rosa spp. essential oil 1 ml
Sandalwood essential oil 1 ml

Mix all of the oils and store in a 4-ounce bottle. Use approximately 1 teaspoon poured into the palm of the hand to rub into the throat, arms, legs, or other area each day after bathing.

Anticancer and Antiaging Effects of Green Tea

Phenolic compounds in green tea (*Camellia sinensis*) are noted to have antioxidant and anti-inflammatory effects; many of the compounds are credited with anticancer effects. Catechins in green tea are a type of phenol for which much research purports to show anticancer activity. Recent investigations are also showing antiaging effects from green tea catechins when used topically. Catechins have free-radical scavenging effects and an ability to protect from UV radiation when used either orally or topically. Green tea catechins can improve type I procollagen synthesis as well as improve water retention.

Antiaging, Antiwrinkle Tincture

This formula can help to build collagen in the skin. *Aloe vera* and *Camellia sinensis* are credited with antiaging effects by antioxidant and tissue-protective abilities, including an ability to protect skin cells from UV radiation when used orally and taken internally. *Actaea* offers isoflavones for hormonal support of the skin, and *Centella asiatica* is a connective tissue repair and regeneration-enhancing herb. *Aloe vera* juice is commercially available.

Aloe vera juice
Centella asiatica tincture
Actaea racemosa tincture
Camellia sinensis tincture

Combine in equal parts. Place 2 or 3 dropperfuls of the combined tincture and 1 teaspoon *Aloe vera* juice in a glass of water or herb tea and consume daily.

Tea to Prevent Radiation Damage

Patients undergoing radiation therapy for lung or other cancer should protect their skin with *Aloe vera* gel before and after each session, while using the following herbs orally. Radiation is injurious to the skin directly but also generates compounds that are harmful systemically. *Hamamelis* tannins may protect dermal fibroblasts, and clinical investigations have shown *Hamamelis* to have a weak but noticeable ability to speed recovery of the skin following exposure to UV light.[129] *Calendula* triterpenes are reportedly anticancer, and *Calendula* is one of the most well-known of all the wound-healing herbs. *Centella* may help protect the liver and tissues from radiation exposure according to modern studies and, like *Calendula*, is wound healing and regenerating. A clinical trial in France showed *Calendula* to prevent radiation-induced dermatitis in breast cancer patients undergoing radiation therapy.[130] In addition to drinking this liquid as a tea, it could also be used as a skin rinse to use at home after undergoing radiation treatments. The herbs may also be macerated in oil or in vinegar to prepare extracts for topical use.

Centella asiatica 2 ounces (60 g)
Hamamelis virginiana 2 ounces (60 g)
Calendula officinalis 2 ounces (60 g)

Use 1 tablespoon of the herb mixture per cup of hot water and bring to a simmer. Remove from heat immediately and let stand, covered, for 20 minutes, and then strain. Drink as much as possible, at least 5 cups a day, while undergoing therapy. Large amounts such as 5 cups a day may be the most effective and worth the effort to prevent permanent skin damage.

Skin-Supportive Effects of Sea Buckthorn

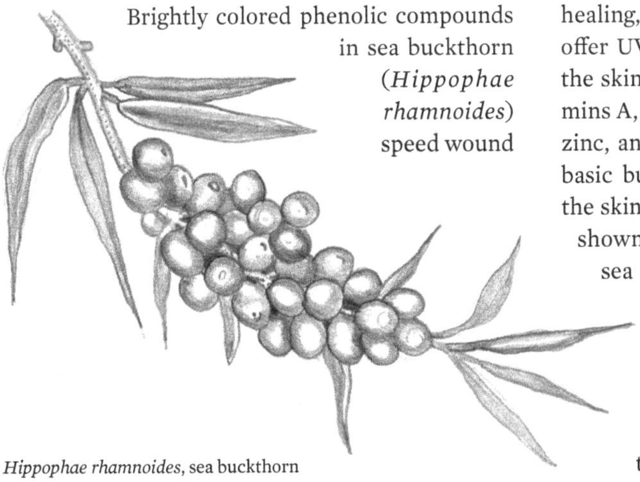
Hippophae rhamnoides, sea buckthorn

Brightly colored phenolic compounds in sea buckthorn (*Hippophae rhamnoides*) speed wound healing, have significant antioxidant activity, and offer UV light protection. Sea buckthorn supports the skin with rich nutrient content including vitamins A, C, and E and the minerals sulfur, selenium, zinc, and copper. Hydroxyproline is an important basic building component of bone, cartilage, and the skin's connective tissue matrix, and it has been shown to increase with the topical application of sea buckthorn oil. Hexosamine, glutathione, vitamin C, and catalase are also all noted to increase in the skin following topical use. Sea buckthorn may also exert anticancer effects. Oral ingestion of the oil is noted to improve fatty acid composition of the skin in patients with atopic dermatitis.

Caffeine and Related Compounds and the Skin

Animal studies on UV-damaged skin have shown that topical application of caffeine to sun-damage lesions may prevent the development of squamous cell cancers. Even drinking green tea or coffee may reduce cancerous changes in cells exposed to UV radiation. These beverages may also reduce the development of actinic keratosis and squamous cell cancers due to their protective effects against UV radiation. While excessive coffee drinking is irritating to the digestive and nervous systems, moderate amounts may offer some health benefits for the skin and possibly the brain.

Herbs Shown to Protect Skin from Ionizing Radiation

The following herbs may be used both topically and internally to help protect the skin from the damaging and inflammatory effects of radiation. Research suggests these herbs may have antiaging effects, protecting collagen from being damaged as well as helping prevent the generation of inflammatory compounds. **Caution:** *Hypericum* is not for topical use prior to radiation or sun exposure, because it photosensitizes skin.

Calendula officinalis
Centella asiatica
Ginkgo biloba
Hamamelis virginiana
Hippophae rhamnoides
Hypericum perforatum
Larrea tridentata
Mentha piperita
Ocimum tenuiflorum (also known as *O. sanctum*)
Panax ginseng
Phyllanthus amarus
Piper longum
Podophyllum peltatum
Syzygium cumini
Zingiber officinale

Skin Solution for Radiation Therapy

Aloe gel can be used as a simple, or may be combined with these oils to prepare a topical formula for use before, after, and between radiotherapy sessions to help protect the skin. Note that *Larrea* oil is not commercially available but is easy to prepare. Also, check the list of "Herbs Shown to Protect Skin from Ionizing Radition," above, which includes herbs that are noted to protect the skin, collagen, and connective tissue from ultraviolet and therapeutic radiation.

Larrea tridentata oil
Hippophae rhamnoides (sea buckthorn) oil
Ricinus communis (castor) oil
Aloe vera gel

Combine the desired or available oils with *Aloe vera* gel. The oils and the *Aloe* are not entirely miscible and will need to be shaken vigorously to homogenize prior to each use. Store in a small glass bottle and apply with the fingertips or a cotton ball.

Formulas for Skin Ulcers and Herpes Lesions

Skin ulcers are moist, raw, open craters and occur most often due to viral infections such as herpes or due to circulatory issues when on the lower legs. Herpes eruptions frequently begin as small vesicles that quickly ulcerate; they often take several weeks or more to resolve. Herbal therapies for herpes lesions may employ antiviral, immune-modulating, pain-relieving, and wound-healing agents.

Skin ulcers on the lower legs are often referred to as venous or stasis ulcers. Poor circulation leads to fluid stasis in the legs, and when combined with poor oxygen delivery, this can result in poorly healing wounds, stasis dermatitis edema, and ulcerations. Stasis ulcers are persistent, enduring for months and even years, becoming quite deep and prone to secondary infections. In diabetics, heavy smokers, and heart failure patients, severe ulcerations may become necrotic and lead to toe, foot, or lower limb amputations. Varicosities, blood clots, and obesity also predispose to stasis ulcers. Elevating the legs, compression dressings, and circulatory-enhancing agents complement the formulas below to best support healing of venous ulcers.

Herbs for Ulcerated Skin with Discharges

A variety of skin conditions may involve ulcers including autoimmune conditions, herpes, chicken pox, and other infections as well as burns and accidental exposure to caustic chemicals. The following herbs are mentioned in the folkloric literature as remedies to consider for ulcerated skin.

Apium graveolens
Betula pendula
Calendula officinalis
Centella asiatica
Echinacea angustifolia, E. purpurea
Galium aparine
Geranium maculatum
Glycyrrhiza glabra
Hamamelis virginiana
Hydrastis canadensis
Phytolacca decandra
Quercus alba, Q. robur
Sambucus nigra

Skin Wash for Ulcerated Skin with Tendency to Bleed

These astringent herbs are noted to help ulcerated and friable skin scab over. Stasis ulcers, herpes lesions, raw tissues from allergic reactivity, and other types of wet, open, ulcerated skin may benefit from a wash such as this. Prepared as an infusion, this formula can be used as a skin wash or prepared into compresses. *Hamamelis* proanthocyanidins have also been shown to have activity against the herpes simplex virus,[131] thus might be included in topicals for cold sores.

Achillea millefolium
Calendula officinalis
Hamamelis virginiana
Equisetum arvense
Quercus alba

Combine equal parts of the dry herbs. Use 1 tablespoon of the herb mixture per cup of water, bring to a simmer, and immediately remove from the heat. Let stand in a covered pan for 15 minutes and strain. Use as a skin wash or to prepare compresses for topical use.

Tea for Ulcerated Skin with Viral Infections

The herpes and the varicella viruses typically cause painful ulcers from cold sores to shingles. *Hypericum* is antiviral and can help address the nerve pain, and *Glycyrrhiza* is antiviral and noted to help heal ulcers through anti-inflammatory effects and by supporting glucosamine glycosides (GAGs molecules). Both *Calendula* and *Centella* can promote healing of ulcerated tissue. Both *Sambucus* flowers and berries have antiviral properties and are mentioned in the folkloric literature for viral exanthems. This formula can also be prepared as a tincture.

Calendula officinalis
Centella asiatica
Glycyrrhiza glabra
Hypericum perforatum
Sambucus nigra leaf and flower
Sambucus nigra berry powder

Combine in equal parts. Use 1 tablespoon of the mixture per cup of boiling hot water for 10 minutes and then strain. Aim to consume as much tea as possible throughout the day in cases of acute herpes outbreaks. This tea can also be used topically.

This formula is offered here as a supportive measure to heal skin eruptions. See also the following tincture and the chapter on infectious illnesses in a separate volume of this series, for ideas on how to better manage the pain of acute shingles.

Oral Anodyne Formula for Acute Shingles

Shingles lesions are extremely sensitive and painful, and this formula may help allay pain. Aconite, a potentially toxic botanical available to licensed clinicians only, is used in the formula. Do not exceed the recommended amount. *Hypericum* is both antiviral and a nerve anti-inflammatory, and *Piper methysticum* and *Scutellaria* both may help reduce the nerve-firing threshold and help to allay neuralgia.

Hypericum perforatum 20 ml
Piper methysticum 20 ml
Scutellaria lateriflora 19 ml
Aconitum napellus 1 ml

Take 1/4 to 1/2 teaspoon of the combined tincture every 2 or 3 hours.

Topical Anodyne for Acute Shingles

As shingles lesions are highly painful, it is difficult to apply salves and compresses. This formula is prepared into a topical spray so as to be as comfortable as possible. Aconite, a potentially toxic botanical available to licensed clinicians only, is used in the formula. Do not exceed the recommended amount.

Hypericum perforatum oil 20 ml
Aloe vera gel 18 ml
Calendula officinalis succus 18 ml
Aconitum napellus tincture 4 ml

Herbs for Crawling and Tingling Sensations

In historical herbals, these herbs are all emphasized as being specific for crawling and tingling sensations in the skin, which are sensations typical of many herpes lesions. Consider including one or more of these herbs in formulas for herpes infections, neuralgias, and other situations causing nerve irritation.

Apium graveolens	*Mentha piperita*
Hypericum perforatum	*Scrophularia nodosa*

Herbs for Hypersensitivity of the Skin

These herbs are listed in folkloric herbals for both topical and systemic use for those with hypersensitivity of the skin. Nerve irritation as well as nervous disorders such as autism or Asperger's can be associated with unusual sensitivity to clothing, bathing, shoes, and touch.

Allium sativum	*Ginkgo biloba*
Apium graveolens	*Hypercium perforatum*
Capsicum annuum	*Syzygium aromaticum*

Combine in a 2-ounce bottle and attach a spray nozzle. Shake well before each use as the ingredients may separate. Spray on the lesions as often as hourly, allowing to air-dry between applications.

Tincture for Herpes Chronic Outbreaks

The use of immune-modulating herbs can help reduce chronic herpes outbreaks. Also, support the immune system, reduce sugar and junk food when necessary, and treat underlying stress and anxiety if present. This formula contains antiviral adrenal and nerve-supportive agents in the form of *Glycyrrhiza*, *Hypericum*, and *Melissa* and general immune-modulating agents *Astragalus* and *Cordyceps*.

Astragalus membranaceus 20 ml
Cordyceps sinensis 10 ml
Glycyrrhiza glabra 10 ml
Hypericum perforatum 10 ml
Melissa officinalis 10 ml

Take ½ to 1 teaspoon of the combined tincture 3 times daily for 3 months. Reduce or stop taking altogether as the occurrence of lesions subsides.

Tincture for Acute Herpes Lesions and Cold Sores

Veratrum is a potentially harmful herb, but when taken in small amounts of just a few drops at any one time, it can reduce acute eruptions and hyperemic skin lesions. It is combined here with antiviral herbs, and the formula may be used both orally and topically. *Veratrum* is usually considered to be a prescription-only botanical, available to licensed clinicians only.

Glycyrrhiza glabra 20 ml
Hypericum perforatum 14 ml
Melissa officinalis 12 ml
Symphytum officinale 12 ml
Veratrum viride or *V. album* 2 ml

Take ½ teaspoon of the combined tincture as often as hourly at the onset, reducing after 24 hours to every 3 or 4 hours. After the lesions have scabbed over, typically at the close of a week, reduce the dose to 3 or 4 times a day. Remove the *Veratrum* from the formula for any formula refills needed. The tincture may also be put on a cotton ball and dabbed on lesions.

Tincture for Stasis Ulcers of the Legs

Because stasis ulcers are due to circulatory insufficiency, systemic measures to support the heart and improve circulation and venous return are important, as are these herbs that can assist wound healing.

Hypericum perforatum
Symphytum officinale
Echinacea angustifolia
Calendula officinalis

Combine in equal parts and take 1 dropperful of the combined tincture as often as hourly while elevating the legs.

Formulas for Cellulite and Stretch Marks

Cellulite involves weakening of the connective tissue structure, resulting in a dimpled appearance of the tissue. It is most common on the thighs and buttocks in women genetically predisposed. Fat deposition, poor circulation, and poor connective tissue repair and regeneration may contribute, as can complex hormonal factors. The fact that the condition is rarely seen in men suggests that estrogen may particularly play a role. Stress and the resultant high catecholamine levels may also contribute. Despite commercial products marketed for cellulite, there is little research to validate their efficacy in eradicating the condition. It is logical that improving circulation and supporting collagen and elastin regeneration below the dermis may be of value, and it would certainly do no harm and would benefit overall health.

Stretch marks, technically referred to as striae, have a very different appearance than cellulite but can also result from connective tissue damage. At times when the skin undergoes rapid stretching, such as the belly and breasts during pregnancy, the ability of the tissues to repair and regenerate collagen and elastin may not keep up with the demand. Rapid weight gain and growth spurts in adolescence are other situations that may result in the formation of stretch marks. In such circumstances the body may be forced to lay down fibrous tissue instead of functional collagen and elastin fibers, resulting in inelastic linear lesions. The use of cortisone may worsen the tendency to stretch marks, thinning the skin and impairing wound repair and new connective tissue formation. Adrenal diseases such as Cushing's syndrome and connective tissue disorders such as Marfan syndrome and Ehlers-Danlos syndrome all involve connective tissue weakness that may result in striae. As with cellulite, connective tissue support may be helpful. *Hypericum perforatum* (St. Johnswort) is a traditional remedy for tissue trauma, strains, and sprains and may support connective tissue repair. St. Johnswort oil may be rubbed into the abdomen and breasts during pregnancy to prevent stretch marks or be part of a broader protocol for treating the striae of Cushing's syndrome.

Tea for Thick, Doughy Skin and Cellulite

Many people may not realize that cellulite has more to do with the circulatory health than one's weight or exercise program. *Ceanothus* can support lymphatic circulation, and *Calendula* supports microcirculation in the dermis, which, along with exercise and systemic cardiovascular support, may help reduce cellulite. Some diabetics also develop thick, doughy skin as a result of poor energy usage and increased blood sugar metabolites being deposited in the tissues. *Stevia* is included here both for flavor and to improve insulin response. This formula can be the starting point of a customized formula for cellulite and can be added to depending on a person's individual circumstances. Address blood pressure, cholesterol, glucose, and basal metabolic rate as individually needed to round out this formula. This may be prepared as a tincture or a tea, adding flavor and specifics as desired.

Ceanothus americanus root, finely ground 2 ounces (60 g)
Calendula officinalis 2 ounces (60 g)
Camellia sinensis 1 ounce (30 g)
Stevia rebaudiana 1 ounce (30 g)
Rosa canina hips, finely chopped 1 ounce (30 g)

Combine and blend thoroughly. Use ½ to 1 tablespoon of the herb mixture per cup of hot water. Bring to a gentle simmer and remove promptly from heat source. Allow to sit in a covered pan for 10 to 15 minutes, and then strain. Drink freely, at least 3 cups per day.

Stretch Mark Oil

Hypericum oil alone may prevent stretch marks when applied topically to the belly and breasts during pregnancy, in my experience. *Equisetum* and *Centella* both support connective tissue strength and integrity, and *Curcuma* has positive effects on fibroblasts, collagen, and elastin, all helping to support dermal elasticity. Placing the herbal oils in a base of *Ricinus* (castor) oil may improve their penetration into the skin. *Hypericum* and castor oil are available commercially, but the other oils will need to be prepared oneself to make this formula.

Hypericum perforatum oil
Equisetum spp. oil
Centella asiatica oil
Curcuma longa oil
Ricinus communis oil

Combine several or all of the herbal oils in equal amounts and use topically twice daily. Applying to skin immediately after getting out of the shower would help facilitate penetration into the skin because the pores are open and circulation is stimulated by the shower's heat.

Formulas for Skin Fissures

Fissures are deep cracks on the skin and may occur due to extreme dryness, eczema, and other skin diseases, such as icthyosis, a genetic disorder that involves scaling and cracking in the skin. Contact dermatitis in atopic individuals may also induce fissures, such as with frequent dishwashing, hand washing, or exposure to chemicals. The first weeks of nursing a newborn baby can result in sore and cracked nipples, a condition that is painful but typically resolves within a month as the nipples become accustomed to the irritation. Anal fissures may occur due to chronic constipation. Cracks at the corners of the mouth, also called cheilosis or angular stomatitis, can occur due to B vitamin deficiency. Cracks on the heels of the feet can occur due to fungal infection such as athlete's foot or due to hypothyroidism.

Treat the various situations as directly as possible, but all cases may improve with the inclusion of quality oils in the diet, including fish oils, flaxseed oil, and evening primrose oil. Innate lipids such as ceramides are important to skin health, and low levels of ceramides are associated with eczema, psoriasis, and easy fissure formation. Rice bran oil is high in naturally occurring ceramides, and commercial creams and supplements are becoming available. Drink adequate water to help keep the skin well hydrated. The use of creams, herbal salves, and lotions may be warranted, and applying oils after bathing can help seal moisture into the skin.

Salve for Nipple Fissures in Nursing Women

Nipple irritation typically occurs in the first week following childbirth as tender nipples are unaccustomed to the rather significant trauma of frequent nursing. The skin will adapt over several weeks, but that period can be very uncomfortable for a new mother as she continues feeding her infant.

Aloe vera gel 15 ml
Calendula officinalis oil 15 ml
Hippophae rhamnoides oil 15 ml
Ricinis communis oil 15 ml

Combine all in a 2-ounce bottle and shake well. Apply after every nursing, and as often as possible in between. The herbs in this salve are not harmful to a nursing baby, but the salve should be gently washed off with soap and water before each nursing.

Oral Tincture for Heel Fissures

Deeply fissured heels can be a sign of hypothyroidism, which should be addressed directly if suspected. Deeply fissured heels can also be a type of chronic eczema, although it is not typically pruritic or vesicular. Eczema should also be addressed systemically if a person displays other evidence of atopy. Those with constitutional dryness and poor circulation may also develop fissured heels. This is an all-purpose starting formula to cover all these bases: *Centella* promotes general healing, *Coleus* may improve both hypothyroidism and chronic inflammatory states, and *Ginkgo* may improve circulation to the feet. Combine this treatment with the Hydrotherapy Treatment for Heel Fissures below.

Centella asiatica
Coleus forskohlii
Ginkgo biloba

Combine in equal parts and take ½ to 1 teaspoon of the combined tincture 3 or more times daily.

Hydrotherapy Treatment for Heel Fissures

A topical and hydrotherapy treatment can help cosmetically as well as promoting circulation and healing of the feet. The treatment would complement an oral therapy chosen to address any underlying issue suspected to contribute to the problem. For this treatment, you'll need two tubs or pans, a pumice stone, castor oil, *Calendula* or *Comfrey* salve, an old pair of cotton socks, and a pair of woolen socks.

1. Immerse the feet in a tub or pan of very hot water for 5 minutes. Then immerse in a tub or pan of icy cold water for 1 minute.
2. Repeat the hot water/cold water soaks three more times.
3. Remove feet from the water and use the pumice stone to remove as much dead skin as possible.
4. Rub a thick coat of castor oil into the heels. Repeat 3 times. Put on the cotton socks and then the wool socks. The double sock covering will drive the oil into the body, and prevent the sticky oil from making a mess.
5. After about an hour, remove the socks and rub in a thick coat of salve. Put the two pairs of socks back on for another hour.

Repeat this treatment daily if possible or at least 3 times a week.

Herbs for Fissures Between the Digits

Fissures on the fingers and between the toes are usually a variation of eczema and associated with atopic dermatitis and therefore will benefit from systemic therapies similar to those appropriate for other types of eczema, addressed in formulas earlier in this chapter. The following herbs are mentioned in the folkloric texts for fissures of the digits. If there is any evidence of a fungal infection in a warm wet climate or circumstance, the condition should instead be treated as athlete's foot.

Apium graveolens
Calendula officinalis
Centella asiatica
Hypericum perforatum
Linum usitatissimum oil
Oenothera biennis oil
Ribes nigrum oil
Syzygium aromaticum
Thuja plicata

Sitz Bath for Anal Fissures with Constipation

Anal fissures can be uncomfortable as well as recurrent and slow to heal. Bowel health and constipation should be addressed (See chapter two, "Formulas for Constipation" on page 46) while this sitz bath therapy begins to provide pain relief and support healing.

Symphytum officinale root, as finely chopped as possible ½ cup (115 g)
Quercus alba, Q. robur bark, in small chunks ½ cup (115 g)
Calendula officinalis flowers ½ cup (115 g)
Hamamelis virginiana leaves ½ cup (115 g)
Epsom salts

Soak the *Symphytum* root and *Quercus* bark in 12 cups of cold water overnight or at least for 3 or 4 hours in a large stockpot (or several smaller ones). Soaking helps dissolve the soothing and healing demulcent components out of the dried *Symphytum* root. Bring the water to a gentle simmer over low heat and hold there for 5 minutes. Remove from the heat and add the *Calendula* flowers and *Hamamelis* leaves. Cover the pan and let steep for 15 to 20 minutes. Obtain a rubber tub large enough to sit in and place it in a bathtub. Strain off the hot liquid into the rubber tub. Add the Epsom salts and agitate the water to help them dissolve. Immersing the rear end, sit in the liquid for 15 or 20 minutes, and follow with a cool shower. Repeat 3 times daily if possible. Apply *Aloe* gel or *Calendula* salve after each shower.

To dispose of the used sitz bath brew, pour it onto a lawn or garden outdoors to avoid putting too much particulate down the household drain.

Herbs for Fissures of the Anus

Chronic constipation can increase pressure and congestion in the bowels and lead to anal fissures as well as hemorrhoids. Alternately, fissures may result with unusually frequent bowel movements passed with straining, urgency, or cramping. Improving intestinal health and elimination is essential to resolving the condition by taking probiotics, supplementing fiber, and optimizing the diet. Irritant and pharmaceutical laxatives should be avoided wherever possible as they typically increase pressure and stress on the tissues. The following herbs can play a role in helping with healing of anal fissures.

Aloe vera. Use aloe juice in juices and teas for its lubricating and gentle laxative effects. Use gel on cotton ball against the anus to support healing.

Aesculus hippocastanum. Include when liver congestion, chronic constipation, hemorrhoids, or varicose veins contributes to anal fissures. *Aesculus* may be included in teas and tinctures as well as prepared into a variety of topical applications.

Calendula officinalis. Use *Calendula* salves directly on the anus and/or include in sitz baths to assist healing of anal fissures.

Hamamelis virginiana. This herb is indicated for tissue irritation, fissures, and hemorrhoids. *Hamamelis* may be used both topically in sitz baths and internally in teas and tinctures.

Hydrastis canadensis. Include in tinctures when poor bowel health or frequent loose stools with mucous or blood are present. *Hydrastis* salves and compresses may help when used topically as well.

Quercus robur, Q. alba. The astringent bark is used for fissures with itching and burning. *Quercus* may be included in both topical and internal formulas.

Symphytum officinale. *Symphytum* is soothing when there is pain with bowel movements and can promote healing of broken tissues when used topically in salves, sitz baths, and compresses.

Thuja plicata. Consider *Thuja* when warts, fungus, or sexually transmitted diseases accompany rectal problems and anal fissures.

Urtica urens, U. dioica. *Urtica* (nettles) can be included in formulas for fissures with burning and itching. Include *Urtica* in teas and tinctures for systemic support.

Formulas for Miscellaneous Dermatologic Conditions

This last section details several topics that don't fit easily into other categories of skin conditions in this chapter. Gangrene is not common in the modern world, but it can occur in diabetics or those with poor circulation. Many skin cancers, such as squamous cell cancers, are slow-growing and slow to metastasize, and herbal therapies can be valuable components of broader therapeutic protocols.

Herbal Vinegars for Weak Fingernails

Weak fingernails that peel and break can be related to inadequate protein intake or more commonly, to poor digestion and/or poor mineral status in the body. These herbs are all excellent sources of organic mineral complexes that are best extracted with vinegar macerations. Neither tinctures nor tea are efficient in extracting minerals.

Centella asiatica
Equisetum spp.
Medicago sativa
Urtica urens, U. dioica
Symphytum officinale

Choose one or more of the herbs and place roughly 1 ounce of each in the bottom of a blender. Cover the dry herbs with apple cider vinegar or other vinegar of choice and liquefy to a small particulate size. Transfer to a glass jar and shake daily for 4 to 6 weeks, then strain, and store in a dark cupboard. Consume this vinegar by adding it to sauces, marinades, and dressings through the week.

However, to get the most minerals, prepare herbal beverages with vinegars, which is more delicious than it might sound. Consuming regularly in beverages allows for a greater quantity to be consumed. Place 1 to 2 tablespoons of the herb(s) in the bottom of a glass, add 1 tablespoon of fruit juice of choice, a lemon wedge, an ice cube or two, and top off with 2 cups of sparkling mineral water. Prepare and drink once or twice a day for many months.

Tincture for Gangrene

Thankfully, herbalists and most physicians in the United States do not see many cases of gangrene. This formula is based on folkloric references to the use of *Echinacea* for decay, necrotic tissue, and gangrene. *Ginkgo* can help perfuse the limbs to help save tissue. *Calendula* may promote microcirculation and healing, while *Phytolacca* has immune-modulating effects and lymph-enhancing actions. However, gangrene requires expert medical care, typically in a hospital setting, and while this tincture may be used in tandem with surgical and pharmaceutical approaches, it is listed here for historical interest.

Echinacea purpurea, E. angustifolia 60 ml
Ginkgo biloba 20 ml
Calendula officinalis 20 ml
Phytolacca decandra 20 ml

General Therapy for Skin Cancer

Basal and squamous cell skin carcinomas are the most common of all cancers. They have a fairly good prognosis because they are usually noticed early and can often be successfully excised. Malignant melanoma is less common and has a poorer prognosis, tending to metastasize aggressively. *Calendula* is a gentle vulnerary with activity against melanoma cells. *Ricinus communis* (castor oil) is believed to have many immune-enhancing mechanisms and might be part of topical application and recovery programs following surgery. Castor oil may also help to pull medicinal compounds deeper into the skin, as the lectins affect cell-to-cell adhesion. *Curcuma longa* (turmeric) has anti-inflammatory, alterative, skin-protective, and anticancer effects and is safe and gentle to include in a variety of protocols, as complementary or standalone therapies. *Podophyllum* is a potentially caustic herb for skilled clinicians only, but it has anticancer properties with topical application, especially in cancer associated with papilloma and condyloma viruses.

Herbs for Poison Oak and Ivy

These herbs are all mentioned in the ethnobotanical literature and pioneer herbals as being effective to remove the inflammatory poisons in the oil of poison ivy and poison oak from the skin and treat the skin inflammation. Tinctures and teas of these herbs can be prepared into skin washes and compresses.

Actaea racemosa
Alnus serrulata
Avena sativa (oatmeal baths)
Grindelia squarrosa
Lavandula spp. (essential oil)
Sanguinaria canadensis

Take 1 teaspoon of the combined tincture every 30 minutes. Seek professional medical attention promptly.

Tincture for Frostbite

Frostbite can lead to tissue death and even loss of digits, and all cases require expert medical attention. Herbal therapies might be complementary to medical treatment and promote healing. *Echinacea* may help save injured tissue and is frequently mentioned historically for gangrene and decaying tissue. *Angelica* may improve circulation in the injured tissue, and *Hypericum* can help allay pain from inflamed nerves.

Herbs for Severe Skin Lesions and Cancers

While skin cancers should be treated by an experienced dermatologist and/or oncologist, these herbs are either of historical note or have enough modern research to warrant mentioning

Conium maculatum. Poison hemlock can be used by skilled clinicians only, in tiny dosages of just several drops at a time internally to help treat pain.

Eucalyptus globulus. Eucalyptus has anti-inflammatory effects and masks the stench of decaying tissue when used topically.

Galium aparine. Cleavers is a traditional remedy for lymphatic cancer and pain, fluid stasis, and enlarged lymph nodes.

Hydrastis canadensis. Goldenseal can be used topically for astringent and antimicrobial effects.

Hypericum perforatum. St. Johnswort can be used topically and internally for inflamed and injured nerve endings.

Podophyllum peltatum. Mayapple contains the caustic resin, podophyllin, that is used to destroy warts and skin growths and shown to have anticancer properties with topical application, especially in cancer associated with papilloma and condyloma viruses.

Thuja plicata. Western red cedar is a traditional anticancer remedy and antiviral specific for warty fleshy growths.

Echinacea angustifolia 30 ml
Angelica sinensis 15 ml
Hypericum perforatum 15 ml

Take 1 teaspoon of the combined tincture every 30 to 60 minutes.

Specific Indications: Herbs for Dermatologic Conditions

Many herbs have an affinity for the skin, and this section of the chapter offers unique or specific indications for which individual herbs would be best used. Some plants may be niche herbs to treat specific diagnoses, such as allergic or infectious conditions, or for specific sensations such as burning or itching.

Achillea millefolium • Yarrow

The young flowering tops are used topically and internally to control bleeding wounds and as antimicrobials, including to treat staphylococcal and streptococcal infections. *Achillea* is useful topically as a poultice, plaster, or skin wash to treat infected wounds, folliculitis, boils, and fungal infections. *Achillea* is specific for skin lesions associated with digestive symptoms and for bleeding hemorrhoids and blood in the stool. *Achillea* has general alterative properties for acne, psoriasis, and eczema and improves concomitant intestinal inflammation, IBS, and liver congestion.

Actaea racemosa • Black Cohosh

Medicines are prepared from the root and are used topically and internally as antidotes for poison ivy. *Actaea* may also be included in formulas when skin or hair changes emerge at menopause or due to female hormonal imbalances. This herb was previously named *Cimicifuga racemosa*, and it may still be referred to by that name in some sources and by vendors.

Allium cepa • Onion

Use fresh raw minced onions topically for burns, to promote healthy wound healing, and to reduce keloid formation. Fresh raw onions may also be used topically to relieve the pain of corns or bunions. Onion poultices may improve seborrheic keratoses, treat burns, and reduce scars, when topically applied.

Allium sativum • Garlic

Garlic bulbs are prepared into medicines to be used orally or topically in formulas and protocols for skin infections, including staphylococcal and streptococcal infections, scabies, athlete's foot, thrush, and other parasitic, fungal, and infectious conditions of the skin. Oils, foot soaks, plasters, and compresses are among the many ways that fresh garlic, garlic powder, and garlic oil can be used topically to treat dermatophytes and various skin fungi, such as athlete's foot. *Allium* may be considered as an ingredient in formulas for warts, both topically and orally. Garlic capsules and tinctures are readily available and may complement topical approaches to treating skin fungus and infection; they may be included in formulas for diabetics who have chronic skin infections. The ingestion of garlic may also deter warts and papillomas.

Alnus serrulata • Alder

Bark or leaf preparations are specific topically as antidotes for poison oak and poison ivy and as skin washes for minor inflammations, pustular eczema, and crops of boils. Alder leaves of many species have tannins and astringent compounds that can help dry up weeping or pustular lesions.

Aloe vera • Aloe

The pulp of the succulent leaves is used topically for burns, to speed healing, and to protect the skin prior to radiation therapy. *Aloe* is specific internally in cases of portal congestion and lymphatic congestion in older, overweight, sedentary people. *Aloe* may improve a variety of skin diseases when taken as a daily drink and be therapeutic to burns, frostbite, wrinkles, stretch marks, pigmentation disorders, and wounds. *Aloe* may reduce skin damage in patients undergoing radiation treatments.

Ammi visnaga • Khella

Khella may be used orally to reduce allergic and inflammatory phenomena. Khella is specific for eczema and concomitant asthma. Tinctures are prepared from leaves and seeds and are used topically to increase photosensitivity in the skin as a therapy for dermatitis and psoriasis. Khella may be used topically in combination with UV light to stimulate melanin production in cases of vitiligo. Consider khella as a foundational herb in formulas for multisystem atopy.

Andrographis paniculata • King of Bitters

This herb, from traditional Ayurvedic medicine, is becoming established in Western herbalism for use as a liver protectant, immunostimulant, anti-infective, and antitoxin. The entire plant has been used as an herbal medicine, especially the roots. Include *Andrographis* as a foundational herb in treating skin conditions that are related to digestive and liver disorders, associated with chronic infectious complaints.

Angelica sinensis • Angelica

The roots of angelica, also called dong quai, are used orally to reduce allergic tendencies in cases of chronic dermatitis and are used topically to induce photosensitivity as part of a protocol for psoriasis or vitiligo. *Angelica* is specific for vascular congestion, pelvic stagnation, menstrual cramps, and allergies.

Apium graveolens • Celery

Tinctures prepared from *Apium* seeds are used to reduce allergic reactivity in cases of dermatitis, urticaria, and ulcerative lesions with profuse discharges. *Apium* seeds are also brewed into teas and may be used topically or in large internal dosages in combination with UV light, to induce photosensitivity for the treatment of psoriasis and vitiligo. *Apium* is specific for itching blotchy skin lesions and a sensation of creeping or crawling on the skin.

Arctium lappa • Burdock

Burdock, also called gobo root, is indicated for acne and all types of chronic skin disorders, such as eczema and psoriasis. *Arctium* is specific for recurrent crops of boils and recurrent styes. *Arctium* is useful as an alterative to improve the liver's processing of hormones in cases of melasma, hyperestrogenism, intestinal dysbiosis, and hyperlipidemia. *Arctium* is especially indicated for mild constipation, acne and skin eruptions, coated tongue, and vague malaise.

Arctostaphylos uva ursi • Uva Ursi

A strong tea of unva ursi, also called kinnickinick, might be used topically to suppress melanocytes in the skin to treat hyperpigmentation disorders. Azelaic acid in uva ursi and other family members is credited with the inhibition of melanogenisis. Azelaic acid may also reduce redness in cases of telangiectasia, rosacea, and postinflammatory hyperpigmentation,[132] though it can be drying and irritating.

Arnica montana • Arnica

Arnica leaves are used to prepare oils and liniments employed topically for bruising, trauma, and soft tissue injury, and are specifically indicated for swollen tissue that one tender to pressure. The plant is also called leopard's bane. Homeopathic preparations of arnica are used orally, more so than full strength *Arnica* tinctures, due to potential irritating effects of *Arnica*. *Arnica* is not to be used on broken skin, unless highly diluted, and is most indicated for strains, sprains, crush injuries, blows, and trauma that do not abrade or lacerate the skin.

Astragalus membranaceus • Milk Vetch

Use root decoctions orally to reduce allergic skin conditions, to reduce the tendency to infections, and for autoimmune conditions. *Astragalus* may be included in formulas to use internally for warts, especially when accompanied by frequent infections, fatigue, and lassitude. *Astragalus* teas are bland and enjoyable, and the crude powder can be worked into smoothies and medicinal truffles. Tinctures and encapsulated products are widely available. *Astragalus* has also been used topically for skin inflammation and lesions, and fermented product recipes are also seen traditional folklore.

Atropa belladonna • Belladonna

Use homeopathically or in small doses botanically for skin that is dry, hot, and bright red, such as with streptococcal infections or erysipelas. Belladonna is specific for sudden onset of high fever with marked vascular congestion, restlessness and throbbing, or burning sensations. Belladonna is also specific for hyperaesthesia of the skin and skin that is red, swollen, and painful, with heat and congestion. *Atropa belladonna* is also indicated for skin complaints accompanied by fever and tender swollen glands, for acne rosacea, for suppurative lesions or wounds, and for rapidly emerging pink, red, or pustular skin eruptions. *Atropa belladonna* is considered a prescription-only botanical, for skilled herbalists only. The dose for belladonna must be low to prevent dryness, visual disturbances, and CNS effects, and the plant is also widely prepared into a homeopathic with similar indications.

Avena sativa • Oats

"Oats," meaning oatmeal, coarsely chopped oats, and oat powders, are used to treat skin complaints topically in baths, plasters, and poultices for pityriasis and to allay the itching of chicken pox, mosquito bites, and other pruritic skin conditions. Oat straw teas and medicinal oat straw vinegars are used long term to supplement minerals in those with malnutrition, nervous digestive disorders, weak fingernails and hair, and poorly healing wounds.

Azadirachta indica • Neem

Most well-known as an antifungal and antiparasitic, neem is also useful topically on boils, blisters, and skin tumors. The diluted oil may be included in skin washes for acne, fungus, and other skin infections. Neem oil may also be used topically undiluted for fungal infections of the skin, on the scalp to kill lice, and in various formulas for scabies, mites, and other skin parasites.

Calendula officinalis • Pot Marigold

Calendula flowers and sometimes the young leaves are used as an all-purpose skin herb for wounds and trauma and to promote healing and treat skin infections. *Calendula* may minimize scarring in those with a tendency to keloids or easy scarring because *Calendula* supports collagen synthesis right below the dermis. Use *Calendula* washes and products topically in cases of erysipelas. Use *Calendula* salves, ointments, and compresses topically on the fissured nipples of nursing mothers. Dilute *Calendula* tea with rosewater for use in the eyes in cases of conjunctivitis. *Calendula* may be used topically and internally simultaneously in the treatment of chronic skin conditions and poorly healing wounds and to help resolve chronic stasis ulcers in the lower legs.

Camellia sinenesis • Green Tea

The leaves of green tea are used to prepare teas, tinctures, capsules, and concentrates and are reported to stimulate hair growth in cases of alopecia when used topically. The stimulating effects of *Camellia* (formerly known as *Thea*) can be used as a synergist in skin formulas where hypothyroidism and general atony and insufficiency accompany.

Cannabis sativa • Marijuana

Marijuana or hemp seed oil may improve eczema, seborrheic dermatitis, psoriasis, lichen planus, and acne rosacea when taken orally or applied topically.

Capsicum annuum • Cayenne

Capsicum tinctures are prepared from the ripe fruits and may be diluted with glycerin and taken orally to promote the healing of extensive ecchymoses. *Capsicum* may be used topically to promote hair growth in cases of alopecia, taking great care to keep it out of the eyes. *Capsicum* may enhance the absorption of other ingredients in herbal tinctures. Due to its strength and intense flavor, *Capsicum* is usually used in lesser proportion to other herbs in formulas. It is also known as *Cayenne frutescens.*

Ceanothus americanus • Red Root

The roots of *Ceanothus*, also called New Jersey tea, are specific for thick doughy skin associated with diabetes and digestive dysfunction. *Ceanothus* may help move fluid through the tissues when liver congestion, pelvic and portal congestion, vascular congestion, diabetes, and hypertension accompany or underlie congestion in the skin.

Centella asiatica • Gotu Kola

The leaves may be dried for teas, prepared into tinctures, or powdered to use topically to treat thickening of the skin, exfoliation, and scales. *Centella*, also known as Pennywort, is most indicated for dry scaly eruptions, as opposed to wet suppurative eruptions. Consider including *Centella* in formulas for acne and papular eruptions with tendency to scale formation, for psoriasis and skin eruptions affecting the soles of the feet, and for excessive sweating. *Centella* may be used as a connective tissue tonic, to promote healing of the skin, joints, mucous membranes, and connective tissues. *Centella* is also indicated following injuries to improve connective tissue integrity and repair and to prevent keloids in those with a prior history.

Chelidonium majus • Celandine

The aerial parts of *Chelidonium* are specific for acne, boils, and skin complaints related to poor digestive and biliary function. *Chelidonium* is also specific for itching and dryness of skin. *Chelidonium* is an effective cholagogue and is indicated for skin disorders accompanied by jaundice, swollen bile ducts, and right upper quadrant pain that radiates to the shoulders and for a full, pale coated tongue.

Cinnamomum species • Cinnamon

The bark can be prepared into teas and tinctures to include in formulas for skin fungus and chronic skin infections in diabetics and those with poor peripheral circulation. Cinnamon essential oil may be diluted with water to use as a topical disinfectant.

Citrus bergamia • Bergamot

The plant contains photosensitizing bergaptens that may be used in small amounts to treat psoriasis vitiligo and other recalcitrant skin lesions.

Cnicus benedictus • Blessed Thistle

The young leaves and stems may be used for skin complaints related to poor digestive and hepatic function. *Cnicus* is useful as an alterative to improve the liver's processing of hormones, carbohydrates, and lipids.

Cnidium officinale, C. monnieri • Snow Parsley

Several species of *Cnidium*, also known as osthole, are used medicinally and contain similar coumarin compounds. Use *Cnidium* orally to reduce allergic tendencies in dermatitis cases. *Cnidium* is a potential photosensitizing agent for cases of vitiligo, psoriasis, and chronic skin complaints. *C. officinale* is sometimes called *Ligusticum officinale.*

Coffea arabica • Coffee

Roasted coffee beans can be used to prepare washes and lotions to use topically on the skin to enhance the penetration and absorption of other herbs or topical agents.

Coleus forskohlii • Coleus

Coleus leaves are specifically indicated for psoriasis and chronic dermatitis, especially when sluggishness of general metabolic functions with excessive inflammatory response are present.

Commiphora mukul • Guggul

Gum from guggul trees is indicated for elevated glucose or lipids due to diabetes, hypothyroidism, or slow metabolism and may be used for doughy skin and susceptibility to skin infections.

Commiphora myrrha • Myrrh

Gum from myrrh trees is processed into tinctures and essential oils. Myrrh tincture prepared with the essential oil as an ingredient can be used topically for infectious complaints and orally for thrush and infections of the oral mucosa, including infectious or inflammatory causes, such as lichen planus and oral manifestations of autoimmune or other systemic illnesses. Myrrh has activity against candida and numerous oral and systemic pathogens. Use myrrh orally in cases of mucosal lesions in the mouth.

Conium maculatum • Poison Hemlock

Conium leaves and seeds are used in very small dosages medically because the herb is potentially toxic. Poison hemlock should be used by experienced clinicians only. The freshly pressed juice is also used. The topical use of *Conium* may allay the pain of cancerous growths and ulcers on the skin, but care must be taken because, even topically, enough of the powerful alkaloid coniine can be absorbed to cause systemic toxicity, resulting in bradycardia and suppression of nerve regulation.

Coptis trifolia, C. chinensis • Goldthread

The bright yellow roots are used for infectious conditions of the skin, including antibiotic-resistant staph infections. *Coptis* is indicated for chronic infections in mucous membranes and for fungal infections of mouth, skin, and vagina. Use *Coptis* orally for thrush in babies, in the mouth for aphthous ulcers, and orally and/or topically for chronic infections in diabetics.

Cordyceps sinensis • Caterpillar Fungus

The entire fungus is prepared into medicine to use orally for streptococcal infections of the skin. *Cordyceps* is specific for immune deficiency and serious infections with difficult recovery.

Coriandrum sativum • Coriander

The leaves of *Coriandrum sativum* are referred to as cilantro, while the ripe seeds are ground into a powder referred to as coriander. Coriand eressential oil or other concentrate may be used topically as a photosensitizing agent in cases of psoriasis or vitiligo.

Crataegus oxyacantha, C. monogyna • Hawthorn

Use *Crataegus* berry products for extensive bruising and skin trauma cases and to improve microcirculation in the skin. *Crataegus* may improve those who tend toward easy contusions with slow recovery, vascular inflammation, and allergic reactivity including rosacea, hives, telangiectasias, and phlebitis. *Crataegus* may improve excessive perspiration. *Crataegus oxyacantha* is also known as *C. rhipidophylla*.

Crocus sativus • Saffron Crocus

The bright orange-red stigmas from the flowers are a prized food coloring. Oral consumption of saffron may exert chemopreventive effects by inducing cellular defense systems.

Curcuma longa and other species • Turmeric

Curcuma roots and the curcuminoids they contain offer significant anti-inflammatory and antimicrobial properties, making them useful internally to promote skin repair and prevent keloid formation. *Curcuma* may reduce excessive dermal proliferation in cases of psoriasis. Use *Curcuma* to protect the liver if using Retin-A and in formulas for acne or melasma related to elevated hormones, because *Curcuma* may assist in hormone clearance via the liver.

Daucus carota • Wild Carrot

Carrot seed oil is useful in skin rejuvenation oils and creams for topical use. It may improve telangiectasias and florid skin and may have chemopreventive effects.

Echinacea angustifolia, E. purpurea • Coneflower

Echinacea roots and in some cases the young stems and leaves are widely used herbal medicines for infections. *Echinacea* is specific for infectious skin conditions, boils, carbuncles, and pustular eruptions and for septic conditions with emaciation and debility. It may also help treat wounds, warts, ulcers, burns, herpes, hemorrhoids, and psoriasis. *Echinacea* can improve chronic folliculitis. *Echinacea* may be used topically as both an anaesthetic and antiseptic. *Echinacea* is specific for skin lesions associated with decay such as tibial ulcers and gangrene and may reduce fetid odor when applied to gangrenous and cancerous ulcerations. Use *Echinacea* orally for bad breath due to dental and gum diseases and topically for frostbite, snakebites, and insect stings. Use *Echinacea* internally to reduce itching in cases of urticaria and to reduce a tendency to abscesses associated with poor digestion and "bad blood."

Eleutherococcus senticosus • Eleuthero

The roots of eleuthero, also called Siberian ginseng, are prepared into medicines to use in skin formulas where underlying allergies or atopic tendencies are present. *Eleutherococcus* is specific for individuals with poor immune response and tendency to infection or excessive immune response and autoimmune skin conditions. Include *Eleutherococcus* in formulas for warts and chronic viruses and as a synergist in formulas for chronic digestive irritability associated with nervous affectations. *Eleutherococcus* is an adrenal tonic that may be used for skin complaints associated with long-term stress resulting in nervous symptoms and fatigue.

Equisetum arvense • Horsetail

The aerial parts of horsetail, also called scouring rush, may be dried to use in teas and capsules for poorly healing wounds, scars, keloids, and skin that is easily injured and friable. Due to its high mineral content, *Equisetum* may help those with fragile fingernails, dry frizzy hair, and chronic strains and sprains. Use *Equisetum* for intestinal ulcerations and to help rebuild connective tissues and mucous membranes. Include *Equisetum* in teas to provide minerals to treat skin conditions where malnutrition and malabsorption may underlie poor healing capacity.

Eucalyptus globulus • Eucalyptus

The young leaves can be prepared into teas and tinctures or distilled to produce concentrated essential oils. Use the essential oil topically for head and body lice. *Eucalyptus* is a powerful antiseptic, expectorant, and diaphoretic. *Eucalyptus* is specific for herpetic skin eruptions with swollen glands; nodular swelling over joints, such as with rheumatoid arthritis; and autoimmune conditions affecting the connective tissues. Use *Eucalyptus* topically to mask the odor of fetid skin lesions such as gangrene and cancerous ulcers as well as to provide pain relief. Include *Eucalyptus* in mouthwash formulas for bad breath and oral lesions. *Eucalyptus* in glycerin may help treat Demodex mites, scabies, and other skin parasites when topically applied.

Eupatorium perfoliatum • Boneset

Eupatorium leaf preparations can be included in formulas for skin complaints associated with viral and bacterial infections. *Eupatorium* is specific for coated tongue and cracks at the corners of the mouth, for aching back and muscles associated with the flu, and for unusual thirst, chilliness, and nausea associated with the flu.

Euphrasia officinalis • Eyebright

Young leaves and flowering tops of *Euphrasia* are specific for inflamed conjunctiva with profuse lachrymation, flushed face, and skin eruptions associated with viral infections.

Foeniculum vulgare • Fennel

Fennel seeds may be dried, prepared into tinctures, or distilled into essential oils. Teas and tinctures may be used orally to reduce allergic tendencies in cases of chronic dermatitis. Include fennel seeds in formulas for dermatitis and chronic inflammation. Fennel essential oil may be used topically to induce photosensitivity as part of a protocol with UV light in cases of psoriasis or vitiligo. Fennel has hormonal effects that may help reduce hirsutism with oral use. Fennel is specific for gas and bloating, peptic distension causing fullness and discomfort, burping, and cramping and gurgling in the intestines and can be considered in any skin formula where these symptoms accompany.

Forsythia suspensa • Forsythia

The small fruits of *Forsythia* are used to treat erysipelas and inflammation or ulceration in the skin as well as to cool fever.

Fucus vesiculosus • Bladderwrack

The fronds of *Fucus* can be orally consumed as a source of iodine and minerals and used topically as a healing and antiaging therapy for the skin. *Fucus* is specific for dermatitis and dry skin in hypothyroid patients. Include *Fucus* in formulas for obesity, goiter, exophthalmos, constipation, and flatulence (symptoms of hypothyroidism). When topically applied *Fucus* may improve elasticity in aging skin. Seaweed wraps are claimed to reduce cellulite but this has not been scientifically validated. Use with caution in those allergic to iodine.

Galium aparine • Cleavers

The aerial parts of *Galium* may be dried, processed into tincture, or juiced. Bedstraw, also called cleavers, is emphasized in the folkloric literature for skin cancer, ulcerative lesions, superficial tumors, and swollen lymph nodes. *Galium* is also specific for chronic skin conditions and the oral manifestations of scurvy. *Galium* may be used both topically and internally for these complaints.

Ganoderma lucidum • Reishi Mushroom

The entire woody mushroom may be sliced and dried (for use in teas), powdered, or tinctured. Include reishi in skin infection and inflammation formulas when frequent or lingering infections, fatigue, and general malaise accompany the skin complaints.

Gentiana lutea • Gentian

Bitter gentian root preparations are specific for skin complaints associated with poor digestion, an acidic stomach with ravenous hunger, nausea, a sense of weight or pain in the stomach, and colicky pain with flatulence and abdominal tenderness. Gentian may be a supportive rather than a primary herb in formulas for skin complaints, often in a lesser proportion than other herbs in a tincture formula.

Geranium maculatum • Wild Geranium

The roots of wild geranium are used to prepare astringent teas and tinctures and are included in pastes/powders to apply topically to reduce bleeding and discharges from the skin and mucous membranes.

Ginkgo biloba • Ginkgo

Ginkgo, which is also called maidenhair tree, is indicated for skin lesions due to vascular insufficiency, such as stasis ulcers, for diabetics with poor healing wounds, or for dermal lesions due to circulatory insufficiency. *Ginkgo* may also exert an antiallergy effect when used systemically for several months.

Glycyrrhiza glabra • Licorice

Licorice may be included in skin formulas where underlying allergies or atopic tendencies are present. Licorice is also indicated for individuals with poor immune response and tendency to infection as well as for excessive immune response and autoimmune skin conditions. Include *Glycyrrhiza* in formulas for skin eruptions due to underlying viral infections as well as in formulas for hives, dermatitis, and chronic inflammation of the skin. Licorice as a solid extract is especially indicated for oral application in cases of oral ulceration, oral lichen planus, oral manifestations of lupus, erythema multiforme, ulcers, canker sores, or other lesions of the oral mucosa. Licorice solid extract can be formulated into a variety of gels, sprays, and lozenges for these purposes. *Glycyrrhiza* may also be valuable in cases of hirsutism and acne due to hormonal imbalances, such as polycystic ovarian syndrome.

Grifola frondosa • Maitake

Medicines are prepared from the entire *Grifola* mushroom and may be included in chronic skin infection formulas as an immunostimulant and antiviral agent, as a supportive measure rather than as a primary herb. The mushroom is edible and is also known as hen of the woods—and may support immune status when eaten regularly.

Grindelia squarrosa • Gumweed

The sticky resinous young flowering tops of gumweed are prepared into herbal medicines that are reported to be antidotes to poison oak and ivy in the folkloric literature. *Grindelia* is specifically indicated topically for burns, blisters, herpetic eruptions, vesicular and papular eruptions with itching and burning, and chronic indolent skin ulcers surrounded by purplish discoloration.

Gymnema sylvestre • Sugar Destroyer

The leaves of *Gymnema*, also called cow plant, are prepared into medicines to include in formulas for chronic fungal infections and skin complaints in those with blood sugar imbalances or in formulas for those with skin lesions due to poor circulation, such as diabetic ulcers. Those individuals with wide swings in blood sugar and reactive hypoglycemia may be sensitive to *Gymnema*; start with a low dose and work up in such cases.

Hamamelis virginiana • Witch Hazel

Young witch hazel stems may be decocted, or the bark and leaves may be dried or tinctured to use medicinally. Use witch hazel tea as a topical and internal anti-inflammatory for the skin and for vascular and mucous membranes. Witch hazel is specific for skin trauma and bruising and is useful to control bleeding in wounds when used as a topical compress. *Hamamelis* may also be used topically to relieve the pain of burns and sunburns and to astringe wounds, dilated capillaries, and shaving nicks and injuries. *Hamamelis* is also specific topically for ulcers, hemorrhoids, bruising, and muscle soreness and for pelvic congestion associated with poor circulation and profuse discharges. Pain and swelling from contusions, hematomas, and wounds can often be allayed by *Hamamelis* soaks.[133]

Hedeoma pulegioides • Pennyroyal

The young leaves and entire aerial parts of pennyroyal may be prepared into medicines or distilled into essential oils. Pennyroyal head washes are effective against head lice, and the essential oil can be used topically but cautiously for skin parasites and scabies and as an ingredient in herbal flea, tick, and mosquito repellents. Like all essential oils, pennyroyal oil must be kept out of the eyes and is not to be used internally, especially during pregnancy as it is a potential abortifacient. The essential oil may also serve as an antidote for poison oak and ivy and may be included in natural mosquito and tick sprays due to its repellent effects.

Helichrysum angustifolium • Curry Plant

The small marigold-like flowers are distilled for their essential oils and used in skin oils and body products. *Helichrysum* is specifically indicated for telangiectasias, florid complexion, and vascular lesions. The oil can be used topically, usually diluted, on acne, wounds, fungal infections, and rough complexions.

Hippophae rhamnoides • Sea Buckthorn

The small berries are pressed to yield a high-quality skin oil, useful for burns, skin lesions, eczema, dryness, and irritation as well as for cosmetic purposes. The oil can be consumed orally as a source of essential fatty acids, but is expensive. Flax and evening primrose oils are more affordable.

Hydrastis canadensis • Goldenseal

The bright yellow roots are prepared into herbal medicines. Goldenseal compresses and poultices may be used topically for bacterial and fungal infections of the skin and for skin cancers and ulcers. *Hydrastis* may also be used as a mouthwash for oral manifestations of lupus or systemic diseases. *Hydrastis* is specific for hyperhydrosis and unhealthy skin, especially when catarrhal states of the stomach and intestines and a bitter taste in the mouth accompany. *Hydrastis* may be included in formulas to use internally for rectal prolapse and both internally and topically for anal fissures.

Hypericum perforatum • St. Johnswort

Use *Hypericum* orally and topically for skin complaints associated with vascular inflammation such as rosacea and telangiectasia. *Hypericum* may help protect and restore connective tissue, and oil-based preparations may be used topically to prevent stretch marks and to promote healing in the skin and joints. It can be used internally as a connective tissue tonic for vascular fragility, chronic hemorrhoids, and easy bruising. *Hypericum* is specific for bruising, strains, sprains, and soft tissue injuries and for injury to highly innervated areas such as fingers and toes. *Hypericum* is also specific for puncture wounds and nail-bed injuries, skin lesions and trauma associated with neuralgia, and shooting, lancinating pain such as from herpes zoster, reflex sympathetic dystrophy, and sciatica. *Hypericum* may also help reduce hyperhidrosis and sweating of the scalp while sleeping. *Hypericum* is also specific for loss of head hair following head trauma and scalp injuries as well as for eczema of the hands and face with intense itching. *Hypericum* is also indicated for sensitive ulcers and sores in the mouth (although I have seen one case where *Hypericum* caused these symptoms). Include *Hypericum* in formulas for warts due to antiviral activity or for skin eruptions due to underlying viral infections.

Iris versicolor • Wild Iris

The roots of wild iris can be powdered or tinctured and used as a complementary ingredient in skin complaints where hypothyroid, digestive insufficiency, poor liver and digestive function, and fat intolerance with steatorrhea accompany. *Iris* is specific for rough, greasy skin, pigmentary changes, and a tendency to sebaceous papules or pustules. Include *Iris* in formulas for herpes zoster, psoriasis, eczema, and chronic pustular eruptions. Because *Iris* is a warming, stimulating herb, small doses of only a few drops at a time, combined with other herbs are needed to gently stimulate the glands. Indicators to choose *Iris* over other common alterative herbs for skin complaints are liver enlargement with sick headache and nausea, burning sensation in the stomach and intestines, liver pain, nausea and vomiting, flatulent colic and diarrhea, greenish loose stools, and skin eruptions with itching. See also the *Iris versicolor* entry on page 85 in chapter 2.

Juglans nigra, J. regia • Walnuts

The tarry inner hulls of black walnut (*Juglans nigra*) are used medicinally and are indicated for acne, comedones, and skin eruptions associated with digestive upset. *Juglans* is specifically indicated for seborrheic dermatitis with crusts behind the ears, and for itching papular and pustular eruptions. *Juglans* is also very useful in formulas where boils and abscesses occur in the axillae. The common name of *J. regia* is English walnut.

Lavandula officinalis • Lavender

Lavender flowers are harvested in the earliest stage of flowering and dried, tinctured, or distilled in essential oil. Lavender essential oil is specific for burns, especially sunburn, providing anodyne as well as anti-inflammatory effects, and to reduce allergic reactions, such as dermatitis and itchy lesions, due to underlying release of histamine from mast cells. Lavender essential oil is one of the better-tolerated essential oils and can often be used undiluted, directly on the skin in cases of scabies, lice, bed bugs, and other parasitic and infectious complaints of the skin. Lavender essential oil is easy to pack in a first aid or emergency kit and can be diluted with water in a spray bottle to use on burns or itchy skin. The calming effect of lavender's aroma also helps to soothe burn patients or children restless from bug bites.

Lobelia inflata • Pukeweed

The young seedpods are used to prepare infusions, vinegars, and tinctures, all of which can be used orally or can be used to prepare compresses to apply to skin rashes due to poison oak or ivy, insect bites, and stings and to allay the pain of bruises and strains. *Lobelia* may be effective topically when used to relieve the pain of erysipelas as mentioned in the folkloric herbals.

Lomatium dissectum • Biscuitroot

Lomatium roots are used medicinally and can be included in oral antiviral formulas to treat chronic warts. Essential oil or other concentrates may be used topically to induce photosensitivity for use with UV light as part of a protocol for psoriasis and vitiligo.

Lonicera japonica • Honeysuckle

Lonicera flowers are traditionally used in skin formulas for inflammation, antiaging, and skin lesions, often combined with *Forsythia*, such as in the classic formula Yin Qiao San. The duo is often claimed to have antitoxin effects.

Mahonia aquifolium • Oregon Grape

The berberine-rich root bark is used orally in teas, tinctures, powders, and encapsulations for skin complaints, acne, boils, and infections related to poor digestion and biliary function. Use *Mahonia* topically on infected wounds or infectious complaints of the skin. *Mahonia* is specific for scaly eczema, acne, blotches, skin eruptions, pimples and general poor complexion, and dry rough skin. *Mahonia* is useful topically as an antimicrobial for skin, eye, and mucous membrane infections. *Mahonia* is especially indicated when skin complaints are accompanied by liver congestion and slow digestion, intestinal dysbiosis, coated tongue, and dyspepsia. *Mahonia* may also be effective for antibiotic-resistant staph infections of the skin. This plant (and many others in the genus) are also known with the genus name *Berberis*.

Melaleuca alternifolia • Tea Tree

Tea tree oil is distilled from the aromatic leaves of the tree and can be used topically for skin infections, scabies, lice, fungal infections, and infected wounds. Tea tree oil soaps, wound salves, antifungal ointments, and acne washes are all commercially available. Tea tree essential oil can also be kept on hand for preparing custom foot soaks and skin washes and to include in minor surgery kits.

Matricaria chamomilla • Chamomile

Young chamomile flowers are used to prepare various medicines including teas, tinctures, and essential oils. Compresses can be useful topically for eczema and allergic dermatitis. Chamomile can be used internally and topically as an anti-inflammatory agent for dermatitis, diaper rash, and nervous irritation. *Matricaria* tea can be used for skin washes and compresses and in large quantities for baths or kitchen sink soaks for babies. Consider *Matricaria* as a supportive herb for when emotional and digestive upsets accompany skin complaints.

Medicago sativa • Alfalfa

Nutritious *Medicago* leaves can be used in mineral tonic teas and encapsulated to consume in cases of poor wound healing, weak fingernails, and connective tissue injury, laxity, or disease. Alfalfa, also called lucerne, is a nutritive herb that may offer nutritional support in cases of malnutrition and may help lower cholesterol. The steroidal saponins it contains may support postmenopausal skin and connective tissue laxity. *Medicago* may improve vitiligo, because it contains genistein, which is shown to promote melanin synthesis.

Melissa officinalis • Lemon Balm

Melissa leaf extracts may be included in formulas for warts and herpes lesions, such as cold sores, due to its antiviral activity. *Melissa* essential oils are quite expensive, but as just a few drops in a lip oil or skin oil can help heal cold sores, it may be worth the investment.

Mentha piperita • Peppermint

Mint leaves can be used to distill an essential oil high in menthol. Mint essential oil may be used topically to allay itching in cases of hives, chicken pox, mosquito bites, and other pruritic skin conditions. The diluted essential oil can be used as a douche for vaginal itching and genital itching such as jock itch and rectal inflammation and pain. The topical application of undiluted oil, although an intense sensation itself, can help reduce acute hemorrhoidal pain.

Mentha spicata • Spearmint

Spearmint may help to reduce elevated androgens in women and may be included in teas for female hirsuitism and skin complaints related to polycystic ovarian syndrome.

Origanum vulgare • Oregano

Oregano leaves are used to produce an essential oil, useful topically for all manner of skin infections including bacterial and fungal infections. Oregano teas, tinctures, and concentrates can be used orally for thrush and oral streptococcal infections as well as for all types of systemic infections. *Origanum* is useful for flatulence and digestive upset and is effective against intestinal parasites and dysbiosis and may be included in skin formulas where poor intestinal health underlies.

Panax ginseng • Ginseng

The roots of *Panax* are prepared into teas, tinctures, and encapsulations and can be used in skin formulas where underlying allergies or atopic tendencies are present. *Panax* is also indicated for individuals with poor immune response and a tendency toward chronic infections. *Panax* may reduce excessive immune response in autoimmune skin conditions as well as improve immunodeficiency states. *Panax* may be included in formulas for warts and chronic skin infections due to its immune-supportive effects. *Panax* is a warming herb best for cold or neutral constitutions.

Phytolacca americana • Pokeroot

Phytolacca roots are used to produce tinctures and oils. Use pokeroot, also called poke, in formulas for chronic boils and skin infections, folliculitis, and abscesses. *Phytolacca* will improve papular and pustular eruptions, dry itching skin with sloughing, and warts. *Phytolacca* is specific for skin eruptions associated with systemic diseases and hard, swollen lymphatic and other glands. *Phytolacca* is also specific for lymphatic congestion, constipation in the elderly, fluid stagnation, and circulatory weakness. Skin complaints accompanied by fluid stasis, tendency to cysts, and a lymphatic constitution will especially respond to *Phytolacca*. *Phytolacca* is often effective for scabies and indolent leg ulcers when taken internally, and according to the historic texts, it may improve skin manifestations of syphilis. Leg ulcers and fluid stasis, enlarged lymph nodes and cysts can also be treated with topical *Phytolacca* preparations. For example, *Phytolacca* with sulfur is a folkloric formula for indolent stasis ulcers.

Picrorhiza kurroa • Kutki

The rhizomes of kutki are a traditional Ayruvedic medicine used to treat allergy, dermatitis, vitiligo, and skin wounds. Kutki is also a traditional remedy for liver disease.

Pilocarpus jaborandi • Jaborandi

Both the leaves and roots are sources of the muscarinic receptor antagonist, pilocarpine. *Pilocarpus* is specific for deficient glandular secretions. *Pilocarpus* is a powerful stimulant to perspiration, salivation, and lacrimation. *Pilocarpus* is used botanically to promote secretions and is used homeopathically to allay hyperhidrosis and excessive heat and sweating. Pilocarpine, an active alkaloid in *Pilocarpus*, is available as a prescription eyedrop used to treat severe dry eyes and as an oral drug to treat dry mouth and deficient secretions as occurs with Sjögren's syndrome. *Pilocarpus* may be mixed with licorice solid extracts for Sjögren's patients, who can use a small amount in the mouth, especially helpful just prior to meals.

Podophyllum peltatum • Mayapple

The resin from the small applelike fruits may be used topically on warts, skin tags, and age spots. *Podophyllum* resin is caustic and is usually placed in a tincture of benzoin or other gummy substance to help a small amount cling to a wart or mole without damaging the surrounding tissues. The herb has antiviral and anticancer effects for superficial epithelial cancers.

Propolis

This is not a plant, but rather a collection of plant resins made by bees and credited with strong antimicrobial and anti-inflammatory effects. The resins in propolis have the physical characteristic of being sticky, waxy, and gummy and are useful to make skin shellacs and disinfectant applications that will dry into a bandagelike covering over wounds and lesions. It is useful for traumatic as well as inflammatory lesions in the skin.

Quercus alba, Q. robur • Oak

Quercus bark may be used topically to reduce bleeding and discharges in the skin and mucous membranes. *Quercus alba* is white oak; *Q. robur* is common oak.

Quercus is specific for indolent skin ulcers and for flabby spongy tissue with suppuration. *Quercus* is also specific for portal congestion, hemorrhoids, chronic liver congestion, intestinal atrophy with mucousy diarrhea, and blood in the stool. When skin complaints or doughy skin are accompanied by liver weakness from chronic alcohol ingestion, splenic enlargement related to portal hypertension, or dependent fluid accumulation due to hepatic and portal stasis, consider *Quercus* as a supportive herb.

Rhamnus purshiana • Cascara

Formerly known as *Cascara sagrada* or sacred bark, *Rhamnus* contains anthraquinone glycosides such as emodin that have been used as irritant laxatives. *Rhamnus* may be included in formulas where constipation and loss of peristaltic muscle function contribute to intestinal dysbiosis, and thereby skin lesions.

Rosmarinus officinalis • Rosemary

Water extracts taken internally and used as skin washes may decrease photo-aging and the risk of skin cancers by preventing cutaneous UV damage. Rosemary essential oil may be effective in treating acne when diluted and applied topically.

Rumex crispus • Yellow Dock

The roots are prepared into teas and tinctures to consider for use when acne, eczema, or psoriasis is accompanied by hypochlorhydria, malabsorption, constipation, or digestive insufficiency. *Rumex* is specific for itchy skin eruptions and hives due to digestive insufficiency. See also the *Rumex* entry on page 87 in chapter 2.

Salix alba • White Willow

The bark of many *Salix* preparations can be used orally to treat hives, itching vesicles, and pustules. *Salix* may also be prepared into a wash for canker sores. The topical application of salicylic acid solutions are known to penetrate the skin and are useful to loosen keratin and promote sloughing of superficial epithelia.[134] *Salix* skin washes for acne are available commercially; however, they are often combined with alcohols and strong soaps, which over time may be more inflaming than helpful.

Salvia miltiorrhiza • Dan Shen

These roots, also called red sage root, are used in teas, tinctures, and encapsulations to treat hyperpigmentation by reducing formation of melanin pigment. Dan shen is traditionally emphasized for blood stasis and vascular inflammation and may improve circulatory congestion and support microcirculation in the skin.

Salvia officinalis • Sage

Aromatic sage leaves are highly effective when used orally for hyperhydrosis, hot flashes, night sweats, and any similar condition to reduce excessive sweating.

Sambucus canadensis • Elderberry

Sambucus leaves and flowers are specific for skin lesions associated with weeping discharges and crust formation and for skin lesions where the epidermis splits and separates, becoming full, flabby, and edematous.

Sanguinaria canadensis • Bloodroot

Sanguinaria roots are potentially caustic and irritating, but may be used in small cautious dosages orally for thrush in adults (not infants), and for lesions and infections of the oral mucosa, oral lichen planus, and oral erythema multiforme. To avoid irritation, place just 10 to 20 drops in water and swish around the mouth. *Sanguinaria* may be effective against tinea and ringworm when applied topically. *Sanguinaria* may also be used as a synergist in formulas when digestive symptoms are associated with burning sensations, hot flashes, and vasomotor symptoms. *Sanguinaria* is specific for hot flashes associated with heat, a red face (especially the cheeks), and burning sensations. *Sanguinaria* is also specific for burning sensations in the palms and soles of the feet and for red blotchy eruptions that burn and itch. It is useful as an antidote for poison oak and ivy. *Sanguinaria* is extremely irritating to the point of being caustic, and some *Sanguinaria* products can burn holes through the dermis and deeper if improperly used. Traditionally remedies for skin lesions, growths, and cancers included *Sanguinaria* prepared into "black salves," capable of destroying tissue when topically applied. When *Sanguinaria* is used topically, knowing the strength and causticity of one's products is essential.

Scrophularia nodosa • Figwort

Scrophularia is specific for dermatitis, seborrhea behind the ears with a tendency to crusting, and prickling and itching sensations in the skin.

Scutellaria baicalensis • Scute

The aerial parts of Chinese scute, also known as huang qin, are used internally to reduce allergic response and treat atopic skin conditions.

Scutellaria lateriflora • Skullcap

Use skullcap leaves or young aerial parts in skin formulas where nervousness underlies or contributes to the condition, such as for neurodermatitis.

Serenoa repens • Saw Palmetto

Ripe *Serenoa* berry preparations can be used in formulas for female hirsutism and acne related to polycystic ovarian syndrome, due to its ability to reduce excessive testosterone-related skin complaints.

Silybum marianum • Milk Thistle

Milk thistle seed preparations may be used in formulas for skin complaints associated with digestive and biliary disorders and to protect the liver when using tretenoin. *Silybum* can help clear drugs and toxins from the system in cases of skin eruptions due to drug side effects.

Smilax ornata • Sarsaparilla

Smilax root is specific for skin eruptions related to hot weather and for seborrhea on the scalp and behind the ears. *Smilax* is also specific for deep fissures of the hands and feet that are worse in spring and summer and for boils and eczema. *Smilax* is an aromatic alterative agent with hormonal and adrenal-supportive properties, helpful in cases of fatigue, muscle weakness, and poor digestion.

Stillingia sylvatica • Queen's Root

The roots of *Stillingia* are specific for chronic dermatitis of the hands and fingers, chronic skin ulcers, chronic skin complaints associated with swollen glands, and what was at one time referred to as "scrophula," an older term for tuberculosis-like symptoms including emaciation, skin lesions, cough, and enlarged lymph nodes. *Stillingia* is especially indicated for skin complaints where liver congestion with jaundice and constipation accompany.

Symphytum officinale • Comfrey

Both the roots and leaves of comfrey are used and can allay the pain of wounds and burns and speed cell division to promote healing when topically applied. Use *Symphytum* topically on lacerations, abrasions, and burns and internally for soft tissue trauma, broken bones, poorly healing wounds, and weak connective tissues and fingernails. Because comfrey contains pyrrolizidine alkaloids, its internal use is avoided during pregnancy or in those with liver disease and is limited to short-term use in everyone. *Symphytum* is also helpful when used topically for pruritis ani (itching rash of the anus), and for trauma to the eyes.

Syzygium aromaticum • Cloves

The aromatic young flower buds may be prepared as teas, powders, or tinctures or processed into concentrated essential oils. Include cloves in tincture or tea formulas for chronic acne with tender pimples, comedones (blackheads), acne rosacea, paronychia, and cracks and fissures about the toes. Clove oil may stimulate saliva flow in cases of dry mouth or appetite in cases of anorexia. The essential oil may be used topically to allay neuralgic pains or to reduce itching in chronic eczema. The essential oil may also kill lice and have a repellent action toward ticks, fleas, and mosquitoes. Care must be taken to keep the essential oil out of the eyes. Cloves have previously been placed in the *Eugenia* genus.

Tabebuia impetiginosa • Pau d'Arco

Tabebuia, also called taheebo, bark preparations can be used topically for fungal infections of the skin and as a mouthwash for thrush. *Tabebuia* may also be taken internally for chronic fungal infections of the skin.

Tanacetum parthenium • Feverfew

The aerial parts of *Tanacetum* may be used to reduce allergic activity in the blood and blood vessels. Include *Tanacetum* in formulas for hives, dermatitis, and vascular inflammation. *Tanacetum* can help reduce skin eruptions due to underlying allergy, drug side effects, or inflammatory reactivity.

Taraxacum officinale • Dandelion

Use *Taraxacum* root and/or leaf preparations for skin complaints related to liver congestion, constipation, or poor nutritional status. *Taraxacum* may be employed as a food, capsule, powder, or tea. The leaf can be prepared as a tea or tincture to serve as a general diuretic and mineral tonic for long-term use in osteoporosis. *Taraxacum* roots can be used as a general alterative when constipation, poor diet, or poor liver and digestive health accompany acne, eczema, and any, and all, skin complaints.

Thuja species • Cedar

Thuja leaves are specific for warts, condyloma, and fleshy vegetative skin growths such as styes and polyps and may be used both internally and topically. *Thuja* has antibacterial and viral properties, and the essential oil may be included in topical formulas for warts. Folkloric literature suggests that *Thuja* is specific for skin eruptions that are better in dry climates and worse in warm humid climates, for an oily complexion with many moles and freckles, as well as for a red, dry scalp, dandruff, hair loss, unhealthy nails, and ano-genital skin lesions. Cedar essential oils may also be included in formulas for warts and skin lesions due to viral or other microbial infections.

Thymus vulgaris • Thyme

Thymus leaves have strong antifungal and antibacterial affects that may improve cellulitis and treat fungus, lice, and scabies when prepared as a skin wash or other topical preparations. Thyme essential oil may also be used but it is very irritating on the skin and must be greatly diluted prior to application.

Trifolium pratense • Red Clover

Red clover blossoms are specific for seborrheic dermatitis with dry, scaly crusts and for tibial ulcers. The plant is traditionally considered to be a "blood mover" with weak hormonal effects. More modern research supports the use of *Trifolium* for postmenopausal bone density, and oral use may be considered in postmenopausal connective tissue laxity, osteoporosis, skin complaints, and weak nails. The genistein content may be useful in hypopigmentation of the skin, protecting melanocytes from damage, and stimulating melanin production.

Urtica dioica, U. urens • Nettles

Urtica leaf preparations are specific for hives and itchy, blotchy skin lesions, genital itching, pruritis ani, pruritis vulvae (itching rashes of the anus and vulva respectively), and itching of the scrotum. Historic literature also suggests *Urtica* for angioedema and for profuse discharges from mucous membranes. Modern research suggests that the root and seed may be useful in formulas for female hirsutism via reducing excessive testosterone stimulation of hormone-sensitive tissues.

Usnea barbata • Old Man's Beard

The entire lichen may be used in teas, tinctures, and topical applications for skin infections and fungal infections.

Vaccinium myrtillus • Blueberry

Use blueberries, also called Bilberry or whortleberry, as a regular food or use ½ teaspoon of solid extract daily. *Vaccinium* leaves in teas and tinctures may be used for bruising and skin trauma and to improve microcirculation in the skin.

Veratrum viride • False Hellebore

This potentially cardiotoxic plant may be used topically for acute skin inflammation or for excessive dermal proliferation, such as psoriasis. *Veratrum* has been traditionally recommended for topical use on boils, abscesses, inflamed cystic acne, and cellulitis, and some early American physicians claimed it to be one of the best remedies for erysipelas when applied topically. Small dosages of *Veratrum* may be included in both topical and internal formulas for herpes labialis and herpes zoster, limiting the quantity to just 2 or 3 milliliters in the entire 60 milliliter (2-ounce) formula. *Veratrum* contains powerful steroidal alkaloids, such as veratridine, that can affect sodium channels and lead to numbness, weakness, bradycardia, cardiac arrhythmias, and finally heart failure. These steroids are abundant during spring and summer growth, and some Native American communities believed that the plant should be harvested only when dormant during the winter. Veratum should be strictly avoided in pregnancy.

Verbena hastata • Vervain

Include *Verbena* in a skin formula where nervousness underlies or contributes to the condition, such as neurodermatitis. *Verbena hastata* is reported to reduce pain in bruising and to promote absorption of extravasated blood.

Viola odorata, V. tricolor • Violets

Viola leaves and flowers can be used as antidotes for stings from venomous insects, topically for seborrhea in children, and for skin lesions with thick scabs and crusts, such as impetigo. Traditional remedies include the use of fresh chopped leaves topically on cradle cap of infants. *Viola odorata* is called wood violet; *V. tricolor* is called Johnny jump up.

Vitex agnus-castus • Vitex

The small peppercorn-like fruits of *Vitex* may be very helpful for adolescent acne when orally consumed. Include *Vitex* in formulas for premenstrual, pubertal, and other cases of hormone-stimulated acne and skin eruptions. *Vitex* is also called chaste tree berry.

Zingiber officinale • Ginger

Zingiber rhizome can be prepared into a variety of medicines internally and topically for staph infections of the skin.

— ACKNOWLEDGMENTS —

I would like to thank my dear friend and colleague, Dr. Mary Bove, for reading these chapters and sharing her insights on the formulas and commentaries, and especially for her vast clinical knowledge and her grounded, firmly moral, and wise-woman ways with me over the years. I would also like to thank Ben Zappin, L.Ac., for helping to enhance the educational value of the text by suggesting many important improvements, for being supportive of my work, and for allowing me to share a small slice of his delightful family's life.

And I would like to thank Mikailah Grover, ND, class of 2019, at the National University of Natural Medicine, for helping with word processing and editing support with a smile and ease.

Thank you to my editor at Chelsea Green, Fern Marshall Bradley, for her professional prowess and for making this book much more polished than it would have been without her guidance. I wish to acknowledge as well Margo Baldwin, the publisher at Chelsea Green, for making this text possible at all, as well as Pati Stone and the entire production team for perfecting the many small details that producing a reference book like this entails.

I extend warm, fond appreciation to my kind, goat-throwing strong, and supportive sweetie, Warren Martin, for his patience and willingness to pitch in with the day-to-day chores of running a household and a business.

I would like to thank and honor my teachers and all people and cultures everywhere who have contributed to herbal healing traditions. And I would like to thank the herbal community for being my beloved tribe, filled with powerful, skillful women and gentle, reverent men, remembering all the times that the healing plants have brought us together in beauty, to share our gifts and wisdom with one another. You are my people!

And last but not least, to Pachamama herself for her mystery, abundance, and amazing gifts. May we all Walk in Beauty and return her gifts tenfold. Blessed Be!

— SCIENTIFIC NAMES — TO COMMON NAMES

The following lists include all of the herbs, medicinal fungi, and homeopathic preparations mentioned in the text of this book.

Abies sibirica	Siberian fir
Achillea millefolium	yarrow
Aconitum napellus	aconite
Acorus calamus	sweet flag
Actaea racemosa (formerly *Cimicifuga racemosa*)	black cohosh
Adiantum capillus-veneris	Southern maidenhair fern, black maidenhair fern
Aesculus hippocastanum	horse chestnut
Agaricus bisporus	white button mushroom
Agathosma betulina (also known as *Barosma betulina*)	buchu
Agrimonia eupatoria, A. pilosa	agrimony
Agropyron repens, A. cristatum (reclassified as *Elymus*)	couch grass
Allium cepa	onion
Allium sativum	garlic
Alnus rubra	red alder
Alnus serrulata	alder
Aloe barbadensis, A. vera, A. ferox, A. arborescens	aloe
Alpinia galanga	Thai galangal
Althaea officinalis	marshmallow
Alcea rosea	hollyhock
Ammi visnaga	khella
Ananas comosus	pineapple
Andrographis paniculata	king of bitters
Anemarrhena asphodeloides	anemarrhena, zhi mu,
Angelica archangelica	garden angelica
Angelica sinensis	angelica, dong quai
Apis mellifica	honey bee
Apium graveolens	celery
Apocynum cannabinum	dogbane
Arctium lappa	burdock, gobo root
Arctostaphylos uva ursi	uva ursi, kinnickinick
Armoracia rusticana (also known as *Cochlearia armoracia*)	horseradish
Arnica montana	arnica, leopard's bane
Artemisia absinthium	wormwood
Artemisia annua	sweet Annie
Artemisia capillaris	capillary wormwood
Artemisia scoparia	redstem wormwood
Artemisia vulgaris	mugwort
Asclepias tuberosa	pleurisy root, butterfly weed
Ascophyllum nodosum	kelp
Asparagus officinalis	asparagus
Asparagus racemosus	shatavari
Astragalus membranaceus	milk vetch
Atractylodes lancea	atractylodes
Atractylodes ovata	white atractylodes, large-headed atractylodes, bai zhu
Atropa belladonna	belladonna
Avena sativa	oats
Avicennia marina	white mangrove, gray mangrove
Azadirachta indica	neem
Baptisia tinctoria	wild indigo, yellow wild indigo
Barosma betulina (also known as *Agathosma betulina*)	buchu

Berberis aquifolium (also known as *Mahonia aquifolium*)	Oregon grape, mahonia
Berberis vulgaris	mahonia, barberry
Bergenia crassifolia	winter-blooming bergenia
Beta vulgaris	beet
Betula pendula	silver birch
Boswellia serrata	frankincense, Indian frankincense
Brassica napus, B. campestris	rapeseed
Bupleurum chinense	Chinese thoroughwax, chai hu
Bupleurum falcatum	Chinese thoroughwax
Cactus grandiflorus (also known as *Selenicereus grandiflorus*)	night-blooming cactus
Calendula officinalis	pot marigold, calendula
Camellia sinensis (formerly *Thea sinensis*)	green tea
Cannabis sativa	marijuana, hemp
Cantharis vesicatoria	Spanish fly
Capsella bursa-pastoris	shepherd's purse
Capsicum annuum (also known as *C. frutescens*)	cayenne
Capsicum spp.	hot peppers
Carbolicum acidum	carbolic acid
Carica papaya	papaya
Carthamus tinctorius	safflower
Carum carvi	caraway
Cassia auriculata (also known as *Senna auriculata*)	avaram senna
Cassia senna (also known as *Senna alexandrina*)	senna
Ceanothus americanus	red root, New Jersey tea
Centella asiatica	gotu kola, pennywort
Ceratonia siliqua	carob
Chelidonium majus	celandine
Chelone glabra	turtlehead
Chenopodium ambrosioides	wormseed, goosefoot

Chimaphila umbellata	pipsissewa, prince's pine
Chionanthus virginicus	fringe tree
Chondrus crispus	Irish moss
Chrysanthemum indicum, C. leucanthemum	chrysanthemum
Cicer arietinum	chickpea, garbanzo bean
Cichorium intybus	chicory
Cimicifuga racemosa (reclassified as *Actaea racemosa*)	black cohosh
Cinchona officinalis, C. pubescens	Peruvian bark, quinine bark, cinchona
Cinnamomum verum, C. zeylanicum	cinnamon
Cistanche tubulosa	rou cong rong
Citrullus colocynthis	bitter apple, bitter cucumber
Citrus aurantium	bitter orange
Citrus bergamia	bergamot
Citrus paradisi	grapefruit
Cnicus benedictus	blessed thistle
Cnidium monnieri	snow parsley, osthole
Cnidium officinale (also known as *Ligusticum officinale*)	snow parsley
Cochlearia armoracia (also known as *Armoracia rusticana*)	horseradish
Codonopsis lanceolata	todok
Codonopsis pilosula	dang shen
Coffea arabica	coffee
Coix lacryma-jobi	Job's tears
Coleus forskohlii	coleus
Collinsonia canadensis	stoneroot
Commiphora mukul	guggul
Commiphora myrrha	myrrh
Conium maculatum	poison hemlock
Convallaria majalis	lily of the valley
Coptis trifolia, C. chinensis	goldthread
Cordyceps sinensis	caterpillar fungus
Coriandrum sativum	coriander, cilantro

Corydalis cava, C. yanhusuo	corydalis, turkey corn
Crataegus monogyna, Crataegus oxyacantha (also known as *C. rhipidophylla*)	hawthorn
Crocus sativus	saffron crocus
Crotalus horridus	rattlesnake
Croton lechleri	dragon's blood
Cucurbita maxima	pumpkin, winter squash
Cucurbita pepo	pumpkin
Curcuma longa	turmeric
Cuscuta reflexa	dodder
Cynara scolymus	artichoke
Cyperus rotundus	nutgrass
Cytisus scoparius	Scotch broom, Scot's broom
Daucus carota	wild carrot, common carrot
Digitalis spp., *D. purpurea*	foxglove, digitalis
Dioscorea villosa, D. rotunda	wild yam
Echinacea angustifolia	coneflower
Echinacea purpurea	coneflower, purple coneflower
Eclipta alba, E. prostata	bhringraj, false daisy, yerba de tago
Elettaria cardamomum	cardamom
Eleutherococcus senticosus	Siberian ginseng, eleuthero
Elymus repens (formerly *Agropyron repens*)	couch grass
Emblica officinalis	amalaki
Ephedra sinica	ephedra, ma-huang
Epimedium pubescens	horny goatweed
Equisetum arvense, E. hymenale	horsetail, scouring rush
Eschscholzia californica	California poppy
Eucalyptus globulus	eucalyptus
Eugenia aromatica, E. caryophyllata (reclassified as *Syzygium aromaticum*)	cloves
Euonymus alatus	burning bush
Eupatorium perfoliatum	boneset
Eupatorium purpureum	gravel root
Euphrasia officinalis	eyebright
Euodia ruticarpa (reclassified as *Tetradium ruticarpum*)	evodia
Filipendula ulmaria	meadowsweet
Foeniculum vulgare	fennel
Forsythia suspensa	forsythia
Fouquieria splendens	ocotillo
Fucus vesiculosus	bladderwrack
Galium aparine	cleavers
Ganoderma lucidum	reishi mushroom
Gardenia jasminoides	gardenia, cape jasmine
Gaultheria procumbens	wintergreen
Gelsemium sempervirens	yellow jessamine, yellow jasmine
Gentiana lutea	gentian
Geranium maculatum, G. robertianum	wild geranium
Ginkgo biloba	ginkgo, maidenhair tree
Glycine max	soybean
Glycyrrhiza glabra	licorice
Glycyrrhiza uralensis	Chinese licorice
Grifola frondosa	maitake, hen of the woods
Grindelia squarrosa	gumweed
Gymnema sylvestre	sugar destroyer, cow plant, gurmar
Hamamelis virginiana	witch hazel
Harpagophytum procumbens	devil's claw
Hedeoma pulegioides	pennyroyal
Helichrysum angustifolium, H. italicum	curry plant
Herniaria glabra	rupturewort
Hibiscus sabdariffa	hibiscus
Hippophae rhamnoides	sea buckthorn
Humulus lupulus	hops, common hop
Hydrangea arborescens, H. macrophylla	hydrangea

Hydrastis canadensis	goldenseal
Hyoscyamus niger	henbane
Hypericum perforatum	St. Johnswort
Iberis amara	candytuft
Iris versicolor, I. tenax	blue flag, wild iris
Juglans nigra	black walnut
Juglans regia	English walnut
Juniperus communis	juniper
Lachesis mutus	bushmaster snake
Laminaria japonica, L. hyperborean	kelp
Larrea tridentata	chaparral, creosote bush
Lavandula angustifolia	English lavender
Lavandula officinalis	lavender
Ledum palustre	marsh tea or wild rosemary
Lentinula edodes	shiitake mushroom
Leonurus cardiaca	motherwort
Lepidium meyenii	maca
Leptandra virginica (also known as *Veronicastrum virginicum*)	Culver's root
Lespedeza capitata	roundhead bushclover
Levisticum officinale	lovage
Ligustrum lucidum	Chinese privet, glossy privet
Ligusticum officinale (also known as *Cnidium officinale*)	snow parsley
Ligusticum porteri	osha, bear root
Linum usitatissimum	flax
Lobelia inflata	pukeweed
Lomatium dissectum	biscuitroot
Lonicera japonica	honeysuckle
Lycium barbarum	goji, wolfberry
Lycopus virginicus	bugleweed
Macrocystis pyrifera	kelp
Magnolia officinalis, M. nervosa	magnolia
Mahonia aquifolium (also known as *Berberis aquifolium*)	Oregon grape, mahonia
Mangifera indica	mango
Matricaria chamomilla, M. recutita	chamomile
Medicago sativa	alfalfa, lucerne
Melaleuca alternifolia	tea tree
Melissa officinalis	lemon balm
Mentha piperita	peppermint
Mentha spicata	spearmint
Myrica cerifera	wax myrtle, bayberry
Myristica fragrans	nutmeg
Nepeta cataria	catnip
Ocimum tenuiflorum (also known as *O. sanctum*)	holy basil, tulsi
Oenothera biennis	evening primrose
Olea europaea	olive
Oplopanax horridus	devil's club
Opuntia ficus-indica	prickly pear
Origanum vulgare, O. marjorana, O. onites	oregano
Paeonia lactiflora	white peony
Paeonia suffruticosa	tree peony, moutan
Panax ginseng	ginseng
Parkia biglobosa	African locust bean
Passiflora incarnata	passionflower
Perilla frutescens	Korean perilla, perilla
Petasites hybridus	butterbur
Petroselinum crispum (also known as *P. sativum*)	parsley
Peumus boldo	boldo
Phaseolus vulgaris	kidney bean, navy bean
Phellodendron spp.	cork tree
Phyllanthus amarus	bahupatra, hurricane weed
Phyllanthus emblica	Indian gooseberry
Phyllanthus niruri	stonebreaker
Phytolacca americana, P. decandra	pokeroot, poke
Picrasma excelsa	quassia

Picrorhiza kurroa	kutki
Pilocarpus jaborandi	jaborandi
Pimpinella anisum	anise, aniseed
Pinellia ternata	crow dipper
Piper cubeba	cubebs
Piper longum	long pepper
Piper methysticum	kava kava
Piper nigrum	black pepper
Piscidia piscipula (also known as *P. erythrina*)	Jamaican dogwood
Plantago lanceolata, P. ovata, P. psyllium	plantain
Podophyllum peltatum	mayapple
Polygonum cuspidatum (reclassified as *Reynoutria japonica*)	Japanese knotweed, donkey rhubarb, Asian knotweed
Polygonum multiflorum (reclassified as *Reynoutria multiflora*)	Chinese knotweed, fo ti, he shou wu
Polyporus umbellatus	lumpy bracket, umbrella polypore
Populus tremuloides	quaking aspen
Poria cocos	hoelen
Propolis	resin gathered from various plants by bees
Prunus persica	peach
Psidium guajava	guava
Pueraria montana, P. mirifica, P. lobata	kudzu
Punica granatum	pomegranate
Pygeum africanum	African plum tree
Quercus alba	white oak
Quercus robur	common oak
Quercus rubra	red oak
Raphanus sativus var. *niger*	Spanish black radish, black Spanish radish
Rauvolfia serpentina, R. vomitoria	Indian snakeroot
Rehmannia glutinosa	di huang, Chinese foxglove

Reynoutria japonica (formerly *Polygonum cuspidatum*)	Japanese knotweed, donkey rhubarb, Asian knotweed
Reynoutria multiflora (formerly *Polygonum multiflorum*)	Chinese knotweed, fo ti, he shou wu
Rhamnus frangula	alder buckthorn, glossy buckthorn
Rhamnus purshiana	cascara
Rheum officinale, R. palmatum	Chinese rhubarb, turkey rhubarb
Rhodiola kirilowii, R. rosea	rhodiola, arctic rose
Rhus aromatica	sweet sumac
Rhus toxicodendron	poison ivy
Ribes nigrum	black currant
Ribes oxyacanthoides	Canadian gooseberry
Ricinus communis	castor bean, castor
Rosa canina	dog rose
Rosa damascena	damask rose
Rosa spp.	rose
Rosmarinus officinalis	rosemary
Rubia tinctorum	common madder
Rubus idaeus	raspberry, red raspberry
Rubus spp.	blackberry
Ruscus aculeatus	butcher's broom
Rumex acetosella	sheep sorrel
Rumex crispus	yellow dock
Salix alba	white willow
Salvia miltiorrhiza	dan shen, red sage
Salvia officinalis	sage
Sambucus nigra, S. canadensis	elderberry
Sanguinaria canadensis	bloodroot
Schisandra chinensis	magnolia vine, five flavor fruit
Scrophularia nodosa	figwort
Scutellaria baicalensis	huang qin, scute, baical
Scutellaria barbata	barbed skullcap
Scutellaria lateriflora	skullcap
Selenicereus grandiflorus (also known as *Cactus grandiflorus*)	night-blooming cactus, queen of the night

Senna alexandrina (also known as *Cassia senna*)	senna
Serenoa repens	saw palmetto
Silybum marianum	milk thistle
Smilax ornata, S. regelii	sarsaparilla
Solidago odora, S. canadensis, S. virgaurea	goldenrod
Spilanthes acmella	paracress
Stellaria media	chickweed
Stevia rebaudiana	stevia, sweet leaf
Stillingia sylvatica	queen's root
Symphytum officinale	comfrey
Syzygium aromaticum (formerly *Eugenia caryophyllata*)	cloves
Syzygium cumini	jambul, jambolan
Tabebuia impetiginosa	pau d'arco, taheebo
Tamarindus indica	tamarind
Tanacetum parthenium	feverfew
Taraxacum officinale	dandelion
Tarentula hispanica	tarentula, wolf spider
Terminalia arjuna	arjuna
Terminalia bellirica	bibhitaki
Terminalia macroptera	kwandari
Terminalis catappa	tropical almond
Terminalis chebula	haritaki
Tetradium ruticarpum (formerly *Euodia ruticarpa*)	euodia
Thea sinensis (reclassified as *Camellia sinensis*)	green tea
Thuja occidentalis	northern white cedar
Thuja plicata	western red cedar
Thymus vulgaris	thyme
Tinospora cordifolia	guduchi
Trifolium pratense	red clover
Trigonella foenum-graecum	fenugreek
Tripterygium wilfordii	thunder god vine
Triticum aestivum	wheat
Tropaeolum majus	nasturtium
Turnera diffusa	damiana
Ulmus fulva, U. rubra	slippery elm
Uncaria tomentosa	uña de gato or cat's claw
Urtica dioica, U. urens	nettles
Usnea barbata	old man's beard
Vaccinium macrocarpon	cranberry
Vaccinium myrtillus	blueberry, bilberry
Valeriana officinalis	valerian
Veratrum album	white hellebore
Veratrum viride	false hellebore, corn lily
Verbascum thapsus	mullein
Verbena hastata	vervain
Veronicastrum virginicum (also known as *Leptandra virginicum*)	Culver's root
Viburnum opulus	crampbark, snowball bush, European cranberry bush
Viburnum prunifolium	blackhaw
Viola odorata	wood violet
Viola tricolor	Johnny jump up
Viscum album	mistletoe
Vitex agnus-castus	vitex, chaste tree berry
Vitis vinifera	common grape vine
Withania somnifera	ashwagandha
Zanthoxylum americanum	northern prickly ash, common prickly ash, toothache tree
Zanthoxylum clava-herculis	southern prickly ash, pepperwood, Hercules' herb, Hercules' club, toothache tree
Zanthoxylum piperitum	Japanese pepper, chopi
Zea mays	corn, cornsilk
Zingiber officinale	ginger
Ziziphus jujuba	Chinese date, jujube

— COMMON NAMES — TO SCIENTIFIC NAMES

The following lists include all of the herbs, medicinal fungi, and homeopathic preparations mentioned in the text of this book.

aconite	*Aconitum napellus*
African locust bean	*Parkia biglobosa*
African plum tree	*Pygeum africanum*
agrimony	*Agrimonia eupatoria, A. pilosa*
alder	*Alnus serrulata*
alder buckthorn	*Rhamnus frangula*
alfalfa	*Medicago sativa*
aloe vera	*Aloe barbadensis, A. vera, A. ferox, A. arborescens*
amalaki	*Emblica officinalis*
anemarrhena	*Anemarrhena asphodeloides*
angelica	*Angelica sinensis*
anise	*Pimpinella anisum*
aniseed	*Pimpinella anisum*
arctic rose	*Rhodiola kirilowii, R. rosea*
arjuna	*Terminalia arjuna*
arnica	*Arnica montana*
artichoke	*Cynara scolymus*
ashwagandha	*Withania somnifera*
Asian knotweed	*Reynoutria japonica* (formerly *Polygonum cuspidatum*)
asparagus	*Asparagus officinalis*
atractylodes	*Atractylodes lancea*
avaram senna	*Cassia auriculata* (also known as *Senna auriculata*)
bahupatra	*Phyllanthus amarus*
bai zhu	*Atractylodes ovata*
baical	*Scutellaria baicalensis*
barbed skullcap	*Scutellaria barbata*
barberry	*Berberis vulgaris*
bayberry	*Myrica cerifera*
bear root	*Ligusticum porteri*
beet	*Beta vulgaris*
belladonna	*Atropa belladonna*
bergamot	*Citrus bergamia*
bhringraj	*Eclipta alba, E. prostata*
bibhitaki	*Terminalia bellirica*
bilberry	*Vaccinium myrtillus*
biscuitroot	*Lomatium dissectum*
bitter apple	*Citrullus colocynthis*
bitter cucumber	*Citrullus colocynthis*
bitter orange	*Citrus aurantium*
black cohosh	*Actaea racemosa* (formerly *Cimicifuga racemosa*)
black currant	*Ribes nigrum*
black maidenhair fern	*Adiantum capillus-veneris*
black pepper	*Piper nigrum*
black Spanish radish	*Raphanus sativus* var. *niger*
black walnut	*Juglans nigra*
blackberry	*Rubus* spp.
blackhaw	*Viburnum prunifolium*
bladderwrack	*Fucus vesiculosus*
blessed thistle	*Cnicus benedictus*
bloodroot	*Sanguinaria canadensis*
blue flag	*Iris versicolor, I. tenax*
blueberry	*Vaccinium myrtillus*
boldo	*Peumus boldo*
boneset	*Eupatorium perfoliatum*

buchu	*Agathosma betulina* (also known as *Barosma betulina*)
bugleweed	*Lycopus virginicus*
burdock	*Arctium lappa*
burning bush	*Euonymus alatus*
bushmaster snake	*Lachesis mutus*
butcher's broom	*Ruscus aculeatus*
butterbur	*Petasites hybridus*
butterfly weed	*Asclepias tuberosa*
calendula	*Calendula officinalis*
California poppy	*Eschscholzia californica*
Canadian gooseberry	*Ribes oxyacanthoides*
candytuft	*Iberis amara*
cape jasmine	*Gardenia jasminoides*
capillary wormwood	*Artemisia capillaris*
caraway	*Carum carvi*
carbolic acid	*Carbolicum acidum*
cardamom	*Elettaria cardamomum*
carob	*Ceratonia siliqua*
cascara	*Rhamnus purshiana*
castor	*Ricinus communis*
castor bean	*Ricinus communis*
cat's claw	*Uncaria tomentosa*
caterpillar fungus	*Cordyceps sinensis*
catnip	*Nepeta cataria*
cayenne	*Capsicum annuum* (also known as *C. frutescens*)
celandine	*Chelidonium majus*
celery	*Apium graveolens*
chai hu	*Bupleurum chinense*
chamomile	*Matricaria chamomilla*, *M. recutita*
chaparral	*Larrea tridentata*
chaste tree berry	*Vitex agnus-castus*
chickpea	*Cicer arietinum*
chickweed	*Stellaria media*
chicory	*Cichorium intybus*
Chinese date	*Ziziphus jujuba*
Chinese foxglove	*Rehmannia glutinosa*
Chinese knotweed	*Reynoutria multiflora* (formerly *Polygonum multiflorum*)
Chinese licorice	*Glycyrrhiza uralensis*
Chinese privet	*Ligustrum lucidum*
Chinese rhubarb	*Rheum officinale*, *R. palmatum*
Chinese thoroughwax	*Bupleurum chinense*, *B. falcatum*
chopi	*Zanthoxylum piperitum*
chrysanthemum	*Chrysanthemum indicum*, *C. leucanthemum*
cilantro	*Coriandrum sativum*
cinchona	*Cinchona officinalis*, *C. pubescens*
cinnamon	*Cinnamomum verum*, *C. zeylanicum*
cleavers	*Galium aparine*
cloves	*Syzygium aromaticum* (formerly *Eugenia caryophyllata*)
coffee	*Coffea arabica*
coleus	*Coleus forskohlii*
comfrey	*Symphytum officinale*
common carrot	*Daucus carota*
common grape vine	*Vitis vinifera*
common hop	*Humulus lupulus*
common madder	*Rubia tinctorum*
common oak	*Quercus robur*
common prickly ash	*Zanthoxylum americanum*
coneflower	*Echinacea angustifolia*, *E. purpurea*
coriander	*Coriandrum sativum*
cork tree	*Phellodendron* spp.
corn	*Zea mays*
corn lily	*Veratrum viride*
cornsilk	*Zea mays*
corydalis	*Corydalis cava*, *C. yanhusuo*
couch grass	*Elymus repens* (formerly *Agropyron repens*)
cow plant	*Gymnema sylvestre*
crampbark	*Viburnum opulus*

cranberry	*Vaccinium macrocarpon*
creosote bush	*Larrea tridentata*
crow dipper	*Pinellia ternata*
cubebs	*Piper cubeba*
Culver's root	*Veronicastrum virginicum* (also known as *Leptandra virginicum*)
curry plant	*Helichrysum angustifolium, H. italicum*
damask rose	*Rosa damascena*
damiana	*Turnera diffusa*
dan shen	*Salvia miltiorrhiza*
dandelion	*Taraxacum officinale*
dang shen	*Codonopsis pilosula*
devil's claw	*Harpagophytum procumbens*
devil's club	*Oplopanax horridus*
di huang	*Rehmannia glutinosa*
digitalis	*Digitalis* spp., *D. purpurea*
dodder	*Cuscuta reflexa*
dog rose	*Rosa canina*
dogbane	*Apocynum cannabinum*
dong quai	*Angelica sinensis*
donkey rhubarb	*Reynoutria japonica* (formerly *Polygonum cuspidatum*)
dragon's blood	*Croton lechleri*
elderberry	*Sambucus nigra, S. canadensis*
eleuthero	*Eleutherococcus senticosus*
English lavender	*Lavandula angustifolia*
English walnut	*Juglans regia*
ephedra	*Ephedra sinica*
eucalyptus	*Eucalyptus globulus*
euodia	*Tetradium ruticarpum* (formerly *Euodia ruticarpa*)
European cranberry bush	*Viburnum opulus*
evening primrose	*Oenothera biennis*
evodia	*Euodia ruticarpa* (reclassified as *Tetradium ruticarpum*)
eyebright	*Euphrasia officinalis*
false daisy	*Eclipta alba, E. prostata*
false hellebore	*Veratrum viride*
fennel	*Foeniculum vulgare*
fenugreek	*Trigonella foenum-graecum*
feverfew	*Tanacetum parthenium*
figwort	*Scrophularia nodosa*
five flavor fruit	*Schisandra chinensis*
flax	*Linum usitatissimum*
fo ti	*Reynoutria multiflora* (formerly *Polygonum multiflorum*)
forsythia	*Forsythia suspensa*
foxglove	*Digitalis* spp., *D. purpurea*
frankincense	*Boswellia serrata*
fringe tree	*Chionanthus virginicus*
garbanzo bean	*Cicer arietinum*
garden angelica	*Angelica archangelica*
gardenia	*Gardenia jasminoides*
garlic	*Allium sativum*
gentian	*Gentiana lutea*
ginger	*Zingiber officinale*
ginkgo	*Ginkgo biloba*
ginseng	*Panax ginseng*
glossy buckthorn	*Rhamnus frangula*
glossy privet	*Ligustrum lucidum*
gobo root	*Arctium lappa*
goji	*Lycium barbarum*
goldenrod	*Solidago odora, S. canadensis, S. virgaurea*
goldenseal	*Hydrastis canadensis*
goldthread	*Coptis trifolia, C. chinensis*
goosefoot	*Chenopodium ambrosioides*
gotu kola	*Centella asiatica*
grapefruit	*Citrus paradisi*
gravel root	*Eupatorium purpureum*
gray mangrove	*Avicennia marina*
green tea	*Camellia sinensis* (formerly *Thea sinensis*)
guava	*Psidium guajava*
guduchi	*Tinospora cordifolia*
guggul	*Commiphora mukul*

gumweed	*Grindelia squarrosa*
gurmar	*Gymnema sylvestre*
haritaki	*Terminalis chebula*
hawthorn	*Crataegus monogyna, Crataegus oxyacantha* (also known as *C. rhipidophylla*)
he shou wu	*Reynoutria multiflora* (formerly *Polygonum multiflorum*)
hemp	*Cannabis sativa*
hen of the woods	*Grifola frondosa*
henbane	*Hyoscyamus niger*
Hercules' club	*Zanthoxylum clava-herculis*
Hercules' herb	*Zanthoxylum clava-herculis*
hibiscus	*Hibiscus sabdariffa*
hoelen	*Poria cocos*
hollyhock	*Alcea rosea*
holy basil	*Ocimum tenuiflorum* (also known as *O. sanctum*)
honey bee	*Apis mellifica*
honeysuckle	*Lonicera japonica*
hops	*Humulus lupulus*
horny goatweed	*Epimedium pubescens*
horse chestnut	*Aesculus hippocastanum*
horseradish	*Armoracia rusticana* (also known as *Cochlearia armoracia*)
horsetail	*Equisetum arvense, E. hymenale*
hot peppers	*Capsicum* spp.
huang qin	*Scutellaria baicalensis*
hurricane weed	*Phyllanthus amarus*
hydrangea	*Hydrangea arborescens, H. macrophylla*
Indian frankincense	*Boswellia serrata*
Indian gooseberry	*Phyllanthus emblica*
Indian snakeroot	*Rauvolfia serpentina, R. vomitoria*
Irish moss	*Chondrus crispus*
jaborandi	*Pilocarpus jaborandi*
Jamaican dogwood	*Piscidia piscipula* (also known as *P. erythrina*)
jambolan	*Syzygium cumini*
jambul	*Syzygium cumini*
Japanese knotweed	*Reynoutria japonica* (formerly *Polygonum cuspidatum*)
Japanese pepper	*Zanthoxylum piperitum*
Job's tears	*Coix lacryma-jobi*
Johnny jump up	*Viola tricolor*
jujube	*Ziziphus jujuba*
juniper	*Juniperus communis*
kava kava	*Piper methysticum*
kelp	*Ascophyllum nodosum, Laminaria japonica, Laminaria hyperborean, Macrocystis pyrifera*
khella	*Ammi visnaga*
kidney bean	*Phaseolus vulgaris*
king of bitters	*Andrographis paniculata*
kinnickinick	*Arctostaphylos uva ursi*
Korean perilla	*Perilla frutescens*
kudzu	*Pueraria montana, P. mirifica, P. lobata*
kutki	*Picrorhiza kurroa*
kwandari	*Terminalia macroptera*
large-headed atractylodes	*Atractylodes ovata*
lavender	*Lavandula officinalis*
lemon balm	*Melissa officinalis*
leopard's bane	*Arnica montana*
licorice	*Glycyrrhiza glabra*
lily of the valley	*Convallaria majalis*
long pepper	*Piper longum*
lovage	*Levisticum officinale*
lucerne	*Medicago sativa*
lumpy bracket	*Polyporus umbellatus*
ma-huang	*Ephedra sinica*
maca	*Lepidium meyenii*
magnolia	*Magnolia officinalis, M. nervosa*
magnolia vine	*Schisandra chinensis*
mahonia	*Berberis aquifolium* (also known as *Mahonia aquifolium*), *B. vulgaris*
maidenhair tree	*Ginkgo biloba*

maitake	*Grifola frondosa*
mango	*Mangifera indica*
marijuana	*Cannabis sativa*
marsh tea or wild rosemary	*Ledum palustre*
marshmallow	*Althaea officinalis*
mayapple	*Podophyllum peltatum*
meadowsweet	*Filipendula ulmaria*
milk thistle	*Silybum marianum*
milk vetch	*Astragalus membranaceus*
mistletoe	*Viscum album*
motherwort	*Leonurus cardiaca*
moutan	*Paeonia suffruticosa*
mugwort	*Artemisia vulgaris*
mullein	*Verbascum thapsus*
myrrh	*Commiphora myrrha*
nasturtium	*Tropaeolum majus*
navy bean	*Phaseolus vulgaris*
neem	*Azadirachta indica*
nettles	*Urtica dioica, U. urens*
New Jersey tea	*Ceanothus americanus*
night-blooming cactus	*Cactus grandiflorus* (also known as *Selenicereus grandiflorus*)
northern prickly ash	*Zanthoxylum americanum*
northern white cedar	*Thuja occidentalis*
nutgrass	*Cyperus rotundus*
nutmeg	*Myristica fragrans*
oats	*Avena sativa*
ocotillo	*Fouquieria splendens*
old man's beard	*Usnea barbata*
olive	*Olea europaea*
onion	*Allium cepa*
oregano	*Origanum vulgare, O. marjorana, O. onites*
Oregon grape	*Berberis aquifolium* (also known as *Mahonia aquifolium*)
osha	*Ligusticum porteri*
osthole	*Cnidium monnieri*
papaya	*Carica papaya*
paracress	*Spilanthes acmella*
parsley	*Petroselinum crispum* (also known as *P. sativum*)
passionflower	*Passiflora incarnata*
pau d'arco	*Tabebuia impetiginosa*
peach	*Prunus persica*
pennyroyal	*Hedeoma pulegioides*
pennywort	*Centella asiatica*
peppermint	*Mentha piperita*
pepperwood	*Zanthoxylum clava-herculis*
perilla	*Perilla frutescens*
Peruvian bark	*Cinchona officinalis, C. pubescens*
pineapple	*Ananas comosus*
pipsissewa	*Chimaphila umbellata*
plantain	*Plantago lanceolata, P. ovata, P. psyllium*
pleurisy root	*Asclepias tuberosa*
poison hemlock	*Conium maculatum*
poison ivy	*Rhus toxicodendron*
poke	*Phytolacca americana, P. decandra*
pokeroot	*Phytolacca americana, P. decandra*
pomegranate	*Punica granatum*
pot marigold	*Calendula officinalis*
prickly pear	*Opuntia ficus-indica*
prince's pine	*Chimaphila umbellata*
pukeweed	*Lobelia inflata*
pumpkin	*Cucurbita maxima, C. pepo*
purple coneflower	*Echinacea purpurea*
quaking aspen	*Populus tremuloides*
quassia	*Picrasma excelsa*
queen of the night	*Selenicereus grandiflorus* (also known as *Cactus grandiflorus*)
queen's root	*Stillingia sylvatica*
quinine bark	*Cinchona officinalis, C. pubescens*
rapeseed	*Brassica napus, B. campestris*
raspberry	*Rubus idaeus*
rattlesnake	*Crotalus horridus*

red alder	*Alnus rubra*
red clover	*Trifolium pratense*
red oak	*Quercus rubra*
red raspberry	*Rubus idaeus*
red root	*Ceanothus americanus*
red sage	*Salvia miltiorrhiza*
redstem wormwood	*Artemisia scoparia*
reishi mushroom	*Ganoderma lucidum*
resin gathered from various plants by bees	*Propolis*
rhodiola	*Rhodiola kirilowii, R. rosea*
rose	*Rosa* spp.
rosemary	*Rosmarinus officinalis*
rou cong rong	*Cistanche tubulosa*
roundhead bushclover	*Lespedeza capitata*
rupturewort	*Herniaria glabra*
safflower	*Carthamus tinctorius*
saffron crocus	*Crocus sativus*
sage	*Salvia officinalis*
sarsaparilla	*Smilax ornata, S. regelii*
saw palmetto	*Serenoa repens*
Scot's broom	*Cytisus scoparius*
Scotch broom	*Cytisus scoparius*
scouring rush	*Equisetum arvense, E. hymenale*
scute	*Scutellaria baicalensis*
sea buckthorn	*Hippophae rhamnoides*
senna	*Cassia senna* (also known as *Senna alexandrina*)
shatavari	*Asparagus racemosus*
sheep sorrel	*Rumex acetosella*
shepherd's purse	*Capsella bursa-pastoris*
shiitake mushroom	*Lentinula edodes*
Siberian fir	*Abies sibirica*
Siberian ginseng	*Eleutherococcus senticosus*
silver birch	*Betula pendula*
skullcap	*Scutellaria lateriflora*
slippery elm	*Ulmus fulva, U. rubra*
snow parsley	*Cnidium monnieri, Cnidium officinale* (also known as *Ligusticum officinale*)
snowball bush	*Viburnum opulus*
southern maidenhair fern	*Adiantum capillus-veneris*
southern prickly ash	*Zanthoxylum clava-herculis*
soybean	*Glycine max*
Spanish black radish	*Raphanus sativus* var. *niger*
Spanish fly	*Cantharis vesicatoria*
spearmint	*Mentha spicata*
St. Johnswort	*Hypericum perforatum*
stevia	*Stevia rebaudiana*
stonebreaker	*Phyllanthus niruri*
stoneroot	*Collinsonia canadensis*
sugar destroyer	*Gymnema sylvestre*
sweet Annie	*Artemisia annua*
sweet flag	*Acorus calamus*
sweet leaf	*Stevia rebaudiana*
sweet sumac	*Rhus aromatica*
taheebo	*Tabebuia impetiginosa*
tamarind	*Tamarindus indica*
tarentula	*Tarentula hispanica*
tea tree	*Melaleuca alternifolia*
Thai galangal	*Alpinia galanga*
thunder god vine	*Tripterygium wilfordii*
thyme	*Thymus vulgaris*
todok	*Codonopsis lanceolata*
toothache tree	*Zanthoxylum americanum, Z. clava-herculis*
tree peony	*Paeonia suffruticosa*
tropical almond	*Terminalis catappa*
tulsi	*Ocimum tenuiflorum* (also known as *O. sanctum*)
turkey corn	*Corydalis cava, C. yanhusuo*
turkey rhubarb	*Rheum officinale, R. palmatum*
turmeric	*Curcuma longa*

turtlehead	*Chelone glabra*
umbrella polypore	*Polyporus umbellatus*
uña de gato	*Uncaria tomentosa*
uva ursi	*Arctostaphylos uva ursi*
valerian	*Valeriana officinalis*
vervain	*Verbena hastata*
vitex	*Vitex agnus-castus*
wax myrtle	*Myrica cerifera*
western red cedar	*Thuja plicata*
wheat	*Triticum aestivum*
white atractylodes	*Atractylodes ovata*
white button mushroom	*Agaricus bisporus*
white hellebore	*Veratrum album*
white mangrove	*Avicennia marina*
white oak	*Quercus alba*
white peony	*Paeonia lactiflora*
white willow	*Salix alba*
wild carrot	*Daucus carota*
wild geranium	*Geranium maculatum, G. robertianum*
wild indigo	*Baptisia tinctoria*
wild iris	*Iris versicolor, I. tenax*
wild yam	*Dioscorea villosa, D. rotunda*
winter squash	*Cucurbita maxima*
winter-blooming bergenia	*Bergenia crassifolia*
wintergreen	*Gaultheria procumbens*
witch hazel	*Hamamelis virginiana*
wolf spider	*Tarentula hispanica*
wolfberry	*Lycium barbarum*
wood violet	*Viola odorata*
wormseed	*Chenopodium ambrosioides*
wormwood	*Artemisia absinthium*
yarrow	*Achillea millefolium*
yellow dock	*Rumex crispus*
yellow jasmine	*Gelsemium sempervirens*
yellow jessamine	*Gelsemium sempervirens*
yellow wild indigo	*Baptisia tinctoria*
yerba de tago	*Eclipta alba, E. prostata*
zhi mu	*Anemarrhena asphodeloides*

— GLOSSARY — OF THERAPEUTIC TERMS

Abortifacient. An agent capable of promoting the expulsion of a developing fetus.

Absorbent. An agent that promotes the absorption of medicinal compounds.

Acidifier. An agent imparting acidity to body fluids, especially blood and urine.

Acute. A condition that has a new onset, comes on suddenly, and is relatively short-lasting in its entire duration.

Aerial parts. The parts of a plant that grow above ground.

Alkalinizer. An agent that increases the alkalinity of bodily fluids, especially the blood and urine.

Allopathic. A term applying to conventional, modern Western medicine. *Allo* refers to "opposite," and in this case, means to oppose pathology. For example: In cases of fever, an antipyretic is used; to treat inflammation, an anti-inflammatory is used; and to treat an infection, antimicrobials are used.

Alterative. An agent that favorably "alters" an individual's health. Alteratives stimulate digestive and absorptive functions while enhancing elimination of wastes. Alteratives are also traditionally said to "purify" the blood and optimize metabolic functions.

Analgesic. An agent that is pain-relieving.

Anaphrodisiac. An agent that diminishes sexual drive or function.

Anesthetic. An agent that diminishes pain and tactile sensations temporarily.

Anhydrotic. An agent that diminishes excessive sweating.

Anodyne. An agent that is pain-relieving.

Antacid. An agent that diminishes stomach acid.

Antagonist. An agent that opposes the action of some other medicine, usually a poison or toxic alkaloid.

Anthelmintic. An agent used to combat intestinal worms.

Antidote. A remedy to counteract the action of poisons or other strong actions.

Antiemetic. An agent that allays nausea and vomiting.

Antigalactogogue. An agent that diminishes lactation.

Antihemorrhagic. An agent that helps control excessive bleeding.

Antilithic. An agent used to reduce the formation of stones and calculi in the body.

Antioxidant. An agent capable of accepting electrons or highly reactive molecules that could damage body membranes if left free in circulation.

Antiperiodic. An agent used to combat the periodic fevers of malaria.

Antiphlogistic. An agent used to reduce fever and inflammation.

Antipyretic. An agent used to reduce fever.

Antiscorbutic. An agent used to provide vitamin C and prevent or treat scurvy.

Antiseptic. An agent having antimicrobial capacity for the prevention of sepsis.

Antisialagogue. An agent capable of reducing salivation.

Antispasmodic. An agent capable of reducing painful spasms in muscles and hollow organs.

Antitussive. An agent used to diminish coughing.

Aperient. A gentle nonirritating laxative.

Aphrodisiac. An agent used to stimulate the libido.

Aromatic. An agent with a strong fragrance to be inhaled or absorbed through the skin.

Astringent. An agent that dries, condenses, and shrinks inflamed or suppurative tissues.

Bitter. An agent that has a bitter flavor and is used to stimulate gastrointestinal tone and secretions. Bitters prepare mucosa for food, stimulate appetite, and enhance digestion.

Cardiotonic. An agent that improves heart function.

Carminative. An agent that reduces gas, bloating, flatulence, and associated pain.

Cathartic. A strong, potentially harsh laxative.

Caustic. An agent having a corrosive action on tissues.

Cholagogue. An agent that increases gallbladder tone and the flow of bile from the gallbladder.

Choleretic. An agent that increases the production of bile.

Chronic. A condition that develops slowly over time and becomes persistent and sometimes permanent.

Corrigent. An agent that balances a harsh or strong action of another agent, a corrective.

Counterirritant. An agent that irritates local tissues to enhance blood flow to the area. Counterirritants are used to induce temporary hyperemia in chronic conditions in an attempt to relieve pain, promote healing, and reduce inflammation.

Dacryagogue. An agent that promotes the flow of tears (lacrimation),

Deficient. Referring to low energy, low vitality, and poor functioning tissues (herbal medicine). In Traditional Chinese Medicine, the term chi deficiency is used when the entire body is in a weakened state. The term may also be used to indicate a poorly functioning organ or biochemical state, such as digestive deficiency, circulatory deficiency, or metabolic deficiency.

Demulcent. A cooling, soothing, mucilagenous substance used internally or topically to emolliate abraded, inflamed, or irritated mucosal tissues.

Depurant. Any agent aimed at purifying, such as a liver depurant, a renal depurant, a blood depurant, and so on. Depurants have a purifying effect by promoting the elimination of wastes from the body.

Diaphoretic. An agent capable of inducing perspiration and often a temporary fever.

Diuretic. An agent that stimulates the production and flow of urine.

Ecbolic. An agent that stimulates childbirth (parturition).

Emetic. An agent that causes vomiting (emesis).

Emmenagogue. An agent that promotes menstrual flow.

Emollient. An agent that soothes and softens the skin and mucosal tissues.

Errhine. An agent that irritates the nasal mucosa and promotes sneezing and secretions.

Escharotic. Any caustic substances applied topically to diseased tissues to kill the cells and promote sloughing away. The word *eschar* means to cast off.

Excess. Indicates a condition with increased or excessive function, such as too much heat, over-stimulated bowels or muscle tone, or other situations of excess in various physiologic functions.

Excitant. An agent that causes excitation of nervous, circulatory, or motor functions; however, in Latin America, the term is more often used to refer to an aphrodisiac, or sexual excitant.

Exhilarant. An agent that causes excitation of psychic functions and promotes euphoria.

Expectorant. An agent that promotes the flow of secretions from the respiratory tract.

Febrifuge. An agent used to bring down the temperature in cases of fever.

Galactogogue. An agent that stimulates lactation.

Hematic, hematinic. An agent that improves the quality of the blood, especially in cases of anemia, but may be used in other situations.

Hemostatic. An agent that reduces blood flow and promotes clotting in cases of hemorrhage, trauma, and internal bleeding.

Hepatic. An agent that improves the function of the liver.

Hydragogue. An agent that promotes watery secretions.

Irritant. An agent applied locally for the purposes of intentionally causing local hyperemia. See Counterirritant.

Laxative. An agent that promotes a mild and painless evacuation of the bowels.

Lithotriptic. An agent aimed at dissolving calculi within the body.

Lipotropic. Literally translates as "fat mover", and used to refer to various alterative and cholagogue herbs, as well as substances such as choline that promote bile flow and biliary function, and thereby improve liver function, and hepatic clearance of lipids, carbohydrates, hormones, toxins, and chemicals.

Material dose. A term used by herbalists and alternative medicine practitioners to distinguish between a highly diluted or homeopathic preparation of a substance and the substance given in a more substantial or "material dose."

Miotic, myotic. An agent that causes the pupil to contract (miosis).

Mydriatic. An agent that promotes dilation of the pupil (mydriasis).

Narcotic. A drug that promotes stupor or sleep and is used to relieve pain or diminish consciousness.

Nervine. An agent having a tonifying effect on the nervous system, usually only used in the context of herbal medicine, with some herbs being referred to nervine herbs.

Nutrient, nutritive. An agent that enhances assimilation, metabolism, and nutrition.

Oxytocic. An agent that promotes uterine contractions and hastens childbirth.

Parturifacient. An agent that facilitates childbirth when taken during labor.

Partus preparator. An agent taken in the last months of pregnancy to tone the uterus and optimize labor and delivery.

Purgative. A strong laxative that may be irritating and cause cramping.

Refrigerant. An agent capable of imparting a cooling sensation when applied topically.

Revulsive. An agent used to enhance the blood flow to a particular body part (hand, foot) in order to draw it away from a congested, engorged area (head, uterus).

Rubefacient. An agent that promotes reddening or hyperemia of the tissues.

Sedative. An agent that calms in cases of nervousness, insomnia, and mania, and may be stronger and less tonifying than a nervine.

Sialagogue, salivant. An agent that increases the flow of saliva.

Simple. A term used to refer to a single herb, not mixed with other herbs or used in a formula, but rather used as "a simple."

Specific. An agent thought to be of specific value for a collection of symptoms.

Sternutatory. An agent that promotes sneezing when inhaled.

Styptic. A strong astringent agent capable of reducing bleeding when applied topically.

Sudorific. An agent capable of inducing perspiration and regarded as being stronger than a diaphoretic.

Synergist. An agent that duplicates, enhances, or pulls together the action of a group of medicinal substances.

Taenicide. An agent that kills or weakens tapeworms.

Tonic. An agent that has a positive effect on the function of an organ or tissue and suggests an ability to restore normal function be it excess or deficient, atonic or hypertonic, overstimulated or understimulated. Tonic supports the optimal physiologic state.

Toxicity. A deranged, inflammatory, or otherwise corrupted or polluted biochemical state in the body.

Vasoconstrictor. An agent that constricts the blood vessels.

Vasodepressant. An agent that slows the pulse rate and lowers the pressure.

Vasodilator. An agent used to dilate the vasculature, usually used in cases of hypertension.

Vermifuge. An agent that promotes the expulsion of intestinal worms.

Vesicant. An agent that promotes blistering or vesication of the skin.

— NOTES —

Chapter 2: Creating Herbal Formulas for Gastrointestinal and Biliary Conditions

1. N. A. Fadeeva et al., "Small Intestinal Bacterial Overgrowth as a Cause of Lactase Deficiency," *Terk Arkh* 87, no. 2 (2015): 20–23, doi:10.17116/terarkh201587220-23.
2. Yanyong Deng et al., "Lactose Intolerance in Adults: Biological Mechanism and Dietary Management," *Nutrients* 7, no. 9 (September 18, 2015): 8020–35, doi:10.3390/nu7095380.
3. Elana Tronconi et al., "The Autoimmune Burden in Juvenile Idiopathic Arthritis," *Italian Journal of Pediatrics* 43 (2017): 56, doi:10.1186/s13052-017-0373-9.
4. Jeong Tyae Jeon et al., "Effect of Ethanol Extract of Dried Chinese Yam (*Dioscorea batatas*) Flour Containing Dioscin on Gastrointestinal Function in Rat Model," *Archives of Pharmacal Research* 29, no. 5 (May 2006): 348–53, doi:10.1007/BF02968583.
5. Ming Ruan et al., "Attenuation of Stress-Induced Gastrointestinal Motility Disorder by Gentiopicroside, from Gentiana Macrophylla Pall," *Fitoterapia* 103 (June 2015): 265–76, doi:10.1016/j.fitote.2015.04.015.
6. Ruan, "Attenuation of Stress-Induced Gastrointestinal Motility Disorder by Gentiopicroside."
7. Michael McMullen, "Bitters: Time for a New Paradigm," *Evidence-Based Complementary and Alternative Medicine* 2015 (2015): doi:10.1155/2015/670504.
8. Mika Yuki et al., "Effects of Daikenchuto on Abdominal Bloating Accompanied by Chronic Constipation: A Prospective, Single-Center Randomized Open Trial," *Current Therapeutic Research* 77 (December 2015): 58–62, doi:10.1016j.curtheres.2015.04.002.
9. Shady Allam et al., "Extracts from Peppermint Leaves, Lemon Balm Leaves and in Particular Angelica Roots Mimic the Pro-Secretory Action of the Herbal Preparation STW 5 in the Human Intestine," *Phytomedicine* 22, no. 12 (November 15, 2015): 1063–70, doi:10.1016/j.phymed.2015.08.008.
10. T. Bokic et al., "Potential Causes and Present Pharmacotherapy of Irritable Bowel Syndrome: An Overview," *Pharmacology* 96, no. 1–2 (2015): 76–85, doi:10.1159/000435816.
11. C. Hasenoehrl et al., "The Gastrointestinal Tract—A Central Organ of Cannabinoid Signaling in Health and Disease," *Neurogastroenterol and Motility* 28, no. 12 (August 26, 2016): 1765–80, doi:10.1111/nmo.12931.
12. Hirotada Akiho and Kazuhiko Nakamura, "Daikenchuto Ameliorates Muscle Hypercontractility in a Murine T-Cell–Mediated Persistent Gut Motor Dysfunction Model," *Digestion* 83, no. 3 (2011): 173–79, doi:10.1159/000321798.
13. G. Cappello et al., "Peppermint Oil (Mintoil®) in the Treatment of Irritable Bowel Syndrome: A Prospective Double Blind Placebo-Controlled Randomized Trial," *Digestive and Liver Disease* 39, no. 6 (June 2007): 530–36, doi:10.1016/j.dld.2007.02.006.
14. Alan C. Logan and Tracey M. Beaulne, "The Treatment of Small Intestinal Bacterial Overgrowth with Enteric-Coated Peppermint Oil: A Case Report," *Alternative Medicine Review* 7, no. 5 (October 2002): 410–17; A. R. Gaby, "Treatment with Enteric-Coated Peppermint Oil Reduced Small-Intestinal Bacterial Overgrowth in a Patient with Irritable Bowel Syndrome," *Alternative Medicine Review* 8, no. 1 (February 2003): 3.
15. Suma Magge and Anthony Lembo, "Low-FODMAP Diet for Treatment of Irritable Bowel Syndrome," *Gastroenterology Hepatology* 8, no. 11 (November 2012): 739–45.
16. Siew C. Ng et al., "Systematic Review: The Efficacy of Herbal Therapy in Inflammatory Bowel Disease," *Alimentary Pharmacology and Therapeutics* 38, no. 8 (October 2013): 854–63, doi:10.1111/apt.12464; J. Langhorst et al., "Systematic Review of Complementary and Alternative Medicine Treatments in Inflammatory Bowel Diseases," *Journal of Crohn's and Colitis* 9, no. 1 (January 2015): 86–106, doi:10.1093/ecco-jcc/jju007; and I. Chakŭrski et al., "Treatment of Chronic Colitis with an Herbal Combination of *Taraxacum officinale, Hipericum perforatum, Melissa officinalis, Calendula officinalis* and *Foeniculum vulgare*," *Vutr Boles* 20, no. 6 (1981): 51–54.
17. S. M. Lim, "Timosaponin AIII and Its Metabolite Sarsasapogenin Ameliorate Colitis in Mice by Inhibiting NF-κB and MAPK Activation and Restoring Th17/Treg Cell Balance," *International Immunopharmacology* 25, no. 2 (April 2015): 493–503, doi:10.1016/j.intimp.2015.02.016.
18. Yasuyuki Deguchi et al., "Curcumin Prevents the Development of Dextran Sulfate Sodium (DSS)–Induced Experimental Colitis," *Digestive Diseases and Sciences* 52, no. 11 (November 2007): 2993–98, doi:10.1007/s10620-006-9038-9; Laura Camacho-Marquero et al., "Curcumin, a *Curcuma longa* Constituent, Acts on MAPK p38 Pathway Modulating COX-2 and iNOS Expression in Chronic Experimental Colitis," *International Immunopharmacology* 7, no. 3 (March 2007): 333–42,

doi:10.1016/j,intimp.2006.11.006; and Anindita Ukil et al., "Curcumin, the Major Component of Food Flavour Turmeric, Reduces Mucosal Injury in Trinitrobenzene Sulphonic Acid–Induced Colitis," *British Journal of Pharmacology* 139, no. 2: 209–18, doi:10.1038/sj.bip.0705241.
19. Tao Wang et al., "Astragalus Saponins Affect Proliferation, Invasion and Apoptosis of Gastric Cancer BGC-823 Cells," *Diagnostic Pathology* 8 (October 24, 2013): 179, doi:10.1186/1746-1596-8-179.
20. Rongli Zhang et al., "Natural Compound Methyl Protodioscin Protects Against Intestinal Inflammation through Modulation of Intestinal Immune Responses," *Pharmacology Research and Perspective* 3, no. 2 (March 2015): e00118, doi:10.1002/prp2.118.
21. Xiao-Dong Wen et al., "*Salvia miltiorrhiza* (Dan Shen) Significantly Ameliorates Colon Inflammation in Dextran Sulfate Sodium Induced Colitis," *American Journal of Chinese Medicine* 41, no. 5: 1097–108, doi:10.1142/S0192415X13500742.
22. Mohammadali Kamali et al., "Efficacy of the *Punica granatum* Peels Aqueous Extract for Symptom Management in Ulcerative Colitis Patients. A Randomized, Placebo-Controlled, Clinical Trial," *Complementary Therapies in Clinical Practice* 21, no. 3 (August 2015): 141–46, doi:10.1016/j.ctcp.2015.03.001.
23. Aikaterini Triantafyllidi et al., "Herbal and Plant Therapy in Patients with Inflammatory Bowel Disease," *Annals of Gastroenterology* 28, no. 2 (April–June 2015): 210–20, doi:10.1002/14651858.CD011223.
24. Triantafyllidi, "Herbal and Plant Therapy in Patients with Inflammatory Bowel Disease."
25. Linjing Zhao et al., "The *In Vivo* and *In Vitro* Study of Polysaccharides from a Two-Herb Formula on Ulcerative Colitis and Potential Mechanism of Action," *Journal of Ethnopharmacology* 153, no. 1 (April 11, 2014): 151–59, doi:10.1016/j.jep.2014.02.008.
26. Bahareh Heidari, "Effect of *Coriandrum sativum* Hydroalcoholic Extract and Its Essential Oil on Acetic Acid–Induced Acute Colitis in Rats," *Avicenna J Phytomed* 6, no. 2 (2016 March/April): 205–14.
27. Natsuko Kageyama-Yahara et al., "The Inhibitory Effect of Ergosterol, a Bioactive Constituent of a Traditional Japanese Herbal Medicine Saireito on the Activity of Mucosal-Type Mast Cells," *Biological and Pharmaceutical Bulletin* 33, no. 1 (2010): 142–45, doi:10.1248/bpb.33.142; and S. J. Kim et al., "Beneficial Effects of the Traditional Medicine Igongsan and Its Constituent Ergosterol on Dextran Sulfate Sodium-Induced Colitis in Mice," *Molecular Medicine Reports* 12, no. 3 (September 2015): 3549–56, doi:10.3892/mmr.2015.3824.
28. Masaaki Kawano et al., "Berberine Is a Dopamine D1- and D2-Like Receptor Antagonist and Ameliorates Experimentally Induced Colitis by Suppressing Innate and Adaptive Immune Responses," *Journal of Neuroimmunology* 289 (December 15, 2015): 43–55, doi:10.1016/j.jneuroim.2015.10.001.
29. Bin Wang et al., "L-Glutamine Enhances Tight Junction Integrity by Activating CaMK Kinase 2-AMP-Activated Protein Kinase Signaling in Intestinal Porcine Epithelial Cells," *Journal of Nutrition* 146, no. 3 (March 2016): 501–8, doi:10.3945/jn.115.224857.
30. Mohammad E. Zohalinezhad et al., "*Myrtus communis* L. Freeze-Dried Aqueous Extract Versus Omeprazol in Gastrointestinal Reflux Disease: A Double-Blind Randomized Controlled Clinical Trial," *Journal of Evidence-Based Complementary and Alternative Medicine* 21, no. 1 (January 2016): 23–9, doi:10.1177/2156587215589403.
31. Arvind K. Giri et al., "Effect of Lycopene against Gastroesophageal Reflux Disease in Experimental Animals," *BMC Complementary and Alternative Medicine* 15 (April 9, 2015): 110, doi:10.1186/s12906-015-0631-6.
32. Flavia Laffleur et al., "Design, Modification and *In Vitro* Evaluation of Pectin's Bucco-Adhesiveness," *Therapeutic Delivery* 7, no. 6 (June 2016): 369–75, doi:10.4155/tde-2016-0019.
33. Manuele Casale et al., "Hyaluronic Acid: Perspectives in Upper Aero-Digestive Tract. A Systematic Review," *PLoS ONE* 10, no. 6 (June 29, 2015): e0130637, doi:10.1371/journal.pone.0130637; and Beniamino Palmieri et al., "Fixed Combination of Hyaluronic Acid and Chondroitin-Sulphate Oral Formulation in a Randomized Double Blind, Placebo Controlled Study for the Treatment of Symptoms in Patients with Non-erosive Gastroesophageal Reflux," *European Review for Medical and Pharmacological Sciences* 17, no. 24 (December 2013): 3272–78.
34. Leo Meunier et al., "Locust Bean Gum Safety in Neonates and Young Infants: An Integrated Review of the Toxicological Database and Clinical Evidence," *Regulatory Toxicology and Pharmacology* 70, no. 1 (October 2014): 155–69, doi:10.1016/j.yrtph.2014.06.023.
35. Khanittha Chawananorasest et al., "Extraction and Characterization of Tamarind (*Tamarind indica* L.) Seed Polysaccharides (TSP) from Three Difference Sources," *Molecules* 21, no. 6 (June 15, 2016): pii: E775, doi:10.3390/molecules21060775; and Amit Kumar Nayak et al., "Development of Calcium Pectinate-Tamarind Seed Polysaccharide Mucoadhesive Beads Containing Metformin HCl," *Carbohydrate Polymers* 101 (January 30, 2014): 220–30, doi:10.1016/j.carbpol.2013.09.024.
36. S. Fujimori, "What Are the Effects of Proton Pump Inhibitors on the Small Intestine?," *World Journal of Gastroenterology* 21, no. 22 (June 14, 2015): 6817–19, doi:10.3748/wjg.v21.i22.6817.
37. Miguel Pérez-Fontan et al., "Inhibition of Gastric Acid Secretion by H2 Receptor Antagonists Associates a Definite Risk of Enteric Peritonitis and Infectious Mortality in Patients Treated with Peritoneal Dialysis," *PLoS ONE* 11, no. 2 (February 12, 2016): e0148806, doi:10.1371/journal.pone.0148806.
38. Daniel E. Freedberg et al., "Use of Acid Suppression Medication Is Associated with Risk for *C. difficile* Infection in Infants and Children: A Population-Based Study," *Clinical*

Infectious Diseases 61, no. 6 (September 15, 2015): 912–17, doi:10.1093/cid/civ432; and Robert MacLaren et al., "Proton Pump Inhibitors and Histamine-2 Receptor Antagonists in the Intensive Care Setting: Focus on Therapeutic and Adverse Events," *Expert Opinion on Drug and Safety* 14, no. 2 (February 2015): 269–80, doi:10.1517/1470338.2015.986456.

39. Katelyn E. Brown et al., "Acid-Suppressing Agents and Risk for *Clostridium difficile* Infection in Pediatric Patients," *Clinical Pediatrics (Phila)* 54, no. 11 (October 2015): 1102–6.
40. John L. Wallace et al., "Toward More GI-Friendly Anti-Inflammatory Medications," *Current Treatment Options in Gastroenterology* 13, no. 4 (December 2015): 377–85, doi:10.1007/s11938-015-0064-9.
41. Sai-Wai Ho et al., "Risk of Stroke-Associated Pneumonia with Acid-Suppressive Drugs: A Population-Based Cohort Study," *Medicine (Baltimore)* 94, no. 29 (July 2015): e1227, doi:10.1097/MD.0000000000001227.
42. Saleem A. Banihani, "Histamine-2 Receptor Antagonists and Semen Quality," *Basic and Clinical Pharmacology and Toxicology* 118, no. 1 (January 2016): 9–13, doi:10.111/bcpt.12446.
43. Victoria S. Benson et al., "Associations between Gastro-Oesophageal Reflux, Its Management and Exacerbations of Chronic Obstructive Pulmonary Disease," *Respiratory Medicine* 109, no. 9 (September 2015): 1147–54, doi:10.1016/j.rmed.2015.06.009.
44. B. Lazarus et al., "Proton Pump Inhibitor Use and the Risk of Chronic Kidney Disease," *JAMA Internal Medicine* 176, no. 2 (February 2016): 238–46, doi:10.1001/jamainternmed.2015.7913.
45. Daniel E. Freedberg et al., "Use of Proton Pump Inhibitors Is Associated with Fractures in Young Adults: A Population-Based Study," *Osteoporos International* 26, no. 10 (October 2015): 2501–7, doi:10.1007/s00198-015-3168-0.
46. Shlomi Cohen et al., "Adverse Effects Reported in the Use of Gastroesophageal Reflux Disease Treatments in Children: A 10 Years Literature Review," *British Journal of Clinical Pharmacology* 80, no. 2 (August 2015): 200–8, doi:10.1111/bcp.12619.
47. Mai Ramadan et al., "Chamazulene Carboxylic Acid and Matricin: A Natural Profen and Its Natural Prodrug, Identified through Similarity to Synthetic Drug Substances," *Journal of Natural Products* 69, no. 7 (July 2006): 1041–45, doi:10.1021/np0601556.
48. Diane L. McKay and Jeffrey B. Blumberg, "A Review of the Bioactivity and Potential Health Benefits of Chamomile Tea (*Matricaria recutita* L.)," *Phytotherapy Research* 20, no. 7: 519–30, doi:10.1002/ptr.1900.
49. Gail B. Mahady et al., "*In Vitro* Susceptibility of *Helicobacter pylori* to Botanical Extracts Used Traditionally for the Treatment of Gastrointestinal Disorders," *Phytotherapy Research* 19, no. 11 (November 2009): 988–91, doi:10.1002/ptr.1776.
50. Fahaid H. Al-Hashem, "Gastroprotective Effects of Aqueous Extract of *Chamomilla recutita* against Ethanol-Induced Gastric Ulcers," *Saudi Medical Journal* 31, no. 11 (November 2010): 1211–16; Mustafa Cemek et al., "Protective Effect of *Matricaria chamomilla* on Ethanol-Induced Acute Gastric Mucosal Injury in Rats," *Pharmaceutical Biology* 48, no. 7 (July 2010): 757–63, doi:10.3109/13880200903296147.
51. S. B. Bezerra et al., "Bisabolol-Induced Gastroprotection against Acute Gastric Lesions: Role of Prostaglandins, Nitric Oxide, and KATP+ Channels," *Journal of Medicinal Food* 12, no. 6 (December 2009): 1403–6, doi:10.1089/jmf.2008.0290.
52. Bezerra, "Bisabolol-Induced Gastroprotection against Acute Gastric Lesions."
53. Nayrton Flávio Moura Rocha et al., "Gastroprotection of (-)-Alpha-Bisabolol on Acute Gastric Mucosal Lesions in Mice: The Possible Involved Pharmacological Mechanisms," *Fundamental and Clinical Pharmacology* 24, no. 1 (February 2010): 63–71, doi:10.1111/j.1472-8206.2009.00726.x.
54. J. von Schönfeld et al., "Oesophageal Acid and Salivary Secretion: Is Chewing Gum a Treatment Option for Gastro-Oesophageal Reflux?," *Digestion* 58, no. 2 (1997): 111–14.
55. Rachel Brown et al., "Effect of GutsyGum™, a Novel Gum, on Subjective Ratings of Gastro Esophageal Reflux Following a Refluxogenic Meal," *Journal of Dietary Supplements* 12, no. 2 (June 2015): 138–45, doi:10.3109/19390211.2014.950783.
56. Manoj K. Rai et al., "Alginate-Encapsulation of Nodal Segments of Guava (*Psidium guajava* L.) for Germplasm Exchange and Distribution," *The Journal of Horticultural Science and Biotechnology* 83, no. 5 (2008): 569–73, doi:10.1080/14620316.2008.11512425.
57. E. Thomas et al., "Randomised Clinical Trial: Relief of Upper Gastrointestinal Symptoms by an Acid Pocket–Targeting Alginate-Antacid (Gaviscon Double Action)—A Double-Blind, Placebo-Controlled, Pilot Study in Gastro-Oesophageal Reflux Disease," *Aliment Pharmacology and Therapies* 39, no. 6 (March 2014): 595–602, doi:10.1111/apt.12640; D. A. Leiman et al., "Alginate Therapy Is Effective Treatment for Gastroesophageal Reflux Disease Symptoms: A Systematic Review and Meta-Analysis," *Diseases of the Esophagus* 30, no. 2 (February 1, 2017): 1–8, doi:10.1111/dote.12535; and Edoardo Savarino et al., "Alginate Controls Heartburn in Patients with Erosive and Nonerosive Reflux Disease," *World Journal of Gastroenterology* 18, no. 32 (August 28, 2012): 4371–78, doi:10.3748/wjg.v18.i32.4371.
58. T. Satapathy et al., "Evaluation of Anti-GERD Activity of Gastro Retentive Drug Delivery System of Itopride Hydrochloride," *Artificial Cells, Blood Substitutes, and Immobilization Biotechnology* 38, no. 4 (August 2010): 200–7, doi:10.3109/10731191003776751.
59. Qiang-Song Wang et al., "Protective Effects of Alginate-Chitosan Microspheres Loaded with Alkaloids from *Coptis chinensis* Franch. and *Evodia rutaecarpa* (Juss.) Benth. (Zuojin Pill) against Ethanol-Induced Acute Gastric Mucosal Injury in Rats," *Drug Design, Development and Therapy* 9 (November 19, 2015): 6151–65, doi:10.2147/DDDT.S96056.

60. Yunes Panahi et al., "Efficacy and Safety of Aloe Vera Syrup for the Treatment of Gastroesophageal Reflux Disease: A Pilot Randomized Positive-Controlled Trial," *Journal of Traditional Chinese Medicine* 35, no. 6 (December 2015): 632–36.
61. Dino Vaira, "How to Proceed in *Helicobacter pylori*-Positive Chronic Gastritis Refractory to First- and Second-Line Eradication Therapy," *Digestive Diseases* 25, no. 3 (2007): 203–5, doi:10.1159/0000103885.
62. L. Zhang et al., "Relation between *Helicobacter pylori* and Pathogenesis of Chronic Atrophic Gastritis and the Research of Its Prevention and Treatment," *Zhongguo Zhong Xi Yi Jie He Za Zhi* (*Chinese Journal of Integrated Traditional and Western Medicine*) 12, no. 9 (September 1992): 521–23, 515–16; George Stamatis et al., "*In Vitro* Anti-*Helicobacter pylori* Activity of Greek Herbal Medicines," *Journal of Ethnopharmacology* 88, no. 2–3 (October 2003): 175–79, doi:10.1016/S0378-8741(03)00217-4; Gail B. Mahady et al., "*In Vitro* Susceptibility of *Helicobacter pylori* to Isoquinoline Alkaloids from *Sanguinaria canadensis* and *Hydrastis canadensis*," *Phytotherapy Research* 17, no. 3 (March 2003): 217–21, doi:10.1002/ptr.1108; and Mahady et al., "*In Vitro* Susceptibility of *Helicobacter pylori* to Botanical Extracts Used Traditionally for the Treatment of Gastrointestinal Disorders."
63. Harathi Yandrapu et al., "A Distinct Salivary Secretory Response Mediated by the Esophago-Salivary Reflex in Patients with Barrett's Esophagus: Its Potential Pathogenetic Implications," *Advances in Medical Sciences* 59, no. 2 (September 2014): 281–87, doi:10.1016/j.advms.2014.08.005.
64. Yan Li et al., "Chemoprotective Effects of *Curcuma aromatica* on Esophageal Carcinogenesis," *Annals of Surgical Oncology* 16, no. 2 (February 2009): 515–23, doi:10.1245/s10434-008-0228-0.
65. Pratibha Singh et al., "Ameliorative Effects of *Panax quinquefolium* on Experimentally Induced Reflux Oesophagitis in Rats," *Indian Journal of Medical Research* 135, no. 3: 407–13.
66. S. H. Sicherer, "Food Allergy," *Mount Sinai Journal of Medicine* 78, no. 5 (September/October 2011): 683–96, doi:10.1002/msj.20292.
67. Ping Bo et al., "Clinical Observations on 46 Cases of Globus Hystericus Treated with Modified *Banxia Houpu* Decoction," *Journal of Traditional Chinese Medicine* 30, no. 2 (June 2010): 103–7.
68. Koh Iwasaki et al., "The Effects of the Traditional Chinese Medicine, 'Banxia Houpo Tang (Hange-Koboku To)' on the Swallowing Reflex in Parkinson's Disease," *Phytomedicine* 7, no. 4 (July 2000): 259–63, doi:10.1016/S0944-7113(0)80042-2.
69. Xue-Gong Feng et al., "Clinical Study on Tongyan Spray for Post-Stroke Dysphagia Patients: A Randomized Controlled Trial," *Chinese Journal of Integrative Medicine* 18, no. 5 (May 2012): 345–49, doi:10.1007/s11655-012-1140-9.
70. Kara M. de Felice et al., "Crohn's Disease of the Esophagus: Clinical Features and Treatment Outcomes in the Biologic Era," *Inflammatory Bowel Diseases* 21, no. 9 (September 2015): 2106–13, doi:10.1097/MIB.0000000000000469.
71. Uday Bandyaopadhyay et al, "Clinical Studies on the Effect of Neem (*Azadirachta indica*) Bark Extract on Gastric Secretion and Gastroduodenal Ulcer," *Life Sciences* 75, no. 24 (October 29, 2004): 2867–78, doi:10.1016/j.lfs.2004.04.050.
72. Mohammad A. Alzohairy, "Therapeutics Role of *Azadirachta indica* (Neem) and Their Active Constituents in Diseases Prevention and Treatment," *Evidence-Based Complementary and Alternative Medicine* 2016 (2016): 7382506, doi:10.1155/2016/7382506.
73. Hyun Lim and Hyun Pyo Kim, "Inhibition of Mammalian Collagenase, Matrix Metalloproteinase-1, by Naturally-Occurring Flavonoids," *Planta Medica* 73, no. 12 (October 2007): 1267–74, doi:10.1055/s-2007-990220.
74. Zhihong Yu and Ping Su, "Effect of β-Aescin Extract from Chinese Buckeye Seed on Chronic Venous Insufficiency," *Pharmazie* 68, no. 6 (June 2013): 428–30, doi:10.1691/ph.2013.2886.
75. J. R. Morling et al., "Rutosides for Prevention of Post-Thrombotic Syndrome," *Cochrane Database of Systematic Review* 4 (April 30, 2013): CD005626, doi:10.1002/14651858.CD005626.pub2.
76. Max H. Pittler et al., "Horse Chestnut Seed Extract for Chronic Venous Insufficiency," *Cochrane Database of Systematic Review* 11 (November 14, 2012): CD003230, doi:10.1002/14651858.CD003230.pub4.
77. Tinja Lääveri et al., "Traveler's Diarrhea, the Most Common Health Problem of Travelers," *Duodecim* 126, no. 4 (2010): 403-10.
78. Charles Darkoh et al., "Bile Acids Improve the Antimicrobial Effect of Rifaximin," *Antimicrobial Agents and Chemotherapy* 54, no. 9 (September 2010): 3618–24, doi:10.1128/AAC.00161-10.
79. Nathan S. Teague et al., "Enteric Pathogen Sampling of Tourist Restaurants in Bangkok, Thailand," *Journal of Travel Medicine* 17, no. 2(March/April 2010): 118–23, doi:10.1111/j.1708-8305.2009.00388.x.
80. Pilar Zamarrón Fuertes et al., "Clinical and Epidemiological Characteristics of Imported Infectious Diseases in Spanish Travelers," *Journal of Travel Medicine* 17, no. 5 (September/October 2010): 303–9, doi:10.1111/j.1708-8305.2010.00433.x.
81. Lei Chen et al., "Bioactivity-Guided Fractionation of an Antidiarrheal Chinese Herb *Rhodiola kirilowii* (Regel) Maxim Reveals (-)-Epicatechin-3-Gallate and (-)-Epigallocatechin-3-Gallate as Inhibitors of Cystic Fibrosis Transmembrane Conductance Regulator," *PLoS ONE* 10, no. 3 (March 6, 2015): e0119122, doi:10.1371/journal.pone.0119122.
82. Yaofang Zhang et al., "Identification of Resveratrol Oligomers as Inhibitors of Cystic Fibrosis Transmembrane Conductance Regulator by High-Throughput Screening of Natural Products from Chinese Medicinal Plants," *PLoS ONE* 9, no. 4 (April 8, 2014): e94302, doi:10.1371/journal.pone.0094302.
83. Nadja Apelt et al., "The Prevalence of Norovirus in Returning International Travelers with Diarrhea," *BMC Infectious Diseases* 10 (May 25, 2010): 131, doi:10.1186/1471-2334-10-131.

84. Marina Soković et al., "Antibacterial Effects of the Essential Oils of Commonly Consumed Medicinal Herbs Using an *In Vitro* Model," *Molecules* 15, no. 11 (October 27, 2010): 7532–46, doi:10.3390/molecules15117532.
85. Dušan Fabian et al., "Essential Oils—Their Antimicrobial Activity against *Escherichia coli* and Effect on Intestinal Cell Viability," *Toxicology in Vitro* 20, no. 8 (December 2006): 1435–45, doi:10.1016/j.tiv.2006.06.012; and Lioa Fang et al., "Experimental Study on the Antibacterial Effect of Origanum Volatile Oil on Dysentery Bacilli *In Vivo* and *In Vitro*," *Journal of Huazhong University Science and Technologogy [Medical Sciences]* 24, no. 4 (August 2004): 400–3.
86. Yong Zhou et al., "A Novel Compound from Celery Seed with a Bactericidal Effect against *Helicobacter pylori*," *Journal of Pharmacy and Pharmacology* 61, no. 8 (August 2009): 1067–77, doi:10.1211/jpp/61.08.0011.
87. Changzhong Wang, "Advances in Study on Traditional Chinese Medicine against Biofilms," *Zhongguo Zhong Yao Za Zhi (Chinese Journal of Chinese Materia Medica)* 35, no. 4 (February 2010): 521–24, doi:10.4268/cjcmm20100426.
88. Julian Andrew Guttman et al., "Gap Junction Hemichannels Contribute to the Generation of Diarrhoea during Infectious Enteric Disease," *Gut* 59, no. 2 (February 2010): 218–26, doi:10.1136/gut.2008.170464.
89. Jean Robert Rapin and Nicolas Wiernsperger, "Possible Links between Intestinal Permeability and Food Processing: A Potential Therapeutic Niche for Glutamine," *Clinics (Sao Paulo)* 65, no. 6 (June 2010): 635–43, doi:10.1590/S1807-59322010000600012.
90. Michael Maes and Jean-Claude Leunis, "Normalization of Leaky Gut in Chronic Fatigue Syndrome (CFS) Is Accompanied by a Clinical Improvement: Effects of Age, Duration of Illness and the Translocation of LPS from Gram-Negative Bacteria," *Neuroendocrinology Letters* 29, no. 6 (December 2008): 902–10.
91. Igor Sukhotnik et al., "Oral Glutamine Prevents Gut Mucosal Injury and Improves Mucosal Recovery Following Lipopolysaccharide Endotoxemia in a Rat," *Journal of Surgical Research* 143, no. 2 (December 2007): 379–84, doi:10.1016/j.jss.2007.02.002.
92. USDA.gov Dr Duke's Phytochemical and Ethnobotanical Databases, https://phytochem.nal.usda.gov/phytochem/chemicals/show/9214?et=
93. A. Malamud and K. T. Wilson, "Treatment of Gastrointestinal Infections," *Current Opinion in Gastroenterology* 16, no. 1 (January 2000): 51–55.
94. Gizachew Yismaw et al., "The Invitro Assessment of Antibacterial Effect of Papaya Seed Extract against Bacterial Pathogens Isolated from Urine, Wound and Stool," *Ethiopian Medical Journal* 46, no. 1 (January 2008): 71–77.
95. M. X. Chang et al., "Bacteriostasis of Rhizoma Coptidis Combined with Trimethoprim (TMPO)," *Zhongguo Zhong Yao Za Zhi (Chinese Journal of Chinese Materia Medica)* 19, no. 12 (December 1994): 748–49, 764.
96. Yihui Yang et al., "Metabolites of Protoberberine Alkaloids in Human Urine Following Oral Administration of Coptidis Rhizoma Decoction," *Planta Medica* 76, no. 16 (November 2010): 1859–63, doi:10.1055/s-0030-1250053.
97. Youvraj R. Sohni et al., "The Antiamoebic Effect of a Crude Drug Formulation of Herbal Extracts against *Entamoeba histolytica* In Vitro and In Vivo," *Journal of Ethnopharmacology* 45, no. 1 (January 1995): 43–52, doi:10.1016/0378-8741(94)01194-5.
98. Jennifer K. Spinler et al., "Probiotics as Adjunctive Therapy for Preventing *Clostridium difficile* Infection—What Are We Waiting For?," *Anaerobe* 41 (October 2016): 51–57, doi:10.1016/j.anaerobe.2016.05.007; and Nynke Postma et al., "The Challenge of *Clostridium difficile* Infection: Overview of Clinical Manifestations, Diagnostic Tools and Therapeutic Options," *International Journal of Antimicrobial Agents* 46, Suppl 1 (December 2015): S47–50, doi:10.1016/j.ijantimicag.2015.11.001.
99. Marie Céline Zanella Terrier et al., "Recurrent *Clostridium difficile* Infections: The Importance of the Intestinal Microbiota," *World Journal of Gastroenterology* 20, no. 23 (June 21, 2014): 7416–23, doi:10.3748/wig.v20.i23.7416.
100. Pieter Van den Abbeele et al., "Prebiotics, Faecal Transplants and Microbial Network Units to Stimulate Biodiversity of the Human Gut Microbiome," *Microbial Biotechnology* 6, no. 4 (July 2013): 335–40, doi:10.1111/1751-7915.12049.
101. Melinda A. Engevik et al., "Human *Clostridium difficile* Infection: Altered Mucus Production and Composition," *American Journal of Physiology-Gastrointestinal and Liver Physiology* 308, no. 6 (March 15, 2015): G510–24, doi:10.1152/ajpgi.00091.2014.
102. Bohyun Yun et al., "Inhibitory Effect of Epigallocatechin Gallate on the Virulence of *Clostridium difficile* PCR Ribotype 027," *Journal of Food Science* 80, no. 12 (December 2015): M2925–31, doi:10.1111/1750-3841.13145.
103. Yuan-Pin Hung et al., "Doxycycline and Tigecycline: Two Friendly Drugs with a Low Association with *Clostridium difficile* Infection," *Antibiotics* 4, no. 2 (June 2015): 216–29, doi: 10.3390/antibiotics4020216; and Nicholas S. Britt et al., "Tigecycline for the Treatment of Severe and Severe Complicated *Clostridium difficile* Infection," *Infectious Diseases and Therapy* 3, no. 2 (December 2014): 321–31, doi:10.1007/s40121-014-0050-x.
104. Fatih Mehmet Birdane et al., "Beneficial Effects of *Foeniculum vulgare* on Ethanol-Induced Acute Gastric Mucosal Injury in Rats," *World Journal of Gastroenterology* 13, no. 4 (January 28, 2007): 607–11; and Ibrahim A. Al Mofleh et al., "Aqueous Suspension of Anise '*Pimpinella anisum*' Protects Rats against Chemically Induced Gastric Ulcers," *World Journal of Gastroenterology* 13, no. 7 (February 21, 2007): 1112–18.
105. Ajay Goel et al., "Curcumin as 'Curecumin': From Kitchen to Clinic," *Biochemical Pharmacology* 75, no. 4, February 15, 2008: 787–809, doi:10.1016/j.bcp.2007.08.016.

106. C. L. Cheng et al., "The Healing Effects of *Centella* Extract and Asiaticoside on Acetic Acid Induced Gastric Ulcers in Rats," *Life Sciences* 74, no. 18 (March 19, 2004): 2237–49.
107. M. Dorababu, "Effect of Aqueous Extract of Neem (*Azadirachta indica*) Leaves on Offensive and Diffensive Gastric Mucosal Factors in Rats," *Indian Journal of Physiology and Pharmacology* 50, no. 3 (July–September 2006): 241–49; and G. P. Garg, "The Gastric Antiulcer Effects of the Leaves of the Neem Tree," *Planta Medica* 59, no. 3 (June 1993): 215–17, doi:10.1055/s-2006-959654.
108. Ekrem Kaya et al., "The Effect of L-Glutamine on Mucosal Healing in Experimental Colitis Is Superior to Short-Chain Fatty Acids," *Turkish Journal of Gastroenterology* 18, no. 2 (June 2007): 89–94.
109. Li Zhang et al., "Inhibition of Urease by Bismuth(III): Implications for the Mechanism of Action of Bismuth Drugs," *BioMetals* 19, no. 5 (October 2006): 503–11, doi:10.1007/s10534-005-5449-0.
110. Sung Eun Kim et al., "Second-Line Bismuth-Containing Quadruple Therapy for *Helicobacter pylori* Eradication and Impact of Diabetes," *World Journal of Gastroenterology* 23, no. 6 (February 14, 2017): 1059–66, doi:10.3748/wjg.v23.i6.1059.
111. Dong-Chan Kim et al., "Isoliquiritigenin Selectively Inhibits H(2) Histamine Receptor Signaling," *Molecular Pharmacology* 70, no. 2 (August 2006): 493–500.
112. Jan van Marle et al., "Deglycyrrhizinised Liquorice (DGL) and the Renewal of Rat Stomach Epithelium," *European Journal of Pharmacology* 72, no. 2–3 (June 19, 1981): 219–25.
113. Donghui Cao et al., "The Protective Effects of 18β-Glycyrrhetinic Acid on *Helicobacter pylori*-Infected Gastric Mucosa in Mongolian Gerbils," *BioMed Research International* 2016 (2016): 4943793, doi:10.115/2016/4943793; and Mannanthendil Kumaran Asha et al., "In Vitro Anti-*Helicobacter pylori* Activity of a Flavonoid Rich Extract of *Glycyrrhiza glabra* and Its Probable Mechanisms of Action," *Journal of Ethnopharmacology* 145, no. 2 (January 30, 2013): 581–86, doi:10.1016/j.jep.2012.11.033.
114. L. Langmead et al., "Antioxidant Effects of Herbal Therapies Used by Patients with Inflammatory Bowel Disease: An In Vitro Study," *Aliment Pharmacology and Therapeutics* 16, no. 2 (February 2002): 197–205.
115. Sei-Jung Lee and Kye-Taek Lim, "Inhibitory Effect of Phytoglycoprotein on Tumor Necrosis Factor-Alpha and Interleukin-6 at Initiation Stage of Colon Cancer in 1,2-Dimethylhydrazine-Treated ICR Mice," *Toxicology and Applied Pharmacology* 225, no. 2, (December 1, 2007): 198–205, doi:10.1016/j.taap.2007.07.010.
116. Sei-Jung Lee and Kye-Taek Lim, "Chemopreventive Effect of Plant Originated Glycoprotein on Colitis-Mediated Colorectal Cancer in A/J Mice," *Journal of Biomedical Science* 15, no. 1 (January 2008): 111–32, doi:10.1007/s11373-007-9169-9.
117. Young Choong Kim et al., "Sesquiterpenes from *Ulmus davidiana* var. *japonica* with the Inhibitory Effects on Lipopolysaccharide-Induced Nitric Oxide Production," *Fitoterapia* 78, no. 3 (April 2007): 196–99, doi:10.1016/j.fitote.2006.11.013.
118. Peng Wang et al., "Paeoniflorin Ameliorates Acute Necrotizing Pancreatitis and Pancreatitis-Induced Acute Renal Injury," *Molecular Medicine Reports* 14, no. 2 (August 2016): 1123–31, doi:10.3892/mmr.2016.5351.
119. Toshiharu Ueki et al., "Prevalence and Clinicopathological Features of Autoimmune Pancreatitis in Japanese Patients with Inflammatory Bowel Disease," *Pancreas* 44, no. 3 (April 2015): 434–40, doi:10.1097/MPA.0000000000000261.
120. Wei Huang et al., "Clinical Study on 100 Cases of Severe Acute Pancreatitis in Aged Patients," *Zhong Xi Yi Jie He Xue Bao (Journal of Chinese Integrative Medicine)* 5, no. 3 (2007): 268–71, doi:10.3736/jcim20070308.
121. Supot Pongprasobchai, "Maldigestion from Pancreatic Exocrine Insufficiency," *Journal of Gastroenterology and Hepatology* 28, Suppl 4 (December 2013): 99–102, doi:10.1111/jgh.12406.
122. Zesi Lin et al., "Anti-fibrotic Effects of Phenolic Compounds on Pancreatic Stellate Cells," *BMC Complementary and Alternative Medicine* 15 (July 30, 2015): 259, doi:10.1186/s12906-015-0789-y.
123. R. K. Jha et al., "Resveratrol Ameliorates the Deleterious Effect of Severe Acute Pancreatitis," *Cell Biochemistry and Biophysics* 62, no. 2 (March 2012): 397–402, doi:10.1007/s12013-011-9313-2.
124. R. Zhang et al., "One Compound of Saponins from *Disocorea zingiberensis* Protected against Experimental Acute Pancreatitis by Preventing Mitochondria-Mediated Necrosis," *Scientific Report* 6 (October 25, 2016): 35965, doi:10.1038/srep35965.
125. Xiao-Wu Xu, "Effect of *Ginkgo biloba* Extract on the Function of Alveolar Polymorphonuclear Neutrophils in Severe Acute Pancreatitis Rats Complicated with Lung Injury," *Zhongguo Zhong Xi Yi Jie He Za Zhi (Chinese Journal of Integrated Traditional and Western Medicine)* 34, no. 4 (April 2014): 460–65.
126. Ravikumar Aruna et al., "Rutin Rich *Emblica officinalis* Geart. Fruit Extract Ameliorates Inflammation in the Pancreas of Rats Subjected to Alcohol and Cerulein Administration," *Journal of Complementary and Integrative Medicine* 11, no. 1 (February 7, 2014): 9–18, doi:10.1515/jcim-2013-0023.
127. L. Zhu et al., "Pharmacological Study on Free Anthraquinones Compounds in Rhubarb in Rats with Experimental Acute Pancreatitis," *Zhongguo Zhong Yao Za Zhi (Chinese Journal of Chinese Materia Medica)* 39, no. 2 (January 2014): 304–8.
128. Cai-Hua Wang et al., "Effect of Emodin on Pancreatic Fibrosis in Rats," *World Journal of Gastroenterology* 13, no. 3 (January 21, 2007): 378–82.
129. W. Y. Yao et al., "Emodin Has a Protective Effect in Cases of Severe Acute Pancreatitis via Inhibition of Nuclear Factor-κB Activation Resulting in Antioxidation," *Molecular Medicine Reports* 11, no. 2 (February 2015): 1416–20, doi:10.3892/mmr.2014.2789.

130. Qingqiang Ni et al., "In Vitro Effects of Emodin on Peritoneal Macrophage Intercellular Adhesion Molecule-3 in a Rat Model of Severe Acute Pancreatitis/Systemic Inflammatory Response Syndrome," *Biomedical Reports* 2, no. 1 (January 2014): 63–68, doi:10.3892/br.2013.178.
131. M. Zhang et al., "Red Peony Root Decoction in Treatment of Severe Acute Pancreatitis: A Randomized Controlled Trial," *Zhong Xi Yi Jie He Xue Bao (Journal of Chinese Integrative Medicine)* 6, no. 6 (June 2008): 569–75, doi:10.3736/jcim20080605.
132. Bing Wan et al., "Efficacy of Rhubarb Combined with Early Enteral Nutrition for the Treatment of Severe Acute Pancreatitis: A Randomized Controlled Trial," *Scandinavian Journal of Gastroenterology* 49, no. 11 (November 2014): 1375–84, doi:10.3109/00365521.2014.958523.
133. K. R. Joo et al., "Effect of Korean Red Ginseng on Superoxide Dismutase Inhibitor–Induced Pancreatitis in Rats: A Histopathologic and Immunohistochemical Study," *Pancreas* 38, no. 6 (August 2009): 661–66, doi:10.1097/MPA.0b013e3181a9eb85.
134. Xi-Ping Zhang et al., "Progress in Research into the Mechanism of *Radix salviae* Miltiorrhizae in Treatment of Acute Pancreatitis," *Hepatobiliary and Pancreatic Diseases International* 5, no. 4 (November 2006): 501–4.
135. Wen Qin Xiao et al., "Catalpol Ameliorates Sodium Taurocholate-Induced Acute Pancreatitis in Rats via Inhibiting Activation of Nuclear Factor Kappa β," *International Journal of Molecular Sciences* 15, no. 7 (July 2014): 11957–72, doi:10.3390/ijms150711957.
136. Shiow Yunn Sheu et al., "Radix Scrophulariae Extracts (Harpagoside) Suppresses Hypoxia-Induced Microglial Activation and Neurotoxicity," *BMC Complementary and Alternative Medicine* 15 (September 14, 2015: 324, doi:10.1186/s12906-015-0842-x.
137. Shi-Feng Zhu et al., "Pharmacokinetic and Pharmacodynamic Comparison of Chinese Herbal Ointment Liu-He-Dan and Micron Liu-He-Dan Ointment in Rats with Acute Pancreatitis," *Evidence-Based Complementary and Alternative Medicine* 2014 (2014): 389576, doi:10.1155/2014/389576; and Yi-Ling Liu et al., "Effect of Acute Pancreatitis on the Pharmacokinetics of Chinese Herbal Micron Liuhe Pill Ointment in Rats," *Chinese Journal of Integrative Medicine* 21, no. 12 (December 2015): 922–27, doi:10.1007/s11655-015-2080-y.
138. Peter R. Holt, "Intestinal Malabsorption in the Elderly," *Digestive Diseases* 25, no. 2 (2007): 144–50, doi:10.1159/000099479.
139. Erika Jensen-Jarolim, "Hot Spices Influence Permeability of Human Intestinal Epithelial Monolayers," *Journal of Nutrition* 128, no. 3 (March 1998): 577–81.
140. Usha N. S. Prakash and Krishnapura Srinivasan, "Beneficial Influence of Dietary Spices on the Ultrastructure and Fluidity of the Intestinal Brush Border in Rats," *British Journal of Nutrition* 104, no. 1 (July 2010): 31–39, doi:10.1016/j.advms.2014.09.001.
141. Angelika Miazga et al., "Current Views on the Etiopathogenesis, Clinical Manifestation, Diagnostics, Treatment and Correlation with Other Nosological Entities of SIBO," *Advances in Medical Sciences* 60, no. 1 (March 2015): 118–24.
142. Ali Rezaie et al., "How to Test and Treat Small Intestinal Bacterial Overgrowth: An Evidence-Based Approach," *Current Gastroenterology Reports* 18, no. 2 (February 2016): 8, doi:10.1007/s11894-015-0482-9.
143. Jordi Serra Pueyo, "Update on Gastroesophageal Reflux Disease," *Gastroenterology and Hepatology* 37, no. 2 (February 2014): 73–82, doi:10.1016/j.gastrohep.2013.11.001.
144. S. Fujimori, "What Are the Effects of Proton Pump Inhibitors on the Small Intestine?" *World Journal of Gastroenterology* 21, no. 22 (June 14, 2015): 6817–19, doi:10.3748/wjg.v21.i22.6817.
145. Anant D. Patil, "Link between Hypothyroidism and Small Intestinal Bacterial Overgrowth," *Indian Journal of Endocrinology and Metabolism* 18, no. 3 (May/June 2014): 307–9, doi:10.4103/2230-8210.131155.
146. Hua Chu et al., "Small Intestinal Bacterial Overgrowth in Patients with Irritable Bowel Syndrome: Clinical Characteristics, Psychological Factors, and Peripheral Cytokines," *Gastroenterology Research and Practice* 2016 (2016): 3230859, doi:10.1155/2016/3230859.
147. Andre Fialho et al., "Association between Small Intestinal Bacterial Overgrowth and Deep Vein Thrombosis," *Gastroenterology Reports* 4, no. 4 (November 2016): 299–303, doi:10.1093/gastro/gow004.
148. Eleni Theocharidou et al., Gastrointestinal Motility Disorders and Their Clinical Implications in Cirrhosis, *Gasteroenterology Research and Practice* 2017 (2017): 8270310, doi:10.1155/2017/8270310.
149. Giuseppina Quartarone, "Role of PHGG as a Dietary Fiber: A Review Article," *Minerva Gastroenterologica e Dietologica* 59, no. 4 (December 20515): 329–40.
150. Manuele Furnari et al., "Clinical Trial: The Combination of Rifaximin with Partially Hydrolysed Guar Gum Is More Effective than Rifaximin Alone in Eradicating Small Intestinal Bacterial Overgrowth," *Alimentary Pharmacology and Therapeutics* 32, no. 8 (October 2010): 1000–6, doi:10.1111/j.1365-2036.2010.04436.x.
151. Victor Chedid et al., "Herbal Therapy Is Equivalent to Rifaximin for the Treatment of Small Intestinal Bacterial Overgrowth," *Global Advances in Health and Medicine* 3, no. 3 (May 2014): 16–24.
152. Adil E. Bharucha, "Epidemiology and Natural History of Gastroparesis," *Gastroenterology Clinics of North America* 44, no. 1 (March 2015): 9–19, doi:10.1016/j.gtc.2014.11.002; and Jessica Chang et al., "Diabetic Gastroparesis-Backwards and Forwards," *Journal of Gastroenterology and Hepatology* 26, Suppl 1 (January 2011): 46–57, doi:10.1111/j.1440-1746.2010.06573.x.

153. William L. Hasler, "Gastroparesis," *Current Opinion in Gastroenterology* 28, no. 6 (November 2012): 621–28.
154. Jing Zhang and J. D. Z. Chen, "Pacing the Gut in Motility Disorders," *Current Treatment Options in Gastroenterology* 9, no. 4 (July 2006): 351–60, doi:10.1007/s11938-006-0017-4.
155. William L. Hasler, "Emerging Drugs for the Treatment of Gastroparesis," *Expert Opinion on Emerging Drugs* 19, no. 2 (June 2014): 261–79, doi:10.1517/14728214.2014.899353; and Jose L. Barboza et al., "The Treatment of Gastroparesis, Constipation and Small Intestinal Bacterial Overgrowth Syndrome in Patients with Parkinson's Disease," *Expert Opinion on Pharmacotherapy* 16, no. 16 (2015): 2449–64, doi:10.1517/14656566.2015.1086747.
156. Fanny Lemarié et al., "Revisiting the Metabolism and Physiological Functions of Caprylic Acid (C8:0) with Special Focus on Ghrelin Octanoylation," *Biochimie* 120 (January 2016): 40–8.
157. Massimo Valoti et al., "Effects of Carnitine and Its Derivatives on Gastric Acid Secretion in Rats," *Pharmacological Research* 34, no. 5–6 (November/December 1996): 219–24, doi:10.1006/phrs.1996.0091.
158. Lixin Wang et al., "Preventive Effect of Rikkunshito on Gastric Motor Function Inhibited by L-Dopa in Rats," *Peptides* 55 (May 2014): 136–44, doi:10.1016/j.peptides.2014.02.011.
159. Teita Asano et al., "Anethole Restores Delayed Gastric Emptying and Impaired Gastric Accommodation in Rodents," *Biochemical and Biophysical Research Communications* 472, no. 1 (March 25, 2016): 125–30, doi:10.1016/j.bbrc.2016.02.078.
160. Ulrike von Arnim et al., "STW 5, a Phytopharmacon for Patients with Functional Dyspepsia: Results of a Multicenter, Placebo-Controlled Double-Blind Study," *American Journal of Gastroenterology* 102, no. 6 (June 2007): 1268–75, doi:10.1111/j.1572-0241.2006.01183.x; J. Melzer et al., "*Iberis amara* L. and Iberogast—Results of a Systematic Review Concerning Functional Dyspepsia," *Journal of Herbal Pharmacotherapy* 4, no. 4 (2004): 51–59; and Sebastian Michael, "Adenosine A2A Receptor Contributes to the Anti-inflammatory Effect of the Fixed Herbal Combination STW 5 (Iberogast®) in Rat Small Intestinal Preparations," *Naunyn-Schmiedeberg's Archive of Pharmacology* 385, no. 4 (April 2012): 411–21, doi:10.1007/s00210-011-0714-y.
161. Miwa Nahata et al., "Impaired Ghrelin Signaling Is Associated with Gastrointestinal Dysmotility in Rats with Gastroesophageal Reflux Disease," *American Journal of Physiology—Gastrointestinal and Liver Physiology* 303, no. 1 (July 2012): G42–53, doi:10.1152/ajpgi.00462.2011.
162. Hiroyuki Kitagawa et al., "Pharmacokinetic Profiles of Active Ingredients and Its Metabolites Derived from Rikkunshito, a Ghrelin Enhancer, in Healthy Japanese Volunteers: A Cross-Over, Randomized Study," *PLoS ONE* 10, no. 7 (July 17, 2015): e0133159, doi:10.1371/journal.pone.0133159.
163. Y. Harada et al., "Ghrelin Enhancer, Rikkunshito, Improves Postprandial Gastric Motor Dysfunction in an Experimental Stress Model," *Neurogastroenterology and Motility* 27, no. 8 (August 2015): 1089–97, doi:10.1111/nmo.12588; Hidekazu Suzuki et al., "Randomized Clinical Trial: Rikkunshito in the Treatment of Functional Dyspepsia—A Multicenter, Double-Blind, Randomized, Placebo-Controlled Study," *Neurogastroenterology and Motility* 26, no. 7 (July 2014): 950–61, doi:10.1111/nmo.12348; and Miwa Nahata et al., "Administration of Exogenous Acylated Ghrelin or Rikkunshito, an Endogenous Ghrelin Enhancer, Improves the Decrease in Postprandial Gastric Motility in an Acute Restraint Stress Mouse Model," *Neurogastroenterology and Motility* 26, no. 6 (June 2014): 821–31, doi:10.1111/nmo.12336.
164. Jing Liu et al., "XiangshaLiujunzi Decoction Alleviates the Symptoms of Functional Dyspepsia by Regulating Brain-Gut Axis and Production of Neuropeptides," *BMC Complementary and Alternative Medicine* 15 (October 27, 2015): 387, doi:10.1186/s12906-015-0913-z.
165. Yuan-Hao Lo et al., "Teaghrelins, Unique Acylated Flavonoid Tetraglycosides in Chin-Shin Oolong Tea, Are Putative Oral Agonists of the Ghrelin Receptor," *Journal of Agricultural and Food Chemistry* 62, no. 22 (June 4, 2014): 5085–91, doi:10.1021/jf501425m.
166. Michael Maws, "Inflammatory and Oxidative and Nitrosative Stress Pathways Underpinning Chronic Fatigue, Somatization and Psychosomatic Symptoms," *Current Opinions in Psychiatry* 22, no. 1 (January 2009): 75–83.
167. Hiroaki Kiyohara et al., "Polysaccharide-Containing Macromolecules in a Kampo (Traditional Japanese Herbal) Medicine, Hochuekkito: Dual Active Ingredients for Modulation of Immune Functions on Intestinal Peyer's Patches and Epithelial Cells," *Evidence-Based Complementary and Alternative Medicine* 2011 (2011): 492691, doi:10.1093/ecam/nep193; and David C. Spray et al., "Proteoglycans and Glycosaminoglycans Induce Gap Junction Synthesis and Function in Primary Liver Cultures," *Journal of Cell Biology* 105, no. 1 (July 1985): 541–51.
168. Hiroaki Kiyohara et al., "Elucidation of Structures and Functions through Peyer's Patches of Responsible Carbohydrate Chains in Intestinal Immune System Modulating Polysaccharides from Japanese Medicinal Herbs," *Yakugaku Zasshi* 128, no. 5 (May 2008): 709–16; Hiroaki Kiyohara et al., "Lignin-Carbohydrate Complexes: Intestinal Immune System Modulating Ingredients in Kampo (Japanese Herbal) Medicine, Juzen-Taiho-To," *Planta Medica* 66, no. 1 (February 2006): 20–4, doi:10.1055/s-2000-11116; and Hiroaki Kiyohara et al., "Intestinal Immune System Modulating Polysaccharides in a Japanese Herbal (Kampo) Medicine," *Phytomedicine* 9, no. 7 (October 2002): 614–24, doi:10.1078/094471102321616427.
169. Cuneyt Temiz, "Effect of Cepea Extract-Heparin and Allantoin Mixture on Epidural Fibrosis in a Rat Hemilaminectomy Model," *Turkish Neurosurgery* 19, no. 4 (October 2009): 387–92.

170. Leandro Silva Costa et al., "Biological Activities of Sulfated Polysaccharides from Tropical Seaweeds," *Biomedicine and Pharmacotherapy* 64, no. 1 (January 2010): 21–28, doi:10.1016/j.biopha.2009.03.005.
171. Valérie Jaulneau et al., "Ulvan, a Sulfated Polysaccharide from Green Algae, Activates Plant Immunity through the Jasmonic Acid Signaling Pathway," *Journal of Biomedicine and Biotechnology* 2010 (2010): 525291, doi:10.1155/2010/525291.
172. Sung Ok Kim and Yung Hyun Choi, "The Ethyl Alcohol Extract of *Hizikia fusiforme* Inhibits Matrix Metalloproteinase Activity and Regulates Tight Junction Related Protein Expression in Hep3B Human Hepatocarcinoma Cells," *Journal of Medicinal Food* 13, no. 1 (February 2010): 31–38.
173. M. Roselli et al., "Effect of Different Plant Extracts and Natural Substances (PENS) against Membrane Damage Induced by Enterotoxigenic *Escherichia coli* K88 in Pig Intestinal Cells," *Toxicology in Vitro* 21, no. 2 (March 2007): 224–29.
174. David R. Hill et al., "Human Milk Hyaluronan Enhances Innate Defense of the Intestinal Epithelium," *Journal of Biological Chemistry* 288, no. 40 (October 4, 2013): 29090–104.
175. Terrence E. Riehl et al., "Hyaluronic Acid Regulates Normal Intestinal and Colonic Growth in Mice," *American Journal of Physiology—Gastrointestinal and Liver Physiology* 303, no. 3 (August 1, 2012): G377–88, doi:10.1152/ajpgio.00034.2012.
176. Xinyu Cao et al., "Resistance of Polysaccharide Coatings to Proteins, Hematopoietic Cells, and Marine Organisms," *Biomacromolecules* 10, no. 4 (April 13, 2009): 907–15, doi:10.1021/bm8014208.
177. Manuele Casale et al., "Hyaluronic Acid: Perspectives in Upper Aero-Digestive Tract. A Systematic Review," *PLoS ONE* 10, no. 6 (June 29, 2015): e0130637, doi:10.1371/journal.pone.0130637.
178. Mei-Fang Hsu and Been-Huang Chiang, "Stimulating Effects of *Bacillus subtilis* Natto-fermented *Radix astragali* on Hyaluronic Acid Production in Human Skin Cells," *Journal of Ethnopharmacology* 125, no. 3 (September 25, 2009): 474–81, doi:10.1016/j.jep.2009.07.011.
179. Kuo-Ching Wen et al., "Comparison of *Puerariae radix* and Its Hydrolysate on Stimulation of Hyaluronic Acid Production in NHEK Cells," *American Journal of Chinese Medicine* 38, no. 1 (2010): 143–55, doi:10.1142/S0192415X10007725.
180. Aline M. B. Pires et al., "Microbial Production of Hyaluronic Acid from Agricultural Resource Derivatives," *Bioresource Technology* 101, no. 16 (August 2010): 6506–9, doi:10.1016/jbiortech.2010.03.074.
181. Man-Hai Liu et al., "Icariin Protects Murine Chondrocytes from Lipopolysaccharide-Induced Inflammatory Responses and Extracellular Matrix Degradation," *Nutrition Research* 30, no. 1 (January 2010): 57–65, doi:10.1016/j.nutres.2009.10.020.
182. Juxian Song et al., "Protective Effect of Bilberry (*Vaccinium myrtillus* L.) Extracts on Cultured Human Corneal Limbal Epithelial Cells (HCLEC)," *Phytotherapy Research* 24, no. 4 (April 2010): 520–24, doi:10.1002/ptr.2974.
183. Delphine Rival et al., "A *Hibiscus abelmoschus* Seed Extract as a Protective Active Ingredient to Favour FGF-2 Activity in Skin," *International Journal of Cosmetic Science* 31, no. 6 (December 2009): 419–26, doi:10.1111/j.1468-2494.2009.00538.x.
184. Jean Robert Apin and Nicolas Wiernsperger, "Possible Links between Intestinal Permeablity and Food Processing: A Potential Therapeutic Niche for Glutamine," *Clinics* 65, no. 6 (June 2010): 635–43, doi:10.1590/S1807-59322010000600012.
185. Zakir Hossain et al., "Docosahexaenoic Acid and Eicosapentaenoic Acid-Enriched Phosphatidylcholine Liposomes Enhance the Permeability, Transportation and Uptake of Phospholipids in Caco-2 Cells," *Molecular and Cellular Biochemistry* 285, no. 1–2 (April 2006): 155–63, doi:10.1007/s11010-005-9074-6.
186. T. Sawai et al., "The Effect of Phospholipids and Fatty Acids on Tight-Junction Permeability and Bacterial Translocation," *Pediatric Surgery International* 17, no. 4 (May 2001): 269–74, doi:10.1007/s003830100592.
187. Elena S. Lysenko et al., "Bacterial Phosphorylcholine Decreases Susceptibility to the Antimicrobial Peptide LL-37/hCAP18 Expressed in the Upper Respiratory Tract," *Infection and Immunity* 68, no. 3 (March 2000): 1664–71, doi:101128/IAI.68.3.1664-1671.2000.

Chapter 3: Creating Herbal Formulas for Liver and Gallbladder Conditions

1. Muniyappan Dhanasekaran et al., "Chemopreventive Potential of Epoxy Clerodane Diterpene from *Tinospora cordifolia* against Diethylnitrosamine-Induced Hepatocellular Carcinoma," *Investigational New Drugs* 27, no. 4 (August 2009): 347–55, doi:10.1007/s10637-008-9181-9; B. Singh et al., "In Vivo Hepatoprotective Activity of Active Fraction from Ethanolic Extract of *Eclipta alba* Leaves," *Indian Journal of Physiology and Pharmacology* 45, no. 4 (October 2001): 435–41; A. K. Saxena et al., "Hepatoprotective Effects of *Eclipta alba* on Subcellular Levels in Rats," *Journal of Ethnopharmacology* 40, no. 3 (December 1993): 155–61, doi:10.1016/0378-8741(93)90063-B; R. Krithika and R. J. Verma, "Mitigation of Carbon Tetrachloride–Induced Damage by *Phyllanthus amarus* in Liver of Mice," *Acta Poloniae Pharmaceutica* 66, no. 4 (July/August 2009): 439–44; and Mani Krishnaveni and Sankaran Mirunalini, "Therapeutic Potential of *Phyllanthus emblica* (Amla): The Ayurvedic Wonder," *Journal of Basic Clinical Physiology and Pharmacology* 21, no. 1 (2010): 93–105.
2. Devjani Charkraborty and Remtej Verma, "Ameliorative Effect of *Emblica officinalis* Aqueous Extract on Ochratoxin-Induced Lipid Peroxidation in the Kidney

and Liver of Mice," *International Journal of Occupational Medicine and Environmental Health* 23, no. 1 (2010): 63–73, doi:10.2478/v10001-010-0009-4; and Vaddi Damodata Reddy et al., "*Emblica officinalis* Protects against Alcohol-Induced Liver Mitochondrial Dysfunction in Rats," *Journal of Medicinal Food* 12, no. 2 (April 2009): 327–33, doi:10.1089/jmf.2007.0694.

3. Mohamed Mohamed Soliman et al., "Immunohistochemical and Molecular Study on the Protective Effect of Curcumin against Hepatic Toxicity Induced by Paracetamol in Wistar Rats," *BMC Complementary and Alternative Medicine* 14 (November 29, 2014): 457, doi:10.1186/1472-6882-14-457.
4. Fang Liu et al., "UPLC/Q-TOFMS-Based Metabolomics Studies on the Protective Effect of *Panax notoginseng* Saponins on Alcoholic Liver Injury," *American Journal of Chinese Medicine* 43, no. 4 (2015): 695–714, doi:10.1142/S0192415X15500433.
5. Muhammad Jahangir Hossen et al., "*Codonopsis lanceolata*: A Review of Its Therapeutic Potentials," *Phytotherapy Research* 30, no. 3 (March 2016): 347–56, doi:10.1002/ptr.5553.
6. Ya. V. Gorina et al., "Evaluation of Hepatoprotective Activity of Water-Soluble Polysaccharide Fraction of *Stellaria media* L.," *Bulletin of Experimental Biology and Medicine* 154, no. 5 (March 2013): 645–48.
7. Hong-Bo Xu et al., "Bioactivity-Guided Isolation of Anti-hepatitis B Virus Active Sesquiterpenoids from the Traditional Chinese Medicine: Rhizomes of *Cyperus rotundus*," *Journal of Ethnopharmacology* 171 (August 2, 2015): 131–40, doi:10.1016/j.jep.2015.05.040.
8. Yi Liu et al., "Baicalein Reduces the Occurrence of Cirrhotic Endotoxemia by Reducing Intestinal Mucosal Apoptosis," *BMC Complementary and Alternative Medicine* 15 (May 29, 2015): 161, doi:10.1186/s12906-015-0682-8.
9. Sheng Liu et al., "In Vitro and in Vivo Anti-hepatitis B Virus Activities of the Lignan Nirtetralin B Isolated from *Phyllanthus niruri* L.," *Journal of Ethnopharmacology* 157 (November 18, 2014): 62–68, doi:10.1016/j.jep.2014.09.019.
10. Yong Zhao et al., "UFLC/MS-IT-TOF Guided Isolation of Anti-HBV Active Chlorogenic Acid Analogues from *Artemisia capillaris* as a Traditional Chinese Herb for the Treatment of Hepatitis," *Journal of Ethnopharmacology* 156 (October 28, 2014): 147–54, doi:10.1016/j.jep.2014.08.043.
11. Yuan-Feng Zou et al., "Complement Activity of Polysaccharides from Three Different Plant Parts of *Terminalia macroptera* Extracted as Healers Do," *Journal of Ethnopharmacology* 155, no. 1 (August 8, 2014): 672–78, doi:10.1016/j.jep.2014.06.017.
12. Wanwisa Waiyaput et al., "Inhibitory Effects of Crude Extracts from Some Edible Thai Plants against Replication of Hepatitis B Virus and Human Liver Cancer Cells," *BMC Complementary and Alternative Medicine* 12 (December 6, 2012): 246, doi:10.1186/1472-6882-12-246; Hye Jin Kim et al., "Antiviral Effect of *Curcuma longa* Linn Extract against Hepatitis B Virus Replication," *Journal of Ethnopharmacology* 124, no. 2 (July 15, 2009): 189–96, doi:10.1016/j.jep.2009.04.046; Maya Mouler Rechtman et al., "Curcumin Inhibits Hepatitis B Virus via Down-regulation of the Metabolic Coactivator PGC-1alpha," *FEBS Letters* 584, no. 11 (June 3, 2010): 2485–90, doi: 10.1016/j.febslet.2010.04.067; and Kyeong Jin Kim et al., "Curcumin Inhibits Hepatitis C Virus Replication via Suppressing the Akt-SREBP-1 Pathway," *FEBS Letters* 584, no. 4 (February 2010): 707–12, doi:10.1016/j.febslet.2009.12.019.
13. Sidra Rehman et al., "Therapeutic Potential of *Taraxacum officinale* against HCV NS5B Polymerase: *In-vitro* and *In silico* Study," *Biomedicine and Pharmacotherapy* 83 (October 2016): 881–91, doi:10.1016/j.biopha.2016.08.002.
14. Yanling Zhao et al., "Large Dose Means Significant Effect—Dose and Effect Relationship of Chi-Dan-Tui-Huang Decoction on Alpha-Naphthylisothiocyanate-Induced Cholestatic Hepatitis in Rats," *BMC Complementary and Alternative Medicine* 15 (April 2, 2015): 104, doi:10.1186/s12906-015-0637-0.
15. Chang-An Geng et al., "Three New Anti-HBV Active Constituents from the Traditional Chinese Herb of Yin-Chen (*Artemisia scoparia*)," *Journal of Ethnopharmacology* 176 (December 24, 2015): 109–17, doi:10.1016/j.jep.2015.10.032.
16. X. Ma et al., "*Paeonia lactiflora* Pall. Protects against ANIT-Induced Cholestasis by Activating Nrf2 via PI3K/Akt Signaling Pathway," *Drug Design, Development and Therapy* 9 (September 2, 2015): 5061–74, doi:10.2147/DDDT.S90030.
17. Maryem Ben Salem et al., "Pharmacological Studies of Artichoke Leaf Extract and Their Health Benefits," *Plant Foods for Human Nutrition* 70, no. 4 (December 2015): 441–53, doi:0.1007/s11130-015-0503-8; and M. Rondanelli, "Health-Promoting Properties of Artichoke in Preventing Cardiovascular Disease by its Lipidic and Glycemic-Reducing Action," *Monaldi Archives for Chest Disease* 80, no. 1 (March 2013): 17–26, doi:4081/mondali.2013.87.
18. V. G. Radchenko et al., "Nonalcoholic Steatohepatitis and Biliary Sludge in People with Metabolic Syndrome," *Ter Arkh* 88, no. 9 (2016): 78–83, doi:10.17116/terarkh201688978-83.
19. Yiqun Liu et al., "Betaine-Enriched Beet Suppresses Hyperhomocysteinemia Induced by Choline Deficiency in Rats," *Wei Sheng Yan Jiu [Journal of Hygiene Research]* 44, no. 2 (March 2015): 279–83.
20. Sakura Kawakami, "Effects of Dietary Supplementation with Betaine on a Nonalcoholic Steatohepatitis (NASH) Mouse Model," *Journal of Nutritional Science and Vitaminology* 58, no. 5 (2012): 371–75.
21. Nilesh K. Jain and Abhay K. Singhai, "Protective Role of *Beta vulgaris* L. Leaves Extract and Fractions on Ethanol-Mediated Hepatic Toxicity," *Acta Poloniae Pharmaceutica* 69, no. 5 (September/October 2012): 945–50.
22. Violetta Krijka-Kuźniak et al., "Beetroot Juice Protects against N-nitrosodiethylamine-Induced Liver Injury in Rats," *Food and Chemical Toxicology* 50, no. 6 (June 2012): 2027–33, doi:10.1016/j.fct.2012.03.062.

23. Shelley C. Lu and José M. Mato, "S-Adenosylmethionine in Liver Health, Injury, and Cancer," *Physiological Reviews* 92, no. 4 (October 2012): 1515–1542, doi:10.1152/physrev.00047.2011.
24. Qin Pan et al., "Fuzheng Huayu Recipe Ameliorates Liver Fibrosis by Restoring Balance between Epithelial-to-Mesenchymal Transition and Mesenchymal-to-Epithelial Transition in Hepatic Stellate Cells," *BioMed Research International* 2015 (2015): 935903, doi:10.1155/2015/935903.
25. Sulochana Priya and P. R. Sudhakaran, "Curcumin-Induced Recovery from Hepatic Injury Involves Induction of Apoptosis of Activated Hepatic Stellate Cells," *Indian Journal of Biochemistry and Biophysics* 45, no. 5 (October 2008): 317–25; and Youcai Tang, "Curcumin Targets Multiple Pathways to Halt Hepatic Stellate Cell Activation: Updated Mechanisms In Vitro and In Vivo," *Digestive Diseases and Sciences* 60, no. 6 (June 2015): 1554–64, doi:10.1007/s10620-014-3487-6.
26. Sseung-Hwa Choi, Keum-Jin Yang, and Dong-Soo Lee, "Effects of Complementary Combination Therapy of Korean Red Ginseng and Antiviral Agents in Chronic Hepatitis B," *Journal of Alternative and Complementary Medicine* 22, no. 12 (December 2016): 964–69, doi:10.1089/acm.2015.0206.
27. I-Shu Chen et al., "Hepatoprotection of Silymarin against Thioacetamide-Induced Chronic Liver Fibrosis," *Journal of the Science of Food and Agriculture* 92, no. 7 (May 2012): 1441–47, doi:0.1002/jsfa.4723.
28. Natasa Milosević et al., "Phytotherapy and NAFLD—from Goals and Challenges to Clinical Practice," *Reviews on Recent Clinical Trials* 9, no. 3 (2014): 195–203.
29. Yuan Xie et al., "Reversing Effects of Silybin on TAA-Induced Hepatic CYP3A Dysfunction through PXR Regulation," *Chinese Journal of Natural Medicines* 11, no. 6 (November 2013): 645–52, doi:10.1016/S1875-5364(13)60075-8; Naglaa M. El-Lakkany et al., "Anti-inflammatory/Anti-fibrotic Effects of the Hepatoprotective Silymarin and the Schistosomicide Praziquantel against *Schistosoma mansoni*-Induced Liver Fibrosis," *Parasites and Vectors* 5 (January 11, 2012): 9, doi:10.1186/1756-3305-5-9; and Ali Reza Ghaffari et al., "The Effects of Milk Thistle on Hepatic Fibrosis Due to Methotrexate in Rat," *Hepatitis Monthly* 11, no. 6 (June 2011): 464–68.
30. Mahmood Bahmani et al., "*Silybum marianum*: Beyond Hepatoprotection," *Journal of Evidence-Based Complementary and Alternative Medicine* 20, no. 4 (October 2015): 292–301, doi:10.1177/2156587215571116.
31. R. B. Badisa et al., "Milk Thistle Seed Extract Protects Rat C6 Astroglial Cells from Acute Cocaine Toxicity," *Molecular Medicine Reports* 10, no. 5 (November 2014): 2287–92, doi:10.3892/mmr.2014.2524.
32. Halimeh Amiridumari et al., "Effects of Milk Thistle Seed against Aflatoxin B1 in Broiler Model," *Journal of Research in Medical Sciences* 18, no. 9 (September 2013): 786–90.
33. Ali Mandegary et al., "Hepatoprotective Effect of Silyamarin in Individuals Chronically Exposed to Hydrogen Sulfide; Modulating Influence of TNF-α Cytokine Genetic Polymorphism," *DARU Journal of Pharmaceutical Sciences* 21, no. 1 (2013): 28, doi:10.1186/2008-2231-21-28.
34. Behjat Moayedi et al., "A Randomized Double-Blind, Placebo-Controlled Study of Therapeutic Effects of Silymarin in β-thalassemia Major Patients Receiving Desferrioxamine," *European Journal of Haematology* 90, no. 3 (March 2013): 202–9, doi:10.1111/ejh.12061.
35. N. Skottová et al., "Phenolics-Rich Extracts from *Silybum marianum* and *Prunella vulgaris* Reduce a High-Sucrose Diet Induced Oxidative Stress in Hereditary Hypertriglyceridemic Rats," *Pharmacological Research* 50, no. 2 (August 2004): 123–30, doi:10.1016/j.phrs.2003.12.013; and Fulvio Cacciapuoti et al., "Silymarin in Non-alcoholic Fatty Liver Disease," *World Journal of Hepatology* 5, no. 3 (March 27, 2013): 109–13, doi:10.4254/wjh.v5.i3.109.
36. Nancy Vargas-Mendoza et al., "Hepatoprotective Effect of Silymarin," *World Journal of Hepatology* 6, no. 3 (March 27, 2014): 144–49, doi:10.4254/wjh.v6.i3.144.
37. Vargas-Mendoza, "Hepatoprotective effect of Silymarin."
38. Y. Chtourou et al., "Therapeutic Efficacy of Silymarin from Milk Thistle in Reducing Manganese-Induced Hepatic Damage and Apoptosis in Rats," *Human and Experimental Toxicology* 32, no. 1 (January 2013): 70–81, doi:10.1177/0960327112455674.
39. Ali McBride et al., "*Silybum marianum* (Milk Thistle) in the Management and Prevention of Hepatotoxicity in a Patient Undergoing Reinduction Therapy for Acute Myelogenous Leukemia," *Journal of Oncology Pharmacy Practices* 18, no. 3 (September 2012): 360–65, doi:10.1177/1078155212438252.
40. Durrin Ozlem Dabak and Nevin Kocaman, "Effects of Silymarin on Methotrexate-Induced Nephrotoxicity in Rats," *Renal Failure* 37, no. 4 (May 2015): 734–39, doi:10.3109/0886022X.2015.1012984.
41. Aleksandar Rašković et al., "The Protective Effects of Silymarin against Doxorubicin-Induced Cardiotoxicity and Hepatotoxicity in Rats," *Molecules* 16, no. 10 (October 12, 2011): 8601–13, doi:10.3390/molecules16108601.
42. Elena J. Ladas et al., "A Randomized, Controlled, Double-Blind, Pilot Study of Milk Thistle for the Treatment of Hepatotoxicity in Childhood Acute Lymphoblastic Leukemia (ALL)," *Cancer* 116, no. 2 (January 15, 2010): 506–13, doi:10.1002/cncr.24723.
43. V. Krecman et al., "Silymarin Inhibits the Development of Diet-Induced Hypercholesterolemia in Rats," *Planta Medica* 64, no. 2 (March 1998): 138–42.
44. Abdullah Aslan and Muhammed Ismail Can, "Milk Thistle Impedes the Development of Carbontetrachloride-Induced Liver Damage in Rats through Suppression of bcl-2 and Regulating Caspase Pathway," *Life Sciences* 117, no. 1 (November 4, 2014): 13–18, doi:10.1016/j.lfs.2014.09.005.

45. El-Lakkany, "Anti-inflammatory/Anti-fibrotic Effects of the Hepatoprotective Silymarin and the Schistosomicide Praziquantel" (see n. 35).
46. Halina Kostek et al., "Silibinin and Its Hepatoprotective Action from the Perspective of a Toxicologist," *Przeglad Lekarski* 69, no. 8 (2012): 541–43; and Ulrich Mengs, Ralf-Torsten Pohl, and Todd Mitchell, "Legalon® SIL: The Antidote of Choice in Patients with Acute Hepatotoxicity from Amatoxin Poisoning," *Current Pharmaceutical Biotechnology* 13, no. 10 (August 2012): 1964–70, doi:10.2174/138920112802273353.
47. Meghna R. Adhvaryu, N. Reddy, and M. H. Parabia, "Effects of Four Indian Medicinal Herbs on Isoniazid-, Rifampicin- and Pyrazinamide-Induced Hepatic Injury and Immunosuppression in Guinea Pigs," *World Journal of Gastroenterology* 13, no. 23 (June 21, 2007): 3199–205; Sreenivasa S. Reddy et al., "Beneficiary Effect of *Tinospora cordifolia* against High-Fructose Diet Induced Abnormalities in Carbohydrate and Lipid Metabolism in Wistar Rats," *Hormone and Metabolic Research* 41, no. 10 (October 2009): 741–46; D. S. Nagarkatti et al., "Modulation of Kupffer Cell Activity by *Tinospora cordifolia* in Liver Damage," *Journal of Postgraduate Medicine* 40, no. 2 (April–June 1994): 65–67; Biswadev Bishayi et al., "Hepatoprotective and Immunomodulatory Properties of *Tinospora cordifolia* in CCl4 Intoxicated Mature Albino Rats," *Journal of Toxicological Sciences* 27, no. 3 (August 2002): 139–46; and T. S. Panchabhai et al., "Protective Effect of *Tinospora cordifolia, Phyllanthus emblica* and Their Combination against Antitubercular Drugs Induced Hepatic Damage: An Experimental Study," *Phytotherapy Research* 22, no. 5 (May 2008): 646–50, doi:10.1002/ptr.2356.
48. Yara Haddad et al., "Antioxidant and Hepatoprotective Effects of Silibinin in a Rat Model of Nonalcoholic Steatohepatitis," *Evidence-Based Complementary and Alternative Medicine* 2011 (2011): nep164, doi:10.1093/ecam/nep164.
49. Moayedi, "A Randomized Double-Blind, Placebo-Controlled Study of Therapeutic Effects of Silymarin in β-Thalassemia Major Patients" (see n. 41).
50. Mohd Farazuddin et al., "Chemotherapeutic Potential of Curcumin-Bearing Microcells against Hepatocellular Carcinoma in Model Animals," *International Journal of Nanomedicine* 9 (March 3, 2014): 1139–52, doi:10.2147/IJN .S34668; Dong-Wei Zhang et al., "Curcumin and Diabetes: A Systematic Review," *Evidence-Based Complementary and Alternative Medicine* 2013 (2013): 636053, doi:10.1155/2013/636053; Mehmet H. Ucisik et al., "Characterization of CurcuEmulsomes: Nanoformulation for Enhanced Solubility and Delivery of Curcumin," *Journal of Nanobiotechnology* 11 (December 6, 2013): 37, doi:10.1186.1477-3155-11-37; O. Naksuriya et al., "Curcumin Nanoformulations: A Review of Pharmaceutical Properties and Preclinical Studies and Clinical Data Related to Cancer Treatment," *Biomaterials* 35, no. 10 (March 2014): 3365–83, doi:10.1016/j.biomaterials.2013.12.090; and Pietro Dulbecco and Vincenzo Savarino, "Therapeutic Potential of Curcumin in Digestive Diseases," *World Journal of Gastroenterology* 19, no. 48 (December 28, 2013): 9256–70, doi:10.3748/wjg.v19.i48.9256.
51. Guido Shoba et al., "Influence of Piperine on the Pharmacokinetics of Curcumin in Animals and Human Volunteers," *Planta Medica* 64, no. 4 (May 1998): 353–56, doi:10.1055/s-2006-957450.
52. N. Rege et al., "Immunotherapy with *Tinospora cordifolia*: A New Lead in the Management of Obstructive Jaundice," *Indian Journal of Gastroenterology* 12, no. 1 (January 1993): 5–8.
53. Matteo Micucci et al., "*Curcuma longa* L. as a Therapeutic Agent in Intestinal Motility Disorders. 2: Safety Profile in Mouse," *PLoS ONE* 8, no. 11 (November 18, 2013): e80925, doi:10.1371/journal.pone.0080925.
54. A. Rasyid et al., "Effect of Different Curcumin Dosages on Human Gall Bladder, " *Asia Pacific Journal of Clinical Nutrition* 11, no. 4 (2002): 314–18.
55. W. Lukita-Atmadja et al., "Effect of Curcuminoids as Anti-inflammatory Agents on the Hepatic Microvascular Response to Endotoxin," *Shock* 17, no. 5 (May 2002): 399–403.
56. Wen-Kuan Huang et al., "The Association of Ursodeoxycholic Acid Use with Colorectal Cancer Risk: A Nationwide Cohort Study," *Medicine* 95, no. 11 (March 2016): e2980, doi:10.1097/MD.0000000000002980.
57. Y. P. Zhang et al., "Systematic Review with Meta-Analysis: Coffee Consumption and the Risk of Gallstone Disease," *Alimentary Pharmacology and Therapy* 42, no. 6 (September 2014): 637–48, doi:10.1111/apt.13328.
58. Ying Guo et al., "Dose-Response Effect of Berberine on Bile Acid Profile and Gut Microbiota in Mice," *BMC Complementary and Alternative Medicine* 16, no. 1 (October 18, 2016): 394, doi:10.1186/s12906-016-1367-7.
59. Paul R. Hanlon, David M. Webber, and David M. Barnes, "Aqueous Extract from Spanish Black Radish (*Raphanus sativus* L. Var. *niger*) Induces Detoxification Enzymes in the HepG2 Human Hepatoma Cell Line," *Journal of Agricultural and Food Chemistry* 55, no. 16 (August 8, 2007): 6439–46, doi:10.1021/jf070530f.
60. Chris Scholl et al., "Raphasatin Is a More Potent Inducer of the Detoxification Enzymes than Its Degradation Products," *Journal of Food Science* 76, no. 3 (April 2011): C504–11, doi:10.1111/j.1750-3841.2011.02078.x.
61. Paul R. Hanlon and David M. Barnes, "Phytochemical Composition and Biological Activity of 8 Varieties of Radish (*Raphanus sativus* L.) Sprouts and Mature Taproots," *Journal of Food Science* 76, no. 1 (January/February 2011): C185–92, doi:10.1111/j.1750-3841.2010.01972.x.
62. Syed Sultan Beevi, Lakshmi Narasu Mangamoori, and Bandi Boje Gowda, "Polyphenolics Profile and Antioxidant Properties of *Raphanus sativus* L.," *Natural Product Research* 26, no. 6 (2012): 557–63, doi:10.1080/14786419.2010.521884.

63. Altaf S. Darvesh, Bahrat B. Aggarwal, and Anupam Bishayee, "Curcumin and Liver Cancer: A Review," *Current Pharmaceutical Biotechnology* 13, no. 1 (January 2012): 218–28.
64. Jungsun Kim et al., "Chemopreventive Effect of *Curcuma longa* Linn on Liver Pathology in HBx Transgenic Mice," *Integrative Cancer Therapies* 10, no. 2 (June 2011): 168–77, doi:10.1177/1534735410380613.
65. Kim, "Chemopreventive Effect of *Curcuma longa* Linn on Liver Pathology."
66. Kim, "Chemopreventive Effect of *Curcuma longa* Linn on Liver Pathology."
67. Kim, "Curcumin Inhibits Hepatitis C Virus Replication" (see n. 20).
68. Bharat B. Aggarwal et al., "Curcumin-Free Turmeric Exhibits Anti-inflammatory and Anticancer Activities: Identification of Novel Components of Turmeric," *Molecular Nutrition and Food Research* 57, no. 9 (September 2013): 1529–42, doi:10.1002/mnfr.201200838; and Leelavinothan Pari, Daniel Tewas, and Juergen Eckel, "Role of Curcumin in Health and Disease," *Archives of Physiology and Biochemistry* 114, no. 2 (April 2008): 127–49, doi:10.1080/13813450802033958.
69. Bharat B. Aggarwal, L. Deb, and S. Prasad, "Curcumin Differs from Tetrahydrocurcumin for Molecular Targets, Signaling Pathways and Cellular Responses," *Molecules* 20, no. 1 (December 24, 2014): 185–205, doi:10.3390/molecules20010185.
70. Aggarwal, "Curcumin-Free Turmeric Exhibits Anti-inflammatory and Anticancer Activities."
71. Jin-Jian Lu et al., "Anti-cancer Properties of Terpenoids Isolated from Rhizoma Curcumae—A Review," *Journal of Ethnopharmacology* 143, no. 2 (September 28, 2012): 406–11, doi:10.1016/j.jep.2012.07.009; and Fazhen Bao, Jie Qiu, and Hong Zhang, "Potential Role of β-elemene on Histone H1 in the H22 Ascites Hepatoma Cell Line," *Molecular Medicine Reports* 6, no. 1 (July 2012): 185–90, doi:10.3892/mmr.2012.891.
72. Sutha Devaraj et al., "Investigation of Antioxidant and Hepatoprotective Activity of Standardized *Curcuma xanthorrhiza* Rhizome in Carbon Tetrachloride-Induced Hepatic Damaged Rats," *Scientific World Journal* 2014 (2014): 353128, doi:10.1155/2014/353128; Shu Ju Wu et al., "Curcumin and Saikosaponin A Inhibit Chemical-Induced Liver Inflammation and Fibrosis in Rats," *American Journal of Chinese Medicine* 38, no. 1 (2010): 99–111, doi:10.1142/S0192415X10007695; and Jittima Weerachayaphorn et al., "Protection of Centrilobular Necrosis by *Curcuma comosa* Roxb. in Carbon Tetrachloride-Induced Mice Liver Injury," *Journal of Ethnopharmacology* 129, no. 2 (May 27, 2010): 254–60, doi:10.1016/j.jep.2010.03.026.
73. Shu-Ye Wang, Jin Xue, and Jin Zhou, "Preliminary Effects of Alcohol and *Curcuma longa* upon CYP2E1 and Hematotoxicity in Benzene-Induced Mice," *Zhonghua Yi Xue Za Zhi [Journal of the Chinese Medical Association]* 89, no. 34 (September 15, 2009): 2429–31.
74. Sujatha Nayak and R. B. Sashidhar, "Metabolic Intervention of Aflatoxin B1 Toxicity by Curcumin," *Journal of Ethnopharmacology* 127, no. 3 (February 17, 2010): 641–44, doi:10.1016/j.jep/2009.12.010; and L. P. Yarru et al., "Effects of Turmeric (*Curcuma longa*) on the Expression of Hepatic Genes associated with Biotransformation, Antioxidant, and Immune Systems in Broiler Chicks Fed Aflatoxin," *Poultry Science* 88, no. 12 (December 2009): 2620–27, doi:10.3382/ps.2009-00204.
75. Yan Li et al., "Hepatic Protection and Anticancer Activity of Curcuma: A Potential Chemopreventive Strategy against Hepatocellular Carcinoma," *International Journal of Oncology* 44, no. 2 (February 2014): 505–13, doi:10.3892/ijo.2013.2184; and Chao Wang et al., "Curcumin Inhibits HMGB1 Releasing and Attenuates Concanavalin A-Induced Hepatitis in Mice," *European Journal of Pharmacology* 697, no. 1–3 (December 15, 2012): 152–57, doi:10.1016/j.ejphar.2012.09.050.
76. Yadira Rivera-Espinoza and Pablo Muriel, "Pharmacological Actions of Curcumin in Liver Diseases or Damage," *Liver International* 29, no. 10 (November 2009): 1457–66, doi:10.1111/j.1478-3231.2009.02086.x.
77. Ehsan Kheradpezhouh et al., "Curcumin Protects Rats against Acetaminophen-Induced Hepatorenal Damages and Shows Synergistic Activity with N-acetyl Cysteine," *European Journal of Pharmacology* 628, no. 1–3): 274–81, doi:10.1016/j.ejphar.2009.11.027; and Rivera-Espinoza, "Pharmacological Actions of Curcumin in Liver Diseases or Damage."
78. Ragaa Hosny Mohamad et al., "The Role of *Curcuma longa* against Doxorubicin (Adriamycin)-Induced Toxicity in Rats," *Journal of Medicinal Food* 12, no. 2 (April 2009): 394–402, doi:10.1089/jmf.2007.0715.
79. A. R. Fetoni et al., "*Curcuma longa* (Curcumin) Decreases In Vivo Cisplatin-Induced Ototoxicity through Heme Oxygenase-1 Induction," *Otology and Neurotology* 35, no. 5 (June 2014): e169–77, doi:10.1097/MAO.0000000000000302.
80. Kim, "Chemopreventive Effect of *Curcuma longa* Linn on Liver Pathology" (see n. 86); and Kim, "Antiviral Effect of *Curcuma longa* Linn Extract against Hepatitis B Virus Replication" (see n. 18).
81. Yildiz Öner-İyidoğan et al., "Effect of Curcumin on Hepatic Heme Oxygenase 1 Expression in High Fat Diet Fed Rats: Is There a Triangular Relationship?" *Canadian Journal of Physiology and Pharmacology* 92, no. 10 (October 2014): 805–12, doi:10.1139/cjpp-2014-0174.
82. D. Bandyopadhyay, "Farmer to Pharmacist: Curcumin as an Anti-invasive and Antimetastatic Agent for the Treatment of Cancer," *Frontiers in Chemistry* 2 (December 23, 2014): 113, doi:10.3389/fchem.2014.00113; and Li, "Hepatic Protection and Anticancer Activity of Curcuma" (see n. 102).
83. U. R. Deshpande, L. J. Jospeh, and A. M. Samuel, "Hepatobiliary Clearance of Labelled Mebrofenin in Normal and D-Galactosamine HCl-Induced Hepatitis Rats and the Protective Effect of Turmeric Extract," *Indian Journal of Physiology and Pharmacology* 47, no. 3 (July 2003): 332–36.

84. Sang-Woo Kim et al., "The Effectiveness of Fermented Turmeric Powder in Subjects with Elevated Alanine Transaminase Levels: A Randomised Controlled Study," *BMC Complementary and Alternative Medicine* 13 (March 8, 2013): 58, doi:10.1186/1472-6882-13-58.
85. Panchabhai, "Protective Effect of *Tinospora cordifolia, Phyllanthus emblica* and Their Combination," (see n. 65).

Chapter 4: Creating Herbal Formulas for Renal and Urinary Conditions

1. G. Marelli, E. Papaleo, and A. Ferrari, "Lactobacilli for Prevention of Urogenital Infections: A Review," *European Review for Medical and Pharmacological Sciences* 8, no. 2 (March–April 2004): 87–95.
2. D. Beaux, J. Fleurentin, and F. Mortier, "Effect of Extracts of *Orthosiphon stamineus* Benth, *Hieracium pilosella* L., *Sambucus nigra* L. and *Arctostaphylos uva-ursi* (L.) Spreng. in Rats," *Phytotherapy Research* 13, no. 3 (May 1999): 222–25, doi:10.1002/(SICI)1099-1573(199905)13:2<333::AID-PTR447>3.0.CO;2-P.
3. Prince Singh et al., "Thiazide Diuretic Prophylaxis for Kidney Stones and the Risk of Diabetes Mellitus," *Journal of Urology* 192, no. 6 (December 2014): 1700–4, doi:10.1016/j.juro.2014.06.078.
4. Jennifer Barber et al., "A Systematic Review and Meta-analysis of Thiazide-Induced Hyponatraemia: Time to Reconsider Electrolyte Monitoring Regimens after Thiazide Initiation?" *British Journal of Clinical Pharmacology* 79, no. 4 (April 2015): 566–77, doi:10.1111/bcp.12499.
5. Chin-Chou Huang et al., "Clinical and Genetic Factors Associated with Thiazide-Induced Hyponatremia," *Medicine* 94, no. 34 (August 2015): e1422, doi:10.1097/MD.0000000000001422.
6. Sirirat Reungiu et al., "Do Thiazides Worsen Metabolic Syndrome and Renal Disease? The Pivotal Roles for Hyperuricemia and Hypokalemia," *Current Opinion in Nephrology and Hypertension* 17, no. 5 (September 2008): 470–76, doi:10.1097/MNH.0b013e328305b9a5.
7. R. O. Pierce, Jr., and A. Perry, "The Effects of Thiazide Diuretics on Bone Mineral Metabolism," *Journal of the National Medical Association* 90, no. 1 (January 1998): 46–50.
8. Marcio L. Griebeler et al., "Thiazide-Associated Hypercalcemia: Incidence and Association with Primary Hyperparathyroidism over Two Decades," *Journal of Clinical Endocrinology and Metabolism* 101, no. 3 (March 2016): 1166–73, doi:10.1210/jc.2015-3964.
9. Betsy Foxman et al., "First-Time Urinary Tract Infection and Sexual Behavior," *Epidemiology* 6, no. 2 (March 1995): 162–68.
10. Danilo Maciel Carneiro et al., "Randomized, Double-Blind Clinical Trial to Assess the Acute Diuretic Effect of *Equisetum arvense* (Field Horsetail) in Healthy Volunteers," *Evidence-Based Complementary and Alternative Medicine* 2014 (2014): 760683, doi:10.1155/2014/760683.
11. Mirian Farinon et al., "Effect of Aqueous Extract of Giant Horsetail (*Equisetum giganteum* L.) in Antigen-Induced Arthritis," *Open Rheumatology Journal* 7 (December 30, 2013): 129–33.
12. Azam Asgharikhatooni et al., "The Effect of Equisetum Arvense (Horse Tail) Ointment on Wound Healing and Pain Intensity after Episiotomy: A Randomized Placebo-Controlled Trial," *Iranian Red Crescent Medical Journal* 17, no. 3 (March 2015): e25637, doi:10.5812/ircmj.25637.
13. Ludek Jahodár, I. Leifertová, and M. Lisá, "Investigation of Iridoid Substances in *Arctostaphylos uva-ursi*," *Pharmazie* 33, no. 8 (August 1978): 536–37.
14. Joachim Quintus et al., "Urinary Excretion of Arbutin Metabolites after Oral Administration of Bearberry Leaf Extracts," *Planta Medica* 71, no. 2 (February 2005): 147–52, doi:10.1055/s-2005-837782; and Gemot Schindler et al., "Urinary Excretion and Metabolism of Arbutin after Oral Administration of *Arctostaphylos uvae ursi* Extract as Film-Coated Tablets and Aqueous Solution in Healthy Humans," *Journal of Clinical Pharmacology* 42, no. 8 (August 2002): 920–27.
15. Sumiko Shiota et al., "Mechanisms of Action of Corilagin and Tellimagrandin I that Remarkably Potentiate the Activity of β-Lactams against Methicillin-Resistant *Staphylococcus aureus*," *Microbiology and Immunology* 48, no. 1 (2004): 67–73.
16. Masato Shimizu et al., "Marked Potentiation of Activity of β-Lactams against Methicillin-Resistant *Staphylococcus aureus* by Corilagin," *Antimicrobial Agents and Chemotherapy* 45, no. 11 (November 2001): 3198–201, doi:10.1128/AAC.45.11.3198-3201.2001.
17. Claudia Siegers et al., "Bacterial Deconjugation of Arbutin by *Escherichia coli*," *Phytomedicine* 10, supplement 4 (2003): 58–60.
18. Hideaki Matsuda et al., "Pharmacological Studies on Leaf of *Arctostaphylos uva-ursi* (L.) Spreng. V. Effect of Water Extract from *Arctostaphylos uva-ursi* (L.) Spreng. (Bearberry Leaf) on the Antiallergic and Antiinflammatory Activities of Dexamethasone Ointment," *Yakugaku Zasshi [Journal of the Pharmaceutical Society of Japan]* 112, no. 9 (September 1992): 673–77.
19. Kathryn Watt, Nick Christofi, and Rodney Young, "The Detection of Antibacterial Actions of Whole Herb Tinctures Using Luminescent *Escherichia coli*," *Phytotherapy Research* 21, no. 12 (December 2007): 1193–99, doi:10.1002/ptr.2238.
20. M. R. S. Zaidan et al., "In Vitro Screening of Five Local Medicinal Plants for Antibacterial Activity Using Disc Diffusion Method," *Tropical Biomedicine* 22, no. 2 (December 2005): 165–70.
21. Meenakshi Singh, Sharad Srivastava, and A. K. S. Rawat, "Antimicrobial Activities of Indian Berberis Species," *Fitoterapia* 78, no.7–8 (December 2007): 574–76, doi:10.1016/j.fitote.2007.03.021.
22. Uwe Albrecht, Karl-Heinz Goos, and Berthold Schneider, "A Randomised, Double-Blind, Placebo-Controlled Trial of a Herbal Medicinal Product Containing *Tropaeoli majoris herba* (Nasturtium) and *Armoraciae rusticanae radix* (Horseradish) for the Prophylactic Treatment of

Patients with Chronically Recurrent Lower Urinary Tract Infections," *Current Medical Research and Opinion* 23, no. 10 (October 2007): 2415–22, doi:10.1185/030079907X233089.

23. Assam M. El-Shazly, S. S. Hafex, and Michael Wink, "Comparative Study of the Essential Oils and Extracts of *Achillea fragrantissima* (Forssk.) Sch. Bip. and *Achillea santolina* L. (Asteraceae) from Egypt," *Pharmazie* 59, no. 3 (March 2004): 226–30.
24. Deborah A. Wing, "Daily Cranberry Juice for the Prevention of Asymptomatic Bacteriuria in Pregnancy: A Randomized, Controlled Pilot Study," *Journal of Urology* 180, no. 4 (October 2008): 1367–72, doi:10.1016/j.juro.2008.06.016.
25. Robert Mittendorf, Michelle A. Williams, and Edward H. Kass, "Prevention of Preterm Delivery and Low Birth Weight Associated with Asymptomatic Bacteriuria," *Clinical Infectious Diseases* 14, no. 4 (April 1992): 927.
26. John W. Warren, "Catheter-Associated Bacteriuria in Long-Term Care Facilities," *Infection Control and Hospital Epidemiology* 15, no. 8 (August 1994): 557–62.
27. Jerry Avorn et al., "Reduction of Bacteriuria and Pyuria after Ingestion of Cranberry Juice," *JAMA: Journal of the American Medical Association* 271, no. 10 (March 1994): 751–54.
28. David T. Bailey et al., "Can a Concentrated Cranberry Extract Prevent Recurrent Urinary Tract Infections in Women? A Pilot Study," *Phytomedicine* 14, no. 4 (April 2007): 237–41, doi:10.1016/j.phymed.2007.01.004; and Marion E. T. McMurdo et al., "Does Ingestion of Cranberry Juice Reduce Symptomatic Urinary Tract Infections in Older People in Hospital? A Double-Blind Placebo-controlled trial," *Age and Ageing* 34, no. 3 (May 2005): 256–61, doi:10.1093/ageing/afi101.
29. Susan A. Oliveria et al., "Estrogen Replacement Therapy and Urinary Tract Infections in Postmenopausal Women Aged 45–89," *Menopause* 5, no. 1 (Spring 1998): 4–8.
30. Raul Raz and Walter E. Stamm, "A Controlled Trial of Intravaginal Estriol in Postmenopausal Women with Recurrant Urinary Tract Infections," *New England Journal of Medicine* 329 (1993): 753–56, doi:10.1056/NEJM199309093291102; and Carla Perrotta et al., "Oestrogens for Preventing Recurrent Urinary Tract Infection in Postmenopausal Women," *Cochrane Database of Systematic Reviews*, no. 2 (April 16, 2008): CD005131, doi:10.1002/14651858.CD005131.pub2.
31. Carolyn Crandall, "Vaginal Estrogen Preparations: A Review of Safety and Efficacy for Vaginal Atrophy," *Journal of Women's Health* 11, no. 10 (December 2002): 857–77, doi:10.1089/154099902762203704.
32. Melinda Zeron Mullins and Konia M. Trouton, "BASIC Study: Is Intravaginal Boric Acid Non-inferior to Metronidazole in Symptomatic Bacterial Vaginosis? Study Protocol for a Randomized Controlled Trial," *Trials* 16 (July 26, 2015): 315, doi:10.1186/s13063-015-0852-5.
33. Andy G. Pinson et al., "Fever in the Clinical Diagnosis of Acute Pyelonephritis," *American Journal of Emergency Medicine* 15, no. 2 (March 1997): 148–51; and Alejandro Hoberman and Ellen R. Wald, "Urinary Tract Infections in Young Febrile Children," *Pediatric Infectious Disease Journal* 16, no. 1 (January 1997): 11–7.
34. Petra Lüthje and Annelie Brauner, "Novel Strategies in the Prevention and Treatment of Urinary Tract Infections," *Pathogens* 5, no. 1 (January 27, 2016): 13, doi:10.3390/pathogens5010013.
35. Shyh-Ming Taso, C. C. Hsu, and M. C. Yin, "Garlic Extract and Two Diallyl Sulphides Inhibit Methicillin-Resistant *Staphylococcus aureus* Infection in BALB/cA Mice," *Journal of Antimicrobial Chemotherapy* 52, no. 6 (December 2003): 974–80, doi:10.1093/jac/dkg476.
36. H. R. Maurer, "Bromelain: Biochemistry, Pharmacology and Medical Use," *Cellular and Molecular Life Sciences* 58, no. 9 (August 2001): 1234–45, doi:10.1007/PL00000936.
37. Vidhya Rathnavelu et al., "Potential Role of Bromelain in Clinical and Therapeutic Applications," *Biomedical Reports* 5, no. 3 (September 2016): 283–88, doi:10.3892/br.2016.720.
38. Valirie Ndip Agbor, Tsi Njim, and Franklin Ngu Mbolingong, "Bladder Outlet Obstruction; a Rare Complication of the Neglected Schistosome, *Schistosoma haematobium*: Two Case Reports and Public Health Challenges," *BMC Research Notes* 9 (2016): 493, doi:10.1186/s13104-016-2303-0.
39. Ebetsam M. Al-Olayan et al., "*Ceratonia siliqua* Pod Extract Ameliorates *Schistosoma mansoni*–Induced Liver Fibrosis and Oxidative Stress," *BMC Complementary and Alternative Medicine* 16 (2016): 434, doi:10.1186/s12906-016-1389-1.
40. Luísa Maria Silveria de Almeida et al., "Flavonoids and Sesquiterpene Lactones from *Artemisia absinthium* and *Tanacetum parthenium* against *Schistosoma mansoni* Worms," *Evidence-Based Complementary and Alternative Medicine* 2016 (2016): 9521349, doi:10.1155/2016/9521349.
41. Olugbenga S. Oile et al., "Detection of Urinary Tract Pathology in Some *Schistosoma haematobium* Infected Nigerian Adults," *Journal of Tropical Medicine* 2016 (2016): 5405207, doi:10.1155/2016/5405207.
42. Welcome M. Wami et al., "Comparative Assessment of Health Benefits of Praziquantel Treatment of Urogenital Schistosomiasis in Preschool and Primary School-Aged Children," *BioMed Research International* 2016 (2016): 9162631, doi:210.1155/2016/9162631.
43. James Monroe McDonald, *Physiologic Medication: A Compend of Physio-Medical Principles and Practices: Containing a Complete Formulary, Dose Book, Dictionary of Therapeutical and Medical Nomenclature* (Chicago, IL: Armstrong, 1900): 90.
44. H. Kaur and A. S. Arunkalaivanan, "Urethral Pain Syndrome and Its Management," *Obstetrical and Gynecological Survey* 62, no. 5 (May 2007): 348–51; quiz 353–54, doi:10.1097/01.ogx.0000261645.12099.2a.
45. Ashley Cox, Nicole Golda, Genevieve Nadeau et al., "CUA Guideline: Diagnosis and Treatment of Interstitial Cystitis/Bladder Pain Syndrome *Canadian Urological Association Journal* 10, no. 5–6 (May–June 2016): E136–E155, doi:10.5489/cuaj.3786.

46. J. Quentin Clemens, Sheila O. Brown, and Elizabeth A. Calhoun, "Mental Health Diagnoses in Patients with Interstitial Cystitis/Painful Bladder Syndrome and Chronic Prostatitis/Chronic Pelvic Pain Syndrome: A Case/Control Study," *Journal of Urology* 180, no. 4 (October 2008): 1378–82, doi:10.1016/j.juro.2008.06.032.
47. Maurice K. Chung, Charles W. Butrick, and Cherie W. Chung, "The Overlap of Interstitial Cystitis/Painful Bladder Syndrome and Overactive Bladder," *JSLS: Journal of the Society of Laproendoscopic Surgeons* 14, no. 1 (January–March 2010): 83–90, doi:10.4293/108680810X12674612014743.
48. John B. Forrest and Sunshine Schmidt, "Interstitial Cystitis, Chronic Nonbacterial Prostatitis and Chronic Pelvic Pain Syndrome in Men: A Common and Frequently Identical Clinical Entity," *Journal of Urology* 172, no. 6, pt. 2 (December 2004): 2561–62.
49. J. Curtis Nickel, Laruel Emerson, and Jillian Cornish, "The Bladder Mucus (Glycosaminoglycan) Layer in Interstitial Cystitis," *Journal of Urology* 149, no. 4 (April 1993): 716–18; and Curtis B. Wilson et al., "Selective Type IV Collagen Defects in the Urothelial Basement Membrane in Interstitial Cystitis," *Journal of Urology* 154, no. 3 (September 1995): 1222–26.
50. R. L. Ochs et al., "Autoantibodies in Interstitial Cystitis," *Journal of Urology* 151, no. 3 (March 1994): 587–92.
51. Jackson E. Fowler, Jr., et al., "Interstitial Cystitis Is Associated with Intraurothelial Tamm-Horsfall Protein," *Journal of Urology* 140, no. 6 (December 1988): 1385–39.
52. Pradeep Tyagi et al., "Functional Role of Cannabinoid Receptors in Urinary Bladder," *Indian Journal of Urology* 26, no. 1 (January–March 2010): 26, doi:10.4103/0970-1591.60440.
53. Merete Holm-Bentzen et al., "A Prospective Double-Blind Clinically Controlled Multicenter Trial of Sodium Pentosanpolysulfate in the Treament of Interstitial Cystitis and Related Painful Bladder Disease," *Journal of Urology* 138, no. 3 (September 1987): 503–7.
54. Stanley Zasalu et al., "*Pentosan polysulfate* (Elmiron): In Vitro Effects on Prostate Cancer Cells Regarding Cell Growth and Vascular Endothelial Growth Factor Production," *American Journal of Surgery* 192, no. 5 (November 2006): 640–43, doi:10.1016/j.amjsurg.2006.08.008.
55. George Chiang et al., "Pentosanpolysulfate Inhibits Mast Cell Histamine Secretion and Intracellular Calcium Ion Levels: An Alternative Explanation of Its Beneficial Effect in Interstitial Cystitis," *Journal of Urology* 164, no. 6 (December 2000): 2119–25.
56. Donald L. Lamm and Dale R. Riggs, "Enhanced Immunocompetence by Garlic: Role in Bladder Cancer and Other Malignancies," *Journal of Nutrition* 131, no. 3s (March 2001): 1067S–70S.
57. C. Manesh and G. Kuttan, "Alleviation of Cyclophosphamide-Induced Urotoxicity by Naturally Occurring Sulphur Compounds," *Journal of Exprimental and Clinical Cancer Research* 21, no. 4 (December 2002): 509–17.
58. Göksel Şener et al., "Chronic Nicotine Toxicity Is Prevented by Aqueous Garlic Extract," *Plant Foods for Human Nutrition* 60, no. 2 (June 2005): 77–86.
59. Rex Munday and Christine M. Munday, "Low Doses of Diallyl Disulfide, a Compound Derived from Garlic, Increase Tissue Activities of Quinone Reductase and Glutathione Transferase in the Gastrointestinal Tract of the Rat," *Nutrition and Cancer* 34, no. 1 (1999): 42–48, doi:10.1207/S15327914NC340106; A. Zeybek et al., "Aqueous Garlic Extract Inhibits Protamine Sulfate-Induced Bladder Damage," *Urologia Internationalis* 76, no. 2 (February 2006): 173–79, doi:10.1159/000090884; Du-Geon Moon et al., "*Allium sativum* Potentiates Suicide Gene Therapy for Murine Transitional Cell Carcinoma," *Nutrition and Cancer* 38, no. 1 (2000): 98–105, doi:10.1207/S15327914NC381_14; Donald L. Lamm and Dale R. Riggs, "The Potential Application of *Allium sativum* (Garlic) for the Treatment of Bladder Cancer" *Urology Clinics of North America* 27, no. 1 (February 2000): 157–62: xi; and Dale R. Riggs, J. I. DeHaven, and Donald L. Lamm, "*Allium sativum* (Garlic) Treatment for Murine Transitional Cell Carcinoma," *Cancer* 79, no. 10 (May 15, 1997): 1987–94.
60. Zeybek, "Aqueous Garlic Extract Inhibits Protamine Sulfate-Induced Bladder Damage."
61. Şener, "Chronic Nicotine Toxicity Is Prevented by Aqueous Garlic Extract."
62. Manesh, "Alleviation of Cyclophosphamide-Induced Urotoxicity by Naturally Occurring Sulphur Compounds."
63. Munday, "Low Doses of Diallyl Disulfide, a Compound Derived from Garlic, Increase Tissue Activities."
64. Jing-Gung Chung, "Effects of Garlic Components Diallyl Sulfide and Diallyl Disulfide on Arylamine N-Acetyltransferase Activity in Human Bladder Tumor Cells," *Drug and Chemical Toxicology* 22, no. 2 (May 1999): 343–58, doi:10.3109/01480549909017839.
65. H. F. Lu et al., "Diallyl Disulfide (DADS) Induced Apoptosis Undergo Caspase-3 Activity in Human Bladder Cancer T24 Cells," *Food and Chemical Toxicology* 42, no. 10 (October 2004): 1543–52, doi:10.1016/j.fct.2003.06.001; Moon, "*Allium sativum* Potentiates Suicide Gene Therapy"; and Lamm, "The Potential Application of *Allium sativum* (Garlic) for the Treatment of Bladder Cancer."
66. Moon, "*Allium sativum* Potentiates Suicide Gene Therapy."
67. Lamm, "Enhanced Immunocompetence by Garlic," (see n. 69).
68. Ian A. Oyama et al., "Modified Thiele Massage as Therapeutic Intervention for Female Patients with Interstitial Cystitis and High-Tone Pelvic Floor Dysfunction," *Urology* 64. no. 5 (November 2004): 862–65, doi:10.1016/j.urology.2004.06.065.
69. Jerome M. Weiss, "Pelvic Floor Myofascial Trigger Points: Manual Therapy for Interstitial Cystitis and the Urgency-Frequency Syndrome," *Journal of Urology* 166, no. 6 (December 2001): 2226–31.
70. Ryuta Inoue et al., "Hydrodistention of the Bladder in Patients with Interstitial Cystitis—Clinical Efficacy and

its Association with Immunohistochemical Findings for Bladder Tissues," *Hinyokika Kiyo [Acta Urologica Japonica]* 52, no. 10 (October 2006): 765–68; and Nasim Zabihi et al., "Bladder Necrosis Following Hydrodistention in Patients with Interstitial Cystitis," *Journal of Urology* 177, no. 1 (January 2007): 149–52, doi:10.1016/j.juro.2006.08.095.

71. I. Binder, G. Rossback, and A. van Ophoven, "The Complexity of Chronic Pelvic Pain Exemplified by the Condition Currently Called Interstitial Cystitis," *Aktuelle Urologie* 39, no. 4 (July 2008): 289–97, doi:10.1055/s-2008-1038199.
72. Payne, "Graded Potassium Chloride Testing in Interstitial Cystitis"; and Parsons, "Potassium Sensitivity Test."
73. "Kava," in *The Review of Natural Products: Facts and Comparisons* (St. Louis, MO: Wolters Kluwer Health, Inc, 2009). June 2009. Accessed May 9, 2012.
74. Georg Boonen and Hanns Häberlein, "Influence of Genuine Kavapyrone Enantiomers on the GABA-A Binding Site," *Planta Medica* 64, no. 6 (August 1998): 504, doi:10.1055/s-2006-957502; L. Davies et al., "Effects of Kava on Benzodiazepine and GABA Receptor Binding," *European Journal of Pharmacology* 183, no. 2 (July 2, 1990): 558, doi:10.1016/0014-2999(90)93467-5; and R. Kretzschmar, H. J. Meyer, and H. J. Teschendorf, "Strychnine Antagonistic Potency of Pyrone Compounds of the Kavaroot (*Piper methysticum* Forst.)," *Experientia* 26, no. 3 (March 1970): 283–84.
75. "Monograph: *Centella asiatica*" (Milan, Italy: Indena S.p.A., 1987); and J. C. Lawrence, "The Morphological and Pharmacological Effects of Asiaticoside upon Skin *In Vitro* and *In Vivo*," *European Journal of Pharmacology* 1, no. 5 (September 1967): 414–24, doi:10.1016/0014-2999(67)90104-5.
76. Beyhan Sağlam et al., "An Aqueous Garlic Extract Alleviates Water Avoidance Stress-Induced Degeneration of the Urinary Bladder," *BJU International* 98, no. 6 (December 2006): 1250–54, doi:10.1111/j.1464-410X.2006.06511.x.
77. Solomon Habtemariam, "Antiinflammatory Activity of the Antirheumatic Herbal Drug, Gravel Root (*Eupatorium purpureum*): Further Biological Activities and Constituents," *Phytotherapy Research* 15, no. 8 (December 2001): 687–90; and Solomon Habtemariam, "Cistifolin, an Integrin-Dependent Cell Adhesion Blocker from the Anti-rheumatic Herbal Drug, Gravel Root (Rhizome of *Eupatorium purpureum*)," *Planta Medica* 64, no. 8 (December 1998): 683–85, doi:10.1055/s-2006-957558.
78. Jean-Sébastien Walczak and Fernando Cervero, "Local Activation of Cannabinoid CB_1 Receptors in the Urinary Bladder Reduces the Inflammation-Induced Sensitization of Bladder Afferents," *Molecular Pain* 7 (May 9, 2011): 31, doi:10.1186/1744-8069-7-31.
79. A. R. Bilia et al., "Effect of Surfactants and Solutes (Glucose and NaCl) on Solubility of Kavain—A Technical Note," *AAPS PharmSciTech* 9, no. 2 (June 2008): 444–48, doi:10.1208/s12249-008-9064-6.
80. Pablo Leitzman et al., "Kava Blocks 4-(Methylnitrosamino)-1-(3-pyridyl)-1-butanone-Induced Lung Tumorigenesis in Association with Reducing O^6-methylguanine DNA Adduct in A/J Mice," *Cancer Prevention Research* 7, no. 1 (January 2014): 86–96, doi:10.1158/1940-6207.CAPR-13-0301; and Y. Tang et al., "Effects of the Kava Chalcone Flavokawain A Differ in Bladder Cancer Cells with Wild-Type versus Mutant p53," *Cancer Prevention Research* 1, no. 6 (November 2008): 439–451, doi:10.1158/1940-6207.CAPR-08-0165.
81. Rajesh Agarwal and Gagan Deep, "Kava, a Tonic for Relieving the Irrational Development of Natural Preventive Agents," *Cancer Prevention Research* 1, no. 6 (November 2008): 409–12, doi:10.1158/1940-6207.CAPR-08-0172.
82. Xuesen Li et al., "Kava Components Down-Regulate Expression of AR and AR Splice Variants and Reduce Growth in Patient-Derived Prostate Cancer Xenografts in Mice," *PLoS ONE* 7, no. 2 (2012): e31213, doi:10.1371/journal.pone.0031213.
83. Xiaoren Tang and Salomon Amar, "Kavain Inhibition of LPS-Induced TNF-α via ERK/LITAF," *Toxicology Research* 5, no. 1 (2016): 188–96, doi:10.1039/C5TX00164A.
84. E. Alramadhan et al., "Dietary and Botanical Anxiolytics," *Medical Science Monitor* 18, no. 4 (April 2012): RA40–48; and Benjamin S. Weeks, "Formulations of Dietary Supplements and Herbal Extracts for Relaxation and Anxiolytic Action: Relarian," *Medical Science Monitor* 15, no. 11 (November 2009): RA256–62.
85. Han Chow Chua et al., "Kavain, the Major Constituent of the Anxiolytic Kava Extract, Potentiates GABAA Receptors: Functional Characteristics and Molecular Mechanism," *PLoS ONE* 11, no. 6 (2016): e0157700.
86. B. M. Dietz and J. L. Bolton, "Biological Reactive Intermediates (BRIs) Formed from Botanical Dietary Supplements," *Chemico-Biological Interactions* 192, no. 1–2 (June 30, 2011): 72–80, doi:10.1016/j.cbi.2010.10.007.
87. Dietz, "Biological Reactive Intermediates (BRIs) Formed from Botanical Dietary Supplements."
88. John M. McPartland, Geoffrey W. Guy, and Vincenzo Di Marzo, "Care and Feeding of the Endocannabinoid System: A Systematic Review of Potential Clinical Interventions that Upregulate the Endocannabinoid System," *PLoS ONE* 9, no. 3 (March 12, 2014): e89566, doi:10.1371/journal.pone.0089566.
89. A. Reirz, C. Fisang, and S. C. Müeller, "Neuromuscular Dysfunction of the Lower Urinary Tract Dysfunction Beyond Spinal Cord Injury and Multiple Sclerosis: A Challenge for Urologists," *Urologe* 47, no. 9 (September 2008): 1097–8, doi:10.1007/s00120-008-1850-y.
90. S. K. Doumouchtsis, S. Jeffery, and M. Fynes, "Female Voiding Dysfunction," *Obstetrical and Gynecological Survey* 63, no. 8 (August 2008): 519–26, doi:10.1097/OGX.0b013e31817f1214.
91. J. Pannek, K. Göcking, and U. Bersch, "'Neurogenic' Urinary Tract Dysfunction: Don't Overlook the Bowel!" *Spinal Cord* 47, no. 1 (January 2009): 93–94, doi:10.1038/sc.2008.79.
92. John R. Beuerle and Fermin Barrueto, Jr., "Neurogenic Bladder and Chronic Urinary Retention Associated with MDMA Abuse," *Journal of Medical Toxicology* 4, no. 2 (June 2008): 106–8.

93. Andrew Ballaro, "The Elusive Electromyogram in the Overactive Bladder: A Spark of Understanding," *Annals of the Royal College of Surgeons of England* 90, no. 5 (July 2008): 362–67, doi:10.1308/003588408X301217.
94. Juergen Pannek, "Treatment of Urinary Tract Infection in Persons with Spinal Cord Injury: Guidelines, Evidence, and Clinical Practice. A Questionnaire-Based Survey and Review of the Literature," *Journal of Spinal Cord Medicine* 34, no. 1 (2011): 11–15, doi:10.1179/107902610X12886261091839.
95. M. De Sèze, "Peripheral Electrical Stimulation in Neurogenic Bladder," *Annals de Réadaptation et de Médecine Physique [Annals of Physical and Rehabilitation Medicine]* 51, no. 6 (July 2008): 473–78, doi:10.1016/j.annrmp.2008.04.003.
96. Haodong Lin et al., "Reconstruction of Reflex Pathways to the Atonic Bladder after Conus Medullaris Injury: Preliminary Clinical Results," *Microsurgery* 28, no. 6 (2008): 429–35, doi:10.1002/micr.20504.
97. R. M. Levin et al., "Low-Dose Tadenan Protects the Rabbit Bladder from Bilateral Ischemia / Reperfusion-Induced Contractile Dysfunction," *Phytomedicine* 12, no. 1–2 (January 2005): 17–24, doi:10.1016/j.phymed.2003.10.002.
98. M. J. Hess et al., "Evaluation of Cranberry Tablets for the Prevention of Urinary Tract Infections in Spinal Cord Injured Patients with Neurogenic Bladder," *Spinal Cord* 46, no. 9 (September 2008): 622–26, doi:10.1038/sc.2008.25.
99. G. Reid et al., "Cranberry Juice Consumption May Reduce Biofilms on Uroepithelial Cells: Pilot Study in Spinal Cord Injured Patients," *Spinal Cord* 39, no. 1 (January 2009): 26–30, doi:10.1038/sj.sc.3101099.
100. Keryn G. Woodman et al., "Nutraceuticals and Their Potential to Treat Duchenne Muscular Dystrophy: Separating the Credible from the Conjecture," *Nutrients* 8, no. 11 (November 9, 2016): pii, E713, doi:10.3390/nu8110713.
101. Xiaoxiao Jiang et al., "Sodium Tanshinone IIA Sulfonate Ameliorates Bladder Fibrosis in a Rat Model of Partial Bladder Outlet Obstruction by Inhibiting the TGF-β/Smad Pathway Activation," *PLoS ONE* 10, no. 6 (2015): e0129655, doi:10.1371/journal.pone.0129655.
102. Weidong Xiao et al., "Ligustilide Treatment Promotes Functional Recovery in a Rat Model of Spinal Cord Injury via Preventing ROS Production," *International Journal of Clinical and Experimental Pathology* 8, no. 10 (2015): 12005–13.
103. M. W. Weinberger, B. M. Goodman, and M. Carnes, "Long-Term Efficacy of Nonsurgical Urinary Incontinence Treatment in Elderly Women," *Journal of Gerontology: Series A, Biological Sciences and Medical Sciences* 54, no. 3 (March 1999): M117–21.
104. Vivek Kumar et al., "Recent Advances in Basic Science for Overactive Bladder," *Current Opinions in Urology* 15, no. 4 (July 2005): 222–26.
105. H. S. Kim, J. C. Kim, and M. S. Choo, "Effects of Nitric Oxide Synthases on Detrusor Overactivity after Removal of Bladder Outlet Obstruction in Rats," *Urologia Internationalis* 81, no. 1 (2008): 107–12, doi:10.1159/000137650.
106. D. M. Schmid et al., "Prospects and Limitations of Treatment with Botulinum Neurotoxin Type A for Patients with Refractory Idiopathic Detrusor Overactivity," *BJU International* 102, supplement 1 (July 25, 2008): 7–10, doi:10.1111/j.1464-410X.2008.07827.x.
107. H. Akino et al., "The Pathophysiology Underlying Overactive Bladder Syndrome Possibly Due to Benign Prostatic Hyperplasia," *Hinyokika Kiyo* 54, no. 6 (June 2008): 449–52.
108. A. Lucioni et al., "The Use of Botulinum Toxin for Treatment of Lower Urinary Tract Symptoms," *Minerva Urologica e Nefrologica [Italian Journal of Urology and Nephrology]* 60, no. 2 (June 2008): 93–103.
109. H. Grunze et al., "Kava Pyrones Exert Effects on Neuronal Transmission and Transmembraneous Cation Currents Similar to Established Mood Stabilizers—A Review," *Progress in Neurō-Psychopharmacology and Biological Psychiatry* 25, no. 8 (November 2001): 1555–70.
110. A. M. Duffield et al., "Identification of Some Human Urinary Metabolites of the Intoxicating Beverage Kava," *Journal of Chromatography* 475, no. 2 (1989): 273–81, doi:10.1016/S0021-9673(01)89682-5.
111. R. F. Weiss, *Herbal Medicine* (New York: Thieme, 2001); and William Boericke, *Homeopathic Materia Medica* (Kessinger Publishing, 2004).
112. Ballaro, "The Elusive Electromyogram in the Overactive Bladder," (see n. 95).
113. M. S. Kava et al., "α1L-Adrenoceptor Mediation of Smooth Muscle Contraction in Rabbit Bladder Neck: A Model for Lower Urinary Tract Tissues of Man," *British Journal of Pharmacology* 123, no. 7 (April 1998): 1359–66, doi:10.1038/sj.bjp.0701748.
114. A. P. Ford et al., "RS-17053 (N-[2-(2-cyclopropylmethoxyphenoxy)ethyl]-5-chloro-α, α-dimethyl-1H-indole-3-ethanamine hydrochloride), a Selective α 1A-Adrenoceptor Antagonist, Displays Low Affinity for Functional α 1-Adrenoceptors in Human Prostate: Implications for Adrenoceptor Classification," *Molecular Pharmacology* 49, no 2 (February 1996): 209–15.
115. J. Mokry and G. Nosalova, "In Vitro Reactivity of Urinary Bladder Smooth Muscle in Rabbits Influenced by Xanthine Derivatives," *Bratislavské Lekárske Listy [Bratislava Medical Journal]* 109, no. 3 (2008): 91–94.
116. Y. Yiangou et al., "Capsaicin Receptor VR1 and ATP-Gated Ion Channel P2X3 in Human Urinary Bladder," *BJU International* 87, no. 9 (June 2001): 774–79.
117. T. Yamanishi, T. Kamai, and K. Yoshida, "Neuromodulation for the Treatment of Urinary Incontinence," *International Journal of Urology* 15, no. 8 (August 2008): 665–72, doi:10.1111/j.1442-2042.2008.02080.x; and N. A. Mungan et al., "Nocturnal Enuresis and Allergy," *Scandinavian Journal of Urology and Nephrology* 39, no. 3 (2005): 237–41.
118. Mungan, "Nocturnal Enuresis and Allergy."
119. J. Egger et al., "Effect of Diet Treatment on Enuresis in Children with Migraine or Hyperkinetic Behavior," *Clinical Pediatrics* 31, no. 5 (May 1992): 302–7.

120. Yamanishi, "Neuromodulation for the Treatment of Urinary Incontinence."
121. Hermann M. Bolt and Klaus Holka, "The Debate on Carcinogenicity of Permanent Hair Dyes: New Insights," *Critical Reviews in Toxicology* 37, no. 6 (2007): 521–36, doi:10.1080/1040844070135671; and J. E. Altwein, "Primary Prevention of Bladder Cancer. What's New?" *Urologe* 46, no. 6 (June 2007): 616–21, doi:10.1007/s00120-007-1348-z.
122. J. H. Lubin et al., "Cigarette Smoking and Cancer: Intensity Patterns in the α-Tocopherol, β-Carotene Cancer Prevention Study in Finnish Men," *American Journal of Epidemiology* 167, no. 8 (April 15, 2008): 970–75, doi:10.1093/aje/kwm392.
123. Daisuke Sato and Masahiro Matsushima, "Preventive Effects of Urinary Bladder Tumors Induced by N-butyl-N-(4-hydroxybutyl)-nitrosamine in Rat by Green Tea Leaves," *International Journal of Urology* 10, no. 3 (March 2003): 160–66.
124. Rayjean J. Hung et al., "Protective Effects of Plasma Carotenoids on the Risk of Bladder Cancer," *Journal of Urology* 176, no. 3 (September 2006): 1192–97, doi:10.1016/j.juro.2006.04.030.
125. Diaa A. Hameed and Tarek H. El-Metwally, "The Effectiveness of Retinoic Acid Treatment in Bladder Cancer: Impact on Recurrence, Survival and TGFα and VEGF as End-point Biomarkers," *Cancer Biology and Therapy* 7, no. 1 (January 2008): 92–100.
126. D. A. Yang, "Inhibitory Effect of Chinese Herb Medicine Zhuling on Urinary Bladder Cancer. An Experimental and Clinical Study," *Zhonghua Wai Ke Za Zhi [Chinese Journal of Surgery]* 29(6):393–5, 399.
127. K. Sheeja and G. Kuttan, "Protective Effect of *Andrographis paniculata* and Andrographolide on Cyclophosphamide-Induced Urothelial Toxicity," *Integrative Cancer Therapy* 5, no. 3 (September 2006): 244–51, doi:10.1177/1534735406291984; and K. Sheeja and G. Kuttan, "Ameliorating Effects of *Andrographis paniculata* Extract against Cyclophosphamide-Induced Toxicity in Mice," *Asian Pacific Journal of Cancer Prevention* 7, no. 4 (October 2006): 609–14.
128. Leemol Davis and Grijia Kuttan, "Effect of *Withania somnifera* on Cyclophosphamide-Induced Urotoxicity," *Cancer Letters* 148, no. 1 (January 1, 2000): 9–17, doi:10.1016/S0304-3835(99)00252-9.
129. C. Manesh and G. Kuttan, "Effect of Naturally Occurring Isothiocyanates in the Inhibition of Cyclophosphamide-Induced Urotoxicity," *Phytomedicine* 12, no. 6–7 (June 2005): 487–93, doi:10.1016/j.phymed.2003.04.004.
130. S.-J. Su et al., "Overexpression of HER-2/neu Enhances the Sensitivity of Human Bladder Cancer Cells to Urinary Isoflavones," *European Journal of Cancer* 37, no. 11 (July 2001): 1413–18, doi:10.1016/S0959-8049(01)00110-1.
131. A. V. Singh et al., "Soy Phytochemicals Prevent Orthotopic Growth and Metastasis of Bladder Cancer in Mice by Alterations of Cancer Cell Proliferation and Apoptosis and Tumor Angiogenesis," *Cancer Research* 66, no. 3 (February 2006): 1851–58.
132. Jin-Rong Zhou et al., "Inhibition of Murine Bladder Tumorigenesis by Soy Isoflavones via Alterations in the Cell Cycle, Apoptosis, and Angiogenesis," *Cancer Research* 58, no. 22 (November 1998): 5231–38.
133. D. A. Yang S. Q. Li, and X. T. Li, "Prophylactic Effects of Zhuling and BCG on Postoperative Recurrence of Bladder Cancer," *Zhonghua Wai Ke Za Zhi [Chinese Journal of Surgery]* 32, no. 7 (July 1994): 433–34.
134. K. Gao et al., "Experimental Study on Decoctum *Agrimonia pilosa* Ledeb-Induced Apoptosis in HL-60 Cells In Vitro," *Zhong Yao Cai [Journal of Chinese Medicinal Materials]* 23, no. 9 (September 2000): 561–62; T. Murayama et al., "Agrimoniin, an Antitumor Tannin of *Agrimonia pilosa* Ledeb., Induces Interleukin-1," *Anticancer Research* 12, no. 5 (September/October 1992): 1471–74; Kenichi Miyamoto, Nobuharu Kishi, and Ryozo Koshioura, "Antitumor Effect of Agrimoniin, a Tannin of *Agrimonia pilosa* Ledeb., on Transplantable Rodent Tumors," *Japanese Journal of Pharmacology* 43, no. 2 (February 1987): 187–95; and Ryozo Koshiura et al., "Antitumor Activity of Methanol Extract from Roots of *Agrimonia pilosa* Ledeb.," *Japanese Journal of Pharmacology* 38, no. 1 (May 1985): 9–16.
135. K. Sheeja, C. Guruvayoorappan, and G. Kuttan, "Antiangiogenic Activity of *Andrographis paniculata* Extract and Andrographolide," *International Immunopharmacology* 7, no. 2 (February 2007): 211–21, doi:10.1016/j.intimp.2006.10.002.
136. Jerry R. Rittenhouse, Paul D. Lui, and Benjamin H. S. Lau, "Chinese Medicinal Herbs Reverse Macrophage Suppression Induced by Urological Tumors," *Journal of Urology* 146, no. 2 (August 1991): 486–90.
137. Li Tang et al., "Potent Activation of Mitochondria-Mediated Apoptosis and Arrest in S and M Phases of Cancer Cells by a Broccoli Sprout Extract," *Molecular Cancer Therapy* 5, no. 4 (April 2006): 935–44, doi:10.1158/1535-7163.MCT-05-0476.
138. Qing-Yi Lu et al., "*Ganoderma lucidum* Extracts Inhibit Growth and Induce Actin Polymerization in Bladder Cancer Cells In Vitro," *Cancer Letters* 216, no. 1 (December 2004): 9–20, doi:10.1016/j.canlet.2004.06.022.
139. Sensuke Konno, "Effect of Various Natural Products on Growth of Bladder Cancer Cells: Two Promising Mushroom Extracts," *Alternative Medicine Review* 12, no. 1 (March 2007): 63–68; and P. C. Griessmayr et al., "Mushroom-Derived Maitake PETfraction as Single Agent for the Treatment of Lymphoma in Dogs," *Journal of Veterinary Internal Medicine* 21, no. 6 (November/December 2007): 1409–12.
140. K. Hostanska et al., "Aqueous Ethanolic Extract of St. John's Wort (*Hypericum perforatum* L.) Induces Growth Inhibition and Apoptosis in Human Malignant Cells In Vitro," *Pharmazie* 57, no. 5 (May 2002): 323–31; Dimitris Skalkos et al., "The Lipophilic Extract of *Hypericum perforatum* Exerts Significant Cytotoxic Activity against T24 and NBT-II Urinary Bladder Tumor Cells,"

Planta Medica 71, no. 11 (November 2005): 1030–35, doi:10.1055/s-2005-873127; and Stavropoulos, "*Hypericum perforatum* L. Extract—Novel Photosensitizer against Human Bladder Cancer Cells" (see n. 138).

141. Rittenhouse, "Chinese Medicinal Herbs Reverse Macrophage Suppression" (see n. 153).
142. Steven Kuan-Hua Huan et al., "Cantharidin-Induced Cytotoxicity and Cyclooxygenase 2 Expression in Human Bladder Carcinoma Cell Line," *Toxicology* 223, no. 1–2 (June 1, 2006): 136–43, doi:10.1016/j.tox.2006.03.012.
143. Xiaolin Zi and Anne R. Simoneau, "Flavokawain A, a Novel Chalcone from Kava Extract, Induces Apoptosis in Bladder Cancer Cells by Involvement of Bax Protein-Dependent and Mitochondria-Dependent Apoptotic Pathway and Suppresses Tumor Growth in Mice," *Cancer Research* 65, no. 8 (April 15, 2005): 3479–86, doi:10.1158/0008-5472.CAN-04-3803.
144. Katarina Hostanska et al., "Evaluation of Cell Death Caused by an Ethanolic Extract of *Serenoae repentis* Fructus (Prostasan) on Human Carcinoma Cell Lines," *Anticancer Research* 27, no. 2 (March/April 2007): 873–81.
145. Jeevan K. Prasain, "Effect of Cranberry Juice Concentrate on Chemically-Induced Urinary Bladder Cancers," *Oncology Reports* 19, no. 6 (June 2008): 1565–70.
146. Ursula Elsässer-Beile, "Adjuvant Intravesical Treatment with a Standardized Mistletoe Extract to Prevent Recurrence of Superficial Urinary Bladder Cancer," *Anticancer Research* 25, no. 6C (November/December 2005): 4733–36.
147. D. Skalkos, et al., "Photophysical Properties of *Hypericum perforatum* L. Extracts—Novel Photosensitizers for PDT," *Journal of Photochemistry and Photobiology B: Biology* 82, no. 2 (February 1, 2006): 146–51, doi:10.1016/j.jphotobiol.2005.11.001.
148. Jing-Cun Zheng et al., "Effect of New Photosensitizer CDHS801-Mediated Photodynamic Therapy on Bladder Cancer: An Experimental Study," *Zhonghua Yi Xue Za Zhi [Chinese Medical Journal]* 85, no. 25 (July 6, 2005): 1762–65; and N. E. Stavropoulos et al., "*Hypericum perforatum* L. Extract—Novel Photosensitizer against Human Bladder Cancer Cells," *Journal of Photochemistry and Photobiology B: Biology* 84, no. 1 (July 3, 2006): 64–69, doi:10.1016/j.jphotobiol.2006.02.001.
149. Hostanska, "Evaluation of Cell Death Caused by an Ethanolic Extract."
150. Anastasia Karioto and Anna Rita Bilia, "Hypericins as Potential Leads for New Therapeutics," *International Journal of Molecular Sciences* 11, no. 2 (2010): 562–94, doi:10.3390/ijms11020562.
151. Chad D. Cole et al., "Hypericin-Mediated Photodynamic Therapy of Pituitary Tumors: Preclinical Study in a GH4C1 Rat Tumor Model," *Journal of Neuro-Oncology* 87, no. 3 (May 2008): 255–61, doi:10.1007/s11060-007-9514-0; and Juergen Berlanda et al., "Comparative In Vitro Study on the Characteristics of Different Photosensitizers Employed in PDT," *Journal of Photochemistry and Photobiology B: Biology* 100, no. 3 (September 2, 2010): 173–80, doi:10.1016/j.jphotobiol.2010.06.004.
152. Ann Huygens et al., "In Vivo Accumulation of Different Hypericin Ion Pairs in the Urothelium of the Rat Bladder," *BJU International* 95, no. 3 (February 2005): 436–41, doi:10.1111/j.1464-410X.2005.05316.x; and A. Kamuhabwa et al., "Hypericin as a Potential Phototherapeutic Agent in Superficial Transitional Cell Carcinoma of the Bladder," *Photochemical and Photobiological Sciences* 3, no. 8 (August 2004): 772–80, doi:10.1039/b315586b.
153. Appolinary A. R. Kamuhabwa et al., "Biodistribution of Hypericin in Orthotopic Transitional Cell Carcinoma Bladder Tumors: Implication for Whole Bladder Wall Photodynamic Therapy," *International Journal of Cancer* 97, no. 2 (January 10, 2002): 253–60; Maike Kober, Kerstin Pohl, and Thomas Efferth, "Molecular Mechanisms Underlying St. John's Wort Drug Interactions," *Current Drug Metabolism* 9, no. 10 (December 2008): 1027–37.
154. A. A. Kamuhabwa et al., "Microscopic Quantification of Hypercin Fluorescence in an Orthotopic Rat Bladder Tumor Model after Intravesical Instillation," *International Journal of Oncology* 22, no. 4 (April 2003): 933–37.
155. Kamuhabwa et al., "Hypericin as a Potential Phototherapeutic Agent in Superficial Transitional Cell Carcinoma of the Bladder."
156. Malini Olivo, Hong-Yan Du, and Boon-Huat Bay, "Hypericin Lights up the Way for the Potential Treatment of Nasopharyngeal Cancer by Photodynamic Therapy," *Current Clinical Pharmacology* 1, no. 3 (September 2006): 217–22.
157. D. Kacerovská et al., "Photodynamic Therapy of Nonmelanoma Skin Cancer with Topical *Hypericum perforatum* Extract—A Pilot Study," *Photochemistry and Photobiology* 84, no. 3 (May/June 2008): 779–85, doi:10.1111/j.1751-1097.2007.00260.x.
158. Johannes Westendorf, Wolfgang Pfau, and Agnes Schulte, "Carcinogenicity and DNA Adduct Formation Observed in ACI Rats after Long-Term Treatment with Madder Root, *Rubia tinctorum* L.," *Carcinogenesis* 19, no. 12 (December 1998): 2163–8.
159. Eduardo De Stefani et al., "Non-alcoholic Beverages and Risk of Bladder Cancer in Uruguay," *BMC Cancer* 7 (2007): 57, doi:10.1186/1471-2407-7-57.
160. K. H. Kurth et al., "Treatment of Superficial Bladder Tumors: Achievements and Needs. The EORTC Genitourinary Group," *European Urology* 37, supplement 3 (2000): 1–9, doi:10.1159/000052386.
161. P. R. Venskutonis, M. Skemaite, and O. Ragazinskiene, "Radical Scavenging Capacity of *Agrimonia eupatoria* and *Agrimonia procera*," *Fitoterapia* 78, no. 2 (February 2007): 166–68, doi:10.1016/j.fitote.2006.10.002.
162. Helena S. Correia, Maria R. Batista, and Teresa C. Dinis, "The Activity of an Extract and Fraction of *Agrimonia eupatoria* L. against Reactive Species," *Biofactors* 29, no. 2–3 (2007): 91–104, doi:10.1002/biof.552029209.

163. H. Correia et al., "Polyphenolic Profile Characterization of *Agrimonia eupatoria* L. by HPLC with Different Detection Devices," *Biomedical Chromatography* 20, no. 1 (January 2006): 88–94, doi:10.1002/bmc.533.
164. Willmann Liang et al., "Inhibitory Effects of *Salviae miltiorrhizae Radix* (Danshen) and *Puerariae lobatae Radix* (Gegen) in Carbachol-Induced Rat Detrusor Smooth Muscle Contractility," *International Journal of Physiology, Pathophysiology and Pharmacology* 4, no. 1 (2012): 36–44.
165. Mayumi Suzuki et al., "Pharmacological Effects of Saw Palmetto Extract in the Lower Urinary Tract," *Acta Pharmacologica Sinica* 30, no. 3 (March 2009): 271–81, doi:10.1038/aps.2009.1.
166. Andrew John Tabner et al., "β-Adrenoreceptor Agonists in the Management of Pain Associated with Renal Colic: A Systematic Review," *BMJ Open* 6, no. 6 (2016): e011315, doi:10.1136/bmjopen-2016-011315.
167. Gianluigi D'Agostino et al., "Prejunctional Muscarinic Inhibitory Control of Acetylcholine Release in the Human Isolated Setrusor: Involvement of the M4 Receptor Subtype," *British Journal of Pharmacology* 129, no. 3 (February 2000): 493–500, doi:10.1038/sj.bjp.0703080.
168. Yuen-Keng Ng, William C. de Groat, and His-Yang Wu, "Muscarinic Regulation of Neonatal Rat Bladder Spontaneous Contractions," *American Journal of Physiology, Regulatory, Integrative and Comparative Physiology* 291, no. 4 (October 2006): R1049–59, doi:10.1152/ajpregu.0023.2006.
169. Linda P. Dwoskin and Peter A. Crooks, "A Novel Mechanism of Action and Potential Use for Lobeline as Treatment for Psychostimulant Abuse," *Biochemical Pharmacology* 63, no. 2 (January 15, 2002): 89–98, doi:10.1016/S0006-2952(01)00899-1; and Nichole M. Neugebauer et al., "Lobelane Decreases Methamphetamine Self-Administration in Rats," *European Journal of Pharmacology* 571, no. 1 (September 24, 2007): 33–38, doi:10.1016/j.ejphar.2007.06.003.
170. A. Bartolini, L. Di Cesare Mannelli, C. Ghelardini, "Analgesic and Antineuropathic Drugs Acting through Central Cholinergic Mechanisms," *Recent Patents on CNS Drug Discovery* 6, no. 2 (May 2011): 119–40.
171. Mazen Zaitouna et al., "Origin and Nature of Pelvic Ureter Innervation," *Neurourology and Urodynamics* 36, no. 2 (February 2017): 271–79, doi:10.1002/nau.22919.
172. Fukashi Yamamichi et al., "β-3 Adrenergic Receptors Could Be Significant Factors for Overactive Bladder-Related Symptoms," *International Journal of Clinical and Experimental Pathology* 8, no. 9 (September 1, 2015): 11863–70.
173. Luis Osorio et al., "Emergency Management of Ureteral Stones: Recent Advances," *Indian Journal of Urology* 24, no. 4 (October–December 2008): 461–66, doi:10.4103/0970-1591.44248.
174. F. Grases et al., "Urolithiasis and Phytotherapy," *International Urology and Nephrology* 26, no. 5 (September 1994): 507–11.
175. McDonald, *Physiologic Medication* (see n. 44): 90.
176. Samad E. J. Golzari et al., "Therapeutic Approaches for Renal Colic in the Emergency Department: A Review Article," *Anesthesiology and Pain Medicine* 4, no. 1 (February 2014): e16222, doi:10.5812/aapm.16222.
177. Verasing Muangman et al., "The Usage of *Andrographis paniculata* Following Extracorporeal Shock Wave Lithotripsy (ESWL)," *Journal of the Medical Association of Thailand* 78, no. 6 (June 1995): 310–13.
178. Grases, "Urolithiasis and Phytotherapy."
179. Akshaya Srikanth Bhagavathula et al., "Ammi Visnaga in Treatment of Urolithiasis and Hypertriglyceridemia," *Pharmacognosy Research* 7, no. 4 (October–December 2015): 397–400, doi:10.4103/0974-8490.167894.
180. Ravindra G. Mali and Avinash S. Dhake, "A Review on Herbal Antiasthmatics," *Oriental Pharmacy and Experimental Medicine* 11, no. 2 (August 2011): 77–90, doi:10.1007/s13596-011-0019-1.
181. Dehong Cao et al., "A Comparison of Nifedipine and Tamsulosin as Medical Expulsive Therapy for the Management of Lower Ureteral Stones without ESWL," *Scientific Reports* 4 (2014): 5254, doi:10.1038/srep05254; and Sam McClinton et al., "Use of Drug Therapy in the Management of Symptomatic Ureteric Stones in Hospitalized Adults (SUSPEND), a Multicentre, Placebo-Controlled, Randomized Trial of a Calcium-Channel Blocker (Nifedipine) and an A-Blocker (Tamsulosin): Study Protocol for a Randomized Controlled Trial," *Trials* 15 (June 2014): 238, doi:10.1186/1745-6215-15-238.
182. P. Vanachayangkul et al., "Prevention of Renal Crystal Deposition by an Extract of *Ammi visnaga* L. and Its Constituents Khellin and Visnagin in Hyperoxaluric Rats," *Urological Research* 39, no. 3 (June 2011): 189–95, doi:10.1186/1745-6215-15-238.
183. Sae Woong Kim, "Phytotherapy: Emerging Therapeutic Option in Urologic Disease," *Translational Andrology and Urology* 1, no. 3 (September 2012): 181–191, doi:10.3978/j.issn.2223-4683.2012.05.10.
184. Dominik Domanski et al., "Molecular Mechanism for Cellular Response to β-Escin and Its Therapeutic Implications," *PLoS ONE* 11, no. 10 (2016): e0164365, doi:10.1371/journal.pone.0164365.
185. N. Tugba Durlu-Kandilci and Alison F. Brading, "Involvement of Rho Kinase and Protein Kinase C in Carbachol-Induced Calcium Sensitization in B-Escin Skinned Rat and Guinea-Pig Bladders," *British Journal of Pharmacology* 148, no. 3 (June 2006): 376–84, doi:10.1038/sj.bjp.0706723.
186. Ines Lindner et al., "β-escin Has Potent Anti-allergic Efficacy and Reduces Allergic Airway Inflammation," *BMC Immunology* 11 (2010): 24, doi:10.1186/1471-2172-11-24.
187. Domanski, "Molecular Mechanism for Cellular Response to β-Escin."
188. Dieter Wetzel et al., "Escin/Diethylammonium Salicylate/Heparin Combination Gels for the Topical Treatment of Acute Impact Injuries: A Randomised, Double Blind, Placebo Controlled, Multicentre Study," *British Journal of Sports Medicine* 36, no. 3 (June 2002): 183–88.

189. Jan F. Stevens et al., "A Novel 2-Hydroxyflavanone from *Collinsonia canadensis*," *Journal of Natural Products* 62, no. 2 (February 1999): 392–94, doi:10.1021/np980421i.
190. Heidari Mahmoud Reza et al., "Effect of Methanolic Extract of *Hyoscymus niger* L. on the Seizure Induced by Picritoxin in Mice," *Pakistan Journal of Pharmaceutical Sciences* 22, no. 3 (July 2009): 308–12.
191. Anahita Alizadeh et al., "Black Henbane and Its Toxicity—A Descriptive Review," *Avicenna Journal of Phytomedicine* 4, no. 5 (September 2014): 297–311.
192. Reza, "Effect of Methanolic Extract of *Hyoscymus niger* L."
193. Anwarul Hassan Gilani et al., "Gastrointestinal, Selective Airways and Urinary Bladder Relaxant Effects of *Hyoscyamus niger* Are Mediated through Dual Blockade of Muscarinic Receptors and Ca2+ Channels," *Fundamental and Clinical Pharmacology* 22, no. 1 (February 2008): 87–99, doi:10.1111/j.1472-8206.2007.00561.x.
194. M. Cenk Gurbuz et al., "Efficacy of Three Different α 1-Adrenergic Blockers and Hyoscine N-Butylbromide for Distal Ureteral Stones," *International Brazilian Journal of Urology* 37, no. 2 (March/April 2011): 195–200, discussion 201–2.
195. Yu-Ming Zhang, Pei Chu, and Wen-Jin Wang, "PRISMA-Combined A-Blockers and Antimuscarinics for Ureteral Stent-Related Symptoms: A Meta-analysis," *Medicine* 96, no. 7 (2017): e6098, doi:10.1097/MD.0000000000006098.
196. Jonathan L. Edwards, "Diagnosis and Management of Benign Prostatic Hyperplasia," *American Family Physician* 77, no. 10 (May 15, 2008): 1403–10; Raj C. Dedhia, Elizabeth Calhoun, and Kevin T. McVary, "Impact of Phytotherapy on Utility Scores for 5 Benign Prostatic Hyperplasia / Lower Urinary Tract Symptoms Health States," *Journal of Urology* 179, no. 1 (January 2008): 220–25, doi:10.1016/j/juro.2007.08.152.
197. T. J. Williams et al., "In Vitro Alpha$_1$-Adrenoceptor Pharmacology of Ro 70-0004 and RS-100329, Novel Alpha$_1$A-Adrenoceptor Selective Antagonists," *British Journal of Pharmacology* 127, no. 1 (May 1999): 252–58, doi:10.1038/sj.bjp.0702541.
198. Yoko Omoto, "Estrogen Receptor-α Signaling in Growth of the Ventral Prostate: Comparison of Neonatal Growth and Post Castration Regrowth," *Endocrinology* 149, no. 9 (September 2008): 4421–27.
199. A. Hurtado et al., "Estrogen Receptor β Displays Cell Cycle-Dependent Expression and Regulates the G1 Phase through a Non-genomic Mechanism in Prostate Carcinoma Cells," *Cell Oncology* 30, no. 4 (2008): 349–65.
200. Stephen J. McPherson, S. J. Ellem, and G. P. Risbridger, "Estrogen-Regulated Development and Differentiation of the Prostate," *Differentiation* 76, no. 6 (July 2008): 660–70, doi:10.1111/j.1432-0436.2008.00291.x.
201. Gail S. Prins, "Endocrine Disruptors and Prostate Cancer Risk," *Endocrine-Related Cancer* 15, no. 3 (September 2008): 649–56, doi:10.1677/ERC-08-0043.
202. F. Branca and S. Lorenzetti, "Health Effects of Phytoestrogens," *Forum of Nutrition*, no. 57 (2005): 100–11; and Sara S. Strom et al., "Phytoestrogen Intake and Prostate Cancer: A Case-Control Study Using a New Database," *Nutrition and Cancer* 33, no. 1 (1999): 20–25, doi:10.1080/01635589909514743.
203. W. A. Fritz et al., "Genistein Alters Growth but Is Not Toxic to the Rat Prostate," *Journal of Nutrition* 132, no. 10 (October 2002): 3007–11.
204. Jonathan L. Edwards, "Diagnosis and Management of Benign Prostatic Hyperplasia," *American Family Physician* 77, no. 10 (May 15, 2008): 1403–10; Drew Keister and Randall Neal, "Managing BPH: When to Consider Surgery," *American Family Physician* 77, no. 10 (Mary 15, 2008): 1375–77.
205. A. Morani, M. Warner, and J. Å. Gustafsson, "Biological Functions and Clinical Implications of Oestrogen Receptors Alfa and β in Epithelial Tissues," *Journal of Internal Medicine* 264, no. 2 (August 2008): 128–42, doi:10.1111/j.1365-2796.2008.01976.x.
206. Risto Santti et al., "Phytoestrogens: Potential Endocrine Disruptors in Males," *Toxicology and Industrial Health* 14, no. 1–2 (January–April 1998): 223–37, doi:10.1177/074823379801400114.
207. Rachel L. Ruhlen et al., "Low Phytoestrogen Levels in Feed Increase Fetal Serum Estradiol Resulting in the 'Fetal Estrogenization Syndrome' and Obesity in CD-1 Mice," *Environmental Health Perspectives* 116, no. 3 (March 2008): 322–28, doi:10.1289/ehp.10448.
208. Claudia Montani et al., "Genistein Is an Efficient Estrogen in the Whole-Body throughout Mouse Development," *Toxicological Science* 103, no. 1 (May 2008): 57–67, doi:10.1093/toxsci/kfn021.
209. J. Drsata, "Enzyme Inhibition in the Drug Therapy of Benign Prostatic Hyperplasiam," *Casopis Lekaru Ceskych [Journal of Czech Physicians]* 141, no. 20 (October 11, 2002): 630–35.
210. Fouad K. Habib et al., "*Serenoa repens* (Permixon) Inhibits the 5α-Reductase Activity of Human Prostate Cancer Cell Lines without Interfering with PSA Expression," *International Journal of Cancer* 114, no. 2 (March 20, 2005): 190–94, doi:10.1002/ijc.20701.
211. Timonthy J. Wilt et al., "*Pygeum africanum* for Benign Prostatic Hyperplasia," *Cochrane Database of Systematic Review*, no. 1 (2002): CD001044, doi:10.1002/14651858.CD001044.
212. Yuan-Shan Zhu and Guang-Huan Sun, "5α-Reductase Isozymes in the Prostate," *Journal of Medical Science* 25, no. 1 (2005): 1–12.
213. Gustavo Francisco Gonzales et al., "Effect of Two Different Extracts of Red Maca in Male Rats with Testosterone-Induced Prostatic Hyperplasia," *Asian Journal of Andrology* 9, no. 2 (March 2007): 245–51, doi:10.1111/j.1745-7262.2007.00228.x; and Gustavo Francisco Gonzales et al., "Antagonistic Effect of *Lepidium meyenii* (Red Maca) on Prostatic Hyperplasia in Adult Mice," *Andrologia* 40, no. 3 (June 2008): 179–85.

214. M. Gasco et al., "Dose-Response Effect of Red Maca (*Lepidium meyenii*) on Benign Prostatic Hyperplasia Induced by Testosterone Enanthate," *Phytomedicine* 14, no. 7–8): 460–64, doi:10.1016/j.phymed.2006.12.003; and Gustavo Francisco Gonzales et al., "Red Maca (Lepidium Meyenii) Reduced Prostate Size in Rats," *Reproductive Biology and Endocrinology* 3 (January 20, 2005): 5, doi:10.1186/1477-7827-3-5.
215. Alan R. Kristal et al., "Dietary Patterns, Supplement Use, and the Risk of Symptomatic Benign Prostatic Hyperplasia: Results from the Prostate Cancer Prevention Trial," *American Journal of Epidemiology* 167, no. 8 (April 15, 2008): 925–34, doi:10.1093/aje/kwm389.
216. Paul F. Englehardt and Claus R. Riedl, "Effects of One-Year Treatment with Isoflavone Extract from Red Clover on Prostate, Liver Function, Sexual Function, and Quality of Life in Men with Elevated PSA Levels and Negative Prostate Biopsy Findings," *Urology* 71, no. 2 (February 2008): 185–90, doi:10.1016/j.urology.2007.08.068.
217. Ikuko Kjima et al., "Grape Seed Extract Is an Aromatase Inhibitor and a Suppressor of Aromatase Expression," *Cancer Research* 66, no. 11 (June 1, 2006): 5960–67, doi:10.1158/0008-5472-CAN-06-0053.
218. Julia E. Chrubasik et al., "A Comprehensive Review on the Stinging Nettle Effect and Efficacy Profiles," *Phytomedicine* 14, no. 7–8 (August 2007): 568–79, doi:10.1016/j.phymed.2007.03.014; Drsata, "Enzyme Inhibition in the Drug Therapy of Benign Prostatic Hyperplasiam" (see n. 209); and J. J. Lichius and C. Muth, "The Inhibiting Effects of *Urtica dioica* Root Extracts on Experimentally Induced Prostatic Hyperplasia in the Mouse," *Planta Medica* 63, no. 4 (August 1997): 307–10, doi:10.1055/s-2006-957688.
219. D. J. Hryb et al., "The Effect of Extracts of the Roots of the Stinging Nettle (*Urtica dioica*) on the Interaction of SHBG with Its Receptor on Human Prostatic Membranes," *Planta Medica* 61, no. 1 (February 1995): 31–32, doi:10.1055/s-2006-957993.
220. Chrubasik, "A Comprehensive Review on the Stinging Nettle Effect and Efficacy Profiles."
221. Ilker Durak et al., "Aqueous Extract of *Urtica dioica* Makes Significant Inhibition on Adenosine Deaminase Activity in Prostate Tissue from Patients with Prostate Cancer," *Cancer Biology and Therapy* 3, no. 9 (September 2004): 855–57.
222. Toshihiko Hirano, M. Homma, and K. Oka, "Effects of Stinging Nettle Root Extracts and Their Steroidal Components on the Na+,K(+)-Atpase of the Benign Prostatic Hyperplasia," *Planta Medica* 60, no. 1 (February 1994): 30–33, doi:10.1055/s-2006-959402.
223. Q. Zhang et al., "Effects of the Polysaccharide Fraction of Urtica Fissa on Castrated Rat Prostate Hyperplasia Induced by Testosterone Propionate," *Phytomedicine* 15, no. 9, (September 2008): 722–27, doi:10.1016/j.phymed.2007.12.005; and Lutz Konrad et al., "Antiproliferative Effect on Human Prostate Cancer Cells by a Stinging Nettle Root (*Urtica dioica*) Extract," *Planta Medica* 66, no. 1 (February 2000): 44–47, doi:10.1055/s-2000-11117.
224. Mohammad Reza Safarinejad, "*Urtica dioica* for Treatment of Benign Prostatic Hyperplasia: A Prospective, Randomized, Double-Blind, Placebo-Controlled, Crossover Study," *Journal of Herbal Pharmacotherapy* 5, no. 4 (2005): 1–11, doi:10.1080/J157v05n04_01.
225. Nikolai A. Lopatkin et al., "Combined Extract of Sabal Palm and Nettle in the Treatment of Patients with Lower Urinary Tract Symptoms in Double Blind, Placebo-Controlled Trial," *Urologiia*, no. 2 (March/April 2006): 12, 14–9; Safarinejad, "*Urtica dioica* for Treatment of Benign Prostatic Hyperplasia"; U. Engelmann et al., "Efficacy and Safety of a Combination of *Sabal* and *Urtica* Extract in Lower Urinary Tract Symptoms. A Randomized, Double-Blind Study versus Tamsulosin," *Arzneimittel Forschung / Drug Research* 56, no. 3 (2006): 222–29, doi:10.1055/s-0031-1296714; Nikolai A. Lopatkin et al., "Long-Term Efficacy and Safety of a Combination of *Sabal* and *Urtica* Extract for Lower Urinary Tract Symptoms—A Placebo-Controlled, Double-Blind, Multicenter Trial," *World Journal of Urology* 23, no. 2 (June 2005): 139–46, doi:10.1007/s00345-005-0501-9; Tommaso Cai et al., "*Serenoa repens* Associated with *Urtica dioica* (Prostamev) and Curcumin and Quercitin (Flogmev) Extracts Are Able to Improve the Efficacy of Prulifloxacin in Bacterial Prostatitis Patients: Results from a Prospective Randomised Study," *International Journal of Antimicrobial Agents* 33, no. 6 (June 2009): 549–53, doi:10.1016/j.ijantimicag.2008.11.012; G. Popa, H. Hägele-Kaddour, and C. Walther, "Efficacy of a Combined *Sabal-Urtica* Preparation in the Symptomatic Treatment of Benign Prostatic Hyperplasia. Results of a Placebo-Controlled Double-Blind Study," *MMW Fortschritter der Medizin* 147, supplement 3 (October 6, 2005): 103–8; and Nikolai A. Lopatkin et al., "Efficacy and Safety of a Combination of *Sabal* and *Urtica* Extract in Lower Urinary Tract Symptoms—Long-Term Follow-Up of a Placebo-Controlled, Double-Blind, Multicenter Trial," *International Urology and Nephrology* 39, no. 4 (2007): 1137–46, doi:10.1007/s11255-006-9173-7.
226. Engelmann, "Efficacy and Safety of a Combination of *Sabal* and *Urtica* Extract in Lower Urinary Tract Symptoms. A Randomized, Double-Blind Study versus Tamsulosin," (see n. 231).
227. Lopatkin, "Efficacy and Safety of a Combination of *Sabal* and *Urtica* Extract in Lower Urinary Tract Symptoms—Long-Term Follow-Up."
228. Canan Aldirmaz Agartan et al., "Protection of Urinary Bladder Function by Grape Suspension," *Phytotherapy Research* 18, no. 12 (December 2004): 1013–18, doi:10.1002/ptr.1620.
229. Rumi Fujita et al., "Anti-androgenic Activities of *Ganoderma lucidum*," *Journal of Ethnopharmacology* 102, no. 1 (October 31, 2005): 107–12, doi:10.1016/j.jep.2005.05.041.
230. Noriko Hirata et al., "Testosterone 5α-Reductase Inhibitory Active Constituents of *Piper nigrum* Leaf," *Biological*

and Pharmaceutical Bulletin 30, no. 12 (December 2007): 2402–5; and Y. C. Kao et al., "Molecular Basis of the Inhibition of Human Aromatase (Estrogen Synthetase) by Flavone and Isoflavone Phytoestrogens: A Site-Directed Mutagenesis Study," *Environmental Health Perspectives* 106, no. 2 (February 1998): 85–92.

231. Jianying Yam, Matthias Heinrich Kreuter, and Jeurgen Drewe, "*Piper cubeba* Targets Multiple Aspects of the Androgen-Signalling Pathway. A Potential Phytotherapy against Prostate Cancer Growth?" *Planta Medica* 74, no. 1 (January 2008): 33–38, doi:10.1055/s-2007-993758.
232. Drsata, "Enzyme Inhibition in the Drug Therapy of Benign Prostatic Hyperplasiam" (see n. 209).
233. Drsata, "Enzyme Inhibition in the Drug Therapy of Benign Prostatic Hyperplasiam" (see n. 209).
234. Drsata, "Enzyme Inhibition in the Drug Therapy of Benign Prostatic Hyperplasiam" (see n. 209).
235. Ezer A. Melo et al., "Evaluating the Efficiency of a Combination of *Pygeum africanum* and Stinging Nettle (*Urtica dioica*) Extracts in Treating Benign Prostatic Hyperplasia (BPH): Double-Blind, Randomized, Placebo Controlled Trial," *International Brazilian Journal of Urology* 28, no. 5 (September/October 2002): 418–25; and Alan D. Edgar et al., "A Critical Review of the Pharmacology of the Plant Extract of *Pygeum africanum* in the Treatment of LUTS," *Neurourology and Urodynamics* 26, no. 4 (2007): 458–63, discussion 464, doi:10.1002/nau.20136.
236. Sonja Schleich et al., "Extracts from *Pygeum africanum* and Other Ethnobotanical Species with Antiandrogenic Activity," *Planta Medica* 72, no. 9 (July 2006): 807–13, doi:10.1055/s-2006-946638; M. Papaioannou et al., "The Natural Compound Atraric Acid Is an Antagonist of the Human Androgen Receptor Inhibiting Cellular Invasiveness and Prostate Cancer Cell Growth," *Journal of Cellular and Molecular Medicine* 13, no. 8B (August 2009): 2210–23, doi:10.1111/j.1582-4934.2008.00426.x; and Sonja Schleich et al., "Activity-Guided Isolation of an Antiandrogenic Compound of *Pygeum africanum*," *Planta Medica* 72, no. 6 (May 2006): 547–51.
237. Yosh Yoshimura et al., "Effect of *Pygeum africanum* Tadenan on Micturition and Prostate Growth of the Rat Secondary to Coadministered Treatment and Post-treatment with Dihydrotestosterone," *Urology* 61, no. 2 (February 2003): 474–78; Papaioannou, "The Natural Compound Atraric Acid Is an Antagonist of the Human Androgen Receptor"; Schleich, "Activity-Guided Isolation of an Antiandrogenic Compound of *Pygeum africanum*"; Schleich, "Extracts from *Pygeum africanum* and Other Ethnobotanical Species"; and Wiebke Hessenkemper et al., "A Natural Androgen Receptor Antagonist Induces Cellular Senescence in Prostate Cancer Cells," *Molecular Endocrinology* 28, no. 11 (November 2014): 1831–40, doi:10.1210/me.2014-1170.
238. Edgar, "A Critical Review of the Pharmacology of the Plant Extract of *Pygeum africanum* in the Treatment of LUTS" (see n. 231); and Delphine Boulbès et al., "*Pygeum africanum* Extract Inhibits Proliferation of Human Cultured Prostatic Fibroblasts and Myofibroblasts," *BJU International* 98, no. 5 (November 2006): 1106–13, doi:10.111/j.1464-410X.2006.06483.x.
239. Anna Santa María Margalef et al., "Antimitogenic Effect of *Pygeum africanum* Extracts on Human Prostatic Cancer Cell Lines and Explants from Benign Prostatic Hyperplasia," *Archivos Españoles de Urología* 56, no. 4 (May 2003): 369–78.
240. Nader S. Shenouda et al., "Phytosterol *Pygeum africanum* Regulates Prostate Cancer In Vitro and In Vivo," *Endocrine* 31, no. 1 (February 2007): 72–81.
241. Boulbès, "*Pygeum africanum* Extract Inhibits Proliferation of Human Cultured Prostatic Fibroblasts and Myofibroblasts."
242. Rosa M. Solano et al., "Effects of *Pygeum africanum* Extract (Tadenan) on Vasoactive Intestinal Peptide Receptors, G Proteins, and Adenylyl Cyclase in Rat Ventral Prostate," *Prostate* 45, no. 3 (November 1, 2000): 245–52.
243. Hyeh-Jean Jeong et al., "Inhibition of Aromatase Activity by Flavonoids," *Archives of Pharmacal Research* 22, no. 3 (June 1999): 309–12; and Shiuan Chen et al., "Structure-Function Studies of Aromatase and Its Inhibitors: A Progress Report," *Journal of Steroid Biochemistry and Molecular Biology* 86, no. 3–5 (September 2003): 231–37.
244. Baiba J. Grube et al., "White Button Mushroom Phytochemicals Inhibit Aromatase Activity and Breast Cancer Cell Proliferation," *Journal of Nutrition* 131, no. 12 (December 2001): 3288–93; and Shiuan Chen et al., "Anti-aromatase Activity of Phytochemicals in White Button Mushrooms (*Agaricus bisporus*)," *Cancer Research* 66, no. 24 (December 15, 2006): 12026–34, doi:10.1158/0008-5472.CAN-06-2206.
245. Monika Papiez, Monika Gancarczyk, and Barbara Bilińska, "The Compounds from the Hollyhock Extract (*Althaea rosea* Cav. var. *Nigra*) Affect the Aromatization in Rat Testicular Cells In Vivo and In Vitro," *Folia Histochemica et Cytobiologica* 40, no. 4 (2002): 353–59.
246. K. Satoh et al., "Inhibition of Aromatase Activity by Green Tea Extract Catechins and Their Endocrinological Effects of Oral Administration in Rats," *Food and Chemical Toxicology* 40, no. 7 (July 2002): 925–33; and Ken-ichi Kimura et al., "Inhibition of 17α-hydroxylase/C17,20-lyase (CYP17) from Rat Testis by Green Tea Catechins and Black Tea Theaflavins," *Bioscience, Biotechnology, and Biochemistry* 71, no. 9 (September 2007): 2325–28, doi:10.1271/bbb.70258.
247. S. M. Hsia et al., "Effects of Adlay (*Coix lachryma-jobi* L. var. *ma-yuen stapf.*) Hull Extracts on the Secretion of Progesterone and Estradiol In Vivo and In Vitro," *Experimental Biology and Medicine* 232, no. 9 (October 2007): 1181–94, doi:10.3181/0612-RM-306.
248. T. K. Lee et al., "Inhibitory Effects of *Scutellaria barbata* D. Don. and *Euonymus alatus* Sieb. on Aromatase Activity of Human Leiomyomal Cells," *Immunopharmacology and Immunotoxicology* 26, no. 3 (August 2004): 315–27.

249. Nam Deuk Kim et al., "Chemopreventive and Adjuvant Therapeutic Potential of Pomegranate (Punica Granatum) for Human Breast Cancer," *Breast Cancer Research and Treatment* 71, no. 3 (February 2002): 203–17.
250. Lee, "Inhibitory Effects of *Scutellaria barbata* D. Don. and *Euonymus alatus* Sieb."
251. C. Morrissey and R. W. Watson, "Phytoestrogens and Prostate Cancer," *Current Drug Targets* 4, no. 3 (April 2003): 231–41.
252. A. Krazeisen et al., "Phytoestrogens Inhibit Human 17β-Hydroxysteroid Dehydrogenase Type 5," *Molecular and Cellular Endocrinology* 171, no. 1–2 (January 22, 2001): 151–62; and J. C. Le Bail et al., "Effects of Phytoestrogens on Aromatase, 3β and 17β-Hydroxysteroid Dehydrogenase Activities and Human Breast Cancer Cells," *Life Sciences* 66, no. 14 (February 25, 2000): 1281–91.
253. Matej Sova et al., "Flavonoids and Cinnamic Acid Esters as Inhibitors of Fungal 17β-Hydroxysteroid Dehydrogenase: A Synthesis, QSAR and Modelling Study," *Bioorganic and Medicinal Chemistry* 14, no. 22 (November 15, 2006): 7404–18, doi:10.1016/j.bmc.2006.07.027.
254. Thomas E. Spires et al., "Identification of Novel Functional Inhibitors of 17β-Hydroxysteroid Dehydrogenase Type III (17β-HSD3)," *Prostate* 65, no. 2 (October 1, 2005): 159–70, doi:10.1002/pros.20279.
255. Maxine Gossell-Williams, A. Davis, and N. O'Connor, "Inhibition of Testosterone-Induced Hyperplasia of the Prostate of Sprague-Dawley Rats by Pumpkin Seed Oil," *Journal of Medicinal Food* 9, no. 2 (Summer 2006): 284–86. doi:10.1089/jmf.2006.9.284; and Yuh-Shyan Tsai et al., "Pumpkin Seed Oil and Phytosterol-F Can Block Testosterone/Prazosin-Induced Prostate Growth in Rats," *Urologia Internationalis* 77, no. 3 (2006): 269–74, doi:10.1159/000094821.
256. M. Friederich, C. Theurer, and G. Schiebel-Schlosser, "Prosta Fink Forte Capsules in the Treatment of Benign Prostatic Hyperplasia. Multicentric Surveillance Study in 2245 Patients," *Forschende Komplementärmedzin und Klassische Naturheilkunde [Research in Complementary and Natural Classical Medicine]* 7, no. 4 (August 2000): 200–4; B.-E. Carbin, B. Larsson, and O. Lindahl, "Treatment of Benign Prostatic Hyperplasia with Phytosterols," *BJU International* 66, no. 6 (December 1990): 639–41; and E. Ryan et al., "Phytosterol, Squalene, Tocopherol Content and Fatty Acid Profile of Selected Seeds, Grains, and Legumes," *Plant Foods for Human Nutrition* 62, no. 3 (September 2007): 85–91, doi:10.1007/s11130-007-0046-8.
257. G. Pourmand et al., "The Risk Factors of Prostate Cancer: A Multicentric Case-Control Study in Iran," *Asian Pacific Journal of Cancer Prevention* 8, no. 3 (July–September 2007): 422–28.
258. Donald T. Wigle, "Role of Hormonal and Other Factors in Human Prostate Cancer," *Journal of Toxicology and Environmental Health B: Critical Reviews* 11, no. 3–4 (March 2008): 242–59, doi:10.1080/10937400701873548.
259. Antonella Dewell et al., "A Very-Low-Fat Vegan Diet Increases Intake of Protective Dietary Factors and Decreases Intake of Pathogenic Dietary Factors," *Journal of the American Dietic Association* 108, no. 2 (February 2008): 347–56, doi:10.1016/j.jada.2007.10.044.
260. Gaelle Fromont et al., "Differential Expression of Genes Related to Androgen and Estrogen Metabolism in Hereditary versus Sporadic Prostate Cancer," *Cancer Epidemiology, Biomarkers and Prevention* 17, no. 6 (June 2008): 1505–9, doi:10.1158/1055-9965.EPI-07-2778; Georges Pelletier, "Expression of Steroidogenic Enzymes and Sex-Steroid Receptors in Human Prostate," *Best Practice and Research: Clinical Endocrinology and Metabolism* 22, no. 2 (April 2008): 223–38; and K. Onsory et al., "Hormone Receptor-Related Gene Polymorphisms and Prostate Cancer Risk in North Indian Population," *Molecular and Cellular Biochemistry* 314, no. 1–2 (July 2008): 25–35, doi:10.1007/s11010-008-9761-1.
261. Joanna M. Day et al., "Design and Validation of Specific Inhibitors of 17β-Hydroxysteroid Dehydrogenases for Therapeutic Application in Breast and Prostate Cancer, and in Endometriosis," *Endocrine-Related Cancer* 15, no. 3 (September 2008): 665–92, doi:10.1677/ERC-08-0042.
262. Schleich, "Extracts from *Pygeum africanum* and Other Ethnobotanical Species" (see n. 232).
263. Margalef, "Antimitogenic Effect of *Pygeum africanum* Extracts on Human Prostatic Cancer Cell Lines and Explants" (see n. 235).
264. Edgar, "A Critical Review of the Pharmacology of the Plant Extract of *Pygeum africanum* in the Treatment of LUTS" (see n. 231); Boulbès, "*Pygeum africanum* Extract Inhibits Proliferation of Human Cultured Prostatic Fibroblasts and Myofibroblasts" (see n. 234); and Shenouda, "Phytosterol *Pygeum africanum* Regulates Prostate Cancer" (see n. 236).
265. Papaioannou, "The Natural Compound Atraric Acid Is an Antagonist of the Human Androgen Receptor" (see n. 232); Schleich, "Activity-Guided Isolation of an Antiandrogenic Compound of *Pygeum africanum*" (see n. 232).
266. Yang Yang et al., "Saw Palmetto Induces Growth Arrest and Apoptosis of Androgen-Dependent Prostate Cancer LNCaP Cells via Inactivation of STAT 3 and Androgen Receptor Signaling," *International Journal of Oncology* 31, no. 3 (September 2007): 593–600; N. A. Lopatkin et al., "Results of a Multicenter Trial of *Serenoa repens* Extract (Permixon) in Patients with Chronic Abacterial Prostatitis," *Urologiia*, no. 5 (September/October 2007): 3–7; and Catherine Ulbricht et al., "Evidence-Based Systematic Review of Saw Palmetto by the Natural Standard Research Collaboration," *Journal of the Society for Integrative Oncology* 4, no. 4 (Fall 2006): 170–86.
267. Hostanska, "Evaluation of Cell Death Caused by an Ethanolic Extract" (see n. 161).
268. Teri L. Wadsworth et al., "Effects of Dietary Saw Palmetto on the Prostate of Transgenic Adenocarcinoma of the Mouse Prostate Model (TRAMP)," *Prostate* 67, no. 6 (May 1, 2007): 661–73, doi:10.1002/pros.20552.

269. Konrad, "Antiproliferative Effect on Human Prostate Cancer Cells by a Stinging Nettle Root" (see n. 226).
270. Rajesh L. Thangapazham et al., "Androgen Responsive and Refractory Prostate Cancer Cells Exhibit Distinct Curcumin Regulated Transcriptome," *Cancer Biology and Therapy* 7, no. 9 (September 2008): 1427–35, doi:10.1016/S0022-5347(08)60553-4.
271. Dong Sheng Ming et al., "Bioactive Compounds from *Rhodiola rosea* (Crassulaceae)," *Phytotherapy Research* 19, no. 9 (September 2005): 740–43, doi:10.1002/ptr.1597; E. Skopińska-Różewska, "The Influence of *Rhodiola quadrifida* 50% Hydro-alcoholic Extract and Salidroside on Tumor-Induced Angiogenesis in Mice," *Polish Journal of Veterinary Science* 11, no. 2 (2008): 97–104; and E. Skopińska-Różewska, "The Effect of *Rhodiola quadrifida* Extracts on Cellular Immunity in Mice and Rats," *Polish Journal of Veterinary Science* 11, no. 2 (2008): 105–11.
272. Robert Michael Hermann et al., "*In Vitro* Studies on the Modification of Low-Dose Hyper-radiosensitivity in Prostate Cancer Cells by Incubation with Genistein and Estradiol," *Radiation Oncology* 3, no. 1 (July 14, 2008): 19, doi:10.1186/1748-717X-3-19.
273. Kentaro Matsumura et al., "Involvement of the Estrogen Receptor β in Genistein-Induced Expression of P21(Waf1/Cip1) in PC-3 Prostate Cancer Cells," *Anticancer Research* 28, no. 2A (March/April 2008): 709–14.
274. Matsumura, "Involvement of the Estrogen Receptor β"; and Nobuyuki Kikuno et al., "Genistein Mediated Histone Acetylation and Demethylation Activates Tumor Suppressor Genes in Prostate Cancer Cells," *International Journal of Cancer* 123, no. 3 (August 1, 2008): 552–60, doi:10.1002/ijc.23590.
275. Nagi B. Kumar et al., "Safety of Purified Isoflavones in Men with Clinically Localized Prostate Cancer," *Nutrition and Cancer* 59, no. 2 (2007): 169–75, doi:10.1080/01635580701432660.
276. John M. Pendleton et al., "Phase II Trial of Isoflavone in Prostate-Specific Antigen Recurrent Prostate Cancer after Previous Local Therapy," *BMC Cancer* 8 (May 11, 2008): 132, doi:10.1186/1471-2407-8-132.
277. M. Weisskopf et al., "A *Vitex agnus-castus* Extract Inhibits Cell Growth and Induces Apoptosis in Prostate Epithelial Cell Lines," *Planta Medica* 71, no. 10 (October 2005): 910–16, doi:10.1055/s-2005-871235.
278. Thomas T. Wang et al., "Differential Effects of Resveratrol on Androgen-Responsive LNCaP Human Prostate Cancer Cells In Vitro and In Vivo," *Carcinogenesis* 29, no. 10 (October 2008): 2001–10, doi:10.1093/carcin/bgn131.
279. Richard B. van Breemen and Natasa Pajkovic, "Multitargeted Therapy of Cancer by Lycopene," *Cancer Letters* 269, no. 2 (October 8, 2008): 339–51, doi:10.1016/j.canlet.2008.05.016.
280. Valeri V. Mossine, P. Chopra, and T. P. Mawhinney, "Interaction of Tomato Lycopene and Ketosamine against Rat Prostate Tumorigenesis," *Cancer Research* 68, no. 11 (June 1, 2008): 4384–91, doi:10.1158/0008-5472.CAN-08-0108.
281. Silke Schwarz et al., "Lycopene Inhibits Disease Progression in Patients with Benign Prostate Hyperplasia," *Journal of Nutrition* 138, no. 1 (June 2008): 49–53.
282. Daniel Peternac et al., "Agents Used for Chemoprevention of Prostate Cancer May Influence PSA Secretion Independently of Cell Growth in the LNCaP Model of Human Prostate Cancer Progression," *Prostate* 68, no. 12 (September 1, 2008): 1307–18, doi:10.1002/pros.20795.
283. Xunxian Liu et al., "Lycopene Inhibits IGF-I Signal Transduction and Growth in Normal Prostate Epithelial Cells by Decreasing DHT-Modulated IGF-I Production in Co-cultured Reactive Stromal Cells," *Carcinogenesis* 29, no. 4 (April 1, 2008): 816–23, doi:10.1093/carcin/bgn011.
284. Timothy Wilt et al., "Cernilton for Benign Prostatic Hyperplasia," *Cochrane Database of Systemic Reviews*, no. 2 (2000): CD001042, doi:10.1002/14651858.CD001042; Roderick MacDonald et al., "A Systematic Review of Cernilton for the Treatment of Benign Prostatic Hyperplasia," *BJU International* 85, no. 7 (May 2000): 836–41; E. W. Rugendorff et al., "Results of Treatment with Pollen Extract (Cernilton N) in Chronic Prostatitis and Prostatodynia," *BJU International* 71, no. 4 (April 1993): 433–38, doi:10.1111/j.1464-410X.1993.tb15988.x; A. Aoki et al., "Clinical Evaluation of the Effect of Tamsulosin Hydrochloride and Cernitin Pollen Extract on Urinary Disturbance Associated with Benign Prostatic Hyperplasia in a Multicentered Study," *Hinyokika Kiyo* 48, no. 5 (May 2002): 259–67; Toshiyuki Kamijo, Shigeru Sato, and Tadaichi Mitamura, "Effect of Cernitin Pollen-Extract on Experimental Nonbacterial Prostatitis in Rats," *Prostate* 49, no. 2 (October 1, 2001): 122–31, doi:10.1002/pros.1126; A. C. Buck et al., "Treatment of Outflow Tract Obstruction Due to Benign Prostatic Hyperplasia with the Pollen Extract, Cernilton. A Double-Blind, Placebo-Controlled Study," *BJU International* 66, no. 4 (October 1990): 398–404, doi:10.1111/j.1464-410X.1990.tb14962.x; and Nadeem Talpur et al., "Comparison of Saw Palmetto (Extract and Whole Berry) and Cernitin on Prostate Growth in Rats," *Molecular and Cellular Biochemistry* 250, no. 1–2 (August 2003): 21–26, doi:10.1023/A:1024988929.
285. Yao-Dong Wu and Yi-Jia Lou, "A Steroid Fraction of Chloroform Extract from Bee Pollen of *Brassica campestris* Induces Apoptosis in Human Prostate Cancer PC-3 Cells," *Phytotherapy Research* 21, no. 11 (November 2007): 1087–91, doi:10.1002/ptr.2235.
286. Xin Zhang et al., "Isolation and Characterization of a Cyclic Hydroxamic Acid from a Pollen Extract, which Inhibits Cancerous Cell Growth In Vitro," *Journal of Medicinal Chemistry* 38, no. 4 (February 17, 1995): 735–38; and Yao-Dong Wu and Yi-Jia Lou, "Brassinolide, a Plant Sterol from Pollen of *Brassica napus* L., Induces Apoptosis in Human Prostate Cancer PC-3 Cells," *Pharmazie* 62, no. 5 (May 2007): 392–95.
287. H. Y. Han et al., "Down-Regulation of Prostate Specific Antigen in LNCaP Cells by Flavonoids from the Pollen of

Brassica napus L.," *Phytomedicine* 14, no. 5 (May 21, 2007): 338–43, doi:10.1016/j.phymed.2006.09.005.

288. Fouad K. Habib et al., "Identification of a Prostate Inhibitory Substance in a Pollen Extract," *Prostate* 26, no. 3 (March 1995): 133–39, doi:10.1002/pros.2990260305.
289. F. K. Habib et al., "In Vitro Evaluation of the Pollen Extract, Cernitin T-60, in the Regulation of Prostate Cell Growth," *BJU International* 66, no. 4 (October 1990): 393–97, doi:10.1111/j.1464-410X.1990.tb14961.x.
290. Wadsworth, "Effects of Dietary Saw Palmetto on the Prostate of Transgenic Adenocarcinoma" (see n. 274).
291. Mark L. Anderson, "A Preliminary Investigation of the Enzymatic Inhibition of 5α-Reduction and Growth of Prostatic Carcinoma Cell Line LNCaP-FGC by Natural Astaxanthin and Saw Palmetto Lipid Extract In Vitro," *Journal of Herbal Pharmacotherapy* 5, no. 1 (2005): 17–26, doi:10.1080/J157v05n01_03.
292. Vittorio Margi et al., "Activity of Serenoa Repens, Lycopene and Selenium on Prostatic Disease: Evidences and Hypotheses," *Archivio Italiano di Urologia e Andrologia [Archives of Italian Urology and Andrology]* 80, no. 2 (June 2008): 65–78.
293. Steven C. Gross et al., "Antineoplastic Activity of *Solidago virgaurea* on Prostatic Tumor Cells in an SCID Mouse Model," *Nutrition and Cancer* 43, no. 1 (2002): 76–81, doi:10.1207/S15327914NC431_9.
294. Michel A. Pontari and Michael R. Ruggieri, "Mechanisms in Prostatitis / Chronic Pelvic Pain Syndrome," *Journal of Urology* 179, no. 5, supplement (May 2008): S61–67, doi:10.1016/j.juro.2008.03.139.
295. W. R. Fair and R. F. Parrish, "Antibacterial Substances in Prostatic Fluid," *Progress in Clinical and Biological Research* 75A (1981): 247–64.
296. H. Zhao et al., "Changes of Seminal Parameters, Zinc Concentration and Antibacterial Activity in Patients with Non-inflammatory Chronic Prostatitis / Chronic Pelvic Pain Syndrome," *Zhonghua Nan Ke Xue [National Journal of Andrology]* 14, no. 6 (June 2008): 530–32.
297. W. Weidner et al., "Ureaplasmal Infections of the Male Urogenital Tract, in Particular Prostatitis, and Semen Quality," *Urologia Internationalis* 40, no. 1 (1985): 5–9, doi:10.1159/000281023.
298. Y. H. Cho et al., "Antibacterial Effect of Intraprostatic Zinc Injection in a Rat Model of Chronic Bacterial Prostatitis," *International Journal of Antimicrobial Agents* 19, no. 6 (June 2002): 576–82.
299. C. Lowell Parsons, "Quantifying Symptoms in Men with Interstitial Cystitis / Prostatitis, and Its Correlation with Potassium-Sensitivity Testing," *BJU International* 95, no. 1 (January 2005): 86–90.
300. C. Lowell Parsons, "Intravesical Potassium Sensitivity in Patients with Prostatitis," *Journal of Urology* 168, no. 3 (September 2002): 1054–57, doi:10.1097/01.ju.0000025143.61057.13.
301. C. Lowell Parsons, "The Role of a Leaky Epithelium and Potassium in the Generation of Bladder Symptoms in Interstitial Cystitis / Overactive Bladder, Urethral Syndrome, Prostatitis and Gynaecological Chronic Pelvic Pain," *BJU International* 107, no. 3 (February 2011): 370–75, doi:10.1111/j.1464-410X.2010.09843.x.
302. C. Lowell Parsons, "Prostatitis, Interstitial Cystitis, Chronic Pelvic Pain, and Urethral Syndrome Share a Common Pathophysiology: Lower Urinary Dysfunctional Epithelium and Potassium Recycling," *Urology* 62, no. 6 (December 2003): 976–82, doi:10.1016/S0090-4295(03)00774-X.
303. Matthias F. Melzig, "Goldenrod—A Classical Exponent in the Urological Phytotherapy," *Wiener Medizinische Wochenschrift* 154, no. 21–22 (November 2004): 523–27.
304. Verena E. Borchert et al., "Extracts from *Rhois aromatica* and *Solidaginis virgaurea* Inhibit Rat and Human Bladder Contraction," *Naunyn-Schmiedeberg's Archives of Pharmacology* 369, no. 3 (March 2004): 281–86, doi:10.1007/s00210-004-0869-x.
305. Florian M. E. Wagenlehner et al., "Pollen Extract for Chronic Prostatitis—Chronic Pelvic Pain Syndrome," *Urologic Clinics of North America* 38, no. 3 (August 2011): 285–92, doi:10.1016/j.ucl.2011.04.004.
306. Florian M. E. Wagenlehner, "A Pollen Extract (Cernilton) in Patients with Inflammatory Chronic Prostatitis–Chronic Pelvic Pain Syndrome: A Multicentre, Randomised, Prospective, Double-Blind, Placebo-Controlled Phase 3 Study," *European Urology* 56, no. 3 (September 2009): 544–51, doi:10.1016/j.eururo.2009.05.046.
307. Gross, "Antineoplastic Activity of *Solidago virgaurea* on Prostatic Tumor Cells" (see n. 296).
308. P. E. Munday and S. Savage, "Cymalon in the Management of Urinary Tract Symptoms," *Genitourinary Medicine* 66, no. 6 (December 1990): 461; and J. B. Spooner, "Alkalinization in the Management of Cystitis," *Journal of International Medical Research* 12, no. 1 (1984): 30–34, doi:10.1177/030006058401200105.
309. J. N. Krieger and D. E. Riley, "Prostatitis: What Is the Role of Infection," *International Journal of Antimicrobial Agents* 19, no. 6 (June 2002): 475–79; I. Szöke et al., "The Possible Role of Anaerobic Bacteria in Chronic Prostatitis," *International Journal of Andrology* 21, no. 3 (June 1998): 163–68, doi:10.1111/j.1365-2605.1998.00110.x; and W. Weidner, H. G. Schiefer, and H. Krauss, "Role of *Chlamydia trachomatis* and Mycoplasmas in Chronic Prostatitis. A Review," *Urologia Internationalis* 43, no. 3 (1988): 167–73, doi:10.1159/000281331.
310. Douglas N. Fish and Larry H. Danziger, "Neglected Pathogens: Bacterial Infections in Persons with Human Immunodeficiency Virus Infection: A Review of the Literature," *Pharmacotherapy* 13, no. 5 (September/October 1993): 415–39, doi:10.1002/j.1875-9114.1993.tb04303.x; and J. N. Krieger and L. A. McGonagle, "Diagnostic Considerations and Interpretation of Microbiological

Findings for Evaluation of Chronic Prostatitis," *Journal of Clinical Microbiology* 27, no. 10 (October 1989): 2240–44.

311. Ghil Suk Yoon et al., "Cytomegalovirus Prostatitis: A Series of 4 Cases," *International Journal of Surgical Pathology* 18, no. 1 (February 2010): 55–59, doi:10.1177/1066896908321182.
312. Zhansong Zhou et al., "Detection of Nanobacteria Infection in Type III Prostatitis," *Urology* 71, no. 6 (June 2008): 1091–95, doi:10.1016/j.urology.2008.02.041; and L. Z. Zegarra Montes et al., "Semen and Urine Culture in the Diagnosis of Chronic Bacterial Prostatitis," *International Brazilian Journal of Urology* 34, no. 1 (January/February 2008): 30–37, discussion 38–40, doi:10.1590/S1677-55382008000100006.
313. Brent C. Taylor et al., "Excessive Antibiotic Use in Men with Prostatitis," *American Journal of Medicine* 121, no. 5 (May 2008): 444–49, doi:10.1016/j.amjmed.2008.01.043.
314. Z. Q. Ye et al., "Tamsulosin Treatment of Chronic Non-bacterial Prostatitis," *Journal of International Medical Research* 36, no. 2 (March/April 2008): 244–52, doi:10.1177/147323000803600205; and Chorok W. Jeong et al., "Treatment for Chronic Prostatitis / Chronic Pelvic Pain Syndrome: Levofloxacin, Doxazosin and Their Combination," *Urologia Internationalis* 80, no. 2 (2008): 157–61, doi:10.1159/000112606.
315. Masanori Noguchi et al., "Effect of an Extract of *Ganoderma lucidum* in Men with Lower Urinary Tract Symptoms: A Double-Blind, Placebo-Controlled Randomized and Dose-Ranging Study," *Asian Journal of Andrology* 10, no. 4 (July 2008): 651–58, doi:10.1111/j.1745-7262.2008.00336.x.
316. D. Wu et al., "Cyclooxygenase Enzyme Inhibitory Compounds with Antioxidant Activities from *Piper methysticum* (Kava Kava) Roots," *Phytomedicine* 9, no. 1 (2002): 41–47, doi:10.1078/0944-7113-00068.
317. Rafikali A. Momin, David L. De Witt, and Muraleedharan G. Nair, "Inhibition of Cyclooxygenase (COX) Enzymes by Compounds drom *Daucus carota* L. Seeds," *Phytotherapy Research* 17, no. 8 (September 2003): 976–79, doi:10.1002/prt.1296.
318. Rafikali A. Momin and M. G. Nair, "Antioxidant, Cyclooxygenase and Topoisomerase Inhibitory Compounds from Apium Graveolens Linn. Seeds," *Phytomedicine* 9, no. 4 (May 2002): 312–18, doi:10.1078/0944-7113-00131.
319. M. A. Kelm et al., "Antioxidant and Cyclooxygenase Inhibitory Phenolic Compounds from *Ocimum sanctum* Linn.," *Phytomedicine* 7, no. 1 (March 2000): 7–13, doi:10.1016/S0944-7113(00)80015-X.
320. Bernd L. Fiebich et al., "Petasites Hybridus Extracts In Vitro Inhibit COX-2 and PGE2 Release by Direct Interaction with the Enzyme and by Preventing P42/44 MAP Kinase Activation in Rat Primary Microglial Cells," *Planta Med* 71, no. 1 (January 2005): 12–19, doi:10.1055/s-2005-837744.
321. Chang Hee Han et al., "Synergistic Effect Between Lycopene and Ciprofloxacin on a Chronic Bacterial Prostatitis Rat Model," *International Journal of Antimicrobial Agents* 31, supplement 1 (February 2008): S102–7, doi:10.1016/j.ijantimicag.2007.07.016.
322. Ken Wojcikowski, David W. Johnson, and Glenda Gobé, "Medicinal Herbal Extracts—Renal Friend or Foe? Part Two: Herbal Extracts with Potential Renal Benefits," *Nephrology* 9, no. 6 (December 2004): 400–5, doi:10.1111/j.1440-1797.2004.00355.x.
323. Z. D. Bao, Z. G. Wu, and F. Zheng, "Amelioration of Aminoglycoside Nephrotoxicity by *Cordyceps sinensis* in Old Patients," *Zhongguo Zhong Xi Yi Jie He Za Zhi [Chinese Journal of Integrated Traditional and Western Medicine]* 14, no. 5 (May 1994): 271–73, 259.
324. X. Zhao and L. Li, "*Cordyceps sinensis* in Protection of the Kidney from Cyclosporine A Nephrotoxicity," *Zhonghua Yi Xue Za Zhi [Journal of the Chinese Medical Association]* 73, no. 7 (July 1993): 410–12, 447.
325. O. Wongmekiat and K. Thamprasert, "Investigating the Protective Effects of Aged Garlic Extract on Cyclosporin-Induced Nephrotoxicity in Rats," *Fundamental and Clinical Pharmacology* 19, no. 5 (October 2005): 555–62, doi:10.1111/j.1472-8206.2005.00361.x.
326. Yin-Wen Zhang, Chao-Yan Wu, and Juei-Tang Cheng, "Merit of Astragalus Polysaccharide in the Improvement of Early Diabetic Nephropathy with an Effect on mRNA Expressions of NF-κb and Iκb in Renal Cortex of Streptozotoxin-Induced Diabetic Rats," *Journal of Ethnopharmacology* 114, no. 3 (December 3, 2007): 387–92, doi:10.1016/j.jep.2007.08.024.
327. H. N. Varzi et al., "Effect of Silymarin and Vitamin E on Gentamicin-Induced Nephrotoxicity in Dogs," *Journal of Veterinary Pharmacology and Therapeutics* 30, no. 5 (October 2007): 477–81, doi:10.1111/j.1365-2885.2007.00901.x.
328. R. Di Paola et al., "Green Tea Polyphenol Extract Attenuates Zymosan-Induced Non-septic Shock in Mice," *Shock* 26, no. 4 (October 2006): 402–9, doi:10.1097/01.shk.0000191379.62897.1d.
329. S. Annie, P. L. Rajagopal, and S. Malini, "Effect of *Cassia auriculata* Linn. Root Extract on Cisplatin and Gentamicin-Induced Renal Injury," *Phytomedicine* 12, no. 8 (August 2005): 555–60, doi:10.1016/j.phymed.2003.11.010.
330. C. Pu, Y. B. Yang, and Q. L. Sun, "Effects of *Salvia miltiorrhiza* on Oxidative Stress and Microinflammatory State in Patients Undergoing Continuous Hemodialysis," *Zhongguo Zhong Xi Yi Jie He Za Zhi [Chinese Journal of Integrated Traditional and Western Medicine]* 26, no. 9 (September 2006): 791–94.
331. Xiao-Hui Sun et al., "Activation of Large-Conductance Calcium-Activated Potassium Channels by Puerarin: The Underlying Mechanism of Puerarin-Mediated Vasodilation," *Journal of Pharmacology and Experimental Therapeutics* 323, no. 1 (October 2007): 391–17, doi:10.1124/jpet.107.125567.
332. L. Y. Wang, A. P. Zhao, and X. S. Chai, "Effects of Puerarin on Cat Vascular Smooth Muscle In Vitro," *Zhongguo Yao Li Xue Bao [Acta Pharmacologica Sinica]* 15, no. 2 (March 1994): 180–82.

333. G. Zhang and S. Fang, "Antioxidation of *Pueraria lobata* Isoflavones (PLIs)," *Zhong Yao Cai [Journal of Chinese Medicinal Materials]* 20, no. 7 (July 1997): 358–60.
334. Zohreh Sedighifard et al., "Silymarin for the Prevention of Contrast-Induced Nephropathy: A Placebo-Controlled Clinical Trial," *International Journal of Preventative Medicine* 7 (2016): 23, doi:10.4103/2008-7802.174762; and Ali Momeni et al., "Effect of Silymarin in the Prevention of Cisplatin Nephrotoxicity, a Clinical Trial Study," *Journal of Clinical and Diagnostic Research* 9, no. 4 (April 2015): OC11–13, doi:10.7860/JCDR/2015/12776.5789.
335. R. M. Montero et al., "Diabetic Nephropathy: What Does the Future Hold?" *International Urology and Nephrology* 48, no. 1 (January 2016): 99–113, doi:10.1007/s11255-015-1121-y; and Mohammad Kazem Fallahzadeh et al., "Effect of Addition of Silymarin to Renin-Angiotensin System Inhibitors on Proteinuria in Type 2 Diabetic Patients with Overt Nephropathy: A Randomized, Double-Blind, Placebo-Controlled Trial," *American Journal of Kidney Disease* 60, no. 6 (December 2012): 896–903, doi:10.1053/j.ajkd.2012.06.005.
336. Su Hee Kim et al., "Effects of *Ginkgo biloba* on Haemostatic Factors and Inflammation in Chronic Peritoneal Dialysis Patients," *Phytotherapy Research* 19, no. 6 (June 2005): 546–48, doi:10.1002/ptr.1633.
337. D. G. Kang et al., "Rehmannia Glutinose Ameliorates Renal Function in the Ischemia/Reperfusion-Induced Acute Renal Failure Rats," *Biological and Pharmaceutical Bulletin* 28, no. 9 (September 2005): 1662–67.
338. Liqiang Meng et al., "A Combination of Chinese Herbs, *Astragalus membranaceus* var. *mongholicus* and *Angelica sinensis*, Enhanced Nitric Oxide Production in Obstructed Rat Kidney," *Vascular Pharmacology* 47, no. 2–3 (August/September 2007): 174–83, doi:10.1016/j.vph.2007.06.002; X. Y. Xu et al., "Adjustment Effect of Radix Astragalus and Radix Angelicae Sinensis on TNF-Alpha and Bfgf on Renal Injury Induced by Ischemia Reperfusion in Rabbit," *Zhongguo Zhong Yao Za Zhi [Chinese Journal of Chinese Materia Medica]* 27, no. 10 (October 2002): 771–73; and Ya-Ni Zhao, Jing-Zi Li, and Ling Yu, "Effect of *Astragalus–Angelica* Mixture on Osteopontin Expression in Rats with Chronic Nephrosclerosis," *Zhongguo Zhong Xi Yi Jie He Za Zhi [Chinese Journal of Integrated Traditional and Western Medicine]* 22, no. 8 (August 2002): 613–17.
339. M. X. Sheng, J. Z. Li, and H. Y. Wang, "Therapeutic Effect of *Astragalus* and *Angelica* on Renal Injury Induced by Ischemia/Reperfusion in Rats," *Zhongguo Zhong Xi Yi Jie He Za Zhi [Chinese Journal of Integrated Traditional and Western Medicine]* 21, no. 1 (January 2001): 43–46.
340. Y. Lu, J. Z. Li, and X. Zheng, "Effect of *Astragalus Angelica* Mixture on Serum Lipids and Glomerulosclerosis in Rats with Nephrotic Syndrome," *Zhongguo Zhong Xi Yi Jie He Za Zhi [Chinese Journal of Integrated Traditional and Western Medicine]* 17, no. 8 (August 1997): 478–80.
341. A. Peng, Y. Gu, and S. Y. Lin, "Herbal Treatment for Renal Diseases," *Annals of the Academy of Medicine, Singapore* 34, no. 1 (January 2005): 44–51.
342. Wei Xiao, Hong-Zhu Deng, and Yun Ma, "Summarization of the Clinical and Laboratory Study on the Rhubarb in Treating Chronic Renal Failure," *Zhongguo Zhong Yao Za Zhi [Chinese Journal of Chinese Materia Medica]* 27, no. 4 (April 2002): 241–44, 262; F. Zheng, "Effect of *Rheum officinal* on the Proliferation of Renal Tubular Cells In Vitro," *Zhonghua Yi Xue Za Zhi [Journal of the Chinese Medical Association]* 73, no. 6 (June 1993): 343–45, 380–81.
343. J. Tian, J. Chen, and J. Li, "Effects Of Cordyceps Sinensis, Rhubarb and Serum Renotropin on Tubular Epithelial Cell Growth. Tian," *Zhong Xi Yi Jie He Za Zhi [Chinese Journal of Modern Developments in Traditional Medicine]* 11, no. 9 (September 1991): 547–49, 518.
344. Heidi H. Ngai, Wai-Hung Sit, and Jennifer M. F. Wan, "The Nephroprotective Effects of the Herbal Medicine Preparation, WH30+, on the Chemical-Induced Acute and Chronic Renal Failure in Rats," *American Journal of Chinese Medicine* 33, no. 3 (2005): 491–500, doi:10.1142/S0192415X05003089.
345. S. M. Chen et al., "Effects of Bupleurum Scorzoneraefolium, Bupleurum Falcatum, and Saponins on Nephrotoxic Serum Nephritis in Mice," *Journal of Ethnopharmacology* 116, no. 3 (March 28, 2008): 397–402, doi:10.1016/j.jep.2007.11.026.
346. Wu Zhao-Long, Wang Xiao-Xia, and Cheng Wei-Ying, "Inhibitory Effect of Cordyceps Sinensis and Cordyceps Militaris on Human Glomerular Mesangial Cell Proliferation Induced by Native LDL," *Cell Biochemistry and Function* 18, no. 2 (June 2000): 93–97, doi:10.1002/(SICI)1099-0844(200006)18:2<93::AID-CBF854>3.0.CO;2-#.
347. Ling-Yu Yang et al., "Efficacy of a Pure Compound H1-A Extracted from *Cordyceps sinensis* on Autoimmune Disease of MRL lpr/lpr Mice," *Journal of Laboratory and Clinical Medicine* 134, no. 5 (November 1999): 492–500, doi:10.1016/S0022-2143(99)90171-3.
348. Lan Lu, "Study on Effect of *Cordyceps sinensis* and Artemisinin in Preventing Recurrence of Lupus Nephritis," *Zhongguo Zhong Xi Yi Jie He Za Zhi [Chinese Journal of Integrated Traditional and Western Medicine]* 22, no. 3 (March 2002): 169–71.
349. Lei-Shi Li, Feng Zheng, and Zhi-Hong Liu, "Experimental Study on Effect of *Cordyceps sinensis* in Ameliorating Aminoglycoside Induced Nephrotoxicity," *Zhongguo Zhong Xi Yi Jie He Za Zhi [Chinese Journal of Integrated Traditional and Western Medicine]* 16, no. 12 (December 1996): 733–37; and Feng Zheng, Jing Tian, and Lei-Shi Li, "Mechanisms and Therapeutic Effect of *Cordyceps sinensis* (CS) on Aminoglycoside Induced Acute Renal Failure (ARF) in Rats," *Zhongguo Zhong Xi Yi Jie He Za Zhi [Chinese Journal of Integrated Traditional and Western Medicine]* 12, no. 5 (May 1992): 288–91, 262.
350. Q. Cheng, "Effect of *Cordyceps sinensis* on Cellular Immunity in Rats with Chronic Renal Insufficiency,"

Zhonghua Yi Xue Za Zhi [Journal of the Chinese Medical Association] 72, no. 1 (January 1992): 27–9, 63.
351. Mi Wang, Jun-Ming Fan, and Xin-Ying Liu, "Effect of Total Saponins of *Panax notoginseng* on Transdifferentiation of Rats' Tubular Epithelial Cell Induced by IL-1α," *Zhongguo Zhong Xi Yi Jie He Za Zhi [Chinese Journal of Integrated Traditional and Western Medicine]* 24, no. 8 (August 2004): 722–25; B. H. Su et al., "Effects of *Panax notoginseng* Saponins on the Process of Renal Interstitial Fibrosis after Unilateral Ureteral Obstruction in Rats," *Sichuan Da Xue Xue Bao Yi Xue Ban [Journal of Suchuan University]* 36, no. 3 (May 2005): 368–71; Y. Wei, J. M. Fan, and L. P. Pan, "Effect of Panax Notoginseng Saponins on Human Kidney Fibroblast," *Zhongguo Zhong Xi Yi Jie He Za Zhi [Chinese Journal of Integrated Traditional and Western Medicine]* 22, no. 1 (January 2002): 47–9.
352. S. J. Liu and S. W. Zhou, "Panax Notoginseng Saponins Attenuated Cisplatin-Induced Nephrotoxicity," *Acta Pharmacologica Sinica* 21, no. 3 (March 2000): 257–60; T. Yokozawa and Z. W. Liu, "The Role of Ginsenoside-Rd in Cisplatin-Induced Acute Renal Failure," *Renal Failure* 22, no. 2 (March 2000): 115–27; H. Y. Kim et al., "Protective Effect of Heat-Processed American Ginseng Against Diabetic Renal Damage in Rats," *Journal of Agriculture and Food Chemistry* 55, no. 21 (October 2007): 8491–7, doi:10.1021/jf071770y.
353. W. C. Cho et al., "Ginsenoside Re of *Panax ginseng* Possesses Significant Antioxidant and Antihyperlipidemic Efficacies in Streptozotocin-Induced Diabetic Rats," *European Journal of Pharmacology* 550, no. 1–3 (November 2006): 173–9.
354. H. J. Han, "Ginsenosides Inhibit EGF-Induced Proliferation of Renal Proximal Tubule Cells via Decrease of c-fos and c-jun Gene Expression In Vitro," *Planta Medica* 68, no.11 (November 2002): 971–4. doi:10.1055/s-2002-35659.
355. T. Yokozawa, H. Oura, and Y. Kawashima, "The Effect of Ginsenoside-Rb2 on Nitrogen Balance," *Journal of Natural Products* 52, no. 6 (November–December 1989): 1350–2.
356. T. Nagasawa, "Effect of Ginseng Extract on Ribonucleic Acid and Protein Synthesis in Rat Kidney," *Chemical and Pharmaceutical Bulletin* 25, no. 7 (July 1977): 1665–70. (Tokyo)
357. X. Y. Zhang, "Prolonged Survival of MRL-lpr/lpr Mice Treated with Tripterygium Wilfordii Hook-F," *Clinical Immunology Immunopathology* 62, no. 1 (January 1992): 66–71.
358. N. S. Lai, "Prevention of Autoantibody Formation and Prolonged Survival in New Zealand Black/New Zealand White F1 Mice with an Ancient Chinese Herb, Ganoderma Tsugae," *Lupus* 10, no. 7 (2001): 461–5.
359. Z. Kang, "Observation of Therapeutic Effect in 50 Cases of Chronic Renal Failure Treated with Rhubarb and Adjuvant Drugs," *Journal of Traditional Chinese Medicine* 13 no. 4 (December 1993): 24952.
360. Y. Sun, B. Chen, and Q. Jia, "Clinical Effect of Xinqingning Combined Low Dose Continuous Gastrointestinal Dialysis in Treating Uremia" *Zhongguo Zhong Xi Yi Jie He Za Zhi [Chinese Journal of Integrated Traditional and Western Medicine]* 20, no. 9 (September 2000): 660–3.
361. N. Yang, X. Liu, and Q. Lin, "Clinical Study on Effect of Chinese Herbal Medicine Combined with Hemodialysis in Treating Uremia," *Zhongguo Zhong Xi Yi Jie He Za Zhi [Chinese Journal of Integrated Traditional and Western Medicine]* 18, no. 12 (December 1998): 712–4.
362. G. Zhang and A. M. el Nahas, "The Effect of Rhubarb Extract on Experimental Renal Fibrosis," *Nephrology, Dialysis, Transplantation: Official Publication of the European Dialysis and Transplant Association* 11, no. 1 (January 1996): 186–90.
363. J. Wei, L. Ni, and J. Yao, "Experimental Treatment of Rhubarb on Mesangio-Proliferative Glomerulonephritis in Rats," *Zhonghua Nei Ke Za Zhi [Chinese Journal of Internal Medicine]* 36, no. 2 (1997): 87–9.
364. Y. J. Wang, L. Xu, and X. X. Cheng, "Clinical Study on Niaodujing in Treating Chronic Renal Failure," *Zhongguo Zhong Xi Yi Jie He Za Zhi* 16, no. 11 (November 1996): 64951.
365. H. Sanada, "Study on the Clinical Effect of Rhubarb on Nitrogen-Metabolism Abnormality Due to Chronic Renal Failure and Its Mechanism" *Nihon Jinzo Gakkai Shi [Japanese Journal of Nephrology]* 38, no. 8 (August 1996): 379–87.
366. Yan Zhou et al., "Artesunate Suppresses the Viability and Mobility of Prostate Cancer Cells through UCA1, the Sponge of miR-184" *Oncotarget* 8, no. 11 (March 2017): 18260–70, doi:10.18632/oncotarget.15353.
367. Wei-Feng Zhong et al., "Eupatilin Induces Human Renal Cancer Cell Apoptosis via ROS-mediated MAPK and PI3K/AKT Signaling Pathways," *Oncology Letters* 12, no. 4 (October 2016): 2894–9, doi:10.3892/ol.2016.4989.
368. B. Li et al., "Astragalus and Angelica Mixture Inhibits the Renal Tubular Epithelial Cell Injury Induced by Aristolochic-acid I," *Beijing Da Xue Xue Bao* 38, no. 4 (August 2006): 381–4.
369. Q. Cai, X. Li, and H. Wang, "Astragali and Angelica Protect the Kidney Against Ischemia and Reperfusion Injury and Accelerate Recovery," *Chinese Medical Journal* 114, no. 2 (February 2001): 119–23.
370. H. Wang et al., "Antifibrotic Effect of the Chinese Herbs, *Astragalus mongholicus* and *Angelica sinensis*, in a Rat Model of Chronic Puromycin Aminonucleoside Nephrosis," *Life Sciences* 74, no. 13 (February 2004): 1645–58; L. Yu et al., "Identification of a Gene Associated with Astragalus and Angelica's Renal Protective Effects by Silver Staining mRNA Differential Display" *Chinese Medical Journal* 115, no. 6 (June 2002): 923–7.
371. V. E. Tyler, Lynn R. Brady, and James E. Robbers, *Pharmacognosy* (London: J&H Churchill, 1967), 160.
372. Harvey Wickes Felter, *Eclectic Materia Medica and Theraputics* (Cincinatti, OH: J. K. Scudder, 1992), 288.

373. Telmo N. Santos et al., "Antioxidant, Anti-Inflammatory, and Analgesic Activities of *Agrimonia eupatoria* L. Infusion," *Evidence Based Complementary Alternative Medicine* (2017): 8309894, https://doi.org/10.1155/2017/8309894.
374. D. L. Lamm and D. R. Riggs, "Enhanced Immunocompetence by Garlic: Role in Bladder Cancer and Other Malignancies" *Journal of Nutrition* 131, no. 3s (March 2001)1067S-70S.
375. M. Papiez, M. Gancarczyk, and B. Bilińska, "The Compounds from the Hollyhock Extract (*Althaea rosea* Cav. var. *nigra*) Affect the Aromatization in Rat Testicular Cells In Vivo and In Vitro," *Folia Histochemica Cytobiologica [Polish Academy of Sciences, Polish Histochemical and Cytochemical Society]* 40, no. 4 (2002): 353–9.
376. K. Sheeja and G. Kuttan, "Protective effect of *Andrographis paniculata* and Andrographolide on Cyclophosphamide-induced Urothelial Toxicity," *Integrative Cancer Therapies* 5, no. 3 (September 2006): 244–51, doi:10.1177/1534735406291984; K. Sheeja and G. Kuttan, "Ameliorating Effects of Andrographis paniculata Extract Against Cyclophosphamide-Induced Toxicity in Mice," *Asian Pacific Journal of Cancer Prevention* 7, no. 4 (October/December 2006): 609–14.
377. W. Mitchell, *Naturopathic Applications of Botanical Remedies*, (Seattle, WA, 1983), p 8; V. Frohne, "Untersuchungen zur frage der harndesifizierenden wirkungen von barentraubenblatt-extracten," *Planta Medica*, 1970, 18.pp1-25; and A. Y. Leung, *Encyclopedia of Common Natural Ingredients in Food, Drugs, and Cosmetics*, (New York: John Wiley and Sons, 1980), 292, 293, and 316–317.
378. M. Shimizu et al., "Marked Potentiation of Activity of Beta-lactams against Methicillin-resistant *Staphylococcus aureus* by Corilagin," *Antimicrobial Agents and Chemotherapy.* 45, no. 11 (November 2001): 3198–201, doi:10.1128/AAC.45.11.3198-3201.2001.
379. D. Sato and M. Matsushima, "Preventive Effects of Urinary Bladder Tumors Induced by N-butyl-N-(4-hydroxybutyl)-nitrosamine in Rat by Green Tea Leaves," *International Journal of Urology* 10, no. 3 (March 2003): 160–6, http://onlinelibrary.wiley.com/journal/10.1111/(ISSN)1442-2042.
380. K. Satoh, "Inhibition of Aromatase Activity by Green Tea Extract Catechins and Their Endocrinological Effects of Oral Administration in Rats," *Food and Chemical Toxicology* 40, no. 7 (July 2002): 925–33.
381. S. K. Huan et al., "Cantharidin-Induced Cytotoxicity and Cyclooxygenase 2 Expression in Human Bladder Carcinoma Cell Line," *Toxicology* 223, no. 1-2 (June 2006): 136–43, doi:10.1016/j.tox.2006.03.012.
382. Herbert Tracy Webster, *Dynamical Theraputics* (Oakland, CA, 1891).
383. Harvey Wickes Felter, *Eclectic Materia Medica and Theraputics* (Cincinatti, OH: J. K. Scudder, 1992) 288.
384. T. Sicilia et al., "Identification and Stereochemical Characterization of Lignans in Flaxseed and Pumpkin Seeds," *Journal of Agricultural and Food Chemistry* 51, no. 5 (February 20013): 1181–8.
385. K. M. Phillips, D. M. Ruggio, and M. Ashraf-Khorassani, "Phytosterol Composition of Nuts and Seeds Commonly Consumed in the United States," *Journal of Agricultural and Food Chemistry* 53, no. 24 (November 2005): 9436–45; J. B. Rodriguez et al., "The Sterols of Cucurbita Moschata ("calabacita") Seed Oil," *Lipids* 31, no. 11 (November 1996): 1205–8.
386. E. Ryan et al., "Phytosterol, Squalene, Tocopherol Content and Fatty Acid Profile of Selected Seeds, Grains, and Legumes," *Plant Foods for Human Nutrition* 62, no. 3 (September 2007): 85–91.
387. Scudder, quoted in Lloyd Bros pharmaceutical label.
388. Harvey Wickes Felter, *Eclectic Materia Medica and Theraputics* (Cincinatti, OH: J. K. Scudder, 1992), 288.
389. Scudder, quoted in Lloyd Bros pharmaceutical label.
390. Scudder, quoted in Lloyd Bros pharmaceutical label.
391. Jonathn Wright, "no title" (presentation, AANP Convention Proceedings, Aspen, CO, October 1995).
392. A. Y. Leung, and S. Foster, *Encyclopedia of Common Natural Ingredients in Food, Drugs, and Cosmetics*, 2nd ed. (New York: John Wiley and Sons, 1996), 213.
393. Piekos, R. u. S. Palslawska: Planta medica 27 (1975) 147.
394. R. Fujita et al., "Anti-androgenic Activities of Ganoderma lucidum," *Journal of Ethnopharmacology* 102, no. 1 (October 2005): 107–12.
395. T. E. Spires et al., "Identification of Novel Functional Inhibitors of 17beta-hydroxysteroid dehydrogenase Type III (17beta-HSD3)," *Prostate* 65, no. 2 (October 1 2005): 159–70.
396. P. C. Griessmayr et al., "Mushroom-derived Maitake PETfraction as Single Agent for the Treatment of Lymphoma in Dogs," *Journal of Veterinary Internal Medicine* 21, no. 6 (November–December 2007): 1409–12.
397. Leibstein, "Article on Urinary Tonics," *Eclectic Medical Association Quarterly* (Cincinatti, OH 1904).
398. M. Takacsov'a, A. Pribela, and M. Faktorova, "Study of the Antioxidative Effects of Thyme, Sage, Juniper, and Oregano," *Nahrung* 39 (1995): 241–3.
399. H. Tunon, C. Olavsdotter, and L. Bohlin, "Evaluation of Anti-Inflammatory Activity of Some Swedish Medicinal Plants. Inhibition of Prostablandin Synthesis and PAF-Induced Exocytosis," *Journal of Ethnopharmacology* 48 (1995): 61–76.
400. G. F. Gonzales, "Antagonistic Effect of *Lepidium meyenii* (red maca) on Prostatic Hyperplasia in Adult Mice," *Andrologia* 40, no. 3 (June 2008): 179–85, doi:10.1111/j.1439-0272.2008.00834.x.
401. G. F. Gonzales, "Antagonistic Effect of *Lepidium meyenii* (red maca) on Prostatic Hyperplasia in Adult Mice."
402. M. Gasco, "Dose-Response Effect of Red Maca (*Lepidium meyenii*) on Benign Prostatic Hyperplasia Induced by Testosterone Enanthate," *Phytomedicine* 14, no. 7–8 (August 2007): 460–4.

403. G. F. Gonzales et al., "Red Maca (*Lepidium meyenii*) Reduced Prostate Size in Rats," *Reproductive Biology and Endocrinology* 3 (January 20 2005): 5, doi:10.1186/1477-7827-3-5.
404. G. F. Gonzales et al., "Effect of Two Different Extracts of Red Maca in Male Rats with Testosterone-Induced Prostatic Hyperplasia," *Asian Journal of Andrology* 9, no. 2 (March 2007): 245–51.
405. N. Hirata et al., "Testosterone 5alpha-reductase Inhibitory Active Constituents of Piper Nigrum Leaf," *Biological and Pharmaceutical Bulletin* 30, no. 12 (December 2007): 2402–5.
406. J. Yam, M. Kreuter, and J. Drewe, "Piper Cubeba Targets Multiple Aspects of the Androgen-Signalling Pathway. A Potential Phytotherapy Against Prostate Cancer Growth?," *Planta Medica* 74, no. 1 (January 2008): 33–8.
407. C. C. Pfeiffer et al., "Effect of Kava in Normal Subjects and Patients," US Public Health Service Pub. No. 1645. Washington, D.C.:U.S. Government Printing Office, 1967
408. N. Hirata et al., "Testosterone 5alpha-Reductase Inhibitory Active Constituents of Piper Nigrum Leaf," *Biological and Pharmaceutical Bulletin* 30, no. 12 (December 2007); and Y. C. Kao et al., "Molecular Basis of the Inhibition of Human Aromatase (Estrogen Synthetase) by Flavone and Isoflavone Phytoestrogens: A Site-Directed Mutagenesis Study," *Environmental Health Perspectives* 106, no. 2 (February 1998): 85–92.
409. G. F. Cheng et al., "Antiinflammatory Effects of Tremulacin, a Salicin-Related Substance Isolated from Populus Tormentosa Leaves" *Phytomedicine* 1 (1994): 209–11; and M. El-Ghazaly, "Study of the Antiinflammatory Activity of *Populus tremula, Solidoago virgqurea,* and *Fraxinus excelsior," Arzneimittel-Forschung* 42 (1992): 333–6; and B. Meyer, W. Schneider, and E. F. Elstner, "Antioxidative Properties of Alcohol Extracts from *Fraxinus excelsior, Populus tremula,* and *Solidago vigaurea," Arzneimittel Forschung Drug Research* 45 (1995): 174–6.
410. A. Tasca et al., "Treatment of Obstructive Symptomatology Caused by Prostatic Adenoma with an Extract of Serenoa Repens. Double Blind Study v. Placebo." *Minerva Urologica e Nefrologice [The Italian Journal of Urology and Nephrology]* 37 (1985): 87–91; and E. Carilla et al., "Binding of Permixon, a New Treatment for Benign Prostatic Hyperplasia, to the Cytosolic Androgen Receptor in the Rat Prostate," *Journal of Steroid Biochemistry* 20 (1984): 521–523; and G. Champault, J. C. Patel, and A. M. Bonnard, "Double Blind Trial of an Extract of the Plant *Serenoa repens* in Benign Prostatic Hyperplasia," *British Journal of Clinical Pharmacology* 20 (1984) 521–3; and C. Boccafoschi and S. Annoscia, "Comparison of *Serenoa repens* Extract with Placebo by Controlled Clinical Trial in Patients with Prostatic Adenomatosis," *Urologia* 50 (1983) 1257–9.
411. T. L. Wadsworth, "Effects of Dietary Saw Palmetto on the Prostate of Transgenic Adenocarcinoma of the Mouse Prostate Model (TRAMP)," *Prostate* 67, no. 6 (May 1 2007): 661–73.
412. F. K. Habib et al., "*Serenoa repens* (Permixon) Inhibits the 5alpha-reductase Activity of Human Prostate Cancer Cell Lines without Interfering with PSA Expression," *International Journal of Cancer* 114, no. 2 (March 20 2005):190–4.
413. R. F. Weiss, *Herbal Medicine* (New York: Thieme, 2001); and William Boericke, *Homeopathic Materia Medica* (Kessinger Publishing, 2004): 241.
414. H. Wagner, B. Kreutzkamp, and K. Jurcic, "Die Alkaloide von Uncaris tormentosa und ihre Phagozytosesteigernde Wirking," *Plant Medica* (1985): 419.
415. R. Aquino et al., "Plant Metabolites: Structure and In Vitro Antiviral Activity of Quinovic Acid Glycosides from *Uncaria tormentosa* and *Guettarda platypoda" Journal of Natural Products* 52, no. 4 (1989): 679; and R. Aquino et al., "New Polyhydroxylated Triterpenes from *Uncaria tormentosa" Journal of Natural Products* 53, no. 3 (1990): 559.
416. M. Harada and Y. Ozaki, "Effect of Indole Alkaloids from Gardneria Genus and Uncaria Genus of Neuromuscular Transmission in the Rat Limb," *Chemical and Pharmaceutical Bulletin* 24, no. 2 (1976): 211.
417. M. Harad et al., "Effects of Indole Alkaloids from Garneria Nutans and *Uncaria rhynochophylla* on a Guinea Pig Urinary Bladder Preparation in Situ," *Chemical Pharmaceutical Bulletin* 27, no. 5 (1979): 1069.
418. M. Moore, *Medicinal Plants of the Desert and Canyon West* (Santa Fe: Museum of New Mexico Press, 1989).
419. U. Elsässer-Beile et al., "Adjuvant Intravesical Treatment with a Standardized Mistletoe Extract to Prevent Recurrence of Superficial Urinary Bladder Cancer," *Anticancer Research* 25, no. 6C (November–December 2005): 4733–6.
420. J. M. McDonald, *Physiologic Medication* (Chicago, IL: 1900).
421. F. Grases et al., "The Influence to *Zea mays* on Urinary Risk Factors for Kidney Stones in Rats,"*Phytotherapy Research* 7 (March 1993), doi:10.1002/ptr.2650070210.
422. Vinay K. Gupta and Seema Malhotra, "Pharmacological Attribute of *Aloe vera*: Revalidation Through Experimental and Clinical Studies," *AYU* 33, no. 2 (April–June 2012): 193–96, doi:10.4103/0974-8520.105237.

Chapter 5: Creating Herbal Formulas for Dermatologic Conditions

1. Soon-Sun Hong et al., "Advanced Formulation and Pharmacological Activity of Hydrogel of the Titrated Extract of *C. asiatica," Archives of Pharmacal Research* 28, no. 4 (April 2005): 502–8.
2. Cheun Lung Cheng et al., "The Healing Effects of Centella Extract and Asiaticoside on Acetic Acid Induced Gastric Ulcers in Rats," *Life Sciences* 74, no. 18 (March 2004): 2237–34, doi:10.1016/j.lfs.2003.09.055.
3. Christopher D. Coldren et al., "Gene Expression Changes in the Human Fibroblast Induced by *Centella asiatica* Triterpenoids," *Planta Medica* 69, no. 8 (August 2003): 725–32, doi:10.1055/s-2003-42791.

4. Amira M. K. Abouelella et al., "Phytotherapeutic Effects of *Echinacea purpurea* in Gamma-Irradiated Mice," *Journal of Veterinary Science* 8, no. 4 (December 2007): 341–51.
5. Steven E. Mansoor et al., "X-ray Structures Define Human P2X3 Receptor Gating Cycle and Antagonist Action," *Nature* 538 (October 2016): 66–71, doi:10.1038/nature19367.
6. M. Yambem, M. Madhale, and D. Bagi, "A Comparative Study to Assess the Effectiveness of Glycerin with Magnesium Sulphate Versus Heparin–Benzyl Nicotinate (Thrombophob) Ointment on Management of Thrombophlebitis among Patients Admitted in Intensive Care Units (ICU) of Selected Hospital in Belgaum, Karnataka." *International Journal of Science and Research* 4, no. 7 (July 2015): 1458–61, http://www.ijsr.net/archive/v4i7/12071501.pdf.
7. R. L. Lv et al., "The Effects of Aloe Extract on Nitric Oxide and Endothelin Levels in Deep-Partial Thickness Burn Wound Tissue in Rat," *Zhonghua Shao Shang Za Zhi* 22, no. 5 (October 2006): 362–65.
8. E. Klouchek-Popova et al., "Influence of the Physiological Regeneration and Epithelialization Using Fractions Isolated from *Calendula officinalis*," *Acta Physiologica et Pharmacologica Bulgarica* 8, no. 4 (1982): 63–7.
9. L. Salas Campos, M. Fernándes Mansilla, and A. M. Martínez de la Chica, "Topical Chemotherapy for the Treatment of Burns," *Revista de Enfermeria* 28, no. 5 (May 2005): 67–70.
10. Erick K. Nishio et al., "Antibacterial Synergic Effect of Honey from Two Stingless Bees: *Scaptotrigona bipunctata* Lepeletier, 1836, and *S. postica Latreille*, 1807," *Scientific Reports* 6 (2016), doi:10.1038/srep21641.
11. Nilgün Oztürk, Seval Korkmaz, and Yusuf Oztürk, "Wound-Healing Activity of St. John's Wort (*Hypericum perforatum* L.) on Chicken Embryonic Fibroblasts," *Journal of Ethnopharmacology* 111, no. 1 (April 2007): 33–9, doi:10.1016/j.jep.2006.10.029.
12. Amala Soumyanath et al., "*Centella asiatica* Accelerates Nerve Regeneration upon Oral Administration and Contains Multiple Active Fractions Increasing Neurite Elongation In-vitro," *Journal of Pharmacy and Pharmacology* 57, no. 9 (September 2005): 1221–9, doi:10.1211/jpp.57.9.0018.
13. Lívia Slobodníková et al., "Antimicrobial Activity of *Mahonia aquifolium* Crude Extract and Its Major Isolated Alkaloids," *Phytotherapy Research* 18, no. 8 (August 2004): 674–6, doi:10.1002/ptr.1517; and Dacheng Wang et al., "Global Transcriptional Profiles of *Staphylococcus aureus* Treated with Berberine Chloride," *FEMS Microbiology Letters* 279, no. 2 (February 2008): 217–25, doi:10.1111/j.1574-6968.2007.01031.x.
14. Hyeon-Hee Yu et al., "Antimicrobial Activity of Berberine Alone and in Combination with Ampicillin or Oxacillin Against Methicillin-Resistant *Staphylococcus aureus*," *Journal of Medicinal Food* 8, no. 4 (December 2005): 454–61, doi:10.1089/jmf.2005.8.454.
15. Yu et al., "Antimicrobial Activity of Berberine."
16. Byeoung-Soo Park et al., "*Curcuma longa* L. Constituents Inhibit Sortase A and *Staphylococcus aureus* Cell Adhesion to Fibronectin," *Journal of Agricultural and Food Chemistry* 53, no. 23 (November 2005): 9005–9, doi:10.1021/jf051765z.
17. Luo Jiaoyang et al., "Multicomponent Therapeutics of Berberine Alkaloids," *Evidence-Based Complementary and Alternative Medicine* (2013), doi:10.1155/2013/545898; and Jesus A. Cuaron et al., "Tea Tree Oil-Induced Transcriptional Alterations in *Staphylococcus aureus*," *Phytotherapy Research* 27, no. 3 (March 2013): 390–96, doi:10.1002/ptr.4738.
18. Oztürk, Korkmaz, and Oztürk, "Wound-Healing Activity of St. John's Wort."
19. Uwe Wollina, Mohamed Badawy Abdel-Naser, and Raj Mani, "A Review of the Microcirculation in Skin in Patients with Chronic Venous Insufficiency: The Problem and the Evidence Available for Therapeutic Options," *The International Journal of Lower Extremity Wounds* 5, no. 3 (September 2006): 169–80, doi:10.1177/1534734606291870; and M. R. Cesarone et al., "Evaluation of Treatment of Diabetic Microangiopathy with Total Triterpenic Fraction of *Centella asiatica*: A Clinical Prospective Randomized Trial with a Microcirculatory Model," supplement 2, *Angiology* 52 (October 2001): S49–54.
20. Hiroshi Tanaka et al., "Fermentable Metabolite of *Zymomonas mobilis* Controls Collagen Reduction in Photoaging Skin by Improving TGF-β/Smad Signaling Suppression," supplement 1, *Archives of Dermatological Research* 300 (November 2007): 57–64, doi:10.1007/s00403-007-0805-2.
21. Z. J. Ming, S. Z. Liu, and L. Cao, "Effect of Total Glucosides of *Centella asiatica* on Antagonizing Liver Fibrosis Induced by Dimethylnitrosamine in Rats," *Zhongguo Zhong Xi Yi Jie He Za Zhi* [*Chinese Journal of Integrated Traditional and Western Medicine*] 24, no. 8 (August 2004): 731–34.
22. Jongsung Lee et al., "Asiaticoside Induces Human Collagen I Synthesis through TGFβ Receptor I Kinase (TβRI Kinase)-Independent Smad Signaling," *Planta Medica* 72, no. 4 (March 2006): 324–8, doi:10.1055/s-2005-916227; and L. Lu et al., "Asiaticoside Induction for Cell-cycle Progression, Proliferation and Collagen Synthesis in Human Dermal Fibroblasts," *International Journal of Dermatology* 43, no. 11 (November 2004): 801–7; and L. Lu et al., "Dermal Fibroblast-Associated Gene Induction by Asiaticoside Shown In Vitro by DNA Microarray Analysis," *British Journal of Dermatology* 151, no. 3 (September 2004): 571–8.
23. T. Nakajima et al., "Inhibitory Effect of Baicalein, a Flavonoid in Scutellaria Root, on Eotaxin Production by Human Dermal Fibroblasts," *Planta Medica* 67, no. 2 (March 2001): 132–5, doi:10.1055/s-2001-11532.
24. K. J. Gohil, J. A. Patel, and A. K. Gajjar, "Pharmacological Review on *Centella asiatica*: A Potential Herbal Cure-all," *Indian Journal of Pharmaceutical Sciences* 72, no. 5 (September–October 2010): 546–56, doi:10.4103/0250-474X.78519.

25. Ilkay Erdogan Orhan et al., "*Centella asiatica* (L.) Urban: From Traditional Medicine to Modern Medicine with Neuroprotective Potential," *Evidence-Based Complementary Alternative Medicine* 2012 (2012): 946259, http://dx.doi.org/10.1155/2012/946259.
26. Y. Lokanathan et al., "Recent Updates in Neuroprotective and Neuroregenerative Potential of *Centella asiatica*," *Malaysian Journal of Medical Sciences* 23, no.1 (January 2016): 4–14.
27. M. Hosnuter et al., "The Effects of Onion Extract on Hypertrophic and Keloid Scars," *Journal of Wound Care* 16, no. 6 (June 2007): 251–4, doi:10.12968/jowc.2007.16.6.27070.
28. V. Duran et al., "Results of the Clinical Examination of an Ointment with Marigold (*Calendula officinalis*) Extract in the Treatment of Venous Leg Ulcers," *International Journal of Tissue Reactions* 27, no. 3 (2005): 101–6.
29. Zhang Luolan, "Observation on the Results of Treatment of Female Infertility in 343 Cases," *Journal of Traditional Chinese Medicine* 6, no. 3 (1986): 175–7; C. P. Sung et al., "Effects of *Angelica polymorpha* on Reaginic Antibody Production," *Journal of Natural Products* 45 (1982): 398–406; and Zhiping H. et al., "Treating Amenorrhea in Vital Energy-Deficient Patients with *Angelica sinensis-Astragalus membranaceus* Menstration Regulating Decoction," *Journal of Traditional Chinese Medicine* 6, no. 3 (1986): 187–90.
30. S. J. Sjeu et al., "Analysis and Processing of Chinese Herbal Drugs; VI The Study of Angelica Radix," *Planta Medica* (1987): 377–8.
31. S. J. Lee et al., "Oral Administration of *Astragalus membranaceus* Inhibits the Development of DNFB-Induced Dermatitis in NC/Nga Mice," *Biological and Pharmaceutical Bulletin* 30, no. 8 (August 2007): 1468–71.
32. Elisabeth Kern et al., "Management des Capecitabin induzierten Hand-Fuß-Syndroms durch locale Phytotherapie" [Management of Capecitabine-Induced Hand-Foot Syndrome by Local Phytotherapy], *Wiener Medizinische Wochenschrift* 157, no. 13–14 (2007): 337–42, doi:10.1007/s10354-007-0435-5.
33. K. Zitterl-Eglseera et al., "Anti-Oedematous Activities of the Main Triterpendiol Esters of Marigold (*Calendula officinalis* L.)," *Journal of Ethnopharmacology* 57, no. 2 (July 1997): 139–44, https://doi.org/10.1016/S0378-8741(97)00061-5.
34. H. Neukirch et al., "Improved Anti-Inflammatory Activity of Three New Terpenoids Derived, by Systematic Chemical Modifications, from the Abundant Triterpenes of the Flowery Plant *Calendula officinalis*," *Chemistry and Biodiversity* 2, no. 5 (May 2005): 657–71, doi:10.1002/cbdv.200590042.
35. M. Saeedi, K. Morteza-Semnani, and M. R. Ghoreishi, "The Treatment of Atopic Dermatitis with Licorice Gel," *Journal of Dermatological Treatment* 14, no. 3 (September 2003): 153–7.
36. Rafikali A. Momin, David L. De Witt, and Muraleedharan G. Nair, "Inhibition of Cyclooxygenase (COX) Enzymes by Compounds from *Daucus carota* L. Seeds," *Phytotherapy Research* 17, no. 8 (September 2003): 976–9, doi:10.1002/ptr.1296.
37. H. M. Kim and S. H. Cho, "Lavender Oil Inhibits Immediate-Type Allergic Reaction in Mice and Rats," *Journal of Pharmacy and Pharmacology* 51, no. 2 (February 1999): 221–6.
38. M. Hou et al., "Topical Apigenin Improves Epidermal Permeability Barrier Homeostasis in Normal Murine Skin by Divergent Mechanisms," *Experimental Dermatology* 22, no. 3 (March 2013): 210–5, doi:10.1111/exd.12102.
39. B. S. Shetty et al., "Effect of *Centella asiatica L* (Umbelliferae) on Normal and Dexamethasone-Suppressed Wound Healing in Wistar Albino Rats," *International Journal of Lower Extremity Wounds* 5, no. 3 (September 2006): 137–43, doi:10.1177/1534734606291313.
40. Shahrooz Rashtak and Joseph A. Murray, "Review Article: Celiac Disease, New Approaches to Therapy," *Alimentary Pharmacology and Therapeutics* 35, no. 7 (April 2012): 768–81, doi:10.1111/j.1365-2036.2012.05013.x.
41. Jonas F. Ludvigsson et al., "Diagnosis and Management of Adult Coeliac Disease: Guidelines from the British Society of Gastroenterology," *Gut* 63, no. 8 (August 2014): 1210–28.
42. G. Gabriel and R. N. Thin, "Treatment of Anogenital Warts. Comparison of Trichloracetic Acid and Podophyllin versus Podophyllin Alone," *British Journal of Venereal Disease* 59, no. 2 (April 1983): 124–6; and P. D. Simmons, "Podophyllin 10% and 25% in the Treatment of Ano-Genital Warts. A Comparative Double-Blind Study," *British Journal of Venereal Disease* 57, no. 3 (June 1981): 208–9.
43. L. Slobodníková et al., "Antimicrobial Activity of Mahonia Aquifolium Crude Extract and Its Major Isolated Alkaloids," *Phytotherapy Research* 18, no. 8 (August 2004): 674–6, doi:10.1002/ptr.1517.
44. H. Dobrev, "Clinical and Instrumental Study of the Efficacy of a New Sebum Control Cream," *Journal of Cosmetic Dermatology* 6, no. 2 (June 2007): 113–8, doi:10.1111/j.1473-2165.2007.00306.x.
45. P. F. Builders et al., "Wound Healing Potential of Formulated Extract from *Hibiscus sabdariffa* Calyx," *Indian Journal of Pharmaceutical Sciences* 75, no. 1 (January–February 2013): 45–52, doi:10.4103/0250-474X.113549.
46. P. F. Builders et al., "Wound Healing Potential of Formulated Extract from *Hibiscus sabdariffa* Calyx"; and A. Bhaskar and V. Nithya, "Evaluation of the Wound-Healing Activity of *Hibiscus rosa sinensis* L (Malvaceae) in Wistar Albino Rats," *Indian Journal of Pharmacology* 44, no. 6 (November–December 2012): 694–8, doi:10.4103/0253-7613.103252.
47. Adolfo Renzi et al., "Myoxinol (Hydrolyzed *Hibiscus esculentus* Extract) in the Cure of Chronic Anal Fissure: Early Clinical and Functional Outcomes," *Gastroenterology Research and Practice* 2015 (2015): 567920, http://dx.doi.org/10.1155/2015/567920.

48. Uwe Wollina, "Recent Advances in the Understanding and Management of Rosacea," *F1000Prime Reports* 6 (2014): 50, doi:10.12703/P6-50.
49. Martin Steinhoff et al., "Topical Ivermectin 10 mg/g and Oral Doxycycline 40 mg Modified-Release: Current Evidence on the Complementary Use of Anti-Inflammatory Rosacea Treatments," *Advances in Therapy* 33, no. 9 (2016): 1481–501.
50. Uwe Wollina, "Recent Advances in the Understanding and Management of Rosacea."
51. Andrzej Slominski, "Steroidogenesis in the Skin: Implications for Local Immune Functions," *Journal of Steroid Biochemistry and Molecular Biology* 137 (September 2013): 107–23, doi:10.1016/j.jsbmb.2013.02.006.
52. Manal Abokwidir and Steven R. Feldman, "Rosacea Management," *Skin Appendage Disorders* 2, no. 1–2 (September 2016): 26–34, doi:10.1159/000446215.
53. Magdalena Działo et al., "The Potential of Plant Phenolics in Prevention and Therapy of Skin Disorders," *International Journal of Molecular Sciences* 17, no. 2 (February 2016): 160, doi:10.3390/ijms17020160; and A. K. Gupta and M. D. Gover, "Azelaic Acid (15% gel) in the Treatment of Acne Rosacea," *International Journal of Dermatology* 46, no. 5 (May 2007): 533–8, doi:10.1111/j.1365-4632.2005.02769.x; and J. Q. Del Rosso et al., "Azelaic Acid Gel 15%: Clinical Versatility in the Treatment of Rosacea," *Cutis* 78, no. S5 (November 2006): 6–19.
54. R. H. Liu et al., "Azelaic Acid in the Treatment of Papulopustular Rosacea: a Systematic Review of Randomized Controlled Trials," *Archives of Dermatology* 142, no. 8 (August 2006): 1047–52, doi:10.1001/archderm.142.8.1047.
55. Tiago H. Silva et al., "Marine Algae Sulfated Polysaccharides for Tissue Engineering and Drug Delivery Approaches," *Biomatter* 2, no. 4 (October 1 2012): 278–89.
56. Tiago H. Silva et al., "Marine Algae Sulfated Polysaccharides for Tissue Engineering and Drug Delivery Approaches," *Biomatter* 2, no. 4 (October 1 2012): 278–89, doi:10.4161/biom.22947.
57. Jianxing Zhang et al., "Novel Sulfated Polysaccharides Disrupt Cathelicidins, Inhibit RAGE and Reduce Cutaneous Inflammation in a Mouse Model of Rosacea," *PLoS ONE* 6, no. 2 (2011): e16658, doi:10.1371/journal.pone.0016658.
58. Sandamali A. Ekanayaka et al., "Glycyrrhizin Reduces HMGB1 and Bacterial Load in *Pseudomonas aeruginosa* Keratitis," *Investigative Ophthalmology and Visual Science* 57, no. 13 (October 2016): 5799–809.
59. A. M. El-Shazly et al., "Treatment of Human Demodex Folliculorum by Camphor Oil and Metronidazole," *Journal of the Egyptian Society of Parasitology* 34, no. 1 (April 2004): 107–16.
60. J. X. Liu, Y. H. Sun, and C. P. Li, "Volatile Oils of Chinese Crude Medicines Exhibit Antiparasitic Activity Against Human Demodex with No Adverse Effects In Vivo," *Experimental and Therapeutical Medicine* 9, no. 4 (April 2015): 1304–8, doi:10.3892/etm.2015.2272; and H. Koo et al., "Ocular Surface Discomfort and Demodex: Effect of Tea Tree Oil Eyelid Scrub in Demodex Blepharitis," *Journal of Korean Medical Science* 27, no. 12 (December 2012): 1574–9, doi:10.3346/jkms.2012.27.12.1574.
61. Verena Katharina Raker, Christian Becker, and Kerstin Steinbrink, "The cAMP Pathway as Therapeutic Target in Autoimmune and Inflammatory Diseases," *Frontiers in Immunology* 7 (2016): 123, doi:10.3389/fimmu.2016.00123.
62. Emmanuel Olorunju Awe and S. Olatunbosun Banjoko, "Biochemical and Haematological Assessment of Toxic Effects of the Leaf Ethanol Extract of *Petroselinum crispum* (Mill) Nyman ex A.W. Hill (Parsley) in Rats," *BMC Complementary Alternative Medicine* 13 (2013): 75, doi:10.1186/1472-6882-13-75.
63. R. E. Klaber, "Phytophotodermatitis," *Archives of Disease in Childhood* 91, no. 5 (May 2006): 385, doi:10.1136/adc.2005.091934; and Jane C. Quinn, Allan Kessell, and Leslie A. Weston, "Secondary Plant Products Causing Photosensitization in Grazing Herbivores: Their Structure, Activity and Regulation," *International Journal of Molecular Sciences* 15, no. 1 (January 2014): 1441–65, doi:10.3390/ijms15011441.
64. Christopher M. Chandler and Owen M. McDougal, "Medicinal History of North American Veratrum," *Phytochemistry Reviews: Proceedings of the Phytochemical Society of Europe* 13, no. 3 (September 2014): 671–94, doi:10.1007/s11101-013-9328-y.
65. Anastasia Karioti and Anna Rita Bilia, "Hypericins as Potential Leads for New Therapeutics," *International Journal of Molecular Sciences* 11, no. 2 (2010): 562–94, doi:10.3390/ijms11020562.
66. Jason Jingjie Yu et al., "Add-On Effect of Chinese Herbal Medicine Bath to Phototherapy for Psoriasis Vulgaris: A Systematic Review," *Evidence-Based Complementary and Alternative Medicine* 2013 (2013): 673078, http://dx.doi.org/10.1155/2013/673078; and Jie Wu, Hong Li, and Ming Li, "Effects of Baicalin Cream in Two Mouse Models: 2,4-dinitrofluorobenzene-Induced Contact Hypersensitivity and Mouse Tail Test for Psoriasis," *International Journal of Clinical and Experimental Medicine* 8, no. 2 (2015): 2128–37.
67. Golnaz Sarafian et al., "Topical Turmeric Microemulgel in the Management of Plaque Psoriasis: A Clinical Evaluation," *Iranian Journal of Pharmaceutical Research* 14, no. 3 (Summer 2015): 865–76; and Jason Jingjie Yu et al., "Add-On Effect of Chinese Herbal Medicine Bath to Phototherapy for Psoriasis Vulgaris: A Systematic Review," *Evidence-Based Complementary and Alternative Medicine* 2013 (2013): 673078, http://dx.doi.org/10.1155/2013/673078.
68. Lihong Yang et al., "Efficacy of Combining Oral Chinese Herbal Medicine and NB-UVB in Treating Psoriasis Vulgaris: A Systematic Review and Meta-Analysis," *Chinese Medicine* 10 (2015): 27, doi:10.1186/s13020-015-0060-y.

69. Gopakumar Ramachandran Nair et al., "Clinical Effectiveness of Aloe Vera in the Management of Oral Mucosal Diseases: A Systematic Review," *Journal of Clinical and Diagnostic Research* 10, no. 8 (August 2016): ZE01–ZE07, doi:10.7860/JCDR/2016/18142.8222; and Neda Babaee et al., "Evaluation of the Therapeutic Effects of Aloe Vera Gel on Minor Recurrent Aphthous Stomatitis," *Dental Research Journal* 9, no. 4 (July–August 2012): 381–5.
70. Seid Javad Kia et al., "Comparative Efficacy of Topical Curcumin and Triamcinolone for Oral Lichen Planus: A Randomized, Controlled Clinical Trial," *Journal of Dentistry* 12, no. 11 (November 2015): 789–96; and V. Singh et al., "Turmeric—A New Treatment Option for *Lichen Planus*: A Pilot Study," *National Journal of Maxillofacial Surgery* 4, no. 2 (July–December 2013): 198–201, doi:10.4103/0975-5950.127651.
71. M. J. Lee et al., "Analysis of the MicroRNA Expression Profile of Normal Human Dermal Papilla Cells Treated with 5α-dihydrotestosterone," *Molecular Medicine Reports* 12, no. 1 (July 2015): 1205–12, doi:10.3892/mmr.2015.3478; and A. Rossi et al., "Conditions Simulating Androgenetic Alopecia," *Journal of European Academy of Dermatology and Venereology* 29, no. 7 (July 2015): 1258–64, doi:10.1111/jdv.12915.
72. B. M. Piraccini and A. Alessandrini, "Androgenetic Alopecia," *Giornale Italiano di Dermatologia e Venereologia* 149, no. 1 (February 2014): 15–24.
73. S. Rinaldi, M. Bussa, and A. Mascaro, "Update on the Treatment of Androgenetic Alopecia," *European Review for Medical and Pharmacological Sciences* 20, no. 1 (January 2016): 54–8.
74. S. Varothai and W. F. Bergfeld, "Androgenetic Alopecia: An Evidence-Based Treatment Update," *American Journal of Clinical Dermatology* 15, no. 3 (July 2014): 217–30, doi:10.1007/s40257-014-0077-5.
75. D. E. Rousso and S. W. Kim, "A Review of Medical and Surgical Treatment Options for Androgenetic Alopecia," *Journal of American Medical Association Facial Plastic Surgery* 16, no. 6 (November–December 2014): 444–50, doi:10.1001/jamafacial.2014.316.
76. Charles J. Ryan and June M. Chan, "Hair, Hormones, and High-Risk Prostate Cancer," *Journal of Clinical Oncology* 33, no. 5 (February 10 2015): 386–7.
77. C. K. Zhou et al., "Male Pattern Baldness in Relation to Prostate Cancer Risks: An Analysis in the VITamins and Lifestyle (VITAL) Cohort Study," *Prostate* 75, no. 4 (March 1, 2015): 415–23, doi:10.1002/pros.22927.
78. X. Chne et al., "Serum Cortisol and Peripheral Blood Mononuclear Cell Glucocorticoid Receptor mRNA Expression in Severe Alopecia Areata with Liver-Kidney Deficiency Syndrome," *Journal of Southern Medical University* 32, no. 2 (February 2012): 230–3.
79. Jiyun Chen, Juhyun Kim, and Jame T. Dalton, "Discovery and Therapeutic Promise of Selective Androgen Receptor Modulators," *Molecular Interventions* 5, no. 3 (June 2005): 173–88, doi:10.1124/mi.5.3.7.
80. Paul Grant and Shamin Ramasamy, "An Update on Plant Derived Anti-Androgens," *International Journal of Endocrinology and Metabolism* 10, no. 2 (Spring 2012): 497–502, doi:10.5812/ijem.3644; and Gita Faghihi et al., "Complementary Therapies for Idiopathic Hirsutism: Topical Licorice as Promising Option," *Evidence-Based Complementary and Alternative Medicine* 2015 (2015): 659041, http://dx.doi.org/10.1155/2015/659041.
81. Susan Arentz et al., "Herbal Medicine for the Management of Polycystic Ovary Syndrome (PCOS) and Associated Oligo/Amenorrhoea and Hyperandrogenism: A Review of the Laboratory Evidence for Effects with Corroborative Clinical Findings," *BMC Complementary and Alternative Medicine* 14 (2014): 511, doi:10.1186/1472-6882-14-511.
82. Susan Arentz et al., "Herbal Medicine for the Management of Polycystic Ovary Syndrome (PCOS) and Associated Oligo/Amenorrhoea and Hyperandrogenism: A Review of the Laboratory Evidence for Effects with Corroborative Clinical Findings."
83. Noel Vinay Thomas and Se-Kwon Kim, "Beneficial Effects of Marine Algal Compounds in Cosmeceuticals," *Marine Drugs* 11, no. 1 (January 2013): 146–64, doi:10.3390/md11010146.
84. Kyungha Shin et al., "Effectiveness of the Combinational Treatment of *Laminaria japonica* and *Cistanche tubulosa* Extracts in Hair Growth," *Laboratory Animal Research* 31, no. 1 (March 2015): 24–32, doi:10.5625/lar.2015.31.1.24; and Joon Seok et al., "Efficacy of *Cistanche tubulosa* and *Laminaria japonica* Extracts (MK-R7) Supplement in Preventing Patterned Hair Loss and Promoting Scalp Health," *Clinical Nutrition Research* 4, no, 2 (April 2015): 124–31, doi:10.7762/cnr.2015.4.2.124.
85. Longyuan Hu et al., "Polysaccharide Extracted from *Laminaria japonica* Delays Intrinsic Skin Aging in Mice," *Evidence-Based Complementary and Alternative Medicine* 2016 (2016): 5137386, http://dx.doi.org/10.1155/2016/5137386.
86. E. Kasumagić-Halilović and A. Prohić, "Association Between Alopecia Areata and Atopy," *Medicinski Arhiv* 62, no. 2 (2008): 82–4.
87. N. Otberg et al., "The Role of Hair Follicles in the Percutaneous Absorption of Caffeine," *British Journal of Clinical Pharmacology* 65, no. 4 (April 2008): 488–92.
88. Mohammad Ali Nilforoushzadeh et al., "The Effects of *Adiantum capillus-veneris* on Wound Healing: An Experimental In Vitro Evaluation," *International Journal of Preventative Medicine* 5, no. 10 (October 2014): 1261–8.
89. Maryam Noubarani et al., "Effect of *Adiantum capillus veneris* Linn on an Animal Model of Testosterone-Induced Hair Loss," supplement *Iranian Journal of Pharmaceutical Research* 13 (Winter 2014): 113–8.
90. T. W. Fischer et al., "Differential Effects of Caffeine on Hair Shaft Elongation, Matrix and Outer Root Sheath Keratinocyte Proliferation, and Transforming Growth Factor-β2/Insulin-

Like Growth Factor-1-Mediated Regulation of the Hair Cycle in Male and Female Human Hair Follicles in Vitro," *British Journal of Dermatology* 171, no. 5 (November 2014): 1031–43.

91. Ozra Akha et al., "The Effect of Fennel (*Foeniculum vulgare*) Gel 3% in Decreasing Hair Thickness in Idiopathic Mild to Moderate Hirsutism, A Randomized Placebo Controlled Clinical Trial," *Caspian Journal of Internal Medicine* 5, no. 1 (Winter 2014): 26–9.
92. E. Loing et al., "A New Strategy to Modulate Alopecia Using a Combination of Two Specific and Unique Ingredients" *Journal of Cosmetic Science* 64, no. 1 (January–February 2013): 45–58.
93. M. H. Kim et al., "Topical Treatment of Hair Loss with Formononetin by Modulating Apoptosis," *Planta Medica* 82, no. 1–2 (January 2016): 65–9, doi:10.1055/s-0035-1557897.
94. A. Rossi et al., "Minoxidil Use in Dermatology, Side Effects and Recent Patents," *Recent Patents on Inflammation and Allergy Drug Discovery* 6, no. 2 (May 2012): 130–6.
95. N. Kumar et al., "5α-Reductase Inhibition and Hair Growth Promotion of Some Thai Plants Traditionally Used for Hair Treatment," *Journal of Ethnopharmacology* 139, no. 3 (February 15 2012): 765–71, doi:10.1016/j.jep.2011.12.010; and K. Murata et al., "Inhibitory Activities of Puerariae Flos Against Testosterone 5α-Reductase and Its Hair Growth Promotion Activities," *Journal of Natural Medicine* 66, no 1 (January 2012): 158–65, doi:10.1007/s11418-011-0570-6.
96. Jain Ruchy et al., "Identification of a New Plant Extract for Androgenic Alopecia Treatment Using a Non-Radioactive Human Hair Dermal Papilla Cell-Based Assay," *BMC Complementary and Alternative Medicine* 16, no. 1 (January 2016): 18, doi:10.1186/s12906-016-1004-5.
97. M. Rondanelli et al., "A Bibliometric Study of Scientific Literature in Scopus on Botanicals for Treatment of Androgenetic Alopecia," *Journal of Cosmetic Dermatology* 15, no. 2 (November 26 2015): 120–30, doi:10.1111/jocd.12198.
98. M. Rondanelli et al., "A Bibliometric Study of Scientific Literature in Scopus on Botanicals for Treatment of Androgenetic Alopecia."
99. Satish Patel et al., "Evaluation of Hair Growth Promoting Activity of *Phyllanthus niruri*," *Avicenna Journal of Phytomedicine* 5, no. 6 (November–December 2015): 51–29.
100. M. Rondanelli et al., "A Bibliometric Study of Scientific Literature in Scopus on Botanicals for Treatment of Androgenetic Alopecia."
101. K. Murata et al., "Promotion of Hair Growth by *Rosmarinus officinalis* Leaf Extract," *Phytotherapy Research* 27, no. 2 (February 2013): 212–7, doi:10.1002/ptr.4712.
102. N. Pazyar, A. Feily, and A. Kazerouni, "Green Tea in Dermatology," *Skinmed* 10, no. 6 (November–December 2012): 352–5.
103. M. H. Kim et al., "Angelica Sinensis Induces Hair Regrowth via the Inhibition of Apoptosis Signaling," *American Journal of Chinese Medicine* 42, no. 4 (2014): 1021–4, doi:10.1142/S0192415X14500645.
104. S. Begum et al., "Exogenous Stimulation with *Eclipta alba* Promotes Hair Matrix Keratinocyte Proliferation and Downregulates TGF-β1 Expression in Nude Mice," *International Journal of Molecular Medicine* 35, no. 2 (February 2015): 496–502, doi:10.3892/ijmm.2014.2022.
105. Begum Shahnaz et al., "Comparative Hair Restorer Efficacy of Medicinal Herb on Nude (Foxn1nu) Mice," *BioMed Research International* 2014 (2014): 319795, http://dx.doi.org/10.1155/2014/319795.
106. L. Zi et al., "*Chrysanthemum zawadskii* Extract Induces Hair Growth by Stimulating the Proliferation and Differentiation of Hair Matrix," *International Journal of Molecular Medicine* 34, no. 1 (July 2014): 130–6, doi:10.3892/ijmm.2014.1768.
107. L. Lin et al., "Traditional Usages, Botany, Phytochemistry, Pharmacology and Toxicology of Polygonum Multiflorum Thunb.: A Review," *Journal of Ethnopharmacology* 159 (January 2015): 158–83, doi:10.1016/j.jep.2014.11.009.
108. Y. N. Sun et al., "Promotion Effect of Constituents from the Root of *Polygonum multiflorum* on Hair Growth," *Bioorganic and Medicinal Chemistry Letters* 23, no. 17 (September 1 2013): 4801–5, doi:10.1016/j.bmcl.2013.06.098.
109. Han Ming-Nuan et al., "Mechanistic Studies on the Use of *Polygonum multiflorum* for the Treatment of Hair Graying," *BioMed Research International* 2015 (2015): 651048.
110. Li Yunfei et al., "Hair Growth Promotion Activity and Its Mechanism of Polygonum multiflorum," *Evidence-Based Complementary Alternative Medicine* 2015 (2015): 517901.
111. C. H. Cho, J. S. Bae, and Y. U. Kim, "5alpha-Reductase Inhibitory Components as Antiandrogens from Herbal Medicine," *Journal of Acupuncture and Meridian Studies* 3, no. 2 (June 2010): 116–8, doi:10.1016/S2005-2901(10)60021-0.
112. H. J. Park, N. Zhang, and D. K. Park, "Topical Application of *Polygonum multiflorum* Extract Induces Hair Growth of Resting Hair Follicles through Upregulating Shh and β-catenin Expression in C57BL/6 Mice," *Journal of Ethnopharmacology* 135, no. 2 (May 2011): 369–75, doi:10.1016/j.jep.2011.03.028.
113. N. Dilek et al., "Cutaneous Drug Reactions in Children: A Multicentric Study," *Postepy Dermatologii i Alergologii* 31, no. 6 (December 2014): 368–71, doi:10.5114/pdia.2014.43881.
114. Gulcan Saylam Kurtipek et al., "Resolution of Cutaneous Sarcoidosis Following Topical Application of *Ganoderma lucidum* (Reishi Mushroom)," *Dermatology and Therapy* 6, no. 1 (March 2016): 105–9, doi:10.1007/s13555-016-0099-4.
115. Tingting Zhou et al., "Beneficial Effects and Safety of Corticosteroids Combined with Traditional Chinese Medicine for Pemphigus: A Systematic Review," *Evidence-Based Complementary Alternative Medicine* 2015 (2015): 815358, http://dx.doi.org/10.1155/2015/815358.
116. Ke Liang et al., "Arbutin Encapsulated Micelles Improved Transdermal Delivery and Suppression of Cellular Melanin Production," *BMC Research Notes* 9 (2016): 254.

117. Debabrata Bandyopadhyay, "Topical Treatment of Melisma," *Indian Journal of Dermatology* 54, no. 4 (October–December 2009): 303–9, doi:10.4103/0019-5154.57602.
118. Nico Smit, Jana Vicanova, and Stan Pavel, "The Hunt for Natural Skin Whitening Agents," *International Journal of Molecular Sciences* 10, no. 12 (December 2009): 5326–49, doi:10.3390/ijms10125326.
119. Barkat Ali Khan et al., "Whitening Efficacy of Plant Extracts Including *Hippophae rhamnoides* and *Cassia fistula* Extracts on the Skin of Asian Patients with Melasma," *Postepy Dermatologii i Alergologii* 30, no. 4 (August 2013): 226–32, doi:10.5114/pdia.2013.37032.
120. Su Jin Kang et al., "Inhibitory Effect of Dried Pomegranate Concentration Powder on Melanogenesis in B16F10 Melanoma Cells; Involvement of p38 and PKA Signaling Pathways," *International Journal of Molecular Sciences* 16, no. 10 (October 2015): 24219–42, doi:10.3390/ijms161024219.
121. Young Chul Kim et al., "Anti-Melanogenic Effects of Black, Green, and White Tea Extracts on Immortalized Melanocytes," *Journal of Veterinary Science* 16, no. 2 (June 2015): 135–43.
122. Huey-Chun Huang et al., "Inhibition of Melanogenesis and Antioxidant Properties of *Magnolia grandiflora* L. Flower Extract," *BMC Complementary and Alternative Medicine* 12 (2012): 72, doi:10.1186/1472-6882-12-72.
123. Mao Lin et al., "Ginsenosides Rb1 and Rg1 Stimulate Melanogenesis in Human Epidermal Melanocytes via PKA/CREB/MITF Signaling," *Evidence-Based Complementary and Alternative Medicine* 2014 (2014): 892073.
124. Chang Taek Oh et al., "Inhibitory Effect of Korean Red Ginseng on Melanocyte Proliferation and Its Possible Implication in GM-CSF Mediated Signaling," *Journal of Ginseng Research* 2013 Oct; 37(4): 389–400.
125. Tagreed Altaei, "The Treatment of Melasma by Silymarin Cream," *BMC Dermatology* 12 (2012): 18, doi:https://doi.org/10.1186/1471-5945-12-18.
126. E. C. Davis and V. D. Callender, "Postinflammatory Hyperpigmentation: A Review of the Epidemiology, Clinical Features, and Treatment Options in Skin of Color," *Journal of Clinical and Aesthetic Dermatology* 3, no. 7 (July 2010): 20–31.
127. K. H. Mou et al., "Combination Therapy of Orally Administered Glycyrrhizin and UVB Improved Active-stage Generalized Vitiligo," *Brazilian Journal of Medical and Biological Research* 49, no 8 (2016): e5354.
128. Orest Szczurko and Heather S. Boon, "A Systematic Review of Natural Health Product Treatment for Vitiligo," *BMC Dermatology* 8 (2008): 2, doi:https://doi.org/10.1186/1471-5945-8-2; and Orest Szczurko et al., "*Ginkgo biloba* for the Treatment of Vitilgo Vulgaris: An Open Label Pilot Clinical Trial," *BMC Complementary and Alternative Medicine* 11 (2011): 21, doi:10.1186/1472-6882-11-21.
129. B. J. Hughes-Formella et al., "Anti-Inflammatory Efficacy of Topical Preparations with 10% Hamamelis Distillate in a UV Erythema Test," *Skin Pharmacology and Applied Skin Physiology* 15, no. 2 (March–April 2002): 125–32; and H. C. Korting et al., "Anti-Inflammatory Activity of Hamamelis Distillate Applied Topically to the Skin," *European Journal of Clinical Pharmacology* 44, no. 4 (1993): 315–8.
130. P. Pommier et al., "Phase III Randomized Trial of *Calendula officinalis* Compared with Trolamine for the Prevention of Acute Dermatitis During Irradiation for Breast Cancer," *Journal of Clinical Oncology* 22, no. 8 (April 2004): 1447–53.
131. C. A. Erdelmeier et al., "Antiviral and Antiphlogistic Activities of *Hamamelis virginiana* Bark," *Planta Medica* 62, no. 3 (June 1996): 241–5, doi:10.1055/s-2006-957868.
132. J. Q. Del Rosso, "The Use of Topical Azelaic Acid for Common Skin Disorders other than Inflammatory Rosacea," Supplement 2 *Cutis* 77 (February 2006): 22–4.
133. H. H. Wolff and M. Kieser, "Hamamelis in Children with Skin Disorders and Skin Injuries: Results of an Observational Study," *European Journal of Pediatrics* 166, no. 9 (September 2007): 943–8. Epub 2006 Dec 20.
134. C. V. Truite et al., "Percutaneous Penetration, Melanin Activation and Toxicity Evaluation of a Phytotherapic Formulation for Vitiligo Therapeutic," *Photochemistry and Photobiology* 83, no. 6 (November–December 2007): 1529–36, doi:10.1111/j.1751-1097.2007.00197.x.

— INDEX —

Note: Page numbers in *italics* refer to figures and illustrations. Page numbers followed by *t* refer to tables.

B

C

E

F

H

J

K

L

S

T

W

X

Y

— ABOUT THE AUTHOR —

Shelly Fry of Battle Ground

Dr. Jill Stansbury is a naturopathic physician with over 30 years of clinical experience. She served as the chair of the Botanical Medicine Department of the National University of Natural Medicine in Portland, Oregon, for more than 20 years. She remains on the faculty, teaching herbal medicine and medicinal plant chemistry and leading ethnobotany field courses in the Amazon. Dr. Stansbury presents numerous original research papers each year and writes for health magazines and professional journals. She serves on scientific advisory boards for several medical organizations. She is the author of *Herbal Formularies for Health Professionals*, Volumes 1, 2, 3, and 4, and is the author of *Herbs for Health and Healing* and coauthor of *The PCOS Health and Nutrition Guide*. Dr. Stansbury lives in Battle Ground, Washington, and is the medical director of Battle Ground Healing Arts. She also runs the Healing Arts Apothecary, offering the best quality medicines from around the world, featuring many of her own custom tea formulas, blends, powders, and medicinal foods.

www.healingartsapothecary.org